D0075088

Graduate Texts in Mathematics

44

Keith Kendig

Elementary Algebraic Geometry

Springer-Verlag
New York Heidelberg Berlin

Dr. Keith Kendig
Cleveland State University
Department of Mathematics
Cleveland, Ohio 44115

AMS Subject Classification 13–01, 14–01

Library of Congress Cataloging in Publication Data

Kendig, Keith, 1938–
Elementary algebraic geometry.
(Graduate texts in mathematics ; 44)
Bibliography: p.
Includes index.
1. Geometry, Algebraic. 2. Commutative algebra.
I. Title. II. Series.
QA564.K46 516′.35 76–22598

Printed in the United States of America.

ISBN 0-387-90199-X Springer-Verlag New York

ISBN 3-540-90199-X Springer-Verlag Berlin Heidelberg

Preface

This book was written to make learning introductory algebraic geometry as easy as possible. It is designed for the general first- and second-year graduate student, as well as for the nonspecialist; the only prerequisites are a one-year course in algebra and a little complex analysis. There are many examples and pictures in the book. One's sense of intuition is largely built up from exposure to concrete examples, and intuition in algebraic geometry is no exception. I have also tried to avoid too much generalization. If one understands the core of an idea in a concrete setting, later generalizations become much more meaningful. There are exercises at the end of most sections so that the reader can test his understanding of the material. Some are routine, others are more challenging. Occasionally, easily established results used in the text have been made into exercises. And from time to time, proofs of topics not covered in the text are sketched and the reader is asked to fill in the details.

Chapter I is of an introductory nature. Some of the geometry of a few specific algebraic curves is worked out, using a tactical approach that might naturally be tried by one not familiar with the general methods introduced later in the book. Further examples in this chapter suggest other basic properties of curves.

In Chapter II, we look at curves more rigorously and carefully. Among other things, we determine the topology of every nonsingular plane curve in terms of the degree of its defining polynomial. This was one of the earliest accomplishments in algebraic geometry, and it supplies the initiate with a straightforward and very satisfying result.

Chapter III lays the groundwork for generalizing some of the results of plane curves to varieties of arbitrary dimension. It is essentially a chapter on commutative algebra, looked at through the eyeglasses of the geometer.

Algebraic ideas are supplied with geometric meaning, so that in a sense one obtains a "dictionary" between commutative algebra and algebraic geometry. I have put this dictionary in the form of a diagram of lattices; this approach does seem to neatly tie together a good many results and easily suggests to the reader a number of possible analogues and extensions.

Chapter IV is devoted to a study of algebraic varieties in $\mathbb{C}^n$ and $\mathbb{P}^n(\mathbb{C})$ and includes a geometric treatment of intersection multiplicity (which we use to prove Bézout's theorem in n dimensions).

In Chapter V we look at varieties as underlying objects upon which we do mathematics. This includes evaluation of elements of the variety's function field (that is, a study of valuation rings), a translation of the fundamental theorem of arithmetic to a nonsingular curve-theoretic setting (the classical ideal theory), some function theory on curves (a generalization of certain basic facts about functions meromorphic on the Riemann sphere), and finally the Riemann–Roch theorem on a curve (which ties in function theory on a curve with the topology of the curve).

After the reader has finished this book, he should have a foundation from which he can continue in any of several different directions—for example, to a further study of complex algebraic varieties, to complex analytic varieties, or to the scheme-theoretic treatments of algebraic geometry which have proved so fruitful.

It is a pleasure to acknowledge the help given to me by various students who have read portions of the book; I also want to thank Frank Lozier for critically reading the manuscript, and Basil Gordon for all his help in reading the galleys. Thanks are also due to Mary Blanchard for her excellent job in typing the original draft, to Mike Ludwig who did the line drawings, and to Robert Janusz who did the shaded figures. I especially wish to express my gratitude to my wife, Joan, who originally encouraged me to write this book and who was an invaluable aid in preparing the final manuscript.

Keith Kendig

Cleveland, Ohio

Contents

CHAPTER I

Examples of curves

1 Introduction

The principal objects of study in algebraic geometry are algebraic varieties. In this introductory chapter, which is more informal in nature than those that follow, we shall define algebraic varieties and give some examples; we then give the reader an intuitive look at a few properties of a special class of varieties, the "complex algebraic curves." These curves are simpler to study than more general algebraic varieties, and many of their simply-stated properties suggest possible generalizations. Chapter II is essentially devoted to proving some of the properties of algebraic curves described in this chapter.

Definition 1.1. Let k be any field.

(1.1.1) The set $\{(x_1, \ldots, x_n) \mid x_i \in k\}$ is called **affine** n**-space over** k; we denote it by k^n, or by $k_{X_1, \ldots, X_n}$. Each n-tuple of k^n is called a **point**.

(1.1.2) Let $k[X_1, \ldots, X_n] = k[X]$ be the ring of polynomials in n indeterminants $X_1, \ldots, X_n$, with coefficients in k. Let $p(X) \in k[X] \backslash k$. The set

$$\mathsf{V}(p) = \{(x) \in k^n \mid p(x) = 0\}$$

is called a **hypersurface** of k^n, or an **affine hypersurface**.

(1.1.3) If $\{p_\alpha(X)\}$ is any collection of polynomials in $k[X]$, the set

$$\mathsf{V}(\{p_\alpha\}) = \{(x) \in k^n \mid \text{each } p_\alpha(x) = 0\}$$

is called an **algebraic variety** in k^n, and **affine algebraic variety**, or, if the context is clear, just a **variety**. If we wish to make explicit reference to the field k, we say **affine variety over** k, k**-variety**, etc.; k is called the **ground field**. We also say $\mathsf{V}(\{p_\alpha\})$ is **defined by** $\{p_\alpha\}$.

(1.1.4) k^2 is called the **affine plane**. If $p \in k[X_1, X_2]\backslash k$, $\mathsf{V}(p)$ is called a **plane affine curve** (or **plane curve**, **affine curve**, **curve**, etc., if the meaning is clear from context)

We will show later on, in Section III,3, that any variety can be defined by only finitely many polynomials p_α.

Here are some examples of varieties in $\mathbb{R}^2$.

EXAMPLE 1.2

(1.2.1) Any variety $\mathsf{V}(aX^2 + bXY + cY^2 + dX + eY + f)$ where $a, \ldots, f \in \mathbb{R}$. Hence all circles, ellipses, parabolas, and hyperbolas are affine algebraic varieties; so also are all lines.

(1.2.2) The "cusp" curve $\mathsf{V}(Y^2 - X^3)$; see Figure 1.

(1.2.3) The "alpha" curve $\mathsf{V}(Y^2 - X^2(X + 1))$; see Figure 2.

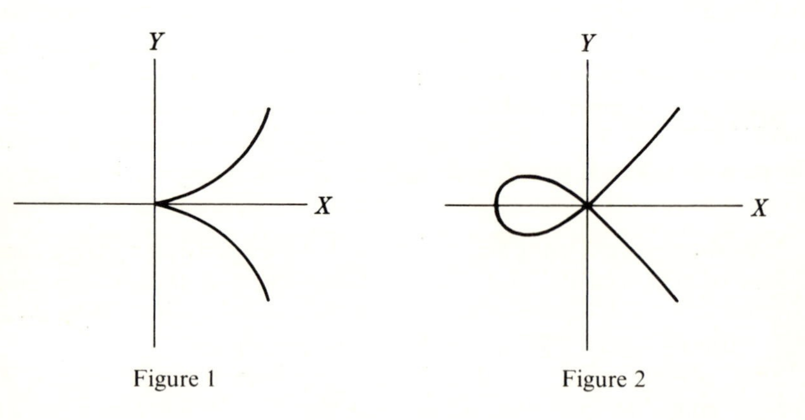

Figure 1 Figure 2

(1.2.4) The cubic $\mathsf{V}(Y^2 - X(X^2 - 1))$; see Figure 3. This example shows that algebraic curves in $\mathbb{R}^2$ need not be connected.

(1.2.5) If $\mathsf{V}(p_1)$ and $\mathsf{V}(p_2)$ are varieties in $\mathbb{R}^2$, then so is $\mathsf{V}(p_1) \cup \mathsf{V}(p_2)$; it is just $\mathsf{V}(p_1 \cdot p_2)$, as the reader can check directly from the definition. Hence one has a way of manufacturing all sorts of new varieties. For instance, $(X^2 + Y^2 - 1)(X^2 + Y^2 - 4) = 0$ defines the union of two concentric circles (Figure 4).

(1.2.6) The graph $\mathsf{V}(Y - p(X))$ in $\mathbb{R}^2$ of any polynomial $Y = p(X) \in \mathbb{R}[X]$ is also an algebraic variety.

(1.2.7) If $p_1, p_2 \in \mathbb{R}[X, Y]$, then $\mathsf{V}(p_1, p_2)$ represents the simultaneous solution set of two polynomial equations. For instance, $\mathsf{V}(X, Y) = \{(0, 0)\} \subseteq \mathbb{R}^2$, while $\mathsf{V}(X^2 + Y^2 - 1, X - Y)$ is the two-point set

$$\left\{\left(\frac{\sqrt{2}}{2}, \frac{\sqrt{2}}{2}\right), \left(-\frac{\sqrt{2}}{2}, -\frac{\sqrt{2}}{2}\right)\right\}$$

in $\mathbb{R}^2$.

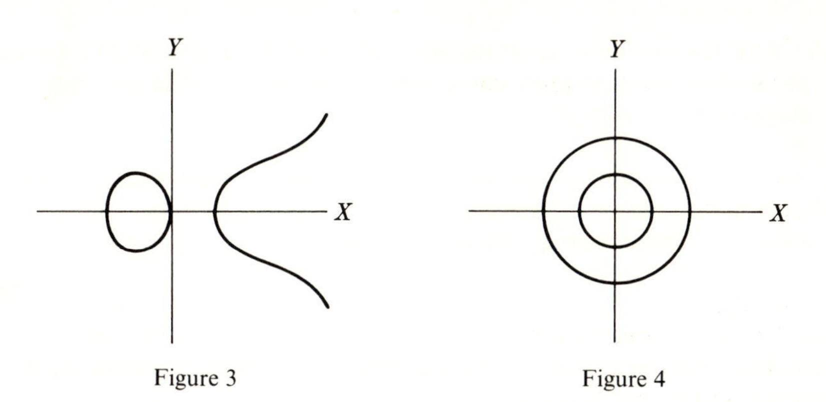

Figure 3

Figure 4

(1.2.8) In $\mathbb{R}^3$, any conic is an algebraic variety, examples being the sphere $\mathsf{V}(X^2 + Y^2 + Z^2 - 1)$, the cylinder $\mathsf{V}(X^2 + Y^2 - 1)$, the hyperboloid $\mathsf{V}(X^2 - Y^2 - Z^2 - 1)$, and so on. A circle in $\mathbb{R}^3$ is also a variety, being represented, for example, as $\mathsf{V}(X^2 + Y^2 + Z^2 - 1, X)$ (geometrically the intersection of a sphere and the (Y, Z)-plane). Any point (a, b, c) in $\mathbb{R}^3$ is the variety $\mathsf{V}(X - a, Y - b, Z - c)$ (geometrically, the intersection of the three planes $X = a$, $Y = b$, and $Z = c$).

Now suppose (still using $k = \mathbb{R}$) that we have written down a large number of sets of polynomials, and that we have sketched their corresponding varieties in $\mathbb{R}^n$. It is quite natural to look for some regularity. How do algebraic varieties behave? What are their basic properties?

First, perhaps a simple "dimensionality property" might suggest itself. For our immediate purposes, we may say that $V \subset \mathbb{R}^n$ has dimension d if V contains a homeomorph of $\mathbb{R}^d$, and if V is the disjoint union of finitely many homeomorphs of $\mathbb{R}^i$ $(i \leqslant d)$. Then in all examples given so far, each equation introduces one restriction on the dimension, so that each variety defined by one equation has dimension one less than the surrounding space—i.e., the variety has *codimension* 1. (In k^n, "codimension" means "n − dimension.") And each variety defined by two (essentially different) equations has dimension two less than the surrounding (or "ambient") space (codimension 2), etc. Hence the sphere $\mathsf{V}(X^2 + Y^2 + Z^2 - 1)$ in $\mathbb{R}^3$ has dimension $3 - 1 = 2$, the circle $\mathsf{V}(X^2 + Y^2 + Z^2 - 1, X)$ in $\mathbb{R}^3$ has dimension $3 - 2 = 1$, and the point $\mathsf{V}(X - a, Y - b, Z - c)$ in $\mathbb{R}^3$ has dimension $3 - 3 = 0$. This same thing happens in $\mathbb{R}^n$ with homogeneous linear equations—each new linearly independent equation cuts down the dimension of the resulting subspace by one.

But if we look down our hypothetical list a bit further, we come to the polynomial $X^2 + Y^2$; $X^2 + Y^2$ defines only the Z-axis in $\mathbb{R}^3$. This one equation cuts down the dimension of $\mathbb{R}^3$ by two—that is, the Z-axis has codimension two in $\mathbb{R}^3$. And further down the list we see $X^2 + Y^2 + Z^2$; the associated variety is only the origin in $\mathbb{R}^3$. And if this is not bad enough, $X^2 + Y^2 + Z^2 + 1$ defines the empty set $\varnothing$ in $\mathbb{R}^3$! Clearly then, one equation does not always cut down the dimension by one.

We might try simply restricting our attention to the "good" sets of polynomials, where the hoped-for dimensional property holds. But one "good" polynomial together with another one may not yield a "good" set of polynomials. For instance, two spheres in $\mathbb{R}^3$ may not intersect in a circle (codimension 2), but rather in a point, or in the empty set.

Though things might not look very promising at this point, mathematicians have often found their way out of similar situations. For instance, mathematicians of antiquity thought that only certain nonconstant polynomials in $\mathbb{R}[X]$ had zeros. But the exceptional status of polynomials having only real roots was removed once the field $\mathbb{R}$ was extended to its algebraic completion, $\mathbb{C}$ = field of complex numbers. One then had a most beautiful and central result, the fundamental theorem of algebra. (Every nonconstant polynomial $p(X) \in \mathbb{C}[X]$ has a zero, and the number of these zeros, when counted with multiplicity, is the degree of $p(X)$.) Similarly, geometers could remove the exceptional behavior of "parallel lines" in the Euclidean plane once they completed it in a geometric way by adding "points at infinity," arriving at the *projective completion* of the plane. One could then say that any two different lines intersect in exactly one point, and there was born a beautiful and symmetric area of mathematics, namely projective geometry.

For us, we may find a way out of our difficulties by using both kinds of completions. We first complete algebraically, using $\mathbb{C}$ instead of $\mathbb{R}$ (each set of polynomials $p_1, \ldots, p_r$ with real or complex coefficients defines a variety $\mathsf{V}(p_1, \ldots, p_r)$ in $\mathbb{C}^n$); and we also complete $\mathbb{C}^n$ *projectively* to *complex projective n-space*, denoted $\mathbb{P}^n(\mathbb{C})$. The variety $\mathsf{V}(p_1, \ldots, p_r)$ in $\mathbb{C}^n$ will be extended in $\mathbb{P}^n(\mathbb{C})$ by taking its topological closure. (We shall explain this further in a moment.) By extending our space and variety this way, we shall see that all exceptions to our "dimensional relation" will disappear, and algebraic varieties will behave just like subspaces of a vector space in this respect.

Hence, although in $\mathbb{R}^2$, $X^2 + Y^2 - 1$ defines a circle but $X^2 + Y^2$ only a point and $X^2 + Y^2 + 1$ the empty set, in our new setting each of these polynomials turns out to define a variety of (complex) codimension one in $\mathbb{P}^2(\mathbb{C})$, independent of what the "radius" of the circle might be. (The "complex dimension" of a variety V in $\mathbb{C}^n$ is just one-half the dimension of V considered as a real point set; we shall see later that as a real point set, the dimension is always even. Also, even though the locus in $\mathbb{C}^2$ of $X^2 + Y^2 = 1$ does not turn out to look like a circle, we shall continue to use this term since the $\mathbb{C}^2$-locus is defined by the same equation. Similarly, we shall use terms like *curve* or *surface* for complex varieties of complex dimension 1 and 2, respectively.)

In general, any nonconstant polynomial turns out to define a point set of complex codimension one in $\mathbb{P}^n(\mathbb{C})$, just as one (nontrivial) linear equation does in any vector space. A generalization of this vector space property is:

If L_1 and L_2 are subspaces of any n-dimensional vector space k^n over k, then

$$\operatorname{cod}(L_1 \cap L_2) \leqslant \operatorname{cod}(L_1) + \operatorname{cod}(L_2)$$

$$(\text{cod} = \text{codimension}).$$

For instance, any two 2-subspaces in $\mathbb{R}^3$ must intersect in at least a line. In $\mathbb{P}^n(\mathbb{C})$ this basic dimension relation holds even for arbitrary complex-algebraic varieties. Certainly nothing like this is true for varieties in $\mathbb{R}^2$. One can talk about disjoint circles in $\mathbb{R}^2$, or disjoint spheres in $\mathbb{R}^3$. These phrases make no sense in $\mathbb{P}^2(\mathbb{C})$ and $\mathbb{P}^3(\mathbb{C})$, respectively; the points missing in $\mathbb{R}^2$ or $\mathbb{R}^3$ simply are not seen because they are either "at infinity," or have complex coordinates. (This will be made more precise soon.) Hence it turns out that what we see in $\mathbb{R}^n$ is just the tip of an iceberg—a rather unrepresentative slice of the variety at that—whose "true" life, from the algebraic geometer's viewpoint, is lived in $\mathbb{P}^n(\mathbb{C})$.

2 The topology of a few specific plane curves

Suppose we have added the missing "points at infinity" to a complex algebraic variety in $\mathbb{C}^n$, thus getting a variety in $\mathbb{P}^n(\mathbb{C})$. It is natural to wonder what the entire "completed" curve looks like. We consider here only curves in $\mathbb{C}^2$ and in $\mathbb{P}^2(\mathbb{C})$; complex varieties of higher dimension have real dimension $\geqslant 4$ and our visual appreciation of them is necessarily limited. Even our complex curves live in real 4-space; our situation is somewhat analogous to an inhabitant of "Flatland" who lives in $\mathbb{R}^2$, when he attempts to visualize an ordinary sphere in $\mathbb{R}^3$. He can, however, see 2-dimensional slices of the sphere. Now in $X^2 + Y^2 + Z^2 = 1$, substituting a specific value Z_0 for Z yields the part of the sphere in the plane $Z = Z_0$. Then if he lets $Z = T =$ time, he can "visualize" the sphere by looking at a succession of parallel plane slices $X^2 + Y^2 = 1 - T^2$ as T varies. He sees a "moving picture" of the sphere; it is a point when $T = -1$, growing to ever larger circles, reaching maximum diameter at $T = 0$, then diminishing to a point when $T = 1$.

Our situation is perhaps even more strictly analogous to his problem of visualizing something like a "warped circle" in 3-space (Figure 5). The

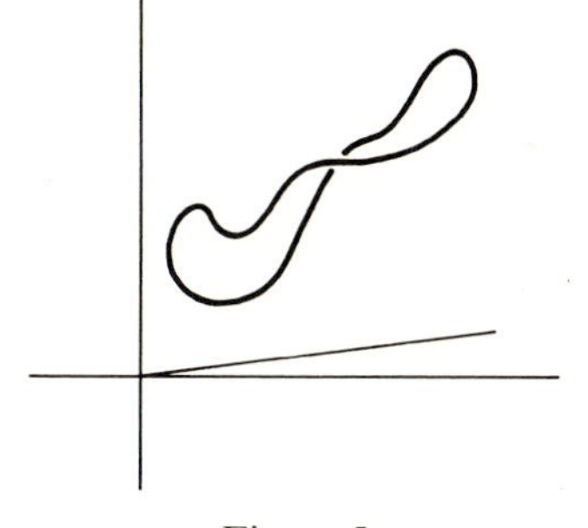

Figure 5

Flatlander's moving picture of the circle's intersections with the planes Z = constant will trace out a topological circle for him. He may not appreciate all the twisting and warping that the circle has in $\mathbb{R}^3$, but he can see its topological structure.

To get a topological look at our complex curves, let us apply this same idea to a hypersurface in complex 2-space. In $\mathbb{C}^2$, we will let the complex X-variable be $X = X_1 + iX_2$; similarly, $Y = Y_1 + iY_2$. We will let X_2 vary with time, and our "screen" will be real (X_1, Y_1, Y_2)-space. The intersection of the 3-dimensional hyperplane X_2 = constant with the real 2-dimensional variety will in general be a real curve; we will then fit these curves together in our own 3-space to arrive at a 2-dimensional object we can visualize. As with the Flatlander, we will lose some of the warping and twisting in 4-space, but we will nonetheless get a faithful topological look, which we will be content with for now.

Since our complex curves will be taken in $\mathbb{P}^2(\mathbb{C})$, we first describe intuitively the little we need here in the way of projective completions. Our treatment is only topological here, and will be made fuller and more precise in Chapter II. We begin with the real case.

$\mathbb{P}^1(\mathbb{R})$: As a topological space, this is obtained by adjoining to the topological space $\mathbb{R}$ (with its usual topology) an "infinite" point, say P, together with a neighborhood system about P. For basic open neighborhoods we take

$$U_N(P) = \{P\} \cup \{r \in \mathbb{R} \,|\, |r| > N\} \qquad N = 1, 2, 3, \ldots.$$

We can visualize this more easily by shrinking $\mathbb{R}^1$ down to an open line segment, say by $x \to x/(1 + |x|)$. We may add the point at infinity by adjoining the two end points to the line segment and identifying these two points. In this way $\mathbb{P}^1(\mathbb{R})$ becomes, topologically, an ordinary circle.

$\mathbb{P}^2(\mathbb{R})$: First note that, except for $\mathbb{R}_X$, the 1-spaces $L_\alpha = \mathsf{V}(X + \alpha Y)$ of $\mathbb{R}_{XY}$ are parametrized by α; a different parametrization, $L_{\alpha'} = \mathsf{V}(\alpha' X + Y)$, includes $\mathbb{R}_X$ (but not $\mathbb{R}_Y$). Then as a topological space, $\mathbb{P}^2(\mathbb{R})$ is obtained from $\mathbb{R}^2$ by adjoining to each 1-subspace of $\mathbb{R}^2$, a point together with a neighborhood system about each such point.

If, for instance, a given line is L_{α_0}, then for basic open neighborhoods about a given P_{α_0} we take

$$U_N(P_{\alpha_0}) = \bigcup_{|\alpha - \alpha_0| < 1/N} (\{P_\alpha\} \cup \{(x, y) \in L_\alpha \,|\, |(x, y)| > N\}) \qquad N = 1, 2, 3, \ldots,$$

where $|(x, y)| = |x| + |y|$.

Similarly for lines parametrized by α'. (When α and α' both represent the same line $L_{\alpha_0} = L_{\alpha'_0}$, the neighborhoods $U_N(P_{\alpha_0})$ and $U_N(P_{\alpha'_0})$ generate the same set of open neighborhoods about $P_{\alpha_0} = P_{\alpha'_0}$.)

Again, we can see this more intuitively by topologically shrinking $\mathbb{R}^2$ down to something small. For instance,

$$(x, y) \to \left(\frac{x}{1 + \sqrt{x^2 + y^2}}, \frac{y}{1 + \sqrt{x^2 + y^2}}\right)$$

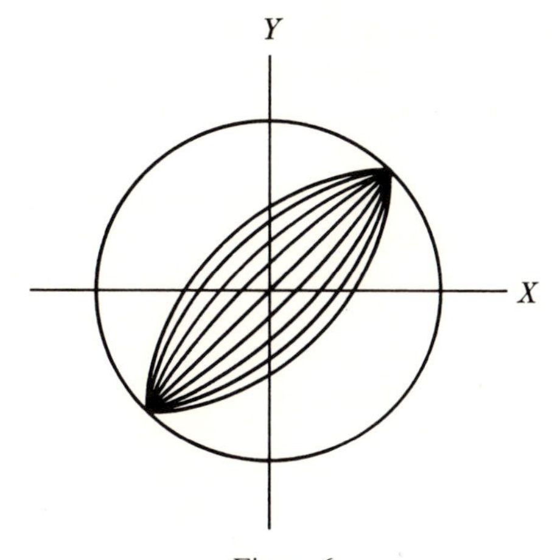

Figure 6

maps $\mathbb{R}^2$ onto the unit open disk. Figure 6 shows this condensed plane together with some mutually parallel lines. (Two lines parallel in $\mathbb{R}^2$ will converge in the disk since distance becomes more "concentrated" as we approach its edge; the two points of convergence are opposite points. If, as in $\mathbb{P}^1(\mathbb{R})$, we identify these points, then any two "parallel" lines in the figure will intersect in that one point. Adding analogous points for every set of parallel lines in the plane means adding the whole boundary of the disk, with opposite (or *antipodal*) points identified. All these "points at infinity" form the "line at infinity," itself topologically a circle, hence a projective line $\mathbb{P}^1(\mathbb{R})$. Since this line at infinity intersects every other line in just one point, it is clear that any two different projective lines of $\mathbb{P}^2(\mathbb{R})$ meet in precisely one point.

$\mathbb{P}^1(\mathbb{C})$: Topologically, the "complex projective line" is obtained by adjoining to $\mathbb{C}$ an "infinite" point P; for basic open neighborhoods about P, take

$$U_N\{P\} = \{P\} \cup \{z \in \mathbb{C} \,|\, |z| > N\} \qquad N = 1, 2, 3, \ldots.$$

Intuitively, shrink $\mathbb{C}$ down so it is an open disk, which topologically is also a sphere with one point missing (just as $\mathbb{R}$ is topologically a circle with one point missing). Adding this point yields a sphere.

$\mathbb{P}^2(\mathbb{C})$: As in the real case, except for the X-axis $\mathbb{C}_X$, the complex 1-spaces of $\mathbb{C}^2 = \mathbb{C}_{XY}$ are parametrized by α:

$$X + \alpha Y = 0 \quad \text{where } \alpha \in \mathbb{C};$$

another parametrization, $\alpha' X + Y = 0$, includes $\mathbb{C}_X$ but not $\mathbb{C}_Y$. Then $\mathbb{P}^2(\mathbb{C})$ as a topological space is obtained from $\mathbb{C}^2$ by adjoining to each complex

1-subspace $L_\alpha = \mathsf{V}(X + \alpha Y)$ (or $L_{\alpha'} = \mathsf{V}(\alpha' X + Y)$) a point P_α (or $P_{\alpha'}$). A typical basic open neighborhood about a given P_{α_0} is

$$U_N(P_{\alpha_0}) = \bigcup_{|\alpha - \alpha_0| < 1/N} (\{P_\alpha\} \cup \{(z_1, z_2) \in L_\alpha \,|\, |(z_1, z_2)| > N\}) \qquad N = 1, 2, 3, \ldots,$$

where $|(z_1, z_2)| = |z_1| + |z_2|$; similarly for neighborhoods about points $P_{\alpha_0'}$.

Intuitively, to each complex 1-subspace and all its parallel translates, we are adding a single "point at infinity," so that all these parallel lines intersect in one point. Each complex line is thus extended to its projective completion, $\mathbb{P}^1(\mathbb{C})$; and all points at infinity form also a $\mathbb{P}^1(\mathbb{C})$. As in $\mathbb{P}^2(\mathbb{R})$, any two different projective lines of $\mathbb{P}^2(\mathbb{C})$ meet in exactly one point.

The reader can easily verify from our definitions that each of $\mathbb{R}$, $\mathbb{R}^2$, $\mathbb{C}$, $\mathbb{C}^2$ is dense in its projective completion; hence the closure of $\mathbb{C}^2$ in $\mathbb{P}^2(\mathbb{C})$ is $\mathbb{P}^2(\mathbb{C})$, and so on. We shall likewise take the projective extension of a complex algebraic curve in $\mathbb{C}^2$ to be its topological closure in $\mathbb{P}^2(\mathbb{C})$.

We next consider some examples of projective curves using the slicing method outlined above.

EXAMPLE 2.1. Consider the circle $\mathsf{V}(X^2 + Y^2 - 1)$. Let $X = X_1 + iX_2$ and $Y = Y_1 + iY_2$. Then $(X_1 + iX_2)^2 + (Y_1 + iY_2)^2 = 1$. Expanding and equating real and imaginary parts gives

$$X_1{}^2 - X_2{}^2 + Y_1{}^2 - Y_2{}^2 = 1, \qquad X_1 X_2 + Y_1 Y_2 = 0. \tag{1}$$

We let X_2 play the role of time; we start with $X_2 = 0$. The part of our complex circle in the 3-dimensional slice $X_2 = 0$ is then given by

$$X_1{}^2 + Y_1{}^2 - Y_2{}^2 = 1, \qquad Y_1 Y_2 = 0. \tag{2}$$

The first equation defines a hyperboloid of one sheet; the second one, the union of the (X_1, Y_1)-plane and the (X_1, Y_2)-plane (since $Y_1 \cdot Y_2 = 0$ implies $Y_1 = 0$ or $Y_2 = 0$). The locus of the equations in (2) appears in Figure 7. It is

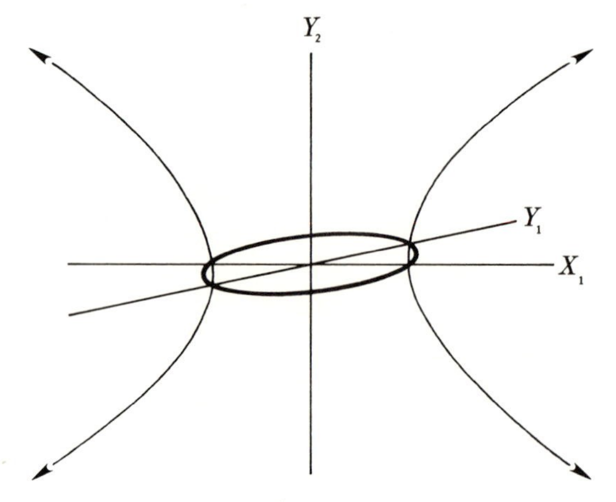

Figure 7

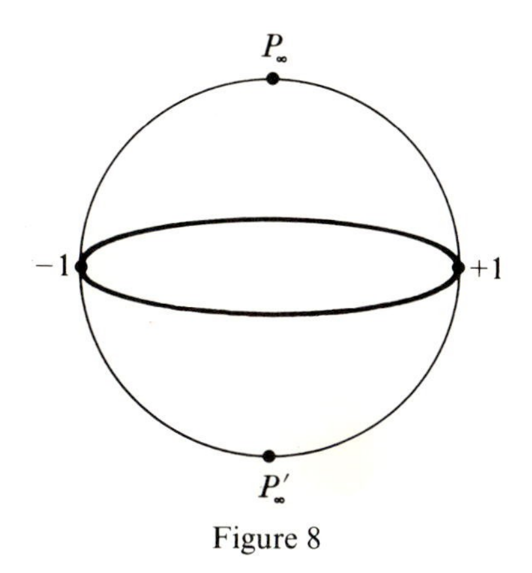

Figure 8

the union of the real circle $X_1{}^2 + Y_1{}^2 = 1$ (when $Y_2 = 0$) and the hyperbola $X_1{}^2 - Y_2{}^2 = 1$ (when $Y_1 = 0$). The circle is, of course, just the real part of the complex circle. The hyperbola has branches approaching two points at infinity, which we call P_∞ and P'_∞.

Now the completion in $\mathbb{P}^2(\mathbb{R})$ of the hyperbola is topologically an ordinary circle. Hence the total curve in our slice $X_2 = 0$ is topologically two circles touching at two points; this is drawn in Figure 8. The more lightly-drawn circle in Figure 8 corresponds to the (lightly-drawn) hyperbola in Figure 7.

Now let's look at the situation when "time" X_2 changes a little, say to $X_2 = \varepsilon > 0$. This defines the corresponding curve

$$X_1{}^2 + Y_1{}^2 - Y_2{}^2 = 1 + \varepsilon^2, \qquad \varepsilon X_1 + Y_1 Y_2 = 0.$$

The first surface is still a hyperboloid of one sheet; the second one, for ε small, in a sense "looks like" the original two planes. The intersection of these two surfaces is sketched in Figure 9. The circle and hyperbola have split into two disjoint curves. We may now sketch these disjoint curves in on Figure 8; they always stay close to the circle and hyperbola. If we fill in all

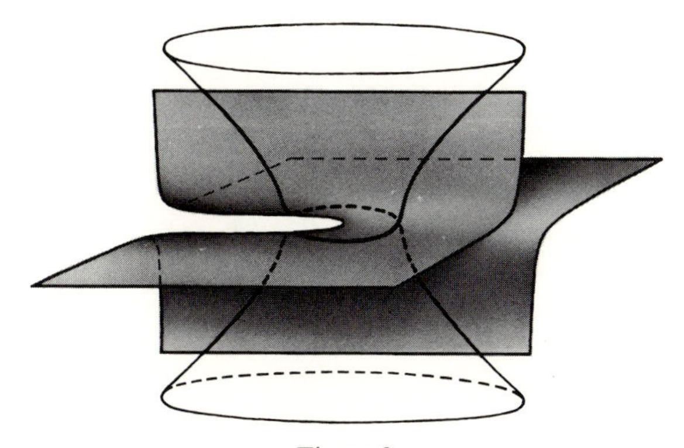
Figure 9

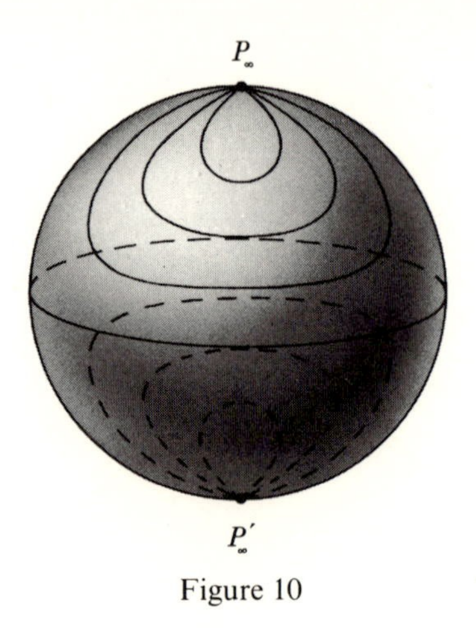

Figure 10

such curves corresponding to X_2 = constant, we will fill in the surface of a sphere. The curves for nonnegative X_2 are indicated in Figure 10.

For $X_2 < 0$, one gets curves lying on the other two quarters of the sphere. We thus see (and will rigorously prove in Section II,10) that all these curves fill out a sphere. We thus have the remarkable fact that *the complex circle* $\mathsf{V}(X^2 + Y^2 - 1)$ *in* $\mathbb{P}^2(\mathbb{C})$ *is topologically a sphere.*

From the complex viewpoint, the complex circle still has codimension 1 in its surrounding space.

EXAMPLE 2.2. Now let us look at a circle of "radius 0," $\mathsf{V}(X^2 + Y^2)$. The equations corresponding to (1) are

$$X_1{}^2 - X_2{}^2 + Y_1{}^2 - Y_2{}^2 = 0, \qquad X_1X_2 + Y_1Y_2 = 0. \tag{3}$$

The part of this variety lying in the 3-dimensional slice $X_2 = 0$ is then given by

$$X_1{}^2 + Y_1{}^2 - Y_2{}^2 = 0, \qquad Y_1Y_2 = 0. \tag{4}$$

The first equation defines a cone; the second one defines the union of two planes as before. The simultaneous solution is the intersection of the cone and planes. This consists of two lines (See Figure 11). The projective closure of each line is a topological circle, so the closure of the two lines in this figure consists of two circles touching at one point. This can be thought of as the limit figure of Figure 8 as the horizontal circle's radius approaches zero.

When $X_2 = \varepsilon$, the saddle-surface defined by $\varepsilon X_1 + Y_1Y_2 = 0$ intersects the one-sheeted hyperboloid given by $X_1{}^2 + Y_1{}^2 - Y_2{}^2 = \varepsilon^2$. As before, their intersection consists of two disjoint real curves, which turn out to be lines (Figure 12); just as in the first example, as X_2 varies, the curves fill out a 2-dimensional topological space which is like Figure 10, except that the radius of the horizontal circle is 0 (Figure 13). To keep the figure simple, only curves for $X_2 \geqslant 0$ have been sketched; they cover the top half of the upper

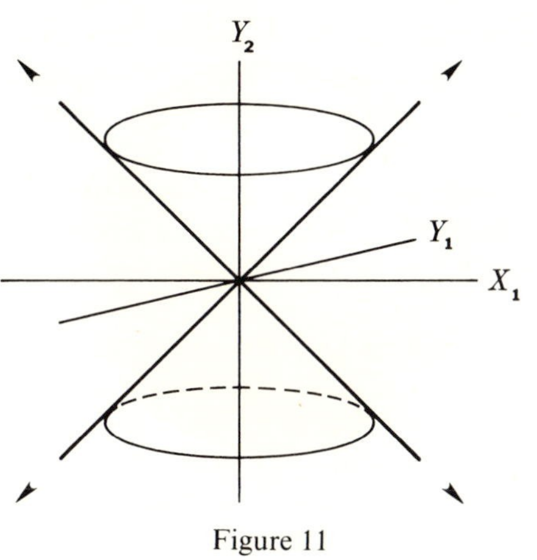

Figure 11

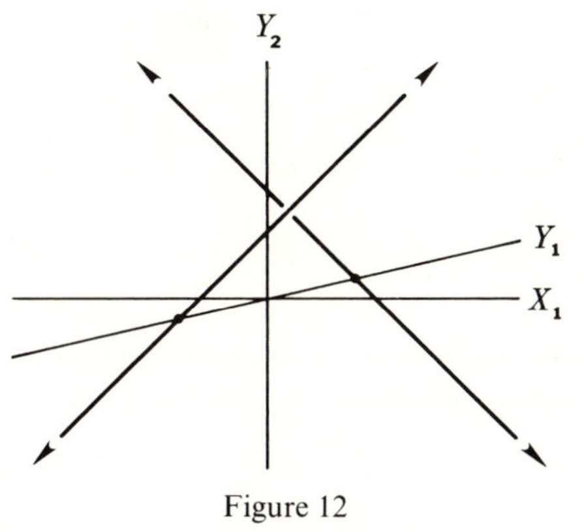

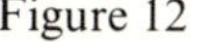
Figure 12

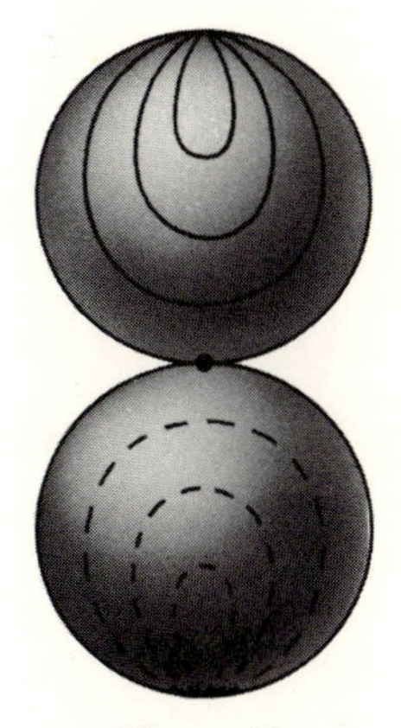

Figure 13

sphere and the bottom half of the lower sphere, the other parts being covered when $X_2 < 0$. Hence: *The complex circle of "zero radius"* $\mathsf{V}(X^2 + Y^2)$ *in* $\mathbb{P}^2(\mathbb{C})$ *is topologically two spheres touching at one point.*

In the complex setting, we see that instead of the dimension changing as soon as the "radius" becomes zero, the complex circle remains of codimension 1, so that one equation $X^2 + Y^2 = 0$ still cuts down the (complex) dimension by one.

Incidentally, here is another fact that one might notice: In Example 2.1, $\mathsf{V}(X^2 + Y^2 - 1)$, the sphere is in a certain intuitive sense "indecomposable," while in Example 2.2, the figure is in a sense "decomposable," consisting of two spheres which touch at only one point. But look at the polynomial $X^2 + Y^2 - 1$; it is "indecomposable" or irreducible in $\mathbb{C}[X, Y]$.[1] And the polynomial $X^2 + Y^2$ is "decomposable," or reducible—$X^2 + Y^2 = (X + iY)(X - iY)$! In fact, $X^2 + Y^2 + \gamma$ is always irreducible in $\mathbb{C}[X, Y]$ if $\gamma \neq 0$. (A proof may be given similar in general spirit to that in Footnote 1.) Hence we should suspect that any complex circle with "nonzero radius" should be somehow irreducible. We shall see later that in an appropriate sense this is indeed true. By the way, $X^2 + Y^2 = (X + iY)(X - iY)$ expresses that $\mathsf{V}(X^2 + Y^2)$ is just the union $\mathsf{V}(X + iY) \cup \mathsf{V}(X - iY)$. Each of these last varieties is a projective line, which is topologically a sphere; and any two projective lines touch in exactly one point in $\mathbb{P}^2(\mathbb{C})$. This is a very different way of arriving at the topological structure of $\mathsf{V}(X^2 + Y^2)$.

Example 2.3. Let us look next at a circle of "pure imaginary radius," $\mathsf{V}(X^2 + Y^2 + 1)$. Separating real and imaginary parts gives

$$X_1{}^2 - X_2{}^2 + Y_1{}^2 - Y_2{}^2 = -1, \qquad X_1X_2 + Y_1Y_2 = 0. \tag{6}$$

At $X_2 = 0$ this defines the part common to a hyperboloid of two sheets and the union of two planes. This is a hyperbola. Its two branches start approaching each other as X_2 increases, finally meeting at $X_2 = 1$ (the hyperboloid of two sheets has become the cone $X_1{}^2 + Y_1{}^2 - Y_2{}^2 = 0$). Then for $X_2 > 1$, we are back to the same kind of behavior as for $\mathsf{V}(X^2 + Y^2 - 1)$ when $X_2 > 0$. Figure 14, analogous to Figures 10 and 13, shows how we end up with a sphere. Later we will supplement this result by proving:

Topologically, $\mathsf{V}(X^2 + Y^2 + \gamma)$ in $\mathbb{P}^2(\mathbb{C})$ is a sphere iff $\gamma \neq 0$.

[1] If $X^2 + Y^2 - 1$ were factorable into terms of lower degree, it would have to be of the form

$$X^2 + Y^2 - 1 = (aX + bY + c)\left(\frac{1}{a}X + \frac{1}{b}Y - \frac{1}{c}\right) \qquad a, b, c \neq 0; \tag{5}$$

this follows from multiplying and equating coefficients. Also, equating X-terms yields $0 = -(a/c) + (c/a)$, or $c^2 = a^2$. Similarly, $c^2 = b^2$, so $a^2 = b^2$, which in turn yields a term $\pm 2XY$ on the right-hand side of (5), a contradiction.

Figure 14

Do other familiar topological spaces arise from looking at curves in $\mathbb{P}^2(\mathbb{C})$? For instance, is a torus (a sphere with one "handle"—that is, the surface of a doughnut) ever the underlying topological space of a complex curve? More generally, how about a sphere with g handles in it (topological manifold of genus g)? Let us consider the following example:

EXAMPLE 2.4. The real part of the curve $\mathsf{V}(Y^2 - X(X^2 - 1))$, frequently encountered in analytic geometry, appears in Figure 3. (The reader will learn, at long last, what happens in those mysterious "excluded regions" $-\infty < X < -1$ and $0 < X < 1$.)

Separating real and imaginary parts in $Y^2 - X(X^2 - 1) = 0$ gives

$$\begin{aligned} Y_1{}^2 - Y_2{}^2 &= X_1{}^3 - 3X_1X_2{}^2 - X_1, \\ 2Y_1Y_2 &= 3X_1{}^2X_2 - X_2{}^3 - X_2. \end{aligned} \tag{7}$$

When $X_2 = 0$, this becomes

$$Y_1{}^2 - Y_2{}^2 = X_1{}^3 - X_1, \qquad Y_1Y_2 = 0.$$

Then either $Y_1 = 0$ or $Y_2 = 0$. When $Y_2 = 0$, the other equation becomes $Y_1{}^2 = X_1{}^3 - X_1$. The sketch of this is of course again in Figure 3—that is, when $X_2 = Y_2 = 0$ we get the real part of our curve. When $Y_1 = 0$, we get a "mirror image" of this in the (X_1, Y_2)-plane. The total curve in the slice $X_2 = 0$ appears in Figure 15.

Note that in the right-hand branch, Y_1 increases faster than X_1 for X_1 large, so the branch approaches the Y_1-axis. Similarly, the left-hand branch approaches the Y_2-axis. But in $\mathbb{P}^2(\mathbb{C})$, exactly one infinite point is added to each complex 1-space, and the (Y_1, Y_2)-plane is the 1-space $Y = 0$. Hence the

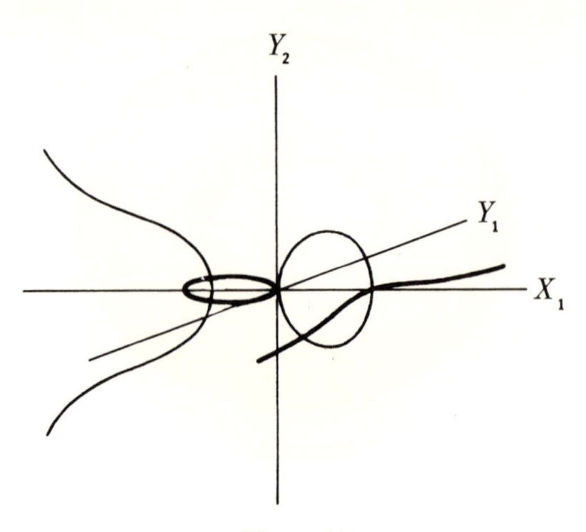

Figure 15

two branches meet at a common point P_∞. We may topologically redraw our curve in the 3-dimensional slice as in Figure 16.

By letting $X_2 = \varepsilon$ in (7) and using continuity arguments, one sees that the curves in the other 3-dimensional slices fill in a torus. In Figure 17, solid lines on top and dotted lines on bottom come from curves for $X_2 \geqslant 0$. The rest of the torus is filled in when $X_2 < 0$. The real part of the graph of $Y^2 = X(X^2 - 1)$ is indeed a small part of the total picture!

We now generalize this example to show we can get as underlying topological space, a "sphere with any finite number of handles"; this is the most general example of a compact connected orientable 2-dimensional manifold. Such a manifold is completely determined by its genus g. (We take this up later on; Figure 19 shows such a manifold with $g = 5$.)

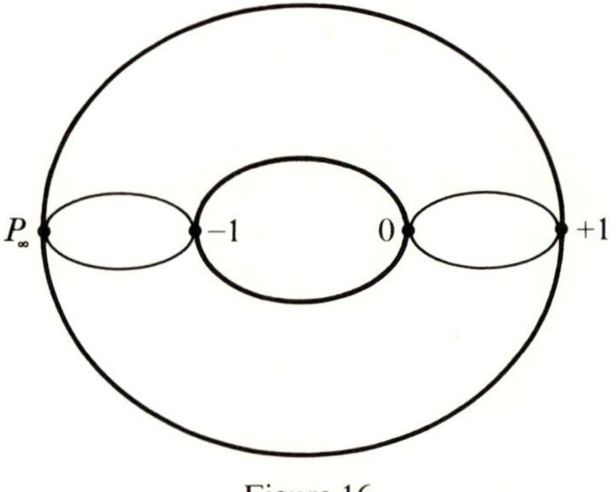

Figure 16

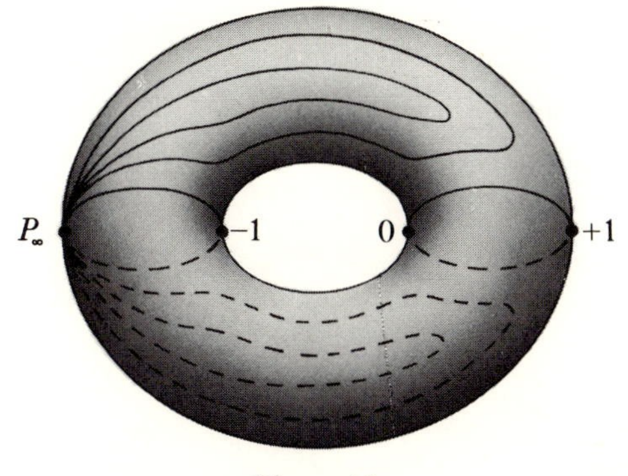

Figure 17

EXAMPLE 2.5. $\mathsf{V}(Y^2 - X(X^2 - 1) \cdot (X^2 - 4) \cdot \ldots \cdot (X^2 - g^2))$. For purposes of illustration we use $g = 5$. The sketch of the corresponding real curve appears in the (X_1, Y_1)-plane of Figure 18. The whole of Figure 18 represents the curve in the slice $X_2 = 0$.

Note the analogy with Figure 15. As before, the branches in Figure 18 meet at the same point at infinity. This may be topologically redrawn as in Figure 19, where also the curves for $X_2 \geqslant 0$ have been sketched in.

We now see that looking at "loci of polynomials" from the complex viewpoint automatically leads us to topological manifolds! Incidentally, these last manifolds of arbitrary genus are intuitively "indecomposable" in a way that the sphere was earlier, so we have good reason to suspect that any polynomial $Y^2 - X(X^2 - 1) \cdot (X^2 - 4) \cdot \ldots \cdot (X^2 - g^2)$ is irreducible in

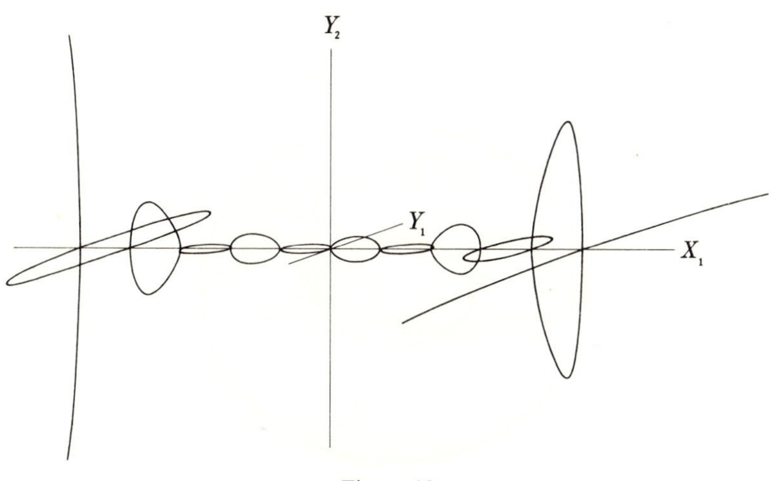

Figure 18

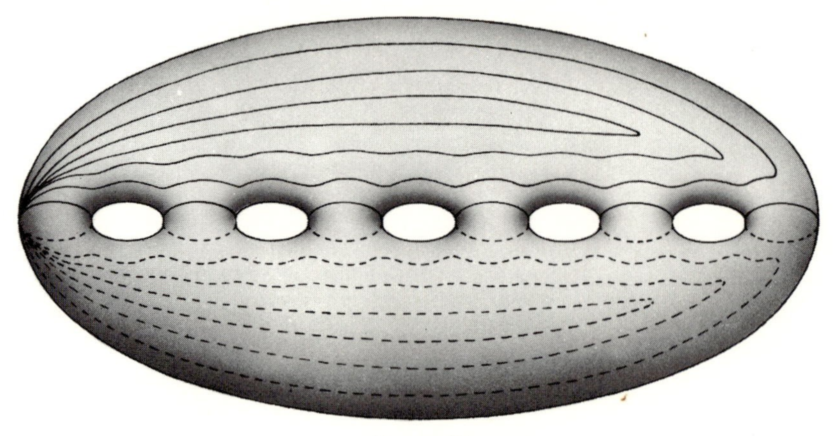

Figure 19

$\mathbb{C}[X, Y]$. This is in fact so. Note, however, that a polynomial having as repeated factors an irreducible polynomial may still define an indecomposable object. (For example, $\mathsf{V}(X - Y) = \mathsf{V}((X - Y)^3)$ is topologically a sphere in $\mathbb{P}^2(\mathbb{C})$.) We also recall that if we take a finite number of irreducible polynomials and multiply them together, the irreducibles' identities are not obliterated, for we can refactor the polynomial to recapture the original irreducibles (by "uniqueness"). The same behavior holds at the geometric level; each topological object in $\mathbb{P}^2(\mathbb{C})$ coming from a (nonconstant) polynomial $p \in \mathbb{C}(X, Y)$ is 2-dimensional, but it turns out that objects coming from different *irreducible* factors of p touch in only a finite number of points, and that removing these points leaves us with a finite number of connected, disjoint parts. These parts are in 1:1 onto correspondence with the distinct irreducible factors of p. For instance, $\mathsf{V}(p)$, with

$$p = (Y^2 - X(X^2 - 1)(X^2 - 4)) \cdot Y \cdot (Y - 1),$$

Figure 20

turns out to look topologically like Figure 20; it falls into three parts, the two spheres corresponding to the factors Y and $(Y - 1)$, and the manifold of genus 2, corresponding to the 5th degree factor. The spheres touch each other in one point, and each sphere touches the third part in 5 points.

EXAMPLE 2.6. We cannot leave this section of examples without at least briefly mentioning curves with singularities; an example is given by the alpha curve $\mathsf{V}(Y^2 - X^2(X + 1))$ (Figure 2). Separating real and imaginary parts of $Y^2 - X^2(X + 1) = 0$ and setting $X_2 = 0$ gives us a curve sketched in Figure 21. The two branches again meet at one point at infinity, P_∞, and the other curves $X_2 =$ constant fit together as in Figure 22. Topologically this is obtained by taking a sphere and identifying two points. Note that $Y^2 - X^2(X + 1)$ is just the limit of $Y^2 - X(X - \varepsilon)(X + 1)$ as $\varepsilon \to 0$. One can think of Figure 22 as being the result of taking the topological circle in Figure 17 between the roots 0 and 1 and "squeezing this circle to a point." Also note that this "squeezing" process not only introduces a singularity, but has the effect of decreasing the genus by one; the genus of $\mathsf{V}(Y^2 - X(X^2 - 1))$ is 1, while $\mathsf{V}(Y^2 - X^2(X + 1))$ is a sphere (genus 0) with two points identified. One may instead choose to squeeze to a point, say, the circle in Figure 17 between roots -1 and 0; this corresponds to $\mathsf{V}(Y^2 - X^2(X - 1))$. Its sketch in real (X_1, Y_1)-space is just the "mirror image" of Figure 2. Squeezing this middle circle to a point gives a sphere with the north and south poles identified to a point; the reader may wish to check that these two different ways of identifying two points on a sphere yield homeomorphic objects.

What if one brings together *all* three zeros of $X(X + 1)(X - 1)$? That is, what does $\mathsf{V}(\lim_{\varepsilon \to 0}[Y^2 - X(X + \varepsilon)(X - \varepsilon)]) = \mathsf{V}(Y^2 - X^3)$ look like? Of course its real part is just the cusp of Figure 1; the origin is again an example of a singular point. As it turns out, $\mathsf{V}(Y^2 - X^3)$ is topologically a sphere (Exercise 2.2).

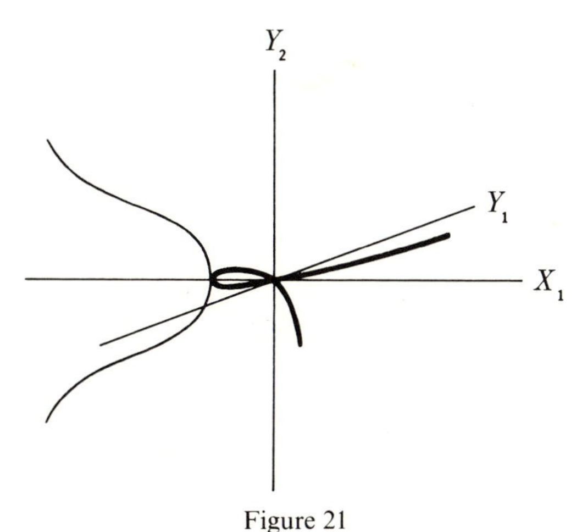

Figure 21

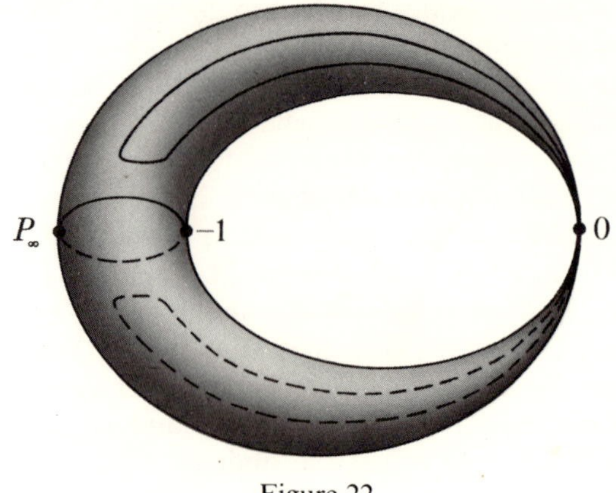

Figure 22

After seeing all these examples, the reader may well wonder:

> What is the most general topological object in $\mathbb{P}^2(\mathbb{C})$ defined by a (nonconstant) polynomial $p \in \mathbb{C}[X, Y]$?

The answer is:

Theorem 2.7. *If $p \in \mathbb{C}[X, Y]\backslash\mathbb{C}$ is irreducible, then topologically $\mathsf{V}(p)$ is obtained by taking a real 2-dimensional compact, connected, orientable manifold (this turns out to be a sphere with $g < \infty$ handles) and identifying finitely many points to finitely many points; for any $p \in \mathbb{C}[X, Y]\backslash\mathbb{C}$, $\mathsf{V}(p)$ is a finite union of such objects, each one furthermore touching every other one in finitely many points.*

We remark that a **(real, topological)** n**-manifold** is a Hausdorff space M such that each point of M has an open neighborhood homeomorphic to an open ball in $\mathbb{R}^n$. For definitions of connectedness and orientability, see Definitions 8.1 and 9.3, and Remark 9.4 of Chapter II.

One of the main aims of Chapter II is to prove this theorem.

EXAMPLE 2.8. In Figure 23 a real 2-dimensional compact, connected, orientable manifold of genus 4 has had 7 points identified to 3 points (3 to 1, 2 to 1, and 2 to 1).

Remark 2.9. We do not imply that every topological object described above actually *is* the underlying space of some algebraic curve in $\mathbb{P}^2(\mathbb{C})$. However, one can, by identifying roots of $Y^2 - X(X^2 - 1)\dots(X^2 - g^2)$, manufacture spaces having any genus, with any number of distinct "2 to 1"

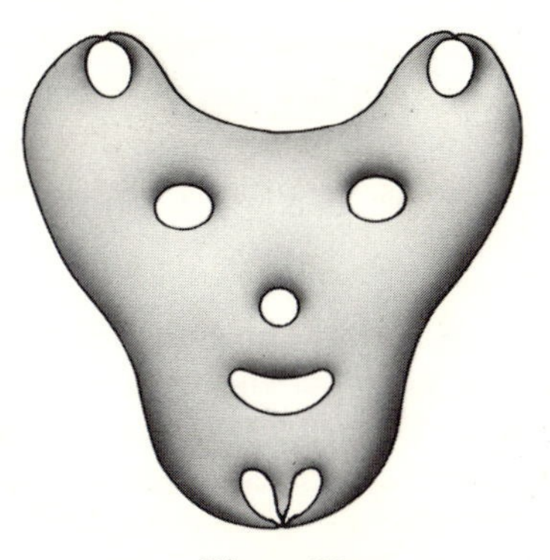

Figure 23

identifications. But how about any number of "3 to 1", or "4 to 1" identifications, etc.? And in just how many points can we make one such "indecomposable" space touch another? Even partial answers to such questions involve a careful study of such things as Bézout's theorem, Plücker's formulas, and the like.

EXERCISES

2.1 Show, using the "slicing method" of this section, that the completion in $\mathbb{P}^2(\mathbb{C})$ of the complex parabola $\mathsf{V}(Y - X^2)$ and the complex hyperbola $\mathsf{V}(X^2 - Y^2 - 1)$ are topologically both spheres.

2.2 Draw figures corresponding to Figures 7–10 to show that the completion in $\mathbb{P}^2(\mathbb{C})$ of $\mathsf{V}(Y^2 - X^3)$ is a topological sphere. Compare your figures with those for $\mathsf{V}(Y^2 - X^2(X + \varepsilon))$, as $\varepsilon > 0$ approaches zero.

2.3 Establish the topological nature of the completion in $\mathbb{P}^2(\mathbb{C})$ of $\mathsf{V}(X^2 - Y^2 + r)$, as r takes on real values in $[-1, 1]$.

3 Intersecting curves

The fact that any two "indecomposable" algebraic curves in $\mathbb{P}^2(\mathbb{C})$ must intersect (as implied by the description in the last section), follows at once from the dimension relation

$$\operatorname{cod}(\mathsf{V}(p_i) \cap \mathsf{V}(p_j)) \leqslant \operatorname{cod} \mathsf{V}(p_i) + \operatorname{cod} \mathsf{V}(p_j),$$

which means, in our case, $\operatorname{cod}(\mathsf{V}(p_i) \cap \mathsf{V}(p_j) \leqslant 2$, or

$$\dim(\mathsf{V}(p_i) \cap \mathsf{V}(p_j) \geqslant 0. \text{ Hence } \mathsf{V}(p_i) \cap \mathsf{V}(p_j) \neq \varnothing \qquad (\dim \varnothing = -1).$$

EXAMPLE 3.1. For two parallel complex lines in $\mathbb{C}^2$, the above amounts to a restatement that these lines must intersect in $\mathbb{P}^2(\mathbb{C})$.

EXAMPLE 3.2. Any complex line and any complex circle in $\mathbb{P}^2(\mathbb{C})$ must intersect. One can actually see, in Figure 7, how any parallel translate of the complex Y-plane along the X_1-axis still intersects the circle $\mathsf{V}(X^2 + Y^2 - 1)$, either in the (X_1, Y_1)-plane (as we usually see the intersection), or in the hyperbola in the (X_1, Y_2)-plane, for $|X_1| > 1$.

EXAMPLE 3.3. Another case may be of some interest. Let us consider one curve which is a complex line, say $\mathsf{V}(Y)$. Let another curve be $\mathsf{V}(Y - q(X))$, where q is a polynomial in X alone. Then the graph of $Y = g(X)$ in $\mathbb{C}_{XY}$ is homeomorphic to $\mathbb{C}_X$; one can then easily check that $\mathsf{V}(Y - q(X))$ is a topological sphere in $\mathbb{P}^2(\mathbb{C})$, as is $\mathsf{V}(Y)$. By our dimension relation, these spheres must intersect, perhaps as in Figure 24.

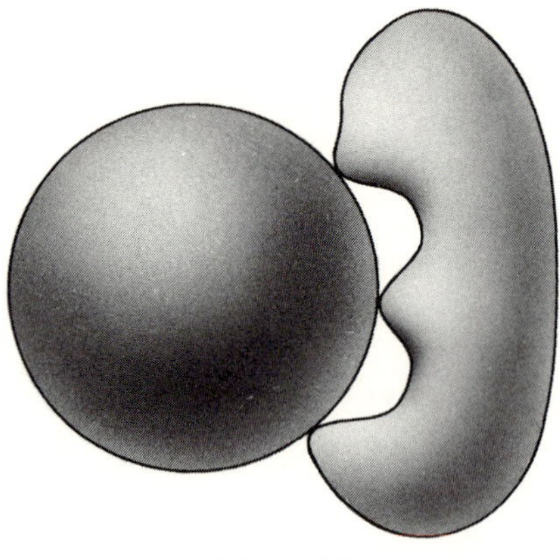

Figure 24

Does this result sound familiar? It is very much like the fundamental theorem of algebra (every nonconstant $q(X) \in \mathbb{C}[X]$ has a zero in $\mathbb{C}$). This famous result can now be put into the $\mathbb{P}^2(\mathbb{C})$ setting. If we do this, then in stating the fundamental theorem of algebra,

(a) There is no need to assume $q(x) \in \mathbb{C}[X]$ is nonconstant; if it is constant, we will have two lines which either coincide or which intersect at one point at infinity.

(b) There is no need to assume the given line is $\mathsf{V}(Y)(=\mathbb{C}_X)$. Any line, in any position, will do as well.

(c) There is no need to assume one polynomial is linear (thus describing a line), or that the other one must be of such a very special form as $Y - q(X)$ (describing the graph of a function: $\mathbb{C} \to \mathbb{C}$). The loci of any two curves in $\mathbb{P}^2(\mathbb{C})$ must intersect.

Of course, the reader might argue that our dimension relation fails to give us certain other information which the fundamental theorem of algebra readily provides. For instance, the fundamental theorem can be stated more informatively as: "Any nonzero polynomial $q(X) \in \mathbb{C}[X]$ has exactly $\deg q(X)$ zeros when counted with multiplicity." However, our dimension relation can be extended in an analogous way.

To see how, first consider $q(X)$ again. We may look at the multiplicity r_i of the zero c_i in $q(X) = (X - c_1)^{r_1} \cdot \ldots \cdot (X - c_k)^{r_k}$ in the following geometric way. Let $a \in \mathbb{C}\backslash\{0\}$, and let L_a be the complex line $Y = a$ in $\mathbb{C}_{XY}$. Then the following holds for all a sufficiently small:

> Of those points in which L_a intersects the graph of $Y = q(X)$, there are exactly r_i of them clustered close to the point $(c_i, 0)$, where r_i is the multiplicity of the zero c_i in $q(X)$.

(A proof of a more general version of this fact will be provided in Section IV,6.)

EXAMPLE 3.4. The point $0 \in \mathbb{C}$ is a double root of the polynomial equation $Y = q(X) = X^2$. For $a \neq 0$, L_a intersects the parabola in two distinct points. (They have complex X-coordinates if $a \notin \mathbb{R}^+$.) See Figure 25.

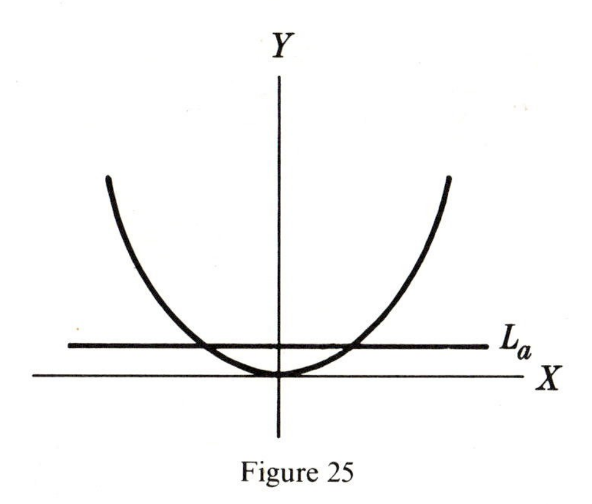

Figure 25

For small a, these two points cluster close to (0, 0); in this way our zero of multiplicity two can be looked at as the limit of two single points which have coalesced. The fundamental theorem says that the sum of all the multiplicities r_i at all zeros $c_1, \ldots, c_k$ of the polynomial $q(X)$ is $\deg q(X)$. The situation of two curves C_1, C_2 in $\mathbb{C}^2$ is very much the same; if P is an isolated point of intersection of C_1 and C_2, there will be a certain integer n_P so that the following holds:

> **(3.5)** For most sufficiently small translates of C_1 or C_2, of those points in which the translated curves intersect, there will be exactly n_P distinct points clustered close to P.

The meaning of "most" above will be made precise in Chapter IV. The reader can get the basic idea via some examples. The integer n_P is called the *intersection multiplicity of C_1 and C_2 at P.*

EXAMPLE 3.6. Consider the intersection at $(0, 0) \in \mathbb{C}^2$ of the two alpha curves $C_1 = \mathsf{V}(Y^2 - X^2(X + 1))$ and $C_2 = \mathsf{V}(X^2 - Y^2(Y + 1))$ (Figure 26). In Figure 27 C_2 has been translated upward a little; there are four distinct points clustered about (0, 0). One can see that "most" translates will yield four points this way. In certain special directions one can get fewer than four points (but never more, nearby). Figure 28 gives an example of this. But in a certain sense the topmost of these clustered points is still "multiple"—that is, a little push up or down of either curve will further separate that point so we again end up with a total of four points.

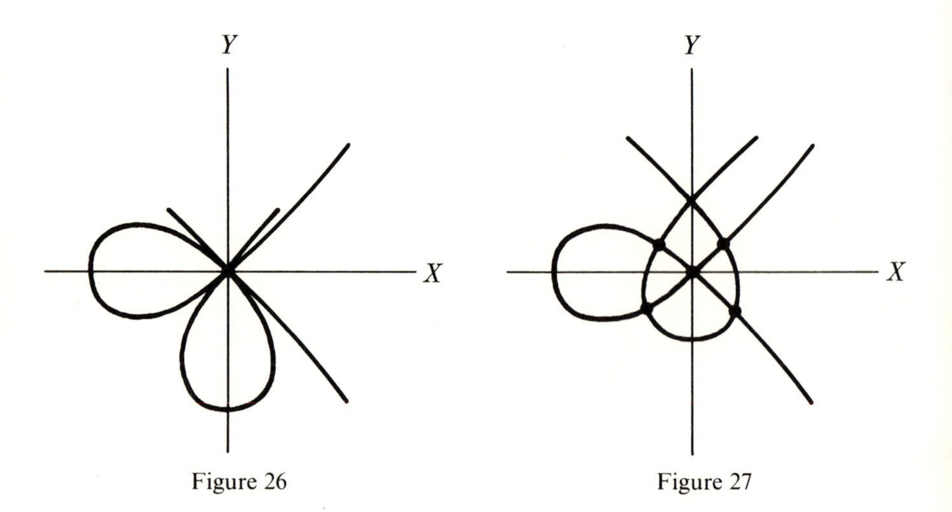

Figure 26

Figure 27

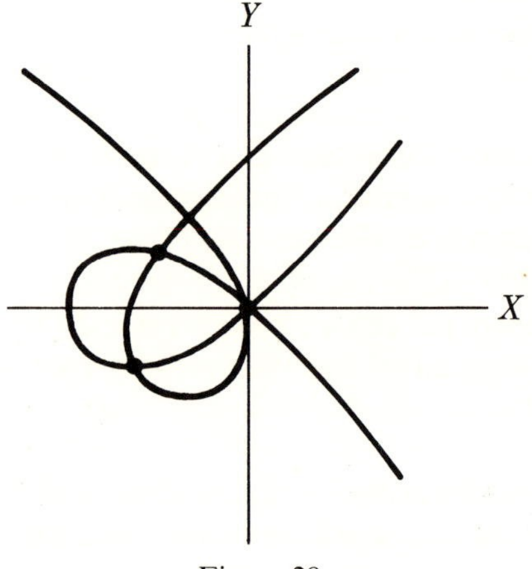

Figure 28

Of course, in most cases, the intersection points have complex coordinates; the above example is quite special in the respect that all intersection points are real. Also, we have not shown by our real pictures that there cannot be more than four points clustered near (0, 0); for this, we need to determine from the equations of the curves all possible simultaneous solutions near (0, 0). But we can translate this geometric idea of "perturbing one curve slightly to separate the points" into algebraic terms; this is done in Chapter IV when we take up intersection multiplicity formally.

An extension of the dimension relation for curves which corresponds to our geometric form of the fundamental theorem of algebra is then the following:

> Let $P_1, \ldots, P_k$ be all the points of intersection of two curves $\mathsf{V}(p_1)$ and $\mathsf{V}(p_2)$ in $\mathbb{P}^2(\mathbb{C})$, where p_1 and p_2 have no repeated factors, and no factor in common. Then the total number of points of intersection of $\mathsf{V}(p_1)$ and $\mathsf{V}(p_2)$, counted with multiplicity, is $(\deg p_1) \cdot (\deg p_2)$. (Often the number of points counted with multiplicity is called the *degree* of the intersection and one writes
>
> $$\deg(\mathsf{V}(p_1) \cap \mathsf{V}(p_2)) = (\deg p_1) \cdot (\deg p_2).)$$

This elegant and central result is known as Bézout's theorem, after its discoverer, the French mathematician E. Bézout (1730–1783).

Remark 3.7. We must assume p_i is a polynomial of lowest degree defining $\mathsf{V}(p_i)$, i.e., that p_i does not have repeated factors. For instance the X and Y axes, which are $\mathsf{V}(Y)$ and $\mathsf{V}(X^2)$, intersect in just one point instead of $1 \cdot 2 = 2$ points. However, using $\mathsf{V}(X)$ yields $1 \cdot 1 = 1$ point. The assumption that p_1 and p_2 have no factor in common is of course needed to make $\mathsf{V}(p_1) \cap \mathsf{V}(p_2)$ finite.

EXAMPLE 3.8. Assuming (3.5), we can now see more precisely the relation between Bézout's theorem and the fundamental theorem of algebra. Assume, in $Y = p(X)$, that p is not constant. Then Y eventually increases like X to a positive power; hence the graph $\mathsf{V}(Y - p(X))$ cannot intersect the X-axis at infinity—that is, all intersections take place in $\mathbb{C}_{XY}$. Secondly, since p is nonconstant, $\deg(Y - p(X)) = \deg p(X)$. Finally the X-axis, i.e., $Y = 0$, has degree 1. Hence the number of intersections of the graph in $\mathbb{C}_{XY}$ with $\mathbb{C}_X$ is $1 \cdot \deg p(X) = \deg p(X)$.

EXAMPLE 3.9. Consider two ellipses as in Figure 29. Each ellipse is defined by a polynomial of degree two, and the total number of intersection points is $2 \cdot 2 = 4$. As the horizontal ellipse is translated upward, we get first one double point at top, and two single points; then two complex points and two real ones; then 2 complexes and 1 real double; and finally, as the ellipses separate entirely in the real plane, we have four complex points of intersection.

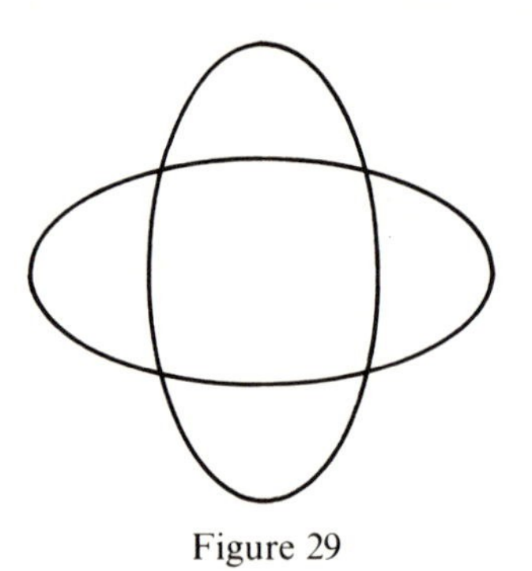

Figure 29

EXAMPLE 3.10. Consider the curves $\mathsf{V}(Y^2 - 5X^2(X + 1))$ and $\mathsf{V}(Y^2 - X - 1)$. Their degrees are 3 and 2, so there should be a total of six intersection points. In Figure 30 four single intersections and one double intersection appear.

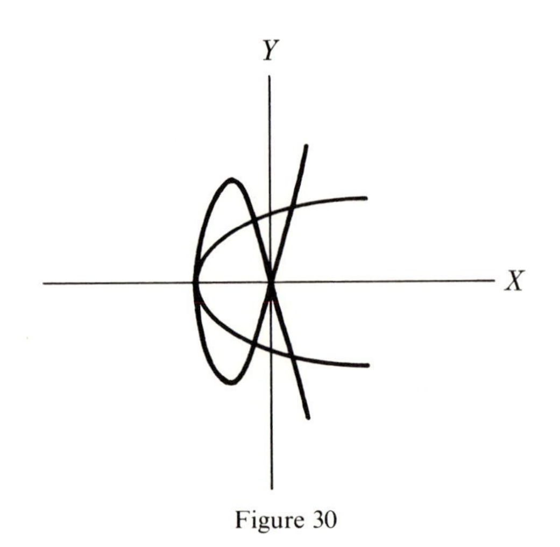

Figure 30

EXERCISES

3.1 Suppose that curves C_1 and C_2 in $\mathbb{P}^2(\mathbb{C})$ have exactly p points of intersection (counting multiplicity), where p is a prime. Show that either C_1 or C_2 must be a topological sphere. (Assume Bézout's theorem.)

3.2 Consider the curves $\mathsf{V}(Y^2 - X^3)$ and $\mathsf{V}(Y^2 - 2X^3)$ in $\mathbb{C}_{XY}$. Find all points of intersection in $\mathbb{C}_{XY}$ near (0, 0), after one curve is given an arbitrarily small nonzero translation. How many points of intersection (counted with multiplicity) are there at the "line at infinity" of $\mathbb{C}_{XY}$?

3.3 Consider the complex circles $\mathsf{V}(X^2 + Y^2 - 1)$ and $\mathsf{V}(X^2 + Y^2 - 4)$ in $\mathbb{P}^2(\mathbb{C})$. Where are their points of intersection? Are all four points of intersection distinct?

4 Curves over $\mathbb{Q}$

Perhaps enough has been said about plane curves to give the reader a first bit of intuition about them. Let us now look in perspective at what we've done so far. We started out using $\mathbb{R}$ as our groundfield, and were led to $\mathbb{C}$ as a ground field where varieties were better able to express an important side of their nature. We then saw topologically, just what certain complex curves look like, and we got a look at how they behave under intersection. This immediately suggests many more questions: What do varieties of *arbitrary* dimension in $\mathbb{P}^n(\mathbb{C})$ look like? How do they behave under intersection? Under union? Can one "multiply" varieties as one does topological spaces to get product varieties? Are there natural maps from one variety to another, as there are continuous maps from one topological space to another? How do they behave? Can one put more structure on an algebraic variety (as one does with topological spaces) to arrive at "algebraic groups," etc.? An important part of algebraic geometry consists in exploring such questions. Also, if going from ground field $\mathbb{R}$ to $\mathbb{C}$ allowed varieties a fuller expression of a certain aspect of their nature, might possibly using other ground fields allow varieties to express a different side of their nature? This is indeed so, and no introductory tour of algebraic varieties would be complete without at least mentioning varieties over ground fields other than $\mathbb{R}$ and $\mathbb{C}$.

For purposes of illustration, let us consider the field $\mathbb{Q}$ of rational numbers. Even asking for the appearance of a few specific curves in $\mathbb{Q}^2$ will show us a totally new aspect of algebraic varieties. For many curves defined by polynomials in $\mathbb{Q}[X, Y]$, their appearance is the "same" as in $\mathbb{R}^2$—that is, if C is the locus in $\mathbb{R}^2$ of $p(X, Y) \in \mathbb{Q}[X, Y]$, then $C \cap \mathbb{Q}^2$ is dense in C. But for other curves the appearances of C in $\mathbb{Q}^2$ versus $\mathbb{R}^2$ are completely different.

EXAMPLE 4.1. $V = \mathsf{V}(Y - X^2) \subset \mathbb{Q}^2$; with the usual topology of $\mathbb{R}^2$, this V is dense in the variety $V' = \mathsf{V}(Y - X^2) \subset \mathbb{R}^2$, so V "looks like" the parabola V' in $\mathbb{R}^2$. The density of V is easily verified because $y = x^2$ is a continuous function, $\mathbb{Q}$ is dense in $\mathbb{R}$, and $x \in \mathbb{Q}$ implies $(x, x^2) \in V \subset \mathbb{Q}^2$.

EXAMPLE 4.2. In a similar way, one sees that if $p(X) \in \mathbb{Q}[X]$, then the "graph-variety" $\mathsf{V}(Y - p(X))$ in $\mathbb{Q}^2$ is dense in the corresponding graph in $\mathbb{R}^2$.

EXAMPLE 4.3. $V = \mathsf{V}(Y^2 - X^3) \subset \mathbb{Q}^2$ is dense in $V' = \mathsf{V}(Y^2 - X^3) \subset \mathbb{R}^2$, so V also looks like the cusp V' in $\mathbb{R}^2$. This is true because the squares of rational numbers form a dense subset of $\mathbb{R}^+$; if s is a square of a rational, then $(s, \pm s^{3/2}) \in V \subset \mathbb{Q}^2$.

EXAMPLE 4.4. The "rational circle" $V = \mathsf{V}(X^2 + Y^2 - 1) \subset \mathbb{Q}^2$ also turns out to be dense in the corresponding real circle of $\mathbb{R}^2$, but this time the reasoning is more subtle. Let $(r, s) \in \mathsf{V}(X^2 + Y^2 - 1) \subset \mathbb{Q}^2$. We may assume without loss of generality that r and s have the same denominator, $r = a/c$, $s = b/c$, a, b, c, integers. Then $r^2 + s^2 = 1$ implies $a^2 + b^2 = c^2$, i.e., that (a, b, c) is a *Pythagorean triple*, meaning that $|a|, |b|, |c|$ form the lengths of the sides of a right triangle (a Pythagorean triangle). Now the number-theoretic problem of finding all Pythagorean triangles was solved already by Euclid's time. The solution says, in essence:

All Pythagorean triples (a, b, c) are obtained from

$$\frac{a}{c} = \frac{v^2 - u^2}{u^2 + v^2} \qquad \frac{b}{c} = \frac{2uv}{u^2 + v^2},$$

where u, v range through all integers (u, v not both zero).

The question of whether the points $(r, s) = (a/c, b/c)$ are dense in the real circle is evidently the same as:

Can the slope

$$\frac{a/c}{b/c} = \frac{a}{b} = \frac{v^2 - u^2}{2uv}$$

be made arbitrarily close to any preassigned slope $m \in \mathbb{R}$?

But

$$\frac{v^2 - u^2}{2uv} \simeq m \quad \text{implies} \quad \frac{v}{u} - \frac{u}{v} \simeq 2m,$$

meaning that for some rational $x = u/v$, $x - (1/x) \simeq 2m$. This implies $x^2 - 2mx - 1 \simeq 0$ which further implies $x \simeq m \pm \sqrt{m^2 + 1}$. But surely any $m \pm \sqrt{m^2 + 1} \in \mathbb{R}$ can be approximated to any degree of accuracy by a rational number.

Geometrically, the density of the rational circle in the ordinary real one says that there are Pythagorean triangles arbitrarily close in shape to any given right triangle.

EXAMPLE 4.5. The last example involved a solution to an honest problem in number theory. Finding out exactly what the rational curve $\mathsf{V}(X^n + Y^n - 1) \subset \mathbb{Q}^2$ looks like for all integers $n > 2$ is probably the most famous unsolved problem in all mathematics; it is equivalent to Fermat's last theorem. This conjecture says that in $\mathbb{Q}$, for any $n > 2$ $\mathsf{V}(X^n + Y^n - 1)$ consists of just the four points $\{(1, 0)\} \cup \{(0, 1)\} \cup \{(-1, 0)\} \cup \{(0, -1)\}$ if n is even, and consists of just the two points $\{(1, 0)\} \cup \{(0, 1)\}$ if n is odd. This is vastly different from the corresponding real curves (Figures 31 and 32). It looks more and more square-shaped as n becomes large and even (Figure 31), and looks like the sketch in Figure 32 for n large and odd.

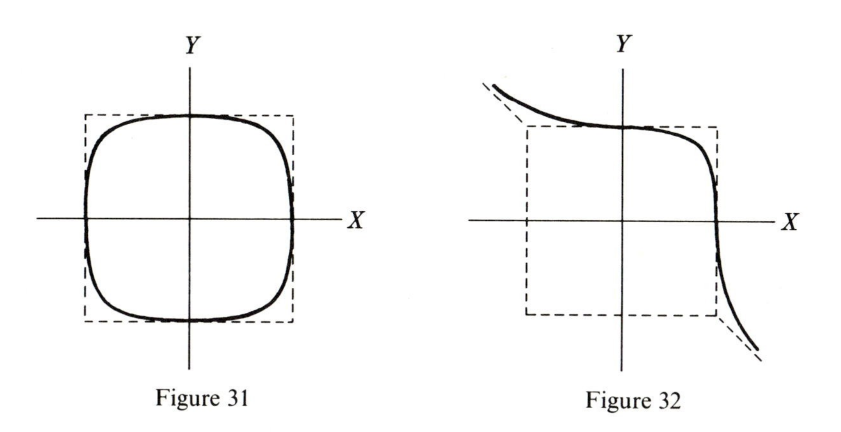

Figure 31

Figure 32

From just these few examples, the reader has perhaps already guessed that by using $\mathbb{Q}$ as ground field, varieties express a strong number-theoretic aspect. (This is true also of algebraic extensions of $\mathbb{Q}$ and finite fields). In fact, much of the modern work in number theory, including looking anew at some of the old classical problems, has been done by looking at number theory geometrically from the vantage point of algebraic geometry.

In this chapter we have tried to give the reader a little feeling for algebraic varieties, mostly by example. A large part of algebraic geometry consists in rigorizing and extending these ideas, not only to arbitrary complex varieties in $\mathbb{P}^n(\mathbb{C})$ or to arbitrary varieties in an analogous $\mathbb{P}^n(k)$, but also to finding appropriate analogues when k is replaced by quite general commutative rings with identity. From our examples of switching from $\mathbb{R}$ to $\mathbb{C}$ or to $\mathbb{Q}$, the reader may be somewhat convinced that such generalization is not just for generalization's sake, but frequently leads to important connections with other areas of mathematics.

Exercises

4.1 Is the set $\mathsf{V}(X^2 - Y^2 - 1) \subset \mathbb{Q}^2$ dense in $\mathsf{V}(X^2 - Y^2 - 1) \subset \mathbb{R}^2$? What about the curves $\mathsf{V}(X^n - Y^n - 1)$, for $n > 2$?

4.2 Is the alpha curve $\mathsf{V}(Y^2 - X^2(X + 1)) \subset \mathbb{Q}^2$ dense in the corresponding curve in $\mathbb{R}^2$?

4.3 Find an ellipse $\mathsf{V}(aX^2 + bY^2 - 1)$ $(a, b \in \mathbb{Q}\backslash\{0\})$ whose graph in $\mathbb{Q}^2$ is the empty set. [*Hint*: Any solution $X_i = x_i$ in $\mathbb{Z}$ of $n_1X_1{}^2 + n_2X_2{}^2 + n^3X_3{}^2 = 0$ $(n_i \in \mathbb{Z})$ implies, for every integer n, a solution in $\mathbb{Z}/(n)$ of $n_1X_1{}^2 + n_2X_2{}^2 + n_3X_3{}^2 \equiv 0 \pmod{n}$.] Can we assume $a = b$?

CHAPTER II

Plane curves

1 Projective spaces

This section is devoted to projective spaces. The question of defining "projective space" is akin to defining "sphere" in that both terms are used at several different levels. By "sphere" one can mean a topological sphere, or sphere as a differential manifold, or, more strictly, as the set $\mathsf{V}(X^2 + Y^2 + Z^2 - 1)$ in $\mathbb{R}^3$. This last object has the most structure of all; for instance, it is also a real algebraic variety. In turn, one can take this specific object, with its multitude of properties, and isolate certain of its properties to get other notions of "sphere." Its equation is defined by a sum of squares, so one has the n-dimensional real sphere $\mathsf{V}({X_1}^2 + \ldots + {X_n}^2 - 1)$. We also have, analogously, complex n-spheres, when $k = \mathbb{C}$; $\mathbb{Q}$-spheres, when $k = \mathbb{Q}$; and so on.

In this book we use a quite specific object in defining projective space. This means there will be quite a bit of structure in our object, and we can bring out various facets of it as needed. To motivate this definition, we start with one of the topological definitions of the last section; by a succession of topologically equivalent definitions we arrive at a definition which will at once suggest a simple definition for any $\mathbb{P}^n(k)$. It will also give us an easy way to pass between affine spaces and projective spaces.

We begin with $\mathbb{P}^2(\mathbb{R})$ as in Chapter I. The disk with antipodal points identified may also be considered, topologically, as an ordinary hemisphere with opposite equatorial points identified (Figure 1). This hemisphere can in turn be looked at as an entire sphere with all pairs of antipodal points identified. This last is very symmetric! In one stroke, the "line at infinity" has lost its special status. The topological space $\mathbb{P}^2(\mathbb{R})$ is just an ordinary sphere with antipodal points identified. But we can go even further and

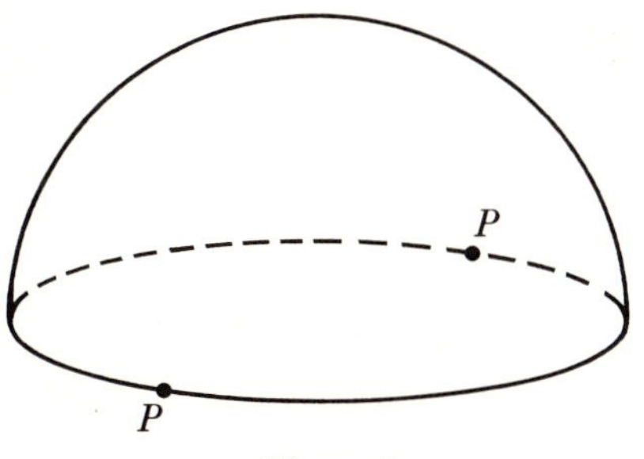

Figure 1

identify to one point all the points on the line-minus-origin through any two antipodal points. And by centering the sphere at the origin of $\mathbb{R}^3$, we see the points of $\mathbb{P}^2(\mathbb{R})$ become identified in a natural way with the 1-subspaces of $\mathbb{R}^3$.

We now make the following general

Definition 1.1. Let k be any field. Then, as a set, **n-dimensional projective space over** k, written $\mathbb{P}^n(k)$, is the set of all 1-subspaces of the vector space k^{n+1}. Each 1-subspace is called a **point** of $\mathbb{P}^n(k)$; the set of all 1-subspaces in an $(r + 1)$-subspace of k^{n+1} comprises an **r-dimensional projective subspace** $\mathbb{P}^r(k)$ **of** $\mathbb{P}^n(k)$. $\mathbb{P}^r(k)$ has **codimension** $n - r$ **in** $\mathbb{P}^n(k)$.

If the field is in addition a topological field then k induces on $\mathbb{P}^n(k)$ a topology as follows: First, k induces on k^{n+1} the usual product topology. Then a typical open set of $\mathbb{P}^n(k)$ consists of all 1-subspaces of k^{n+1} intersecting an arbitrary open set of k^{n+1}.

Remark 1.2. If $\mathbb{R}$ and $\mathbb{C}$ are given their usual topologies, then $\mathbb{P}^n(\mathbb{R})$ and $\mathbb{P}^n(\mathbb{C})$ are compact. See Exercises 1.1–1.3.

One advantage of Definition 1.1 is that it at once gives us the general n-dimensional analogue of the basic relation: *Any two projective lines in* $\mathbb{P}^2(\mathbb{R})$ *must intersect*. This may be equivalently expressed as: If L_1, L_2 are projective lines in $\mathbb{P}^2(\mathbb{R})$, then $\text{cod}(L_1 \cap L_2) \leqslant \text{cod } L_1 + \text{cod } L_2$. (This is the "projective" analogue of the subspace theorem for vector spaces, noted in Chapter I.) The generalization is as follows:

Lemma 1.3. *Let S_1 and S_2 be any two projective subspaces of $\mathbb{P}^n(k)$. Then*

$$\text{cod}(S_1 \cap S_2) \leqslant \text{cod } S_1 + \text{cod } S_2.$$

PROOF. Any subspace k^{r+1} has codimension $n - r$ in k^{n+1}; therefore the associated subspace $\mathbb{P}^r(k)$ has the same codimension $n - r$ in $\mathbb{P}^n(k)$. Then apply the corresponding vector space theorem. □

Hence any two projective 2-spaces in $\mathbb{P}^3(\mathbb{R})$ intersect in at least a line, and so on. We show later that this dimension relation holds for any two algebraic varieties in $\mathbb{P}^n(\mathbb{C})$ (Section IV,3); we prove it for projective curves in $\mathbb{P}^2(\mathbb{C})$ in Section II,6.

Another advantage of Definition 1.1 is that it allows us to define coordinates on $\mathbb{P}^n(k)$. This will be extremely useful later on.

Definition 1.4. Let P be a point of $\mathbb{P}^n(k)$, and L_P, the corresponding 1-subspace of k^{n+1}. The $(n + 1)$-tuple of coordinates $(a_1, \ldots, a_{n+1})$ of any nonzero point in L_P is called a **coordinate set** of P. More informally, we say that $(a_1, \ldots, a_{n+1})$ are **coordinates of** P.

Remark 1.5. Coordinate sets of P are never uniquely determined, unless k is the two-element field; however, any two coordinate sets of P differ by a scalar multiple.

Definition 1.6. Two nonzero $(n + 1)$-tuples $(a_1, \ldots, a_{n+1})$ and $(b_1, \ldots, b_{n+1})$ of k^{n+1} are **equivalent** if

$$(b_1, \ldots, b_{n+1}) = (ca_1, \ldots, ca_{n+1})$$

for some nonzero $c \in k$. (Hence all coordinate sets of any $P \in \mathbb{P}^n(k)$ are equivalent.)

Now let us relate, in a more direct way, Definition 1.1 to our definition of $\mathbb{P}^2(\mathbb{R})$ in Chapter I. The same kind of arguments we use now will enable us to see how the general definition reduces to those in Chapter I for the special cases considered there.

First, using Definition 1.1 and an (X_1, X_2, X_3)-coordinate system of $\mathbb{R}^3$, we see the points of $\mathbb{P}^2(\mathbb{R})$ fall into two classes—those having zero as last coordinate, and those with nonzero last coordinate. If a triple (a_1, a_2, a_3) satisfies $a_3 \neq 0$, then dividing by a_3 yields a triple $(b_1, b_2, 1)$. All such points of $\mathbb{R}^3$ constitute the 2-plane $\mathsf{V}(X_3 - 1)$, and this establishes a 1:1 correspondence between a part of $\mathbb{P}^2(\mathbb{R})$ and the plane $\mathsf{V}(X_3 - 1)$ (since two triples $(b_1, b_2, 1)$ and $(b'_1, b'_2, 1)$ are equivalent iff $b_1 = b'_1$ and $b_2 = b'_2$.

Now if a point of $\mathbb{P}^2(\mathbb{R})$ has last coordinate zero, say $(b_1, b_2, 0)$, then all scalar multiples of it form a line L in $\mathbb{R}_{X_1X_2}$. Then $L + (0, 0, 1)$ is a line in our hyperplane through $(0, 0, 1)$. Hence the points with zero last coordinate may be identified in a natural way with the set of all *lines* through $(0, 0, 1)$ within the plane $\mathsf{V}(X_3 - 1)$, while the points with nonzero last coordinates correspond to the *points* of the plane $\mathsf{V}(X_3 - 1)$. Hence the set $\mathbb{P}^2(\mathbb{R})$, according to Definition 1.1, is in 1:1 onto correspondence with the points of $\mathbb{R}^2$, together with all 1-subspaces of $\mathbb{R}^2$. But this is precisely the way $\mathbb{R}^2$ was completed in Chapter I—to $\mathbb{R}^2$ we added one new element for each different 1-subspace of $\mathbb{R}^2$. It is straightforward to check that either definition yields the same open sets on $\mathbb{P}^2(\mathbb{R})$; hence both definitions yield the same topological space. We can now apply precisely the same kind of reasoning to show that

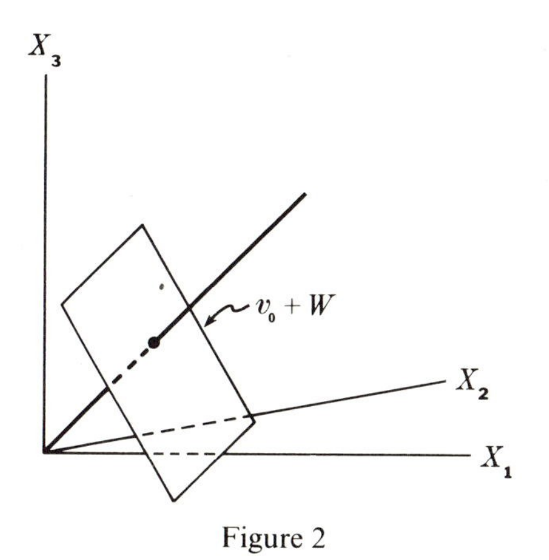

Figure 2

Definition 1.1 reduces to the ones in Chapter I for the special cases there.

There is yet another advantage of Definition 1.1—it gives a very nice way of passing between the "affine" and the "projective." Let $\mathbb{P}^n(k)$ be as above; any n-dimensional subspace W of k^{n+1} then defines an $(n-1)$-dimensional projective subspace $\mathbb{P}^{n-1}(k)$ of $\mathbb{P}^n(k)$. We may choose this subspace to play the role of "projective hyperplane at infinity," $\mathbb{P}_\infty{}^{n-1}(k) \subset \mathbb{P}^n(k)$. What does the set of remaining points of $\mathbb{P}^n(k)$ look like? If we parallel-translate the subspace W through a fixed vector v_0 in $k^{n+1}\backslash W$, obtaining

$$v_0 + W = \{v_0 + w \,|\, w \in W\},$$

then each 1-subspace in $k^{n+1}\backslash W$ meets $v_0 + W$ in exactly one point. This sets up a 1:1 onto correspondence between the points of $\mathbb{P}^n(k)\backslash\mathbb{P}_\infty{}^{n-1}(k)$ and the points of k^n; Figure 2 indicates a typical situation for $k = \mathbb{R}$, $n = 2$.

Definition 1.7. We call the set $\mathbb{P}^n(k)\backslash\mathbb{P}_\infty{}^{n-1}(k)$ **the affine part of** $\mathbb{P}^n(k)$ **relative to the hyperplane at infinity** $\mathbb{P}_\infty{}^{n-1}(k)$.

Any affine n-space may be regarded as the affine part of a $\mathbb{P}^n(k)$ relative to some $\mathbb{P}_\infty{}^{n-1}(k)$ by taking a parallel translate of an n-dimensional subspace W of k^{n+1}, and identifying each point P of this parallel translate with the 1-subspace of k^{n+1} through P.

There are $n + 1$ particularly simple choices of $\mathbb{P}_\infty{}^{n-1}(k)$, namely the projective hyperplanes defined by each of the $n + 1$ hyperplanes $X_i = 0$ where $i = 1, \ldots, n + 1$. The following important observation will be particularly useful in the sequel:

The corresponding $n + 1$ affine parts of $\mathbb{P}^n(k)$ completely cover $\mathbb{P}^n(k)$.

This is true since the affine part corresponding to $X_1 = 0$ covers all of $\mathbb{P}^n(k)$ except those points represented by the 1-subspaces contained in $X_1 = 0$;

the union of the affine parts corresponding to $X_1 = 0$ and $X_2 = 0$ then covers all of $\mathbb{P}^n(k)$ except those points represented by the 1-subspaces in the intersection of $X_1 = 0$ and $X_2 = 0$; and so on. Clearly there are no 1-subspaces in k^{n+1} common to $X_1 = X_2 = \ldots = X_{n+1} = 0$, so the union of all these $n + 1$ affine parts covers all of $\mathbb{P}^n(k)$.

We have seen how an arbitrary $(n - 1)$-dimensional subspace can be the hyperplane at infinity of $\mathbb{P}^n(k)$. It is fair to ask why one would want to do this in the first place. Why not just stick to one standard affine n-space, letting its points be the finite points, and the added points always be *the* points at infinity? One answer is this: There is often much important geometry going on "at infinity," and many times one needs to know more precisely what is happening there. It is helpful in this to be able to "move" the line at infinity so that the infinite points become finite; these points then become points in an ordinary affine variety, where methods developed for affine varieties can be applied.

EXAMPLE 1.8. Consider the cubic curve $\mathsf{V}(Z - X^3)$ in $\mathbb{P}^2(\mathbb{C})$. It is of degree 3, so by Bézout's theorem any projective line must intersect this projective curve in 3 points, counted with multiplicity. Now it happens that if the projective line contains a line in the real part of $\mathbb{C}_{XZ}$, then all three of these points are "real"—that is, they are all in the projective completion $\mathbb{P}^2(\mathbb{R})$ of the real part of $\mathbb{C}_{XZ}$. Figure 8e shows the part of this curve in $\mathbb{P}^2(\mathbb{R})$. Since Z increases much faster than X for large X, the branches approach the Z-axis, and both meet at the infinite point of the Z-axis. The completed Z-axis should intersect the cubic in 3 points. The origin is clearly one point, the point at infinity, another. So far we have two points. But let's look more closely at what happens at infinity—in fact, let us try to get an explicit equation describing the curve near this infinite point.

First, represent the points of $\mathsf{V}(Z - X^3)$ in $\mathbb{P}^2(\mathbb{R})$ by 1-subspaces of $\mathbb{R}^3$. The usual (X, Z)-coordinate system is naturally induced in the plane $Y = 1$; that is, a point (x, z) corresponds to $(x, 1, z)$. Through each point of this translate $Y = 1$ of the subspace $Y = 0$, there is a unique 1-subspace of $\mathbb{R}^3$. The corresponding picture of $Z = X^3$, when looked at as a collection of 1-subspaces, is a surface in $\mathbb{R}^3$; it is a kind of generalized "cone through the origin" (cf. Figure 7).

To determine the equation of this surface, let $(a, 1, a')$ be a typical point of the curve as embedded in the hyperplane $Y = 1$. Now the symmetric equations of the 1-space through $(a, 1, a')$ are

$$\frac{X}{a} = \frac{Y}{1} = \frac{Z}{a'};$$

hence

$$a = \frac{X}{Y} \quad \text{and} \quad a' = \frac{Z}{Y}.$$

But we know $a' - a^3 = 0$, so

$$\frac{Z}{Y} - \left(\frac{X}{Y}\right)^3 = 0,$$

that is,

$$ZY^2 - X^3 = 0.$$

Hence each point of our surface satisfies this equation. Furthermore, no other points of $\mathbb{R}^3 \backslash \mathbb{R}_{XZ}$ satisfy it, because if (b, b', b'') is any point satisfying it, then clearly so does (cb, cb', cb'') for each $c \in \mathbb{R}$. Hence the intersection of the 1-space through (b, b', b'') with the plane $Y = 1$ satisfies it. Since this point is on the original curve $Z - X^3$ $(Y = 1)$, (b, b', b'') is on one of the lines of our surface.

Hence the whole algebraic variety $\mathsf{V}(ZY^2 - X^3) \subset \mathbb{R}^3$ coincides precisely with our surface on the "affine part," that is, on the 1-subspaces of $\mathbb{R}^3$ in $\mathbb{R}^3 \backslash \mathbb{R}_{XZ}$. (So far, we've "homogenized" $\mathsf{V}(Z - X^3)$ to the variety $\mathsf{V}(ZY^2 - X^3)$. We take this up formally in the next section.)

Now let us choose a new line at infinity so the original point at infinity is no longer infinite; we can do this by letting the hyperplane $Z = 0$ of $\mathbb{R}^3$ define the new line at infinity. We can get a picture of the new affine part by translating this hyperplane to $Z = 1$ and looking in this plane at the intersections with the 1-subspaces of $\mathbb{R}^3$. Such intersections of 1-subspaces in our surface $ZY^2 - X^3 = 0$ are then determined by setting $Z = 1$ in this equation. This yields the curve $Y^2 - X^3 = 0$, as in Figure I,1. Note that the infinite point in Figure 8e is now the new origin. One can see that the set of points constituting the original Z-axis in Figure 8e now corresponds to the new Y-axis.

But this Y-axis intersects the cusp with multiplicity two! Hence we do indeed have a total of 3 points of intersection, as Bézout's theorem promises.

This example perhaps gives the reader some feeling for why changing the roles of hyperplanes is important—sometimes essentially all the interesting geometry of a curve or variety takes place "at infinity," and one must be able to deal with this situation.

Exercises

1.1 Let S be a set and let T be a collection of mutually disjoint subsets of S whose union is S. If $x \in S$, denote by $\pi(x)$ that set of T containing x. (This is an identification map.) Now if S has a topology, define the corresponding identification topology on T as follows: An open set of T is any set $\mathcal{O} \subset T$ such that $\pi^{-1}(\mathcal{O})$ is open in S. Note that π is continuous with respect to this topology.

Let S be the real sphere $\mathsf{V}(X_1{}^2 + \ldots + X_{2n+2}{}^2 - 1)$ in $\mathbb{R}_{X_1, \ldots, X_{2n+2}} = \mathbb{C}^{n+1}$. Show that the intersections with S of the complex 1-subspaces of $\mathbb{C}^{n+1}$ form a mutually disjoint set of subsets (circles) of S whose union is S. Using the continuity of the corresponding identification topology and the compactness of S relative to the usual topology on $\mathbb{R}^{2n+2}$, conclude that $\mathbb{P}^n(\mathbb{C})$ is compact.

1.2 Prove that $\mathbb{P}^n(\mathbb{R})$ is compact.

1.3 Let B be a closed unit n-ball in $\mathbb{R}^n$, supplied with the usual topology. Identify to a point each pair of antipodal points on B's boundary. Show that the resulting space is topologically the same as $\mathbb{P}^n(\mathbb{R})$. Can you analogously identify appropriate subsets to points on the surface of a real $2n$-ball to arrive at $\mathbb{P}^n(\mathbb{C})$?

2 Affine and projective varieties; examples

In this section we look at projective varieties and some general facts about passing from the "affine" to the "projective" and back again; we illustrate these facts with some examples.

Let k be an arbitrary field.

Definition 2.1. A **homogeneous subset of** k^n is any subset S of k^n satisfying

$$x \in S \text{ implies } cx \in S \quad \text{for all } c \in k.$$

Hence a set is homogeneous iff it is $\varnothing$, $\{0\}$, or consists of a nonempty union of 1-subspaces of k^n. In view of Definition 2.1, a homogeneous set in k^n can be regarded as a set of points in $\mathbb{P}^{n-1}(k)$. (One regards $\{0\} \subset k^n$ as $\varnothing \subset \mathbb{P}^{n-1}(k)$.)

Definition 2.2. A homogeneous set in k^n **represents** a set in $\mathbb{P}^{n-1}(k)$. Any subset of a projective space is a **projective set**; a projective set in $\mathbb{P}^{n-1}(k)$ is **represented by** the corresponding homogeneous set in k^n.

Definition 2.3. A nonzero polynomial $q = q(X_1, \ldots, X_n) \in k[X_1, \ldots, X_n]$ is **homogeneous** (and q is called a **form**) if all its terms have the same total degree; if this degree is d, then q is **homogeneous of degree** d. Any polynomial p of degree s can be written as a sum of polynomials $p_0 + p_1 + \ldots + p_s$, where p_i is 0 or a homogeneous polynomial of degree i; if p_i is nonzero, then p_i is called the **homogeneous component of degree** i **of** p.

EXAMPLE 2.4. $q(X, Y) = X^5 + 2X^2Y^3 - 3XY^4 + Y^5$ is homogeneous of degree 5; $q(X, X_2, \ldots, X_n) \equiv 1 \in k[X_1, \ldots, X_n]$ is homogeneous of degree 0; by convention, the zero polynomial in $k[X_1, \ldots, X_n]$ has degree ∞.

Definition 2.5. A **homogeneous variety** in k^n is an algebraic variety which is a homogeneous set.

Theorem 2.6. *Let k be infinite. An algebraic variety V in k^n is homogeneous iff it is defined by a set of homogeneous polynomials. (We agree that the variety defined by the empty set of polynomials is k^n.)*

Proof. Since the theorem is trivial if $V = k^n$, assume $V \subsetneqq k^n$.

$\Leftarrow$: Let the variety be $V = \mathsf{V}(q_1, \ldots, q_r)$, where each $q_i \in k[X_1, \ldots, X_n]$ is homogeneous of degree d_i. Now $x \in V$ iff $q_i(x) = 0$ for $i = 1, \ldots, r$. But for any $t \in k$, $q_i(tx) = t^{d_i} q_i(x)$, so $x \in V$ implies $tx \in V$ for all $t \in k$.

$\Rightarrow$: Suppose $V = \mathsf{V}(q_1, \ldots, q_r)$ is homogeneous. Now V may be homogeneous without every (or even any) q_i being homogeneous. (Example: $\{0\} \subset \mathbb{R}^2$ is homogeneous, yet it is the intersection of two parabolas, $\{0\} = \mathsf{V}(Y + X^2, Y - X^2)$.) However, we shall manufacture from $\{q_1, \ldots, q_r\}$ a set of homogeneous polynomials defining V; this set is just *the set of all homogeneous components of all the* q_i. (Thus, $\{0\} = \mathsf{V}(Y, X^2, Y, -X^2) = \mathsf{V}(Y, X^2)$.)

Let x_0 be a fixed point in V; then each $q_i(x_0) = 0$. Now let t be an arbitrary element of k, and write $q_i = \sum q_{ij}$, where q_{ij} is the homogeneous component of degree j of q_i. Then

$$q_i(tx_0) = \sum t^j q_{ij}(x_0); \tag{1}$$

since x_0 is fixed and t is arbitrary, the polynomial in (1) may be looked at as a polynomial in an indeterminant T, namely $q_i(Tx_0) \in k[T]$. Since V is homogeneous, $q_i(Tx_0)$ is 0 for each $T = t$; because k is infinite, $q_i(Tx_0)$ is the zero polynomial in $k[T]$. Hence each coefficient of $q_i(Tx_0)$ is 0—that is, each $q_{ij}(x_0) = 0$. Hence $x_0 \in V$ implies $q_{ij}(x_0) = 0$ for each q_{ij} above, or q_{ij} is 0 at each point of V. Hence $V \subset \mathsf{V}(\{q_{ij}\})$. But obviously each $q_{ij}(x_0) = 0$ implies each $q_i(x_0) = 0$, so $\mathsf{V}(\{q_{ij}\}) \subset V$. Therefore $V = \mathsf{V}(\{q_{ij}\})$, so "$\Rightarrow$" is proved. □

Definition 2.7. A **projective variety** V in $\mathbb{P}^n(k)$ is a subset of $\mathbb{P}^n(k)$ represented by a homogeneous variety in k^{n+1}. If $q_i \in k[X_1, \ldots, X_{n+1}]$, where $i = 1, \ldots, r$, are homogeneous polynomials, then by abuse of language, $\mathsf{V}(q_1, \ldots, q_r)$ denotes the projective variety in $\mathbb{P}^n(k)$ represented by the homogeneous variety $\mathsf{V}(q_1, \ldots, q_r)$ in k^{n+1}.

It is clear that the intersection of any number of homogeneous varieties in k^{n+1} is a homogeneous variety; likewise for projective varieties in $\mathbb{P}^n(k)$. Hence for any subset of $\mathbb{P}^n(k)$ there is a smallest projective variety in $\mathbb{P}^n(k)$ containing that subset.

Definition 2.8. Let $k^n \subset \mathbb{P}^n(k)$, and let V be a variety in k^n. The smallest projective variety in $\mathbb{P}^n(k)$ containing V is called the **projective completion of V in $\mathbb{P}^n(k)$** and is denoted by V^c; sometimes notationally it is preferable to refer to the homogeneous variety $\mathsf{H}(V)$ in k^{n+1} representing V^c, and we then also denote V^c by $\mathsf{H}(V)$.

Definition 2.9. Let $p(X_1, \ldots, X_n) \in k[X_1, \ldots, X_n]$ be of degree d, and write $p = p_0 + p_1 + \ldots + p_d$ as in Definition 2.3. Then

$$p_0 X_{n+1}{}^d + p_1 X_{n+1}{}^{d-1} + \ldots + p_d \in k[X_1, \ldots, X_n, X_{n+1}]$$

is homogeneous of degree d and is called the **homogenization of** p; we denote it by $\mathsf{H}_{X_{n+1}}(p)$, $\mathsf{H}_{n+1}(p)$, or by just $\mathsf{H}(p)$, depending on context. If $p \in k[X_1, \ldots, \hat{X}_i, \ldots, X_{n+1}]$, the **homogenization** $\mathsf{H}_{X_i}(p) = \mathsf{H}_i(p)$ **of** p **at** X_i is defined analogously.

Remark 2.10. Suppose $k = \mathbb{C}$. It turns out that if $\mathsf{V}(p_1, \ldots, p_r) \subset \mathbb{C}^n$, then

V^c is represented by the homogeneous variety

$$\mathsf{V}(\mathsf{H}_{n+1}(p_1), \ldots, \mathsf{H}_{n+1}(p_r)) \subset \mathbb{C}^{n+1}. \tag{2}$$

Also we shall see that V^c is the topological closure of V in $\mathbb{P}^n(\mathbb{C})$. Neither of these statements is true in general for varieties over $\mathbb{R}$ (see Example 2.22). However, any projective variety $V \subseteq \mathbb{P}^n(\mathbb{R})$ is topologically closed. $H(V)$ is the intersection of hypersurfaces, and each hypersurface is the inverse image under a polynomial function of the closed set $\{0\} \subset \mathbb{R}$; hence each hypersurface is closed. (We note, of course, that our canonical map from $\mathbb{R}^{n+1}$ to $\mathbb{P}^n(\mathbb{R})$ sends closed sets to closed sets.) Thus if the topological closure of V in $\mathbb{P}^n(\mathbb{R})$ is a variety, it is the projective completion of V; this will be of use to us in the examples of this section.

We have homogenized both polynomials and varieties; those operations, when $k = \mathbb{C}$, are related in (2). We can also reverse the process:

Definition 2.11. Let $V \subset \mathbb{P}^n(k)$ be a projective variety, and $\mathbb{P}_\infty{}^{n-1}(k)$ a choice of hyperplane at infinity. The part of V in $\mathbb{P}^n(k)\backslash\mathbb{P}_\infty{}^{n-1}(k)$ is called the **dehomogenization of** V **at** $\mathbb{P}_\infty{}^{n-1}(k)$, or the **affine part of** V **relative to** $\mathbb{P}_\infty{}^{n-1}(k)$.

The $n + 1$ canonical choices of hyperplanes described in the last section (defined by $X_1 = 0, \ldots, X_{n+1} = 0$ in k^{n+1}) induce $n + 1$ canonical dehomogenizations of $\mathbb{P}^n(k)$, and also of any projective variety V in $\mathbb{P}^n(k)$. As before, V is covered by the $n + 1$ corresponding affine parts of V.

Notation 2.12. We denote the dehomogenization of V at $\mathbb{P}_\infty{}^{n-1}$ by $\mathsf{D}_{\mathbb{P}_\infty{}^{n-1}}(V)$, or by $\mathsf{D}(V)$ if the hyperplane $\mathbb{P}_\infty{}^{n-1}$ is clear from context. We denote the above canonical dehomogenizations of any V by $\mathsf{D}_1(V), \ldots, \mathsf{D}_{n+1}(V)$.

Just as there are $n + 1$ canonical dehomogenizations of $\mathbb{P}^n(k)$, there are also $n + 1$ canonical dehomogenizations of any homogeneous polynomial $p \in k[X_1, \ldots, X_{n+1}]$.

Definition 2.13. Let $q(X_1, \ldots, X_{n+1}) \in k[X_1, \ldots, X_{n+1}]$ be a homogeneous polynomial. The polynomial

$$q(X_1, \ldots, X_{i-1}, 1, X_{i+1}, \ldots, X_{n+1})$$

is called the **dehomogenization of** $q(X_1, \ldots, X_{n+1})$ **at** X_i; we denote it by $\mathsf{D}_{X_i}(q)$, by $\mathsf{D}_i(q)$, or by $\mathsf{D}(q)$ if clear from context.

Lemma 2.14. *Let* $q_1, \ldots, q_r \in k[X_1, \ldots, X_{n+1}]$ *be homogeneous; let* $\mathsf{V}(q_1, \ldots, q_r) \subset \mathbb{P}^n(k)$ *be the projective variety defined by* $q_1, \ldots, q_r$. *Then*

$$\mathsf{D}_i(\mathsf{V}(q_1, \ldots, q_r)) = \mathsf{V}(\mathsf{D}_i(q_1), \ldots, \mathsf{D}_i(q_r)). \tag{3}$$

PROOF. The variety $\mathsf{V}(\mathsf{D}_i(q_1), \ldots, \mathsf{D}_i(q_r))$ can be looked at as the intersection of the variety $\mathsf{V}(q_1, \ldots, q_r)$ with the plane given by $X_i = 1$ in k^{n+1}. □

Here are some relations between D and H:

Lemma 2.15. *Let* $p \in k[X_1, \ldots, X_n]$. *Then*

$$\mathsf{D}_i(\mathsf{H}_i(p)) = p.$$

PROOF. Obvious from the definitions of D_i and H_i. □

Lemma 2.16. *Let* q *be a homogeneous polynomial in* $k[X_1, \ldots, X_n]$. *Then for any* $i = 1, \ldots, n$, *it can happen that*

$$\mathsf{H}_i(\mathsf{D}_i(q)) \neq q.$$

PROOF. Let $q(X_1, X_2) = X_1X_2$. Then $\mathsf{D}_1(q) = X_2$, and $\mathsf{H}_1(\mathsf{D}_1(q)) = X_2 \neq X_1X_2$. Similarly, $\mathsf{H}_2(\mathsf{D}_2(q)) = X_1 \neq X_1X_2$. □

Lemma 2.17. *Let* $\mathbb{P}_\infty{}^{n-1}(k)$ *be a hyperplane at infinity of* $\mathbb{P}^n(k)$, *and let* $V \subset k^n = \mathbb{P}^n(k)\setminus\mathbb{P}_\infty{}^{n-1}(k)$. *Let* $\mathsf{H}(V)$ *be the projective completion of* V, *and* D *the operation of dehomogenizing* $\mathsf{H}(V)$ *at* $\mathbb{P}_\infty{}^{n-1}$. *Then*

$$\mathsf{D}(\mathsf{H}(V)) = V. \tag{4}$$

But if V *is a variety in* $\mathbb{P}^n(k)$, *then it can happen that*

$$\mathsf{H}(\mathsf{D}(V)) \neq V. \tag{5}$$

PROOF. We leave verification of (4) as an easy exercise. (5) follows from Lemma 2.16 by letting V be $\mathsf{V}(X_1X_2)$. More generally, if V is any variety in $\mathbb{P}^n(k)$ not containing $\mathbb{P}_\infty{}^{n-1}(k)$, then (5) holds for the variety $V \cup \mathbb{P}_\infty{}^{n-1}(k)$. □

We now give some illustrations of the above ideas. Many of the essential features can be brought out using real varieties; in fact we can learn much from real curves in $\mathbb{R}^2$ and $\mathbb{P}^2(\mathbb{R})$. The reader will see that various ways of looking at $\mathbb{P}^2(\mathbb{R})$ (1-subspaces in $\mathbb{R}^3$, sphere with identified antipodal points, disk with antipodal boundary points identified) will all be valuable in understanding the nature of projective curves, of homogenization, and of dehomogenization.

In the first four examples (Examples 2.18–2.21) we start with an affine variety $V = \mathsf{V}(p) \subset \mathbb{R}_{XY}$, $p \in \mathbb{R}[X, Y]$. The homogenized polynomial $\mathsf{H}_Z(p)$ then defines a homogeneous variety $\mathsf{H}_Z(V) = \mathsf{H}(V)$ in $\mathbb{R}_{XYZ}$, the original affine part in $\mathbb{R}_{XY}$ being represented by the 1-spaces of $\mathsf{H}(V)$ in $\mathbb{R}_{XYZ} \backslash \mathbb{R}_{XY}$. In each of these examples we note that

$$[\mathsf{H}(V) \cap (\mathbb{R}_{XYZ} \backslash \mathbb{R}_{XY})]^{-} = \mathsf{H}(V).$$

(The bar denotes topological closure in $\mathbb{R}_{XYZ}$.) Since $\mathsf{H}(V)$ is a variety, by our earlier observation (Remark 2.10), $\mathsf{H}(V)$ represents the projective completion V^c of V.

EXAMPLE 2.18. Consider the real circle $\mathsf{V}(X^2 + Y^2 - 1) \subset \mathbb{R}^2$. The homogenized polynomial $X^2 + Y^2 - Z^2 \in \mathbb{R}[X, Y, Z]$ determines the cone in Figure 3 as well as the circles in Figure 4. (Since antipodal points are identified, this is just one circle.)

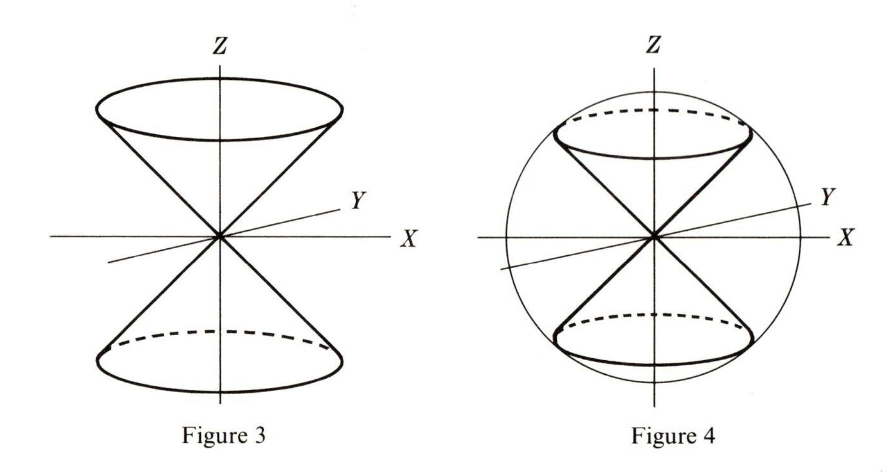

Figure 3

Figure 4

We may dehomogenize at an arbitrary $\mathbb{P}_\infty{}^1(\mathbb{R})$ by choosing an appropriate 2-space in $\mathbb{R}_{XYZ}$. Since the intersection of the cone with a parallel translate of this 2-space yields a copy of the affine part of the curve with respect to $\mathbb{P}_\infty{}^1(\mathbb{R})$, we see the various affine parts of the circle in $\mathbb{P}^2(\mathbb{R})$ are conic sections. Thus, dehomogenizing at the plane in $\mathbb{R}_{XYZ}$ with Z-axis as normal gives a circle, and as we vary the normal, we get ellipses, a parabola, and hyperbolas. Likewise, dehomogenizing $X^2 + Y^2 - Z^2$ to $X^2 + Y^2 - 1$, to $X^2 + 1 - Z^2$, and to $1 + Y^2 - Z^2$ yields a circle and two hyperbolas, respectively.

We may also get specific equations for affine curves induced in 2-spaces other than in the above canonical way. For example, let the 2-space in $\mathbb{R}^3$ given by $X = Z$ define the line at infinity; this subspace intersects our cone in just one 1-subspace L. Hence in $\mathbb{P}^2(\mathbb{R})$ the curve touches this line at infinity in exactly one point. An affine representative with respect to this infinite line is obtained by intersecting the cone with a parallel translate of the plane $X = Z$, say $X = Z + 1$. What is the polynomial describing this affine representative? It may easily be found by choosing new coordinates X', Y', Z' of $\mathbb{R}^3$ so the new Z'-axis is the 1-subspace L. This may be done by setting

$$X = X' + Z', \qquad Y = Y', \qquad Z = Z'.$$

The plane $X = Z + 1$ then becomes $X' = 1$; the equation of the cone in these coordinates becomes

$$(X')^2 + 2X'Z' + (Y')^2 = 0.$$

In the affine plane $X' = 1$ this equation becomes

$$1 + 2Z' + (Y')^2 = 0$$

which is a parabola. The sketches of the affine curve in $\mathsf{V}(X' - 1)$ appears in Figure 5. (We identify $\mathsf{V}(X' - 1)$ with $\mathbb{R}_{Y'Z'}$.) The sketch of the entire projective curve appears in Figure 6.

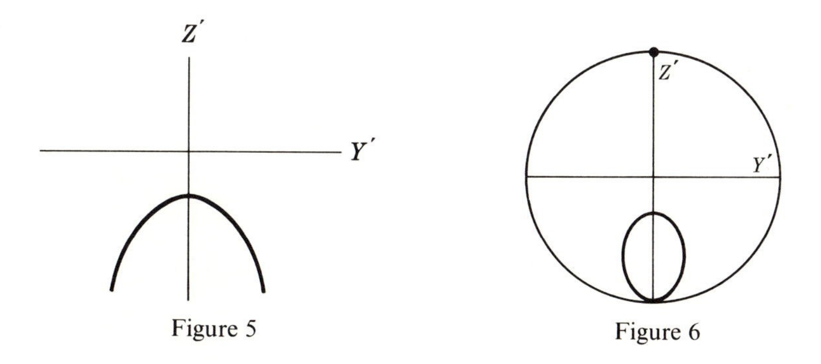

Figure 5 Figure 6

Our projective circle now touches the line at infinity in just one point, but by making this point finite, we can easily show it does so with multiplicity two.

From what we have said so far, the reader can see that from a projective viewpoint, the difference between circles, ellipses, parabolas and hyperbolas is simply a matter of where the line at infinity is chosen, and these affine curves all correspond to one projective curve. Hence, from the complex viewpoint, the extension to $\mathbb{P}^2(\mathbb{C})$ of any real conic section is still topologically a sphere.

EXAMPLE 2.19. We return to the curve $\mathsf{V}(Y^2 - X^3)$. Homogenizing $Y^2 - X^3$ gives $Y^2Z - X^3$; Figure 7 shows the homogeneous surface $\mathsf{V}(Y^2Z - X^3)$. The surface is, of course, the union of lines through the origin. In Figure 8a–c we sketch the curves in $\mathbb{R}^2$ after dehomogenizing at the planes $Z = 0$, $Y = 0$, and $X = 0$; the reader can see these are just the intersections of the surface of Figure 7 with parallel translates of the (X, Y), (X, Z), and (Y, Z) planes, respectively. Figure 8d–f show the corresponding completions in the disk with opposite boundary points identified. The points P and Q correspond to two points on the projective curve, the *cusp point* and *flex point*.

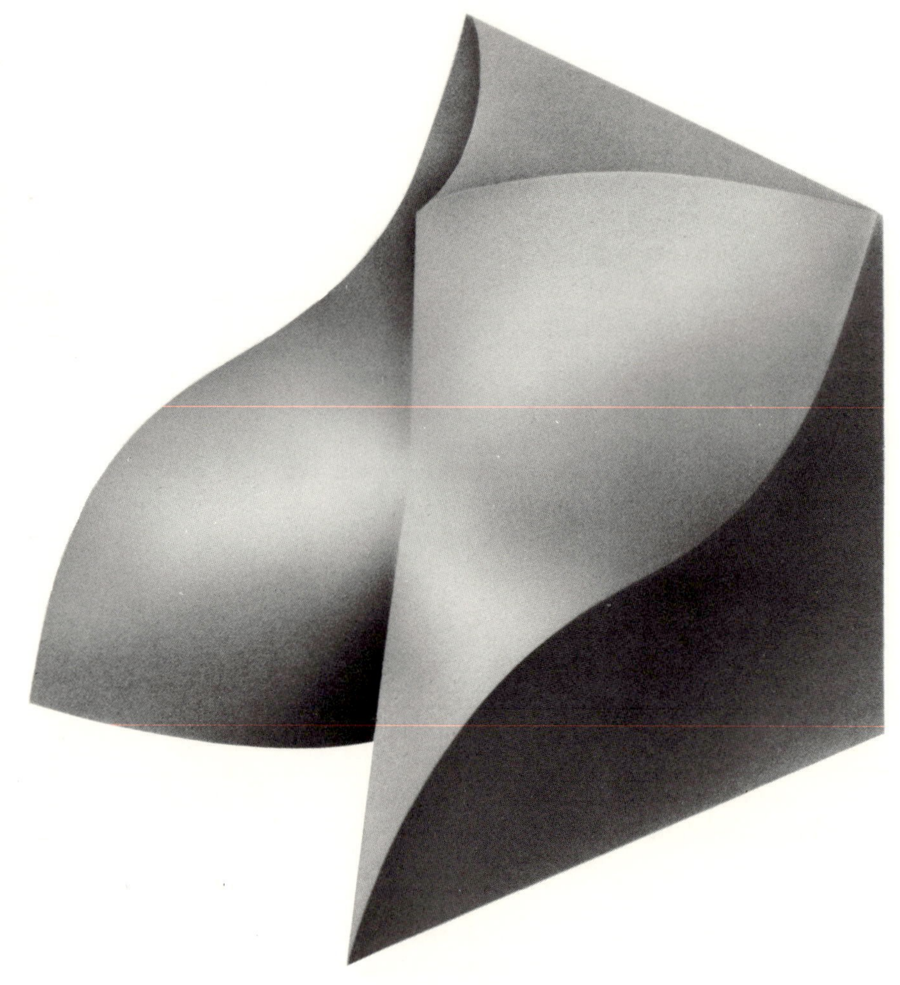

Figure 7

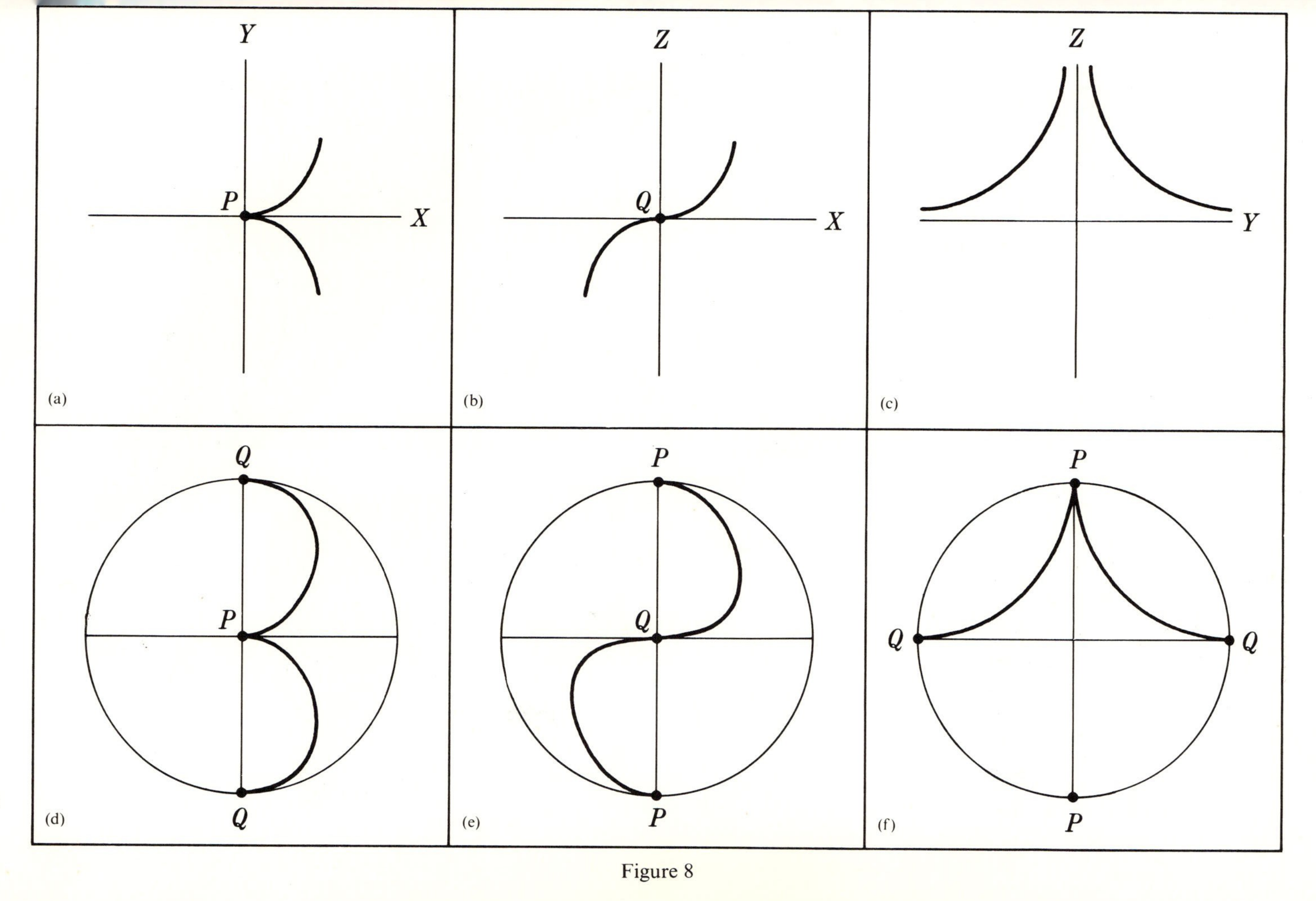

Figure 8

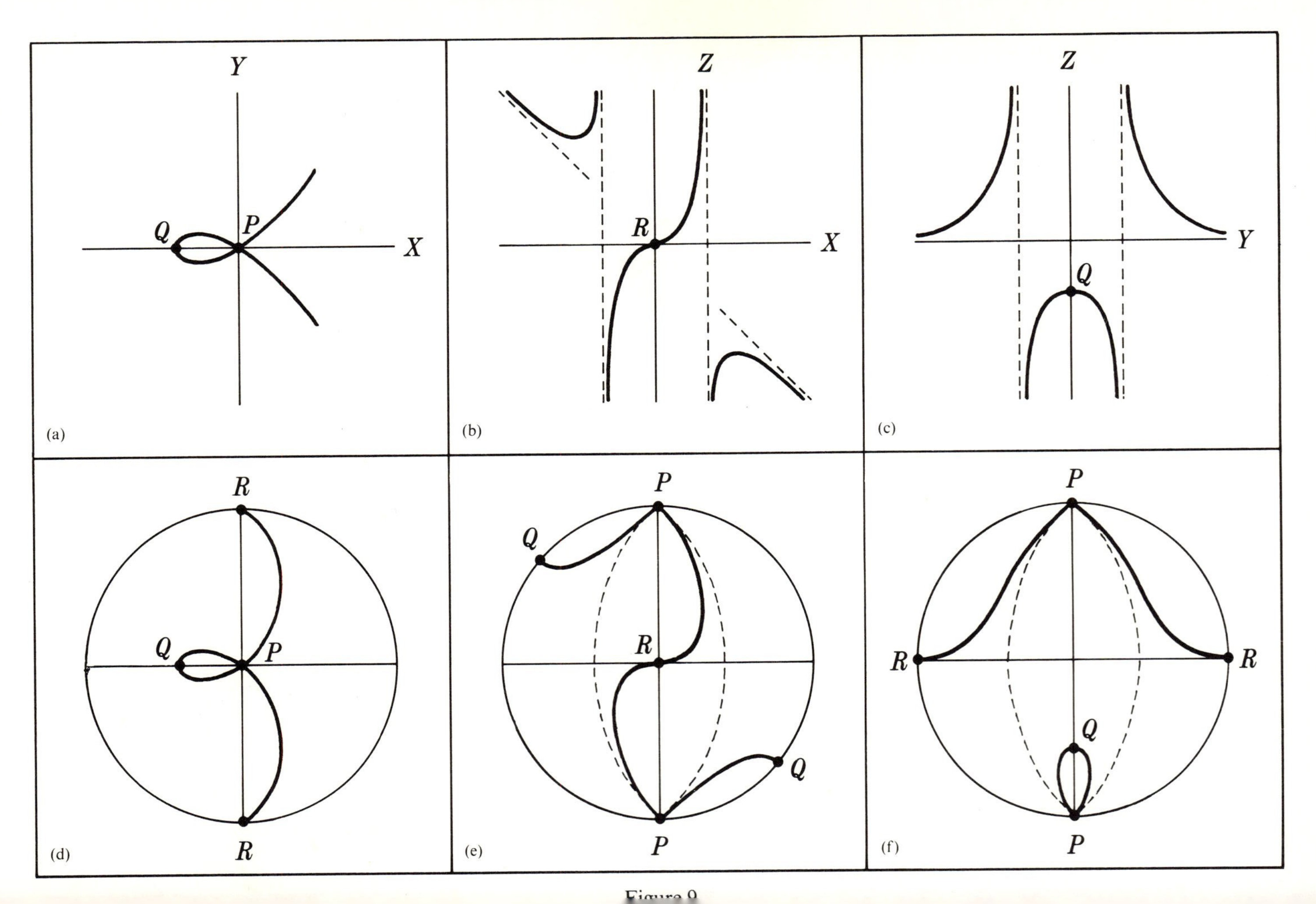

Figure 9

EXAMPLE 2.20. We next consider the alpha curve $V(Y^2 - X^2(X + 1)) \subset \mathbb{R}^2$. Homogenizing the polynomial gives

$$Y^2Z - X^2(X + Z) = 0;$$

the intersection of $V(Y^2Z - X^2(X + Z))$ with a sphere centered at (0, 0, 0) is shown in Figure 10. Figure 9a–c show the affine parts after dehomogenizing at Z, Y, and X, respectively; Figure 9d–f are the corresponding completions in the disk. Three points, P, Q, R of the projective curve are indicated. The loop between P and Q forms a topological circle, as do the two arcs between P and R. It is instructive to trace out a cycle around each of these two loops, as well as the figure 8 pattern, in the six sketches and on the sphere. These two real loops are the ones between -1 and 0, and between 0 and P_∞ appearing on the pinched torus of Figure I,22.

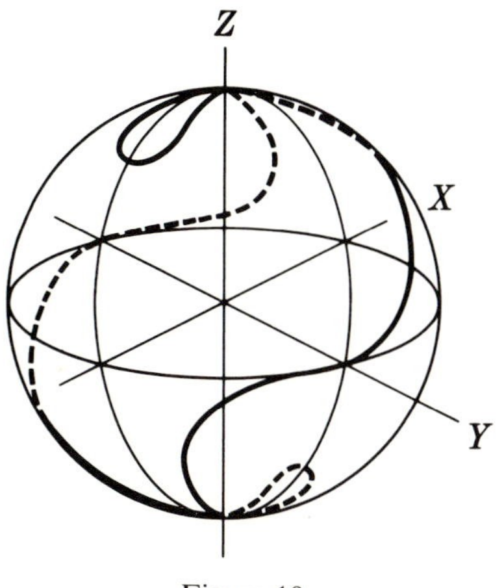

Figure 10

EXAMPLE 2.21. Let $V \subset \mathbb{R}^2$ consist of n distinct parallel lines. If the lines are $L_1, \ldots, L_n$, given by, say, $Y = 1$, $Y = 2, \ldots, Y = n$, then the union of these lines is given by $p(X, Y) = (Y - 1)(Y - 2) \cdot \ldots \cdot (Y - n) = 0$. Since the projective completion of each line intersects any other projective line in one point, it might be guessed that the union of the n projective lines should intersect any other distinct line in n points, counted with multiplicity. This is obvious except when the line is the line at infinity, or if the line is parallel to the X-axis. To explore these cases, we homogenize and then dehomogenize so the intersection point (at infinity) becomes the new origin.

Homogenizing $p(X, Y)$ gives us

$$H_Z(p) = (Y - Z) \cdot (Y - 2Z) \cdot \ldots \cdot (Y - nZ) = 0$$

this describes the union of n planes containing the X-axis in $\mathbb{R}_{XYZ}$; its intersection with a sphere centered at (0, 0, 0) consists of n great circles (Figure 11a). Dehomogenizing at $X = 0$ yields Figure 11b; the original line at

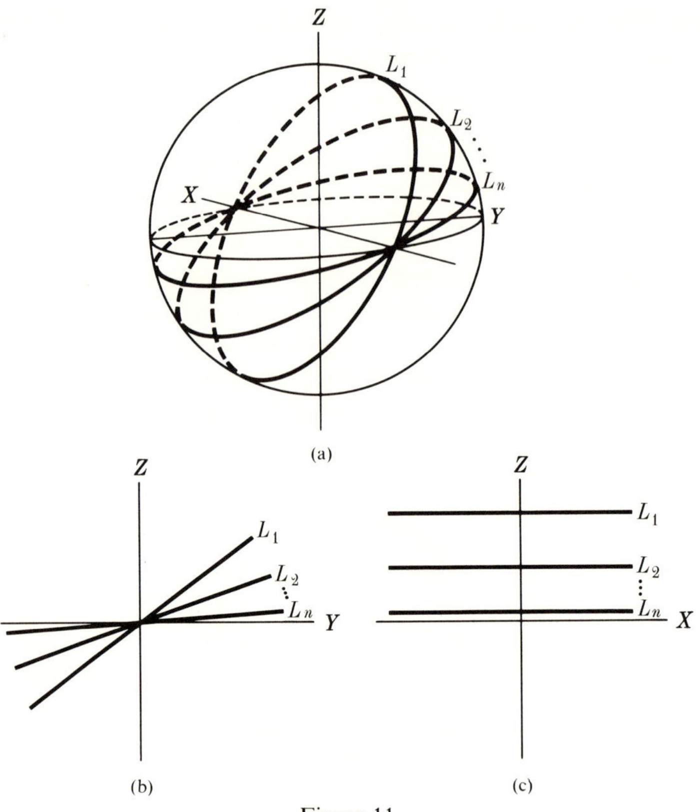

Figure 11

infinity is the great circle corresponding to $Z = 0$, which appears as the Y-axis in Figure 11b; the n lines intersect the Y-axis at the origin with multiplicity n. And any other distinct line parallel to the X-axis in the original (X, Y)-plane appears as a distinct line through the origin in the (Y, Z)-plane, so again will intersect the n lines in one point with multiplicity n. Of course since the degree of $p(X, Y) = (Y - 1) \cdot (Y - 2) \cdot \ldots \cdot (Y - n)$ is n, Bézout's theorem tells us that in the extension to $\mathbb{P}^2(\mathbb{C})$, the n complex projective lines intersect any other line in n points, counted with multiplicity. All these points turn out to be real in our example.

The affine part after dehomogenizing at $Y = 0$ appears in Figure 11c.

In all our examples so far the projective completion described by the homogenization of the polynomial has turned out to be just the topological closure in $\mathbb{P}^2(\mathbb{R})$ of the original affine variety. As mentioned earlier, this does not always hold when $k = \mathbb{R}$. We now illustrate this exceptional behavior.

EXAMPLE 2.22. If an isolated point is part of the real curve and this point is at infinity in $\mathbb{P}^2(\mathbb{R})$, then dehomogenizing at any line through this point yields an affine curve whose topological closure leaves that point out. We can easily manufacture a curve in $\mathbb{R}_{XY}$ having an isolated point. Consider the curve

$$\mathsf{V}(Y^2 - (X + 1)(X)(X - 1))$$

(see Figure I,3); if we let the zero -1 in the polynomial approach 0, the topological circle between -1 and 0 is squeezed to a point, and the curve's equation becomes

$$\lim_{\varepsilon \to 0}(Y^2 - (X + \varepsilon)(X)(X - 1)) = Y^2 - X^2(X - 1) = 0;$$

see Figure 12.

We can now make the origin an infinite point. First, the homogenized polynomial is $Y^2Z - X^2(X - Z)$; the reader can easily sketch the resulting surface by letting the plane of Figure 12 be the plane $Z = 1$, drawing

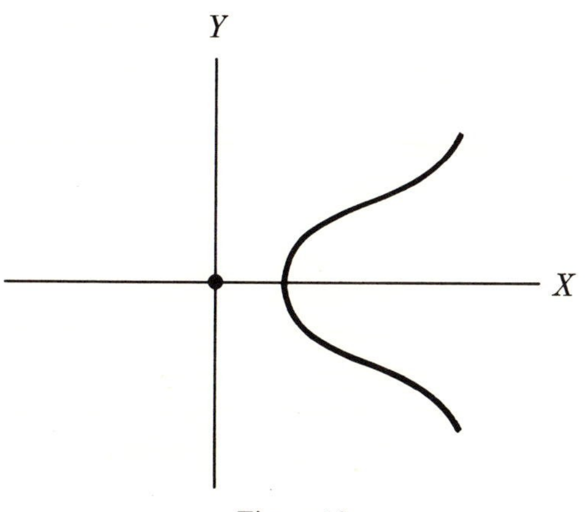

Figure 12

1-subspaces of $\mathbb{R}^3$ through this curve, and looking separately at the 1-subspaces in the plane $Z = 0$. Note that the Z-axis is in a sense an "isolated" line.

Now dehomogenizing at X gives us the desired equation of our curve:

$$Y^2Z - (1 - Z) = 0$$

The curve in $\mathbb{R}^2$ is sketched in Figure 13; Figure 14 shows the whole projective curve in the disk.

Next note that $\mathsf{H}_X(Y^2Z - (1 - Z))$ is irreducible in $\mathbb{R}[X, Y, Z]$. (First, $Y^2 - X^2(X - 1)$ is irreducible in $\mathbb{R}[X, Y]$ for since it has no Y-term, any factorization would have to be of the form $(Y + f(X))(Y - f(X)) = Y^2 - f^2(X)$. However, $X^2(X - 1)$ is not the square of any polynomial. Second, one can easily check that $Y^2 - X^2(X - 1)$ irreducible implies $\mathsf{H}_Z(Y^2 - X^2(X - 1))$ irreducible, so $\mathsf{H}_X(Y^2Z - (1 - Z))$, which equals

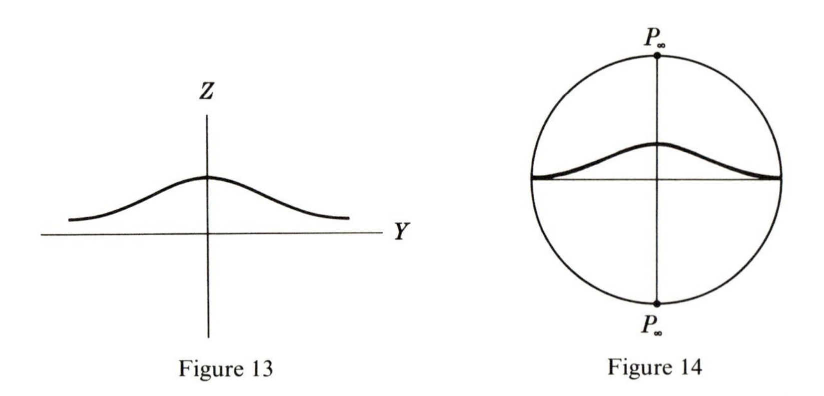

Figure 13

Figure 14

$\mathsf{H}_Z(Y^2 - X^2(X - 1))$, is irreducible.) We show in Exercise 4.3 of Chapter III that if p is irreducible, then $\mathsf{V}(p)$ is irreducible in the sense that it is never the union of two properly smaller varieties. Hence the part in $\mathbb{P}^2(\mathbb{R})$ excluding this isolated point is the topological closure of the affine part, but is not a variety in $\mathbb{P}^2(\mathbb{R})$.

Topologically, the extension to $\mathbb{P}^2(\mathbb{C})$ of this curve turns out to be the limit of Figure I,16 as the circle between -1 and 0 shrinks to a point; this is a "pinched sphere." The point where, say, the north and south poles of a sphere are identified, is the isolated point in the real curve, and the equator can be taken to correspond to the branch. Recall (Remark 2.10) that we stated that for $k = \mathbb{C}$ the topological closure always gives the projective completion. But our isloated point is no longer isolated in the curve's complex extension to $\mathbb{P}^2(\mathbb{C})$! Hence the topological closure in $\mathbb{P}^2(\mathbb{C})$ of the pinched sphere without this one point is again the whole pinched sphere.

Exercises

2.1 Sketch six figures corresponding to the six parts of Figure 8 or 9 for the curve $\mathsf{V}(XY^2 - Y - X)$. How many points of the curve are on $\mathbb{R}_{XY}$'s line at infinity? Are there additional points of the curve on $\mathbb{C}_{XY}$'s line at infinity?

2.2 Do the same as above for the curve $\mathsf{V}(X^2Y^2 + X^2 - Y^2)$.

2.3 For any positive integer n, find an algebraic curve C_n in $\mathbb{R}_{XY}$ whose topological closure in $\mathbb{P}^2(\mathbb{R})$ omits n points of C_n's real projective completion.

3 Implicit mapping theorems

In Chapter I we stated that a topological copy of any complex-algebraic curve can be obtained by taking a compact connected orientable 2-manifold and identifying a finite number of points to a finite number of points. Part of the proof of this fact (given in the next section) uses an "implicit mapping theorem." Since implicit mapping theorems are important in their own right and are used frequently in algebraic geometry, we devote a section to them.

There are quite a number of implicit mapping theorems; they occur at various levels; for instance in the real case there are "differentiable" and "analytic" implicit mapping theorems. The ones of greatest importance in algebraic geometry are "complex-analytic." To gain a little perspective, we state some of these mapping theorems in various forms; we then prove an analytic version needed in the next section.

We begin by recalling the following standard

Definition 3.1. Let U be an open set of $\mathbb{R}_{X_1,\dots,X_n}$. A function $f\colon U \to \mathbb{R}_Y$ is **differentiable at** $(a) = (a_1, \dots, a_n) \in U$ provided there is a real n-plane through $(a_1, \dots, a_n, f(a))$ given by, say,

$$Y = f(a) + c_1(X_1 - a_1) + \dots + c_n(X_n - a_n) \tag{6}$$

such that

$$\lim_{(x)\to(a)} \frac{f(x) - [f(a) + c_1(x_1 - a_1) + \dots + c_n(x_n - a_n)]}{|x_1 - a_1| + \dots + |x_n - a_n|} = 0,$$

where $(x) = (x_1, \dots, x_n) \in U \backslash (a)$.

The function f is called **differentiable on** U if it is differentiable at each point of U, and a map $f = (f_1 \dots f_m)\colon U \to \mathbb{R}^m$ is **differentiable at a point of** U, or **on** U, if each f_i is. If all partial derivatives

$$\frac{\partial^n f_i}{\partial X_{j_1} \dots \partial X_{j_n}}$$

of each f_i exist and are continuous at (a) or on U, then we say f is **smooth at** (a), or **smooth on** U.

Remark 3.2. The hyperplane in (6) is the "tangent hyperplane to the graph of f at $(a, f(a))$"; by letting $(x) \to (a)$ in the direction of the coordinate axis $\mathbb{R}_{X_i}$, we see that in (6),

$$c_i = (\partial f/\partial X_i)(a).$$

For more than one variable ($n > 1$), the definition of real-differentiability represents a great difference from only requiring the partials $\partial f/\partial X_i$ to exist. Definition 3.1 says that $f\colon U \to \mathbb{R}$ is "uniformly close" to the n-plane. Hence knowledge of the partials of f at $(a) \in \mathbb{R}^n$ is enough to determine its directional derivatives in every direction.

The definition of *complex-differentiability* for a function $f(X_1, \dots, X_n)$: $U \to \mathbb{C}_{Y_1,\dots,Y_n}$ (U open in $\mathbb{C}_{X_1,\dots,X_n}$) may be taken to be verbatim the same as in Definition 3.1, except everywhere we replace "real" and "$\mathbb{R}$" by "complex" and "$\mathbb{C}$", respectively, and $| \ |$ becomes the usual complex norm. For instance, in the special case of $f: \mathbb{C}_X \to \mathbb{C}_Y$ the definition is

Definition 3.3. $f(X): \mathbb{C}_X \to \mathbb{C}_Y$ is **complex-differentiable at** $a = a_1 + ia_2 \in \mathbb{C}_X$ if there exists $c \in \mathbb{C}$ such that

$$\lim_{x \to a} \frac{f(x) - [f(a) + c(x - a)]}{|x - a|} = 0.$$

Remark 3.4. If a function $f: \mathbb{R}_{X_1X_2} \to \mathbb{R}$ is real-differentiable at (a), then the derivatives along *two* different real lines through (a) determine the derivative in *any* real direction through (a). But in the complex case, the derivative along *one* real line through (a) determines the same derivative in *all* real directions through (a). This is a stringent condition, and leads at once to the Cauchy–Riemann equations. Also, this definition of complex-differentiability for f at all points in a neighborhood of $(a) \in \mathbb{C}_{X_1, \ldots, X_n}$ implies in particular that at (a) all partials $\partial^n f/(\partial X_i)^n$ exist, which means f is analytic in each variable separately—that is, for each $i, f(b_1 \ldots b_{i-1}, X_i, b_{i+1} \ldots b_n)$ is analytic at a_i for each b_j near a_j, $j = 1, \ldots i - 1, i + 1, \ldots n$. This in turn implies f itself is analytic; this is a central result due to Hartogs. (See, e.g., [Bochner and Martin, Chapter VII, Section 4].) (Recall that if U is an open set in $\mathbb{C}_{X_1, \ldots, X_n}$, then a complex-valued function $f(X_1, \ldots, X_n)$ on U is defined to be **complex-analytic**, or **analytic**, **at** $(a) = (a_1, \ldots, a_n) \in U$ provided that f is represented at all points of U in some neighborhood of (a) by a power series in $X_1 - a_1, \ldots, X_n - a_n$. The function is **analytic in** U if it is analytic at each point of U.) Thus complex-differentiability is equivalent to analyticity; in developing the theory of several complex variables, one usually simply starts with the concept of analyticity.

Now let us look at implicit function theorems. They are often useful in investigating the local nature of zero-sets. Essentially, they give conditions under which a zero-set may be considered as the graph of an appropriate function. This is important, because since a function is essentially its graph, differentiability or analyticity of a function are perfectly reflected in its graph. In contrast to this, the "niceness" of one or several functions often has little to do with the "niceness" of the corresponding zero-set. For example, any closed set in $\mathbb{R}^n$, no matter how "wild," is nonetheless the zero-set of some differentiable (even infinitely differentiable!) function $f: \mathbb{R}^n \to \mathbb{R}$ (see Exercise 3.1). Thus although $f(X, Y) = Y^2 - X^2 \in \mathbb{R}[X, Y]$ is smooth, the corresponding zero-set consists of two intersecting lines. However, we can write $V(Y^2 - X^2) = V(Y - X) \cup V(Y + X)$; each part (being the graph of a nice function) is therefore smooth.

We now state a general implicit mapping theorem at the complex level. All these theorems are *local*—they make a statement about points near an arbitrary but fixed point. Without loss of generality we let this point be the origin.

Theorem 3.5 (Implicit complex-analytic mapping theorem). *Suppose*:

(3.5.1) *$\mathbb{C}$-valued functions $f_1, \ldots, f_q$ are complex-analytic in a neighborhood of* $(0) \in \mathbb{C}_{X_1, \ldots, X_n} = \mathbb{C}_X$;

(3.5.2) $f_1(0) = \ldots = f_q(0) = 0$ (*i.e., the origin of $\mathbb{C}^n$ is in the zero-set of* $\{f_1, \ldots, f_q\}$);

(3.5.3) *The $q \times n$ Jacobian matrix*

$$J(f)_x = J(f_1, \ldots, f_q)_x = \begin{pmatrix} \dfrac{\partial f_1}{\partial X_1} \cdots \dfrac{\partial f_1}{\partial X_n} \\ \vdots \qquad \vdots \\ \dfrac{\partial f_q}{\partial X_1} \cdots \dfrac{\partial f_q}{\partial X_n} \end{pmatrix}$$

has constant rank r throughout some $\mathbb{C}^n$-open neighborhood of (0).

Then there exist subspaces $\mathbb{C}^{n-r}$ and $\mathbb{C}^r$ $(\mathbb{C}^{n-r} \cap \mathbb{C}^r = (0))$, *neighborhoods $U^{n-r} \subset \mathbb{C}^{n-r}$ and $U^r \subset \mathbb{C}^r$ about* (0), *and a unique complex-analytic map*

$$\phi = (\phi_1, \ldots, \phi_r)\colon U^{n-r} \to U^r$$

such that within $U^{n-r} \times U^r$, the graph of ϕ coincides with the zero-set of $\{f_1, \ldots, f_q\}$.

We will prove this theorem in Section IV,2. In fact we will show more precisely that if the independent variables X_i are remembered so that the last r columns of $J(f)_x$ are linearly independent for each (x) in a neighborhood of (0), then $\mathbb{C}^{n-r}$ may be taken to be $\mathbb{C}_{X_1, \ldots, X_{n-r}}$, and $\mathbb{C}^r$ to be $\mathbb{C}_{X_{n-r+1}, \ldots, X_n}$. Then the conclusion of Theorem 3.5 says that within $U^{n-r} \times U^r$, the equations

$$f_i(X_1, \ldots, X_{n-r}, Y_1, \ldots, Y_r) = 0 \qquad (i = 1, \ldots, q)$$

may be "solved" for $Y_1, \ldots, Y_r$ in a unique way, say

$$\begin{aligned} Y_1 &= \phi_1(X_1, \ldots, X_{n-r}), \\ &\vdots \\ Y_r &= \phi_r(X_1, \ldots, X_{n-r}). \end{aligned}$$

Special cases of the above theorem

1. Often the rank of $J(f)_x$ is the same as the number of functions f_i—that is, $r = q$. (Geometrically this means that the zero-sets of $f_1, \ldots, f_q$ intersect "transversally.") Let the last r columns of $J(f)$ be linearly independent at (0), and denote the determinant of the associated $r \times r$ matrix by $\det(J_r(f))$. Since this determinant is continuous in X at (0), (3.5.3) can be simplified to

$$\det(J_r(f))_{X=(0)} \neq 0.$$

2. Sometimes there is given just one function f (that is, $q = 1$). Then if at (0) not every partial of f is zero (say $(\partial f/\partial X_i)(0) \neq 0$), the implicit mapping theorem tells us that near (0), the zero-set in $\mathbb{C}_{X_1,\ldots,X_n}$ of f forms the graph of an analytic function $\phi:\mathbb{C}_{X_1,\ldots,X_{i-1},X_{i+1},\ldots,X_n} \to \mathbb{C}_{X_i}$. (Thus, about $(0) \in \mathbb{R}_{XY}$, the zero-set of $Y^2 - X$ does not form a graph of a function from $\mathbb{C}_X$ to $\mathbb{C}_Y$, but it does from $\mathbb{C}_Y$ to $\mathbb{C}_X$. But about (0) there exists no coordinate system relative to which the zero-set of, say, $Y^2 - X^2$, is the graph of a function.)

3. The implicit mapping theorem can be used to give a condition for the existence of inverse mappings. Let $f = (f_1, \ldots, f_m)$ map $(0) \in \mathbb{C}_{X_1,\ldots,X_m} = \mathbb{C}_X$ to $(0) \in \mathbb{C}_{Y_1,\ldots,Y_m} = \mathbb{C}_Y$ and let f be complex-analytic at (0). First, look at the graph of f as the zero-set of m analytic functions $h_1, \ldots, h_m$ on a neighborhood of (0) in $\mathbb{C}^{2m} = \mathbb{C}_{XY}$, where

$$h_i(X, Y) = Y_i - f_i(X).$$

We know the common zero-set of the h_i forms the graph of a function having domain in $\mathbb{C}_X$. (And of course the last m columns (the "Y-columns") of $J(h) = J(h_1, \ldots, h_m)$ are linearly independent, forming the identity matrix.) Now f has an inverse at $(0) \in \mathbb{C}_X$ if the same zero-set forms the graph of a function with a neighborhood in $\mathbb{C}_Y$ as domain, instead. This will be satisfied if the *first* m columns (the "X columns") of the $m \times 2m$ Jacobian of $h_1, \ldots, h_m$, are linearly independent. Since $\partial h_i/\partial X_j = \partial f_i/\partial X_j$, the implicit mapping theorem becomes in this case:

> If f is analytic at $(0) \in \mathbb{C}_{X_1,\ldots,X_m}$ and $f(0) = 0 \in \mathbb{C}_{Y_1,\ldots,Y_m}$, then f has a unique analytic inverse in a neighborhood of $(0) \in \mathbb{C}_{Y_1,\ldots,Y_m}$ provided
>
> $$\det\left(\frac{\partial f_i}{\partial X_j}\right)_{X=(0)} \neq 0.$$

A very simple case of this is when $f:\mathbb{C} \to \mathbb{C}$ is analytic at, say, $x_0 \in \mathbb{C}$. If $f'(x_0) \neq 0$, then f has an analytic inverse in a neighborhood of $f(x_0)$.

We now prove the following case of the implicit mapping theorem which is used in the next section. The proof readily extends to one for the full Theorem 3.5 which we present when needed, in Section IV,2. To keep formulas compact, we will use subscripts for partial differentiation, e.g., p_Y for $\partial p/\partial Y$.

Theorem 3.6. *Let* $p(X, Y) \in \mathbb{C}[X, Y]$ *satisfy*

(3.6.1) $p(0, 0) = 0$, *and*
(3.6.2) $p_Y(0, 0) \neq 0$.

Then within some neighborhood of (0, 0), *those points* (x, y) *satisfying* $p(x, y) = 0$ *form the graph of a function* $Y = \phi(X)$ *analytic at* $(0) \in \mathbb{C}_X$.

In proving this theorem we assume the following standard integral theorems of complex variables. For our purposes it suffices to state them "in a disk." (See, e.g., [Ahlfors, Chapter IX] for fuller statements and generalizations.)

Definition 3.7. Let $f(X)$ be analytic at a point $a \in \mathbb{C}_X$. Then a is a **zero of multiplicity** n, or a **zero of order** n, if $f(X) = (X - a)^n h(X)$, where $h(X)$ is analytic at a, and $h(a) \neq 0$; we also say f **has order** n **at** a.

Theorem 3.8 (Two basic integral theorems). *Let $f(X)$ be a function analytic at each point of an open set containing a closed disk $\bar{\Delta}$ in $\mathbb{C}$ with boundary $\partial\Delta$, and suppose that within $\Delta = \bar{\Delta}\backslash\partial\Delta$ there are exactly N zeros of $f(X)$, counted with multiplicity. Then:*

(3.8.1) Cauchy integral formula:
For any point $b \in \Delta$,

$$\frac{1}{2\pi i}\int_{\partial\Delta}\frac{f(X)}{X - b}\,dX = f(b);$$

(3.8.2) Argument principle:
If $f(X) \neq 0$ on $\partial\Delta$, then

$$\frac{1}{2\pi i}\int_{\partial\Delta}\frac{f'(X)}{f(X)}\,dX = N.$$

PROOF OF THEOREM 3.6. That the zero-set of $p(X, Y)$ near $(0, 0)$ forms the graph of some function $Y = \phi(X)$ will follow easily from the argument principle; it will then be our task to prove that ϕ is analytic.

To show the zero-set forms a graph, we first note that the definition of multiplicity of a zero shows that the hypotheses $p(0, 0) = 0$ and $p_Y(0, 0) \neq 0$ together form a way of expressing that the polynomial $p(0, Y) \in \mathbb{C}(Y)$ has $Y = 0$ as a zero of multiplicity 1. Hence if Δ is a sufficiently small open disk about $(0) \in \mathbb{C}_Y$ with boundary $\partial\Delta$, then

$$\frac{1}{2\pi i}\int_{\partial\Delta}\frac{p_Y(0, Y)}{p(0, Y)}\,dY = 1. \tag{7}$$

Now p and p_Y are continuous in X and Y; since $p(0, Y)$ is bounded away from 0 on the compact set $\partial\Delta$, the values on $\partial\Delta$ of the above integrand vary continuously in X. Thus for all sufficiently small $c \in \mathbb{C}_X$,

$$\frac{1}{2\pi i}\int_{Y \in \partial\Delta}\frac{p_Y(c, Y)}{p(c, Y)}\,dY \tag{8}$$

is close to 1. But by the argument principle the value of the integral is always an integer, so the expression in (8) always equals 1 for all c sufficiently small. Since this integral counts the number of zeros in $\partial\Delta$, for each c near $0 \in \mathbb{C}_X$, $p(c, Y) = 0$ has *exactly one* solution near $0 \in \mathbb{C}_Y$; we denote this unique solution by $\phi(c)$. Hence the zero-set of $p(X, Y)$ near $(0, 0)$ does indeed form the graph of a function $Y = \phi(X)$.

We now show that $\phi(X)$ is analytic at $X = 0$. First, the assumption $p_Y(0, 0) \neq 0$ easily implies the existence of an open disk $\Delta \subset \mathbb{C}_Y$ and a neighborhood U of $0 \in \mathbb{C}_X$ such that for each $c \in U$, $p(c, Y)$ has $\phi(c) \in \Delta$ as a zero of multiplicity 1. Expanding $p_Y(c, Y)/p(c, Y)$ about $Y = \phi(c)$ then gives us

$$\frac{p_Y(c, Y)}{p(c, Y)} = \frac{1 + a_1(Y - \phi(c)) + a_2(Y - \phi(c))^2 + \ldots}{Y - \phi(c)} \qquad (a_i \in \mathbb{C}).$$

Therefore

$$\frac{1}{2\pi i}\int_{\partial\Delta} \frac{Y p_Y(c, Y)}{p(c, Y)}\,dY = \frac{1}{2\pi i}\int_{\partial\Delta} \frac{Y + a_1 Y(Y - \phi(c)) + \ldots}{Y - \phi(c)}\,dY \tag{9}$$

By the Cauchy integral formula, the value of the right-hand integral in (9) is just the numerator of the integrand evaluated at $\phi(c)$, which of course is $\phi(c)$.

Now $p(X, Y) \neq 0$ for all X sufficiently small and for all $Y \in \partial\Delta$. Hence for U sufficiently small, we may represent $Yp_Y(X, Y)/p(X, Y)$ as a power series $\sum_{n=0}^{\infty} g_n(Y)X^n$, where each coefficient $g_n(Y)$ is analytic on $\partial\Delta$ and where this power series converges uniformly on $U \times \partial\Delta$. We may therefore integrate termwise: For each $c \in U$, we have

$$\phi(c) = \int_{\partial\Delta} \frac{Y p_Y(c, Y)}{p(c, Y)}\,dY = \sum_{n=0}^{\infty}\left(\int_{\partial\Delta} g_n(Y)dY\right)c^n = \sum_{n=0}^{\infty} b_n c^n \qquad (b_n \in \mathbb{C}).$$

Thus $\phi(X)$ is indeed analytic at $0 \in \mathbb{C}_X$. □

The theorem we have just proved tells us something important about the nature of a complex algebraic curve: Since we can just as well state Theorem 3.6 with the roles of X and Y reversed, we have at once

Corollary 3.9. *At any point (x_0, y_0) of $C = \mathrm{V}(p(X, Y))$ where either $p_X(x_0, y_0) \neq 0$ or $p_Y(x_0, y_0) \neq 0$, C is locally the graph of an analytic function.*

Remark 3.10. We say that C is **locally an analytic manifold** at such a point.

We have now proved a first fact about the structure of an arbitrary complex curve. We shall see in the next section that for a given curve C, $p(X, Y)$ may without loss of generality be chosen so that there are only finitely many points where the hypothesis of Corollary 3.9 fail to hold.

One can give a "real-variable" proof that at any (x_0, y_0) where $p_X(x_0, y_0) \neq 0$ or $p_Y(x_0, y_0) \neq 0$, C is locally the graph of a real-analytic mapping. (That is, C is a **real-analytic manifold** at (x_0, y_0).) We do this in Lemma 3.11, next. In proving Lemma 3.11, we shall assume that the real analogue of Theorem 3.5 has been proved. (Actually, it suffices for our purposes to know that $\phi_1, \ldots, \phi_r$ in Theorem 3.5's conclusion have first order partial derivatives. In this form, the result is found in most books on advanced calculus.)

Lemma 3.11. *Let $p(X, Y) \in \mathbb{C}[X, Y] \setminus \mathbb{C}$. Then $C = \mathsf{V}(p(X, Y))$ is a real-analytic manifold at any point (x_0, y_0) where either $p_X(x_0, y_0) \neq 0$ or $p_Y(x_0, y_0) \neq 0$.*

PROOF. Suppose without loss of generality that $p_Y(x_0, y_0) \neq 0$. Writing $p(X, Y)$ as $p_1(X, Y) + ip_2(X, Y)$ gives

$$p_1(X_1 + iX_2, Y_1 + iY_2) = p_2(X_1 + iX_2, Y_1 + iY_2) = 0;$$

for convenience we write this as

$$p_1(X_1, X_2, Y_1, Y_2) = 0, \qquad p_2(X_1, X_2, Y_1, Y_2) = 0. \tag{10}$$

Then $(p_1, p_2): \mathbb{R}^4 \to \mathbb{R}^2$ plays the role of f in the real-analytic form of Theorem 3.5 with $n = 4$ and $q = r = 2$.

Let us now look at the determinant of our Jacobian

$$\det(J(p_1, p_2)) = \begin{pmatrix} \dfrac{\partial p_1}{\partial Y_1} & \dfrac{\partial p_1}{\partial Y_2} \\[2ex] \dfrac{\partial p_2}{\partial Y_1} & \dfrac{\partial p_2}{\partial Y_2} \end{pmatrix}$$

This is

$$\frac{\partial p_1}{\partial Y_1} \cdot \frac{\partial p_2}{\partial Y_2} - \frac{\partial p_1}{\partial Y_2} \cdot \frac{\partial p_2}{\partial Y_1};$$

using the Cauchy–Riemann equations

$$\frac{\partial p_1}{\partial Y_1} = \frac{\partial p_2}{\partial Y_2}, \qquad \frac{\partial p_1}{\partial Y_2} = -\frac{\partial p_2}{\partial Y_1},$$

we obtain

$$\det(J(p_1, p_2)) = \frac{\partial p_1}{\partial Y_1} \cdot \frac{\partial p_1}{\partial Y_1} + \frac{\partial p_2}{\partial Y_1} \cdot \frac{\partial p_2}{\partial Y_1} = |p_{Y_1}|^2 = |p_Y|^2.$$

By our initial assumption, this is nonzero at (x_0, y_0). Therefore the zero-set of $p(X, Y) = 0$ in a neighborhood of $(0) \in \mathbb{R}^4$ is described by unique real-analytic functions

$$Y_1 = \phi_1(X_1, X_2), \qquad Y_2 = \phi_2(X_1, X_2). \qquad \square$$

One can even push the above real-variable approach a little further to show that C in Lemma 3.11 is complex-analytic at (x_0, y_0). (See Exercise 3.5.)

EXERCISES

3.1 Show that any closed subset S of $\mathbb{R}^n$ is the zero-set of an infinitely-differentiable real function. [*Hint*: Cover $\mathbb{R}^n \setminus S$ with closed n-cubes (of various sizes) whose interiors are mutually disjoint. Then for each cube construct an infinitely-differentiable function which is never zero on that cube, but which attains nonzero values on only finitely many cubes. (Remember the e^{-1/x^2}-type function).]

3.2 Find functions $f_1, \ldots, f_q$ satisfying hypotheses (3.5.1) and (3.5.2), but not satisfying the conclusion of Theorem 3.5.

3.3 In what sense is Gauss elimination for linear systems over $\mathbb{C}$ a special case of Theorem 3.5?

3.4 Show that there isn't an "implicit complex polynomial mapping theorem"—that is, if $f_1, \ldots, f_q$ in Theorem 3.5 are polynomials, one cannot in general conclude that each ϕ_i is a polynomial.

3.5 Show that C in Lemma 3.11 is actually *complex-analytic* at (x_0, y_0) by showing that if $Y_1 = \phi_1(X_1, X_2)$ and $Y_2 = \phi_2(X_1, X_2)$ are the real-analytic functions of Lemma 3.11, then $Y = Y_1 + iY_2$ is a complex-analytic function of $X = X_1 + iX_2$. [Hint: Verify the Cauchy–Riemann equations

$$\frac{\partial Y_1}{\partial X_1} = \frac{\partial Y_2}{\partial X_2}, \qquad \frac{\partial Y_1}{\partial X_2} = -\frac{\partial Y_2}{\partial X_1}$$

by differentiating the equations in (10) partially with respect to X_1 and X_2.]

4 Some local structure of plane curves

In Chapter I we stated Theorem I,2.7 which says in part that a topological copy of any complex algebraic projective curve may be obtained by identifying finitely many points in some appropriate compact connected orientable 2-manifold.

In this section we prove part of this result for curves in $\mathbb{P}^2(\mathbb{C})$. First, the topological space obtained by taking finitely many open disks, selecting one point in each disk and then identifying these selected points to one point will be called **the one-point union of finitely many open disks**. Figure 15 shows the one-point union of three (topological) disks.

We shall prove the following part of Theorem 2.7 of Chapter I.

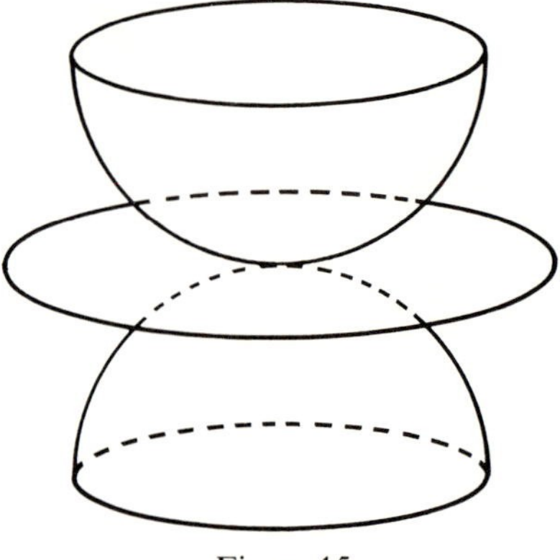

Figure 15

Lemma 4.1

(4.1.1) *Any complex algebraic curve C in $\mathbb{P}^2(\mathbb{C})$ is compact.*
(4.1.2) *Let U_P be a neighborhood of $P \in C$. Then at all but finitely many points P of C, for a sufficiently small U_P, $C \cap U_P$ is topologically an open disk.*
(4.1.3) *At each of the remaining points of C, for a sufficiently small U_P, $C \cap U_P$ is the one-point union of finitely many open disks.*

We shall devote this section to a proof of this lemma. The proof of compactness is immediate; proofs of the other two statements are longer. Actually, our proofs of (4.1.2) and (4.1.3) lead in a natural way to some concepts and results which are important in their own right. In proving (4.1.2) we meet the notions of resultant and discriminant, used throughout algebraic geometry; in proving (4.1.3) we meet fractional-power series. Rather than presenting these new ideas separately and in isolation, we show how a working mathematician might naturally meet them in determining the structure of curves in $\mathbb{P}^2(\mathbb{C})$. Thus, instead of striving for the shortest proofs, we will take a little time along the way to present these new notions.

PROOF OF (4.1.1). Recall that any curve in $\mathbb{P}^2(\mathbb{C})$ is definable in $\mathbb{C}^3$ by some nonconstant homogeneous polynomial, say $q(X, Y, Z)$, and that the curve is covered by the three affine representatives defined by $q(1, Y, Z)$, $q(X, 1, Z)$, and $q(X, Y, 1)$. Since each of $q(1, Y, Z)$, $q(X, 1, Z)$ and $q(X, Y, 1)$ is continuous, each affine representative is closed in its affine space; since these three open subsets of $\mathbb{P}^3(\mathbb{C})$ cover $\mathbb{P}^2(\mathbb{C})$, the whole curve is closed in $\mathbb{P}^2(\mathbb{C})$. Thus, since $\mathbb{P}^2(\mathbb{C})$ is compact (Exercises 1.1 and 1.2), so is the curve. □

Since (4.1.2) and (4.1.3) involve only small neighborhoods, it clearly suffices to prove these local statements in each of the three affine representatives. Without loss of generality, we work in $\mathbb{C}_{XY}$ with the polynomial $q(X, Y, 1)$ (which we henceforth denote by $p(X, Y)$), and with $C = \mathsf{V}(p(X, Y)) \subset \mathbb{C}_{XY}$. Though it may happen that p is a nonzero constant (when $q(X, Y, Z) = Z^n$), p would then define $\emptyset \subset \mathbb{C}_{XY}$, and there would be nothing to prove in $\mathbb{C}_{XY}$. We therefore assume in this section that p is nonconstant.

Now it is immediate from Lemma 3.9 that topologically an affine curve $C = \mathsf{V}(p(X, Y))$ is locally a disk at any point $(x_0, y_0) \in C$ which satisfies either $p_X(x_0, y_0) \neq 0$ or $p_Y(x_0, y_0) \neq 0$ (or both). Thus (4.1.2) follows from

(4.1.2′) $p(X, Y)$ may be chosen so there are only finitely many points $(x_0, y_0) \in \mathbb{C}_{XY}$ where $p(x_0, y_0) = p_Y(x_0, y_0) = 0$.

And (4.1.3) becomes

(4.1.3′) About any such (x_0, y_0) there is some neighborhood $U \subset \mathbb{C}_{XY}$ such that $C \cap U$ is the one-point union of finitely many open disks.

We now begin the proof of (4.1.2′). If $C = \mathsf{V}(p)$, let $p = p_1^{n_1} \cdot \ldots \cdot p_r^{n_r}$ be p's factorization into irreducibles in $\mathbb{C}[X, Y]$; then

$$\mathsf{V}(p) = \mathsf{V}(p_1^{n_1}) \cup \ldots \cup \mathsf{V}(p_r^{n_r}) = \mathsf{V}(p_1) \cup \ldots \cup \mathsf{V}(p_r) = \mathsf{V}(p_1 \cdot \ldots \cdot p_r)$$

Therefore from those polynomials defining C, we may choose one which is a product of distinct irreducible factors. It is easily proved that for a given C, this polynomial is uniquely determined up to a nonzero constant multiple (Exercise 4.1); henceforth by p in $C = \mathsf{V}(p)$, we shall mean this unique polynomial.

Now clearly the topological structure of $C \subset \mathbb{C}_{XY}$ is an intrinsic property in the sense that a coordinate change in $\mathbb{C}_{XY}$ does not alter the topology of C. As a matter of convenience, throughout the remainder of the proof of (4.1.2′) we make without loss of generality the following

Assumption 4.2. Coordinates (X, Y) in $\mathbb{C}_{XY}$ have been chosen so that if $\deg p(X, Y) = n$, then p is of the form

$$p(X, Y) = Y^n + a_1(X)Y^{n-1} + \ldots + a_n(X),$$

where $a_i(X) \in \mathbb{C}[X]$, and where either $\deg a_i(X) \leqslant i$, or $a_i(X) = 0$.

(If this assumption is not already satisfied in $\mathbb{C}_{XY}$, new coordinates defined by

$$X = X' + cY', \qquad Y = Y'$$

can be chosen so that the coefficient of the new $(Y')^n$-term is a nonzero polynomial in c; hence the coefficient is nonzero for all but finitely many choices of c. We shall continue to denote these new coordinates by (X, Y).)

To prove (4.1.2′), we need a condition telling just where the two polynomials p_X and p_Y can have common zeros. As it turns out, one can easily answer a much more general question. First, one can look at p_X or p_Y as belonging to $D[X]$, where D is the unique factorization domain $\mathbb{C}[Y]$. One can then ask, given any two polynomials $f, g \in D[X]$, is there an $a \in D$ such that $f(a) = g(a) = 0$?

We now begin our first side trip. Recall that for any $f(X) \in D[X]$ and $a \in D$, $f(a) = 0$ iff $(X - a)$ is a factor of f. Therefore one answer to the above question is: $f, g \in D[X]$ have a common zero iff they have a common factor of the form $X - a$.

But one can generalize this question even further: When do f and g have *any* factor in common? The answer to this (Theorem 4.4) is not very hard to come by; it will be of use to us several times throughout the book and will at once yield (4.1.2′).

A preliminary form of the criterion is the following

Lemma 4.3. *Let D be any unique factorization domain. Let two polynomials in $D[X]$ be*

$$f(X) = a_0 X^m + \dots + a_m,$$
$$g(X) = b_0 X^n + \dots + b_n.$$

We assume that at least one of a_0, b_0 is nonzero. Then $f(X)$ and $g(X)$ have a nonconstant factor in common iff there are polynomials $F(X), G(X) \in D[X]$ such that

$$fG = gF, \tag{11}$$

where $\deg F < m$ *and* $\deg G < n$.

PROOF. Since D is a unique factorization domain, so is $D[X]$, by Gauss' lemma. Let the unique factorizations of f and g be

$$f = d \cdot f_1{}^{m_1} \cdot \ldots \cdot f_r^{m_r}, \qquad g = e \cdot g_1{}^{n_1} \cdot \ldots \cdot g_s{}^{n_s}, \tag{12}$$

$d, e \in D$, f_i, g_i irreducible in $D[X]$.

Suppose that f and g have a nonconstant common factor, say f_1. Then $F = f/f_1$ and $G = g/f_1$ satisfy the equations in (12), and $\deg F < \deg f$, $\deg G < \deg g$.

Conversely, suppose f and g have no nonconstant factors in common, and suppose $a_0 \neq 0$ (i.e., $\deg f = m$). If $fG = gF$, then by the uniqueness of factorizations in (12), every $f_i^{m_i}$ must appear in F's decomposition, hence $\deg F \geqslant m$, a contradiction. If $a_0 = 0$, then $b_0 \neq 0$ and one similarly derives $\deg G \geqslant n$. □

Lemma 4.3 may easily be translated into a statement about the coefficients of f and g. Write

$$f(X) = a_0 X^m + a_1 X^{m-1} + \dots + a_m,$$
$$g(X) = b_0 X^n + b_1 X^{n-1} + \dots + b_n,$$

where $a_0 \neq 0$ or $b_0 \neq 0$. If $\deg F < m$ and $\deg G < n$, then F and G may be written as

$$F(X) = A_0 X^{m-1} + \dots + A_{m-1},$$
$$G(X) = B_0 X^{n-1} + \dots + B_{n-1}.$$

Hence f and g have a common nonconstant factor precisely when one can find coefficients $A_0, \ldots, A_{m-1}, B_0, \ldots, B_{n-1}$ (not all of them zero) such that

$$(a_0 X^m + \dots + a_m)(B_0 X^{n-1} + \dots + B_{n-1}) = (b_0 X^n + \dots + b_n)(A_0 X^{m-1} + \dots + A_{m-1}). \tag{13}$$

Now two polynomials are equal iff their coefficients are equal. Hence multiplying out each side of (13) and equating coefficients yields

$$\begin{aligned} a_0 B_0 &= b_0 A_0 \\ a_1 B_0 + a_0 B_1 &= b_1 A_0 + b_0 A_1 \\ &\vdots \\ a_m B_{n-1} &= b_n A_{m-1}. \end{aligned}$$

This is a homogeneous linear system of $m + n$ equations in the $m + n$ unknowns $A_0, \ldots, A_{m-1}, B_0, \ldots, B_{n-1}$. This system has a nonzero solution in D iff the determinant of the coefficient matrix is zero, i.e., iff

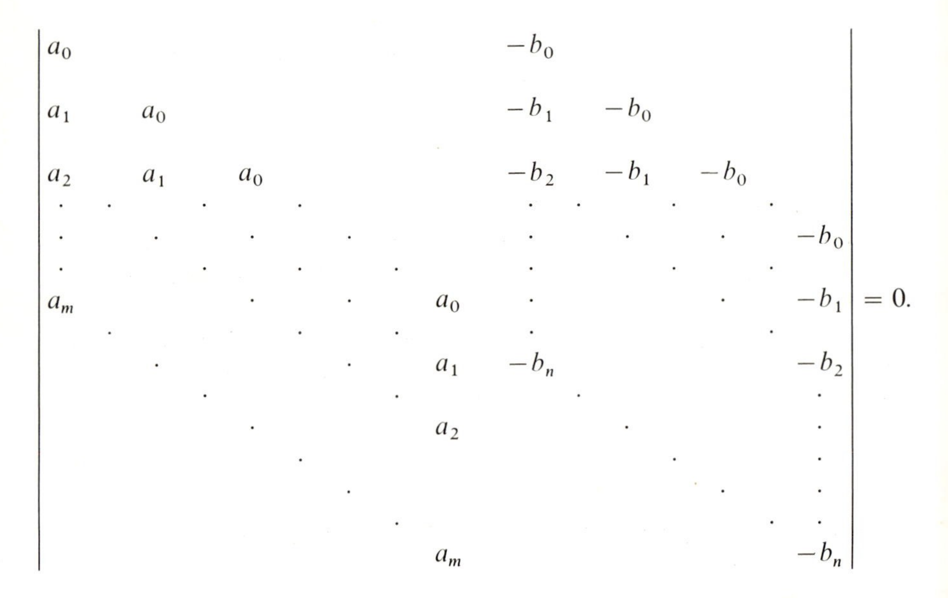

(The blank entries are understood to be 0). If we multiply each of the "b" columns by -1 and interchange rows with columns, we may express this result as

Theorem 4.4. *Let D be a unique factorization domain, and let*

$$\begin{aligned} f(X) &= a_0 X^m + a_1 X^{m-1} + \ldots + a_m, \\ g(X) &= b_0 X^n + b_1 X^{n-1} + \ldots + b_n \end{aligned}$$

be two polynomials in $D[X]$. Assume that the leading coefficients a_0 and b_0 of $f(X)$ and $g(X)$ are not both zero. Then $f(X)$ and $g(X)$ have a nonconstant common factor iff the following determinant is zero:

$$\begin{vmatrix}
a_0 & a_1 & a_2 & \cdots & a_m & & & \\
 & a_0 & a_1 & \cdots & \cdots & a_m & & \\
 & & a_0 & \cdot & & & a_m & \\
 & & & \ddots & \ddots & & & \ddots \\
 & & & & a_0 & a_1 & a_2 & \cdots & a_m \\
b_0 & b_1 & b_2 & \cdots & b_n & & & \\
 & b_0 & b_1 & \cdots & \cdots & b_n & & \\
 & & b_0 & \cdot & & & b_n & \\
 & & & \ddots & \ddots & & & \\
 & & & b_0 & b_1 & b_2 & \cdots & \cdots & b_n
\end{vmatrix} \tag{14}$$

where there are n rows of "a" entries and m rows of "b" entries.

Definition 4.5. The determinant in (14) is called the **resultant of** f **and** g; we denote it by $\mathscr{R}(f, g)$. If $f, g \in D[X_1, \ldots, X_t]$, then for any i,

$$f, g \in D[X_1, \ldots, X_{i-1}, X_{i+1}, \ldots, X_t][X_i] = D'[X_i];$$

the corresponding resultant is called the **resultant of** f **and** g **with respect to** X_i, denoted by $\mathscr{R}_{X_i}(f, g)$. For any $f \in D[X]$, one can define the formal derivative $df/dX \in D[X]$ using the relations

$$\frac{d(au)}{dX} = a\frac{du}{dX}, \qquad \frac{d(uv)}{dX} = u\frac{dv}{dX} + v\frac{du}{dX} \qquad (a \in D,\ u, v \in D[X]).$$

Then the resultant $\mathscr{R}(f, f')$ of $f \in D[X]$ and its derivative $df/dX = f' \in D[X]$ is called the **discriminant of** f, denoted $\mathscr{D}(f)$; if $f \in D[X_1, \ldots D_t]$, then $\mathscr{R}_{X_i}(f, \partial f/\partial X_i)$ is called the **discriminant of** f **with respect to** X_i, denoted $\mathscr{D}_{X_i}(f)$. If $f \in \mathbb{C}[X_1, \ldots, X_t]$, then the variety

$$\mathsf{V}(\mathscr{D}_{X_i}(f)) \subset \mathbb{C}_{X_1, \ldots, X_{i-1}, X_{i+1}, \ldots, X_t} = \mathbb{C}^{t-1}$$

is called the **discriminant variety of** $\mathscr{D}_{X_i}(f)$.

Remark 4.6. It is easily checked that $\mathscr{D}_X(aX^2 + bX + c)$ is essentially the familiar "$b^2 - 4ac$" (Exercise 4.2).

The following will be used frequently in the sequel:

Lemma 4.7. *Let D be any unique factorization domain of characteristic zero. Then $f \in D[X]$ has a repeated (nonconstant) factor iff f and f' have a common factor. Thus*

$$f \text{ has a repeated factor iff } \mathscr{D}(f) = 0.$$

In particular:

If $D = \mathbb{C}[X_1, \ldots, X_{i-1}, X_{i+1}, \ldots, X_t]$, then $f(X_i) \in D[X_i]$ has a repeated factor (involving X_i) iff $\mathscr{D}_{X_i}(f) = 0$.

PROOF. First, suppose that f has no repeated factors. Then $f = p_1 p_2, \ldots, p_r$, where the p_i are distinct irreducible polynomials. Differentiating, we obtain

$$f' = p_1' p_2, \ldots, p_r + p_1 p_2', \ldots, p_r + \ldots + p_1 p_2, \ldots, p_r'.$$

All terms except the ith are divisible by p_i, but the ith term is not divisible by p_i. Indeed, $p_i \nmid p_i'$ in characteristic zero, since $p_i' \neq 0$ and $\deg p_i' < \deg p_i$. Hence $p_i \nmid f'$, so f and f' have no common factors.

Conversely, suppose that f has a repeated factor, say $f = g^s h$, where $s \geqslant 2$. Then $f' = sg^{s-1}g'h + g^s h'$, so g is a common factor of f and f'. □

Lemma 4.8. *Suppose $p(X, Y) \in \mathbb{C}[X, Y]$ satisfies Assumption 4.2, p having (total) degree n. Then the points $x_0 \in \mathbb{C}_X$ at which $p(x_0, Y)$ has fewer than n zeros are precisely the zeros of the polynomial $\mathscr{D}_Y(p) \in \mathbb{C}[X]$.*

PROOF. Let $x_0 \in \mathbb{C}_X$. Then $\deg p(x_0, Y) = n$, and from the form of the resultant in (14) it is evident that

$$\mathscr{D}_Y(p(X, Y))_{X = x_0} = \mathscr{D}(p(x_0, Y)).$$

This, together with Lemma 4.7, gives the result. □

Remark 4.9. Note that the conclusion of Lemma 4.8 need not hold if Assumption 4.2 on $p(X, Y)$ is not satisfied. For instance, $p(X, Y) = Y - X^2$ does not satisfy the condition, and

$$\mathscr{D}_Y(p) = \mathbb{R}_Y[1Y + (-X^2)Y^0, 1Y^0] \equiv 1.$$

And for each $X = x_0$, $p(x_0, Y)$ has only one zero ($Y = x_0{}^2$), not two. (One can think of "the other zero" as lying on the line at infinity.)

This completes our detour into resultants and discriminants. We now return to the proof of (4.1.2′). We are almost done.

Let us write, in accordance with Assumption 4.2,

$$p(X, Y) = Y^n + \ldots + a_n(X),$$

$$\frac{\partial p}{\partial Y}(X, Y) = p_Y(X, Y) = b_0(X)Y^{n-1} + \ldots + b_{n-1}(X),$$

where $a_i(X), b_i(X) \in \mathbb{C}[X]$, $\deg a_i(X) \leqslant i$ (or $a_i(X) = 0$), and $b_0(X) = n \neq 0$.

If p and $\partial p/\partial Y$ have a common zero at $X = x_0$, the determinant $\mathscr{D}_Y(p) \in \mathbb{C}[X]$ must vanish at $X = x_0$. If this discriminant polynomial is not the zero polynomial, there are, of course, only finitely many values x_0 for which $p(x_0, Y)$ and $(\partial p/\partial Y)(x_0, Y)$ could possibly possess a common zero. Thus, to prove there are only finitely many points $(x_0, y_0) \in \mathbb{C}_{XY}$ satisfying

$$p(x_0, y_0) = \frac{\partial p}{\partial Y}(x_0, y_0) = 0,$$

there remain only these two things to clear up:

(4.10) $\mathscr{D}_Y(p)$ is not the zero polynomial.
(4.11) At any zero $x_0 \in \mathbb{C}_X$ of $\mathscr{D}_Y(p)$, there are not infinitely many solutions to $p(x_0, Y) = (\partial p/\partial Y)(x_0, Y) = 0$.

First, (4.10) follows at once from the assumption that p has no repeated irreducible factors (Lemma 4.7).

Second, (4.11) holds since for any x_0, $p(x_0, Y)$ is a nonzero polynomial in Y having at most n zeros.

We have thus completed the proof of (4.1.2′). □

We now turn to the proof of (4.13′). First recall the following standard fact from complex analysis:

Theorem 4.12 (Riemann extension theorem). *Let Ω be a nonempty open subset of $\mathbb{C}$, let c be an arbitrary point of Ω and let $h(X)$ be single-valued and analytic at each point of $\Omega\backslash\{c\}$. Then if h is bounded at c (i.e., if there is an $M \in \mathbb{R}$ such that $|h(X)| < M$, for all X near c), h may be uniquely extended to a function holomorphic on all of Ω (i.e., there is a unique h^*, analytic on Ω with restriction $h^*|\Omega\backslash\{c\} = h$.)*

In proving (4.1.3′), we continue to assume that p is a product of distinct factors.

Let (x_0, y_0) be a point of $\mathbb{C}_{XY}$ satisfying, without loss of generality, $p(x_0, y_0) = (\partial p/\partial Y)(x_0, y_0) = 0$. Then y_0 is a multiple root of $p(x_0, Y) = 0$. Let $r > 1$ be its multiplicity, and let $\Delta = \Delta(y_0, \varepsilon)$ be a disk in $\mathbb{C}_Y$ centered at y_0, whose closure contains no other y_{0i}. By the argument principle (see Theorem 3.8), we have

$$\frac{1}{2\pi i}\int_{\partial\Delta} \frac{p_Y(x_0, Y)}{p(x_0, Y)}\, dY = r;$$

we now reason as before in the proof of Theorem 3.6.

Since $p(x_0, Y)$ is never zero on $\partial\Delta$, a small change in x_0 to x_1, yields a small change in the integrand, hence in the integral. Thus the integral

$$\frac{1}{2\pi i}\int_{\partial\Delta} \frac{p_Y(x_1, Y)}{p(x_1, Y)}\, dY$$

has value r for all $x_1 \in \mathbb{C}_X$ sufficiently near x_0. We then see that for a sufficiently small disk $\Delta(y_0, \varepsilon) \subset \mathbb{C}_Y$ centered at y_0, there is a sufficiently small disk $\Delta'(x_0, \delta) \subset \mathbb{C}_X$ about x_0 so that for each $x_1 \in \Delta'(x_0, \delta)\backslash\{x_0\}$ there are exactly r zeros of $p(x_1, Y)$ in $\Delta(y_0, \varepsilon)$, counted with multiplicity. But for $\Delta'(x_0, \delta)$ sufficiently small, $x_1 \in \Delta'(x_0, \delta)\backslash\{x_0\}$ is never in the discriminant variety $\mathrm{V}(\mathscr{D}_Y(p(X, Y))) \subset \mathbb{C}_X$; hence each zero is of multiplicity one. Thus, for $x' \in \Delta'(x_0, \delta)\backslash\{x_0\}$, *there are exactly r distinct zeros of $p(x_1, Y)$ in $\Delta(y_0, \varepsilon)$*. Let these distinct zeros be $y_{11}, y_{12}, \ldots, y_{1r}$.

We now fix our attention on one fixed but arbitrary zero, say y_{11}. Starting at x_1, let us travel once around the circle in $\mathbb{C}_X$ centered at x_0. Since $p_Y(X, Y) \neq 0$ on $\Delta'(x_0, \delta)\backslash\{x_0\}$, the implicit function theorem (Theorem 3.6) tells us that the part of C in a $\mathbb{C}_{XY}$-neighborhood about any point in $(\Delta'(x_0, \delta)\backslash\{x_0\}) \times \{y_0\}$ is the graph of a holomorphic function. Thus, y_{11} depends holomorphically on X, and as X moves around the circle, y_{11} varies continuously, always staying within $\Delta(y_0, \varepsilon)$. It therefore must return to one of $y_{11}, y_{12}, \ldots, y_{1r}$ (not necessarily to y_{11}). If this new y_{1i} is not y_{11}, let us go around the circle a second time; we will end up at another one of the zeros. Obviously the process of starting with each y_{1j} and following its image as we make exactly one revolution, defines a permutation of $\{y_{11}, y_{12}, \ldots, y_{1r}\}$. Hence after some number $m(\leqslant r)$ of trips around the circle, we must return to y_{11} for the first time.

Now set $X - x_0 = T^m$. The part of C about any (x_1, y_{1i}) forms the graph of a function, and this function, considered as a function of T, extends to a single-valued function Y_{1i} throughout some neighborhood of $\mathbb{C}_T$ about $0 \in \mathbb{C}_T$. This is because values of X traverse a circle about x_0 m times when values of T go once around a circle centered at $0 \in \mathbb{C}_T$. This function is of course analytic at each point of some neighborhood of $0 \in \mathbb{C}_T$, except possibly at 0 itself. But since this function is bounded, by the Riemann extension theorem (Theorem 4.12), it has a unique analytic extension to a neighborhood of $0 \in \mathbb{C}_T$. We still denote this function by Y_{1i}; its value at 0 is y_0.

Let this function's expansion about 0 be

$$Y_{1i} = y_0 + a_1 T + a_2 T^2 + \ldots.$$

Now since $T^m = X - x_0$, all the m roots of $X - x_0$ are given by

$$\varepsilon^i (X - x_0)^{1/m} \tag{15}$$

where ε is a primitive m^{th} root of unity, say $\varepsilon = e^{2i\pi/m}$ $(i^2 = -1)$. We thus get m different corresponding fractional-power series

$$\begin{aligned} Y_{1i} &= y_0 + \varepsilon^i a_1 (X - x_0)^{1/m} + \varepsilon^{2i} a_2 (X - x_0)^{2/m} + \ldots, \\ i &= 0, 1, \ldots, m - 1. \end{aligned} \tag{16}$$

For any given $x \in \Delta'(x_0, \delta)$, δ sufficiently small, the m y-values when $X = x$, are m distinct zeros of $p(x, Y)$. These zeros again form a *cyclic set*, i.e., are cyclically permuted by going around the circle centered at x_0 and containing

x. The equations in (16) for $i = 0, \ldots, m-1$ each describe the same set $S_i \subset \mathbb{C}_{XY}$ in a neighborhood of (x_0, y_0). We denote by Y_1 some arbitrarily chosen Y_{1i} in (16).

Now set $m = m_1$. We may denote by y_{1,m_1+1} one of the zeros of $p(x_1, Y)$ not in the cyclic set $y_{11}, \ldots, y_{1m_1}$. Repeating the above argument produces another cyclic set of m_2 distinct zeros; it is clear from that argument that the sets containing y_{11} and y_{1,m_1+1} are disjoint. Continuing in this way, we get a finite number N of disjoint cyclic sets of zeros; if each set contains m_i zeros, we have

$$m_1 + m_2 + \ldots + m_N = r.$$

For each of these cyclic sets there is a corresponding set of fractional-power series like (16); selecting one fixed series from each of these N sets gives representatives $Y_1, \ldots, Y_N$. This result is a very central one in the theory of plane algebraic curves, and represents a generalization of the implicit function theorem (Theorem 3.6). Let us state it formally:

Theorem 4.13. *Let* $p(X, Y) = Y^n + a_1(X)Y^{n-1} + \ldots + a_n(X) \in \mathbb{C}[X, Y]$ $(n > 0)$ *have no repeated nonconstant factors. Let* (x_0, y_0) *be a point of* $C = \mathsf{V}(p(X, Y)) \subset \mathbb{C}_{XY}$. *Then the set of points of* C *lying in a sufficiently small open neighborhood* U *of* (x_0, y_0) *is the union of* N *different point sets* S_j, *where* S_j *is the set of points in* U *satisfying the fractional-power series*

$$Y_j = y_0 + a_{j1}(X - x_0)^{1/m_j} + a_{j2}(X - x_0)^{2/m_j} + \ldots, \tag{17}$$

where $m_1 + \ldots + m_N = r =$ *multiplicity of the zero* y_0 *in* $p(x_0, Y)$. *For* U *sufficiently small,* $S_i \cap S_j = \{(x_0, y_0)\}$ *if* $i \neq j$.

Remark 4.14. If $(\partial p/\partial Y)(x_0, y_0) \neq 0$, then by the implicit function theorem there is just one S_j (i.e., $N = 1$), the fractional-power series becomes an ordinary power series (i.e., $m_i = m_1 = 1$), and $r = 1$.

Theorem 4.13 can now be used to give us the topological structure of C at each of the finitely many points $(x_0, y_0) \in C$ where $(\partial p/\partial X)(x_0, y_0) = (\partial p/\partial Y)(x_0, y_0) = 0$, for the topology of C about (x_0, y_0) is known once we know the topology of each S_j. Since (x_0, y_0) is an isolated point of intersection of the S_j through (x_0, y_0), we at once have Statement 4.13′ (that C is topologically locally the one-point union of finitely many disks) if we can show:

> **(4.15)** For some neighborhood U about $(x_0, y_0) \in C$, each S_j is homeomorphic to a disk.

We first observe that any fractional-power series

$$Y = y_0 + a_1(X - x_0)^{1/m} + a_2(X - x_0)^{2/m} + \ldots$$

is the composition $h = h_2 \circ h_1$ of

$$h_1(X) = T = (X - x_0)^{1/m} \tag{18}$$

$$h_2(T) = Y = y_0 + a_1 T + a_2 T^2 + \dots. \tag{19}$$

Let us look at the "graph" in $\mathbb{C}^2$ defined by h_1. We may just as well look at the inverse $X - x_0 = T^n$, or, by translation of the X-coordinate, at $X = T^n$.

What is the effect of this map on an open disk Δ about $0 \in \mathbb{C}_T$? As a point travels once around a circle centered at $0 \in \mathbb{C}_T$, the image point goes n times around a circle about $0 \in \mathbb{C}_X$. Hence the part of the graph over Δ, namely $\{(T, T^n) | T \in \Delta\} \subset \mathbb{C}_{TX}$, can then be looked at this way: Consider Δ as being made of rubber. Slit Δ along the positive real axis, and, keeping the lower edge fixed, rotate the upper edge n times about 0 (this forms a kind of "spiral ramp"); and then sew the slit edges back together. This particular construction (which sets up a 1 : 1-onto map between the disk and "ramp") cannot be realized in $\mathbb{R}^3$ without self-intersections, but Figure 16 gives the idea for

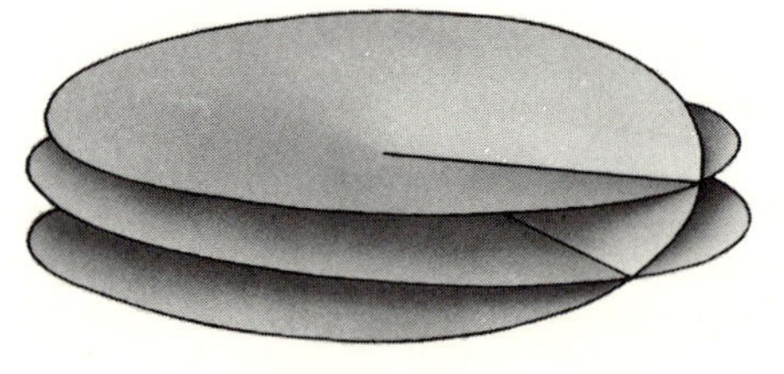

Figure 16

$n = 3$. If we perform our sewing so that the same points are identified before as after the slit, one can then define a topology on this image by taking as open sets the images of the open sets of Δ. The "ramp" is thus homeomorphic to a disk.

Now to see the topological nature of S_j itself, note that h_1 sends a spiral ramp into an ordinary disk, thus setting up a homeomorphism between two topological disks. As for h_2, let a_M be the first nonzero coefficient in (19). Then

$$Y = a_M T^M(1 + \text{higher powers of } T).$$

Within a very small disk about $0 \in \mathbb{C}_T$, the contribution of the higher powers is very small compared with the T^M term. Hence as a point goes around $0 \in \mathbb{C}_T$ one time, the image point goes around $0 \in \mathbb{C}_Y$ M times, and we end up with a spiral ramp as before. Thus h_2 also sets up a homeomorphism from a sufficiently small disk to a disk; hence so does $h = h_2 \circ h_1$. But the graph of a homeomorphism between two disks is surely itself topologically a disk. Hence each S_j is a disk, as desired. Hence we have proved (4.15), therefore (4.1.3)′, and therefore (4.1.3). Since we have also established (4.1.1) and (4.1.2), Lemma 4.1 is proved in its entirety. □

We close this section by stating an important generalization of Theorem 4.13, namely Theorem 4.16. Since the proof of Theorem 4.16 is a little long, and since a nice account appears in [Walker, Chapter IV, Theorem 3.1], we do not reproduce it here. First, let $\mathbb{C}(X)^*$ denote the set of all fractional-power series $\sum_{i=i_0}^{\infty} a_i X^{i/n}$ with arbitrary coefficients from $\mathbb{C}$ (that is, "formal" fractional-power series); for a given such fractional-power series, n is an arbitrary but *fixed* positive integer, and $i_0 \in \mathbb{Z}$. (Thus, for example, $X + X^{1/2} + X^{1/4} + X^{1/8} + \ldots$ is *not* in $\mathbb{C}(X)^*$.) We define $X^{1/1} = X$, and, for integers a, c and positive integers b, d, we define $X^{a/b} = X^{c/d}$ iff $a/b = c/d$. Two such series are equal if they are equal termwise; similarly, one adds and subtracts these formal series termwise. (The sum $\sum_i a_i X^{i/n} + \sum_j b_j X^{j/m}$ is a series $\sum_k c_k X^{k/mn}$.) Multiplication is similar to multiplication of polynomials. The quotient A/B ($B \neq 0$) is the series C such that $A = BC$, C's coefficients being (uniquely) determined by equating terms of like degree in C and AB. With these definitions, it is easily seen that $\mathbb{C}(X)^*$ forms a field. We note that if $x_0 \in \mathbb{C}$, then $\mathbb{C}(X) = \mathbb{C}(X - x_0)$, and all the above considerations apply equally well to fractional-power series in $(X - x_0)$. The basic result about these series is

Theorem 4.16. *$\mathbb{C}(X)^*$ is algebraically closed.*

The proof of Theorem 4.16 in [Walker] is actually constructive—that is, it supplies a general algorithm for constructing the power series factors of any polynomial over not only $\mathbb{C}(X)^*$, but over the analogous field $k(X)^*$, where k is any field.

Corollary 4.17. *Let $p(X, Y)$ be any polynomial in $\mathbb{C}[X, Y]$ of degree n in Y and monic in Y. Then for any fixed $x_0 \in \mathbb{C}_X$, $p(X, Y)$ factors into a product*

$$p(X, Y) = \prod_{k=1}^{n} \left(Y - \left(\sum_i a_{ik}(X - x_0)^{i/m_k} \right) \right); \tag{20}$$

this factorization is unique up to order of the factors.

Corollary 4.18. *Let x_0 be an arbitrary point of $\mathbb{C}_X$. Each of the n series in (20) converges in a neighborhood of x_0.*

Proof. Each of the m series in (16) converges in a neighborhood of x_0. Of course these m series are only the ones corresponding to an m-fold ramp at (x_0, y_0). Considering now the totality of all the series analogous to (16) corresponding to *all* the roots of $p(x_0, Y) = 0$, we see that there are altogether $n = \deg p$ of them. By the uniqueness of the factorization in (20), we see that the factors in (20) must be just these n convergent series. □

From the uniqueness of the factors in (20), we see that any method yielding the formal factors in (20) supplies a method of obtaining the Y's of (16). We look at some specific examples of this in Exercises 4.3 and 4.4.

EXERCISES

4.1 Show that the product of distinct irreducible polynomials defining a given curve $C \subset \mathbb{C}_{XY}$ is unique.

4.2 Compare $\mathscr{D}_X(aX^2 + bX + c)$ with $b^2 - 4ac$.

4.3 In (a) and (b), find the unique factorization of the form given in (20).

(a) $Y^2 - X^2(X - 1)$, $x_0 = (0)$.
(b) $Y^2 - X(X^2 - 1)$, $x_0 = (0)$.

4.4 If one knows $m_1, \ldots, m_n$ in (20), then one can substitute $\sum_i d_{ik} X^{i/m_k}$ for Y in $p(X, Y) = 0$ and solve for the a_{ik}. Find the first few fractional-power series terms in each factor of:

(a) $(X^2 + Y^2)^2 + 3X^2Y - Y^3$, $x_0 = (0)$. This polynomial defines a *three-leaved rose* in $\mathbb{R}_{XY}$ (see Figure 17). [*Hint*: Note that all four branches through points on $\mathbb{C}_Y$ appear, in $\mathbb{R}_{XY}$, to possibly be described by functions analytic in X. Hence try $m_1 = m_2 = m_3 = m_4 = 1$.]

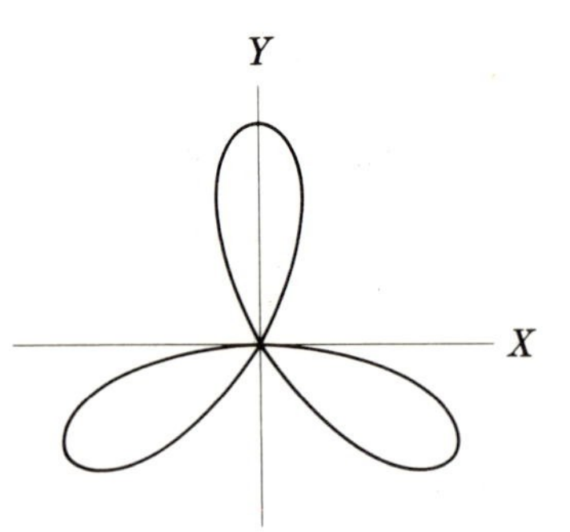

Figure 17

(b) $(X^2 + Y^2)^2 + 3XY^2 - X^3$, $x_0 = (0)$. (This is the result of interchanging axes in (a).) [*Hint*: Only two arcs through points on $\mathbb{R}_Y$ appear possibly analytic in X. Near $(0, 0) \in \mathbb{R}_{XY}$, the other one looks something like a parabola tangent at (0) to $\mathbb{R}_Y$. This suggests trying $m_1 = m_2 = 1$, $m_3 = m_4 = \frac{1}{2}$.]

Note: The real part of the curve does not always geometrically suggest what the values of $m_1, \ldots, m_n$ are. For further discussion, see [Walker, Chapter IV, Sections 3.2 and 3.3].

5 Sphere coverings

In the last section we looked at the local structure of general plane curves. We may use these ideas to help establish the overall, or *global*, structure of a plane curve. In Chapter I we got a look at the overall appearance of a few curves using the "slicing" method; this is very direct and gives some information about how the curve lies in its surrounding space. However, even for quite simple curves this method can become very involved; furthermore, our treatment was more on an intuitive level—in a more rigorous treatment one would, for instance, have to make sure that the same curves couldn't somehow fit together in a different way to form a different topological object.

In this section we look at the question of the global topology of a curve from the new viewpoint of sphere coverings. This approach, though less informative about the precise way the curve lies in an affine representative, does give us purely topological information quite easily; this method also readily generalizes, and has important applications. For instance, we use it in establishing the genus formula in Section 10 of this chapter. The present section consists mostly of definitions and examples.

The general idea is this: Somewhat as the graph of a function projects onto its domain (the graph thus forming a "one-valued" or "one-sheeted" cover of the domain), the points of a curve $C \subset \mathbb{P}^2(\mathbb{C})$ will form an "s-sheeted cover" of a subspace $\mathbb{P}^1(\mathbb{C})$ or $\mathbb{P}^2(\mathbb{C})$, except possibly over finitely many points of $\mathbb{P}^1(\mathbb{C})$. The s sheets are attached at these finitely many points in a way suggested by the last section—as "ramps," or as the one point union of disks; from this one can then derive the topological structure of a given curve.

We next make some definitions. **Disk** will mean "topological image of an open disk in $\mathbb{R}^2$"; a **connected component** of a topological space is any maximal connected subset of that topological space. (See Definition 8.1.) A topological space is **locally compact** if for each point in the space, there is an open neighborhood of that point whose closure is compact. Recall (Section I,2) that a topological 2-manifold M is a Hausdorff space in which each point has a disk as an open neighborhood.

Definition 5.1. Let M be a topological 2-manifold and let A be a locally compact topological space. Suppose that there is a continuous map $\pi : A \to M$ satisfying these properties:

(5.1.1) π is onto;

(5.1.2) For each point $p \in M$, there is some disk $\Delta(p) \subset M$ about p such that each connected component of $\pi^{-1}(\Delta(p))$ is a disk $\Delta_\alpha(p)$. Furthermore, each such disk is open in A;

(5.1.3) For each disk $\Delta_\alpha(p)$, the restricted map $\pi | \Delta_\alpha(p)$ is a homeomorphism between $\Delta_\alpha(p)$ and $\Delta(p)$.

Then A is called a **covering of** M, the triple (A, M, π) is called a **cover**, and π, the **covering map**. A cover (A, M, π) is an s-**sheeted cover** if for each $p \in M$, there is some disk $\Delta(p) \subset M$ about p such that $\pi^{-1}(\Delta(p))$ consists of exactly s disjoint disks.

A triple (A, M, f) is a "near cover" if it is a cover except over finitely many points of M. More precisely,

Definition 5.2. Let A, M be as above, let $P_1, \ldots, P_r$ be finitely many points of M, and let $f: A \to M$ be a (not necessarily onto) map. Then the triple (A, M, f) is called a **near cover** and A, a **near covering of** M if

$$(A \backslash f^{-1}(\{P_1, \ldots, P_r\}), M \backslash (\{P_1, \ldots, P_r\}), f \mid A \backslash f^{-1}(\{P_1, \ldots, P_r\}))$$

is a cover. If this last triple is an s-sheeted cover, (A, M, f) is a **near s-sheeted cover**.

Notation 5.3. Let S and T be sets. The map $\pi_T: S \times T \to S$ is defined by:

$$\pi_T((s, t)) = s \quad \text{for each } (s, t) \in S \times T.$$

It is called the **projection of $S \times T$ on S along T**. Similarly we define π_S by

$$\pi_S((s, t)) = t.$$

If A is a subset of $S \times T$, and if no confusion can arise, we also denote the restrictions $\pi_S | A$ and $\pi_T | A$ by π_S and π_T respectively. The map $\pi_{\mathbb{C}_X}: \mathbb{C}_{XY} \to \mathbb{C}_Y$ is denoted by π_X, and $\pi_{\mathbb{C}_Y}: \mathbb{C}_{XY} \to \mathbb{C}_X$, by π_Y, etc.

In the literature, π's subscript usually denotes the space into which we project, instead of along which we project. For the purposes of this book, our notation will result in somewhat smoother exposition later on.

Example 5.4. Let Δ be an open disk of $\mathbb{R}_{XY}$, let $\mathbb{R}$ have the usual topology, and let $\mathbb{Z}$ (integers) have the discrete topology (every point is open). Then $(\Delta \times \mathbb{Z}, \Delta, \pi_{\mathbb{Z}})$ is a cover. However, $(\Delta \times \mathbb{R}, \Delta, \pi_{\mathbb{R}})$ is *not* a cover, since $\pi_{\mathbb{R}}^{-1}(\Delta)$ is itself connected; since it is of dimension 3, it is not a disk.

Example 5.5. If $A = \mathsf{V}(Y^2 - X) \subset \mathbb{C}_{XY}$, then $(\mathsf{V}(Y^2 - X), \mathbb{C}_X, \pi_Y)$ is a near 2-cover.

Example 5.6. If $A = \mathsf{V}(Y(X^2 - Y^2)) \subset \mathbb{C}_{XY}$, then $(\mathsf{V}(Y(X^2 - Y^2)), \mathbb{C}_X, \pi_Y)$ is a near 3-cover.

Example 5.7. A very important general example is expressed in the following:

Lemma 5.8. *Let $p(X, Y)$ be a polynomial with no repeated factors,*

$$p(X, Y) = a_0(X)Y^n + a_1(X)Y^{n-1} + \ldots + a_n(X), \tag{21}$$

where $a_i(X) \in \mathbb{C}[X]$, $a_0 \neq 0$, and $n \geqslant 1$. Then

$$(\mathsf{V}(p), \mathbb{C}_X, \pi_Y)$$

is a near n-sheeted cover.

Proof. As we saw in the last section, the discriminant $\mathscr{D}_Y(p) \in \mathbb{C}[X]$ is not the zero-polynomial since p has no repeated factors; hence the discriminant variety $\mathsf{V}(\mathscr{D}_Y(p)) \subset \mathbb{C}_X$ consists of only finitely many points. There are thus n

distinct zeros $y_{01}, \ldots, y_{0n}$ of $p(x_0, Y)$ at all but finitely many values x_0. When these n zeros are distinct at x_0, then $(\partial p/\partial Y)(x_0, y_{0i}) \neq 0, i = 1, \ldots, n$. Hence by the implicit function theorem (Theorem 3.6), the part of $\mathsf{V}(p)$ near each such (x_0, y_{0i}) forms the graph of an analytic function $Y = h(X)$. Thus the connected components of C lying above a sufficiently small disk $\Delta(x_0)$ about x_0 in $\mathbb{C}_X$ are all disks $\Delta_\alpha(x_0)$, and π induces on each $\Delta_\alpha(x_0)$ a homeomorphism with $\Delta(x_0)$. This being true at all but finitely many points of $\mathbb{C}_X$, one sees that $(\mathsf{V}(p), \mathbb{C}_X, \pi_Y)$ is a near n-sheeted cover. □

Now, in what sense may a curve in $\mathbb{P}^2(\mathbb{C})$ be regarded as a sphere covering? To answer this, let $\mathbb{P}^1(\mathbb{C})$ denote any fixed projective 1-subspace of $\mathbb{P}^2(\mathbb{C})$, and let P_∞ be any point of $\mathbb{P}^2(\mathbb{C})$ not on $\mathbb{P}^1(\mathbb{C})$. Clearly each point of $\mathbb{P}^2(\mathbb{C})$ is contained in some line through P_∞; also, any two distinct lines in the set of all lines through P_∞ intersect in exactly P_∞, so they are disjoint in $\mathbb{P}^2(\mathbb{C})\backslash P_\infty$, so the parts of these lines within $\mathbb{P}^2(\mathbb{C})\backslash P_\infty$ are disjoint. Finally, each line in this set intersects $\mathbb{P}^1(\mathbb{C})$ in just one point, and distinct lines through P_∞ intersect $\mathbb{P}^1(\mathbb{C})$ in distinct points. There is thus defined in a natural way a projection $\pi : \mathbb{P}^2(\mathbb{C})\backslash\{P_\infty\} \to \mathbb{P}^1(\mathbb{C})$ mapping any point $P \in \mathbb{P}^2(\mathbb{C})\backslash\{P_\infty\}$ to that point in which the line through P and P_∞ intersects $\mathbb{P}^1(\mathbb{C})$ (see Figure 18).

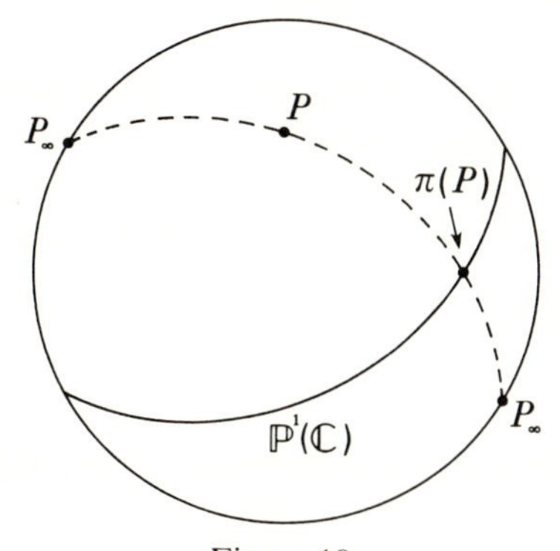

Figure 18

Now let C be any curve in $\mathbb{P}^2(\mathbb{C})$. We may without loss of generality suppose that coordinates in $\mathbb{C}^3$ have been chosen so that after dehomogenizing at Z, C is the completion of the curve in $\mathbb{C}_{XY}$ defined by a polynomial of the form given in (21) (no repeated factors, $a_0 \neq 0$, and $n \geqslant 1$). By Lemma 5.8, $(\mathsf{V}(p), \mathbb{C}_X, \pi_Y)$ is a near n-sheeted cover. If $\mathbb{P}^1(\mathbb{C})$ is the projective 1-subspace of $\mathbb{P}^2(\mathbb{C})$ containing $\mathbb{C}_X \subset \mathbb{C}_{XY}$, if P_∞ is the point completing $\mathbb{C}_Y$, and if π is as above, then

$$(C\backslash\{P_\infty\}, \mathbb{P}^1(\mathbb{C}), \pi)$$

is also a near n-sheeted cover.

If $P_\infty \notin C$, we have succeeded in representing all of C as a near covering of the sphere $\mathbb{P}^1(\mathbb{C})$. If $P_\infty \in C$, this point is not part of the covering, but we can put back this missing point by taking the "one-point compactification" of $C\backslash\{P_\infty\}$.

Definition 5.9. Let T be a topological space, and let P be an abstract point not in T. The **one-point compactification** T^* **of** T is the space described as follows:

(5.9.1) The underlying set of T^* is $T \cup \{P\}$;
(5.9.2) A basis for the open sets is given by:
  (a) the open sets of T;
  (b) subsets U of $T \cup \{P\}$ such that $(T \cup \{P\})\backslash U$ is a closed compact set of T.

EXAMPLE 5.10
(5.10.1) The one-point compactification $\mathbb{R}^*$ of $\mathbb{R}$ (with the usual topology) is a real circle (that is, the topological space $\mathbb{P}^1(\mathbb{R})$).
(5.10.2) $(\mathbb{R}^2)^* =$ sphere.
(5.10.3) The one-point compactification of a sphere with finitely many points $P_1, \ldots, P_n$ missing is the sphere with $P_1, \ldots, P_n$ all identified to one point.
(5.10.4) The one point compactification of a compact set T is T together with an extra closed, isolated point.

Lemma 5.11. *Let T be a compact Hausdorff space, and let P be any point of T. Then*

$$(T\backslash\{P\})^* = T.$$

The proof is a strightforward exercise and is left to the reader.

One can now see the following:

If C is any curve in $\mathbb{P}^2(\mathbb{C})$ and if $\mathbb{P}^1(\mathbb{C})$ any subspace of $\mathbb{P}^2(\mathbb{C})$, then C is either a near s-sheeted covering of $\mathbb{P}^1(\mathbb{C})$, or the one-point compactification of such a covering.

Remark 5.12. If we dehomogenize $\mathbb{P}^2(\mathbb{C})$ at any 1-subspace through P_∞ and choose linear coordinates X, Y in the resulting affine space, then the part of C in this $\mathbb{C}_{XY}$ is $\mathsf{V}(p)$, for some $p(X, Y) \in \mathbb{C}[X, Y]$. If $\deg_Y p = n$, then $(C, \mathbb{C}_X, \pi_Y)$ is a near s-sheeted cover, where $s \leqslant n$. If $\deg_Y p = 0$, then $p(X, Y)$ is in $\mathbb{C}[X]$, and C is simply the completion of finitely many parallel lines $X =$ a constant in $\mathbb{C}_{XY}$.

Now if we are given any such representation of C as a near cover, and if we know the nature of C about each of the finitely many exceptional (discriminant) points, then in practice it is fairly easy to determine the topology of the whole curve. We now illustrate this with a few specific examples; in Section II,10 we use sphere coverings to obtain a more general result, the topological nature of an important class of curves.

EXAMPLE 5.13. We first reconsider from this new viewpoint the circle $C \subset \mathbb{P}^2(\mathbb{C})$ defined by $\mathsf{V}(X^2 + Y^2 - Z^2) \subset \mathbb{C}_{XYZ}$. Let $\mathbb{P}^1(\mathbb{C})$ and P_∞ be represented in $\mathbb{C}_{XYZ}$ by $\mathbb{C}_{XZ}$ and $\mathbb{C}_Y$, respectively. (Hence relative to the affine part $\mathbb{C}_{XY}$, $\mathbb{P}^1(\mathbb{C})$ contains $\mathbb{C}_X \subset \mathbb{C}_{XY}$ and P_∞ completes $\mathbb{C}_Y$.) Now $X^2 + Y^2 - Z^2$ evaluated at $(0, 1, 0) \in \mathbb{C}_Y \subset \mathbb{C}_{XYZ}$ is nonzero, so $P_\infty \notin C$. And by looking at affine representatives of C in dehomogenizations at Z and X, we see that C is a near 2-covering of $\mathbb{P}^1(\mathbb{C})$. There are two exceptional points of $\mathbb{P}^1(\mathbb{C})$ above which there are fewer than two points of $C \backslash \{P_\infty\}$; these are both in $\mathbb{C}_X$, at $X = \pm 1$.

What is the nature of C above each of these two points? Let us first expand $X^2 + Y^2 - 1$ about the point $X = 1$, $Y = 0$, or, what is the same, set $X' = X - 1$ and $Y' = Y$ and expand about $X' = 0$, $Y' = 0$. This gives $(X' + 1)^2 + (Y')^2 - 1 = 0$, or

$$(Y')^2 = -X'(2 + X').$$

What is the effect of going once around a small circle in $\mathbb{C}_{X'}$ centered at $X' = 0$? Set $X' = re^{i\theta}$, r small. Then

$$Y' = \pm\sqrt{r}\, e^{i\theta/2}(2 + re^{i\theta})^{1/2};$$

as θ increases from 0 to 2π, $e^{i\theta/2}$ changes from $+1$ to -1, while for r sufficiently small, the factor $(2 - re^{i\theta})^{1/2}$ remains the same. Hence one circuit about a circle of small radius r cannot lead us from one zero of $(Y') + r(2 + r)$ (when $\theta = 0$) back to itself, so one circuit must lead to a different zero. However two circuits obviously do lead back to the original zero. Thus the part of C near $(X', Y') = (0, 0)$ behaves like $(Y')^2 = -2X'$, and one gets a 2-ramp about $(X, Y) = (1, 0)$. Similarly, there is another such 2-ramp about $(X, Y) = (-1, 0)$.

One can construct a double covering of $\mathbb{P}^1(\mathbb{C})$ having 2-ramps above any two distinct points $P_1, P_2 \in \mathbb{P}^1(\mathbb{C})$ as follows: Take two concentric spheres and make two slits, one above the other. Let the edges of the cut inner sphere be E_1 and E_2, and of the outer sphere be E_3 and E_4, where E_3 lies above E_1, and E_4 above E_2. Now sew E_1 to E_4 and E_2 to E_3. (This amounts to first switching the edges, then sewing.) This construction gives us a 2-ramp at each of P_1 and P_2.

At the top of Figure 19 we have separated the two cut spheres. We may easily see the topology of our curve C if we perform the sewing as indicated in the rest of Figure 19. We thus see from this new viewpoint that the complex circle C is topologically a sphere.

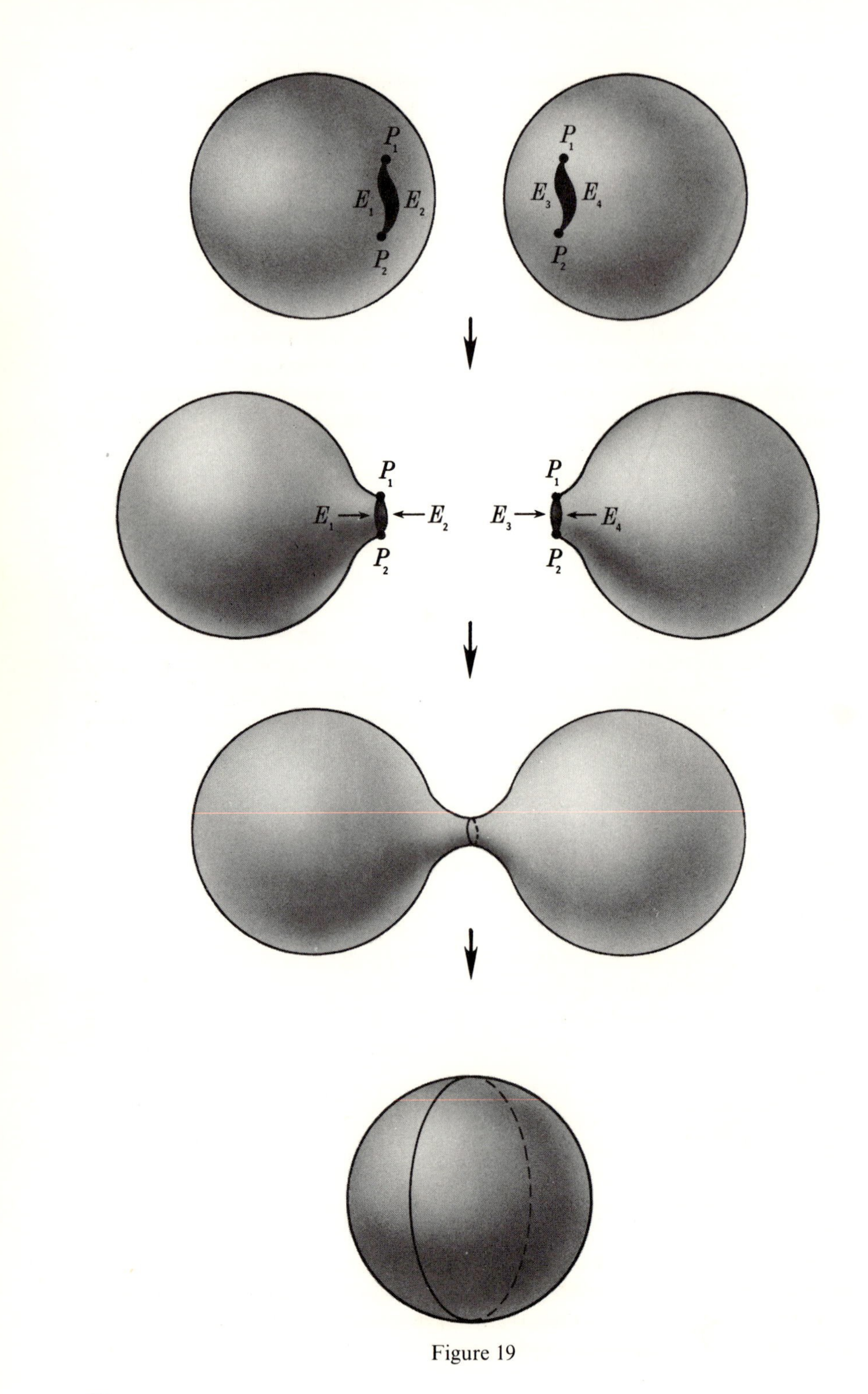

Figure 19

EXAMPLE 5.14. The representation of a given curve C as a near covering can change markedly as we vary $\mathbb{P}^1(\mathbb{C})$ and P_∞. For instance in Example 5.13, one might choose for P_∞ a point in the circle. This can be done, for example, by picking coordinates in $\mathbb{C}_{XYZ}$ so that dehomogenizing at Z gives, in affine space $\mathbb{C}_{XY}$, the complex parabola $\mathsf{V}(Y - X^2)$. Let $\mathbb{P}^1(\mathbb{C})$ be the 1-subspace containing $\mathbb{C}_X \subset \mathbb{C}_{XY}$, and let P_∞ be the point of $\mathbb{P}^2(\mathbb{C})$ completing $\mathbb{C}_Y$ ($\subset \mathbb{C}_{XY}$). Then (0, 1, 0) is a point in the 1-space $\mathbb{C}_Y$ of $\mathbb{C}_{XYZ}$ ($\mathbb{C}_Y$ represents P_∞), and $YZ - X^2$ evaluated at (0, 1, 0) is zero, so $P_\infty \in C$.

Now $\mathsf{V}(Y - X^2)$ is a near 1-covering of $\mathbb{P}^1(\mathbb{C})$ and a 1-covering of $\mathbb{C}_X$ since there is exactly one point of C over each point of $\mathbb{C}_X$. Thus C is topologically the one-point compactification of a 1-sheeted covering of $\mathbb{C}_X$. It is easily seen that a 1-sheeted cover of $\mathbb{C}_X$ is itself homeomorphic to $\mathbb{C}_X$; since the one-point compactification of $\mathbb{C}$ is a sphere, we again end up with a sphere as underlying topological space of C.

EXAMPLE 5.15. Choosing the same $\mathbb{P}^1(\mathbb{C})$ and P_∞ as in Example 5.14 but writing the parabola as $\mathsf{V}(Y^2 - X)$ again represents a change in the relative position of $\mathbb{P}^1(\mathbb{C})$, P_∞, and the curve. The variety V now describes a near 2-sheeted cover of $\mathbb{C}_X$; there are two distinct points above each point of $\mathbb{C}_X$ except at 0. The homogenization $Y^2 - XZ$ evaluated at (0, 1, 0) is nonzero, so $P_\infty \notin V$. The graph in $\mathbb{C}_{XY}$ of $Y^2 = X$ near (0, 0) is, of course, a 2-ramp. What about above the infinite point $\mathbb{P}^1(\mathbb{C})\backslash\mathbb{C}_X$? Dehomogenizing $Y^2 - XZ$ at $X = 1$ places this point at the origin, the new affine representative being given by $Y^2 - Z = 0$; in $\mathbb{C}_{YZ}$ it is $\mathbb{C}_Z$ whose completion is $\mathbb{P}^1(\mathbb{C})$. Then $Y^2 = Z$ describes another 2-ramp (a "ramp at infinity" from the viewpoint of our original $\mathbb{C}_{XY}$). We thus have a near 2-sheeted covering of a sphere, with two ramps at the exceptional points, thus bringing us back to the situation of Example 5.13. Hence we again get a sphere as the underlying space.

EXAMPLE 5.16. One can derive the topology of a curve in $\mathbb{P}^2(\mathbb{C})$ by looking at any affine part of it (though if the line at infinity is in the curve, one must put it back again after analyzing the topological closure of the affine part). The alpha curve C defined in $\mathbb{C}_{XYZ}$ by $Y^2Z - X^2(X + Z)$ is an interesting example of how looking at different affine representatives leads to the same topological result.

Dehomogenizing $Y^2Z - X^2(X + Z)$ at Y gives $Z - X^3 - X^2Z$. Equating this to zero yields

$$Z = \frac{X^3}{1 - X^2}. \tag{22}$$

The real part of the graph of (22) appears in Figure 9b. Let $\mathbb{P}^1(\mathbb{C})$ be the projective completion of $\mathbb{C}_X$ in the affine part $\mathbb{C}_{XZ}$, and let P_∞ be the point P in Figure 9e. Then (22) describes, in the obvious way, a near 1-sheeted cover of $\mathbb{P}^1(\mathbb{C})$. Exceptional points are at $X = \pm 1$, above which there are no points of $C \backslash \{P_\infty\}$. Dehomogenizing $Y^2Z - X^2(X + Z)$ at X shows there is also one point (Q, in Figure 9e) of $C \backslash \{P_\infty\}$ above $\mathbb{P}^1(\mathbb{C}) \backslash \mathbb{C}_X$. Hence the part of C in $\mathbb{P}^2(\mathbb{C}) \backslash \{P_\infty\}$ is represented as a sphere with two points missing. Moreover $P_\infty \in C$, since the homogeneous polynomial is zero at $(0, 0, 1) \in \mathbb{C}_{XYZ}$. (Note that P_∞ is represented in $\mathbb{C}_{XYZ}$ by the subspace $\mathbb{C}_Z$.) Thus C is topologically the one-point compactification of a sphere with two points missing. This is just a sphere with two points identified, which agrees with our earlier result (Figure I,22).

Now let us look at the part of our projective curve in $\mathbb{C}_{XY}$, namely $\mathsf{V}(Y^2 - X^2(X + 1))$. Choose $\mathbb{P}^1(\mathbb{C})$ and P_∞ as in Example 5.13. This time our covering is quite different. First, $P_\infty \in C$, since $Y^2Z - X^2(X + Z)$ is zero at $(0, 1, 0)$, but now we have a near 2-sheeted covering of $\mathbb{P}^1(\mathbb{C})$. Exceptional points occur when $Y = 0$ (that is, when $X = -1$ and when $X = 0$ in $\mathbb{C}_X$), and possibly at $\mathbb{P}^1(\mathbb{C}) \backslash \mathbb{C}_X$. There is only one point of $C \backslash P_\infty$ above $X = -1$ and above $X = 0$; the points of $\mathbb{P}^2(\mathbb{C}) \backslash P_\infty$ above $\mathbb{P}^1(\mathbb{C}) \backslash \mathbb{C}_X$ are represented in $\mathbb{C}_{XYZ}$ by 1-subspaces through $(1, a, 0)$ ($a \in \mathbb{C}$), and $Y^2Z - X^2(X + Z)$ is never zero at any such point. Therefore there are no points of $C \backslash P_\infty$ above the point $\mathbb{P}^1(\mathbb{C}) \backslash \mathbb{C}_X$.

Now let us look at the part of C about $(0, 0)$. We argue as in Example 5.13, setting $X = re^{i\theta}$, r small. Then

$$Y = \pm re^{i\theta}\sqrt{e^{i\theta} + 1}.$$

As θ increases from 0 to 2π, $e^{i\theta}$ varies continuously, starting and ending up at 1; one then sees that starting from either of the zeros $\pm\sqrt{2}r$ (corresponding to $\theta = 0$), we end up at the same zero after one revolution about the circle. Hence the two disks of the near cover do not attach in a ramp fashion, but rather as a one-point union.

Next, expanding $Y^2 = X^2(X + 1)$ about $(X, Y) = (-1, 0)$ (or expanding about $(X', Y') = (0, 0)$ after setting $X' = X + 1$ and $Y' = Y$), we get

$$(Y')^2 = X'(X' - 1)^2.$$

As in Example 5.13, the part of C near $(X, Y) = (-1, 0)$ then forms a 2-ramp.

Hence $C \backslash P_\infty$, as a near 2-cover of *this* $\mathbb{P}^1(\mathbb{C})$, consists of two concentric spheres with points missing above the point $\mathbb{P}^1(\mathbb{C}) \backslash \mathbb{C}_X$, attached by means of a 1-point union above $X = 0$, and attached rampwise above $X = -1$. But constructing a ramp involves making a slit. How shall we do this? It can be done in a way consistent with all our requirements on a near cover by cutting the sheets leftward to infinity, starting from the point $X = -1$, and then reattaching the edges after switching them, as before. Then C is topologically the one-point compactification of this construction. It is easily seen that this construction induces a ramp at infinity, too, for the infinite point is just the other "end" of the slit. Hence we have two slit spheres touching at one point, these slits to be attached in a way similar to what was done in Figure 19. One thus gets, again, a single sphere with two points identified to one.

One could also look at the projective alpha curve as a 3-sheeted covering by using, in the same affine part $\mathbb{C}_{XY}$, the completion of $\mathbb{C}_Y \subset \mathbb{C}_{XY}$ as $\mathbb{P}^1(\mathbb{C})$, and the point at infinity of $\mathbb{C}_X$ as P_∞. We leave this as an exercise for the reader.

EXERCISES

5.1 Prove Lemma 5.11.

5.2 Derive the topology of the completion of $\mathsf{V}(Y^2 - X^2(X + 1))$ in $\mathbb{P}^2(\mathbb{C})$ by choosing $\mathbb{P}^1(\mathbb{C})$ and P_∞ as suggested at the end of Example 5.16.

5.3 Show that the completion C in $\mathbb{P}^2(\mathbb{C})$ of the curve $\mathsf{V}(Y^2 - X^3)$ is topologically a sphere by looking at C as a near 2-sheeted covering of some $\mathbb{P}^1(\mathbb{C})$, and then as a near 3-sheeted covering of some $\mathbb{P}^1(\mathbb{C})$.

5.4 Determine the topology of the completion in $\mathbb{P}^2(\mathbb{C})$ of $\mathsf{V}(XY^2 - Y - X)$. (Cf. Exercise 2.1.)

5.5 Determine the topology of the completion in $\mathbb{P}^2(\mathbb{C})$ of $\mathsf{V}(X^4 + Y^4 - 1)$.

5.6 Using sphere coverings, show that the completion in $\mathbb{P}^2(\mathbb{C})$ of

$$\mathsf{V}(Y^2 - X(X^2 - 1)\cdot \ldots \cdot (X^2 - g^2))$$

has genus g.

6 The dimension theorem for plane curves

In Chapter I we stated that for any two varieties V_1 and V_2 in $\mathbb{P}^n(\mathbb{C})$,

$$\operatorname{cod}(V_1 \cap V_2) \leqslant \operatorname{cod} V_1 + \operatorname{cod} V_2.$$

In this section we prove this fundamental dimension relation for curves. Our proof will be of importance, for it points the way to a proof for arbitrary dimension (Section IV,3).

We shall give definitions of dimension for arbitrary varieties in Section IV,2. For now, we briefly describe the situation for plane curves. First, note that at all but finitely many points of a curve C in $\mathbb{C}^2$ or $\mathbb{P}^2(\mathbb{C})$, C is locally homeomorphic to an open set of $\mathbb{C}^1$; in this sense we say *C has complex dimension one*. Similarly $\mathbb{C}^2$ and $\mathbb{P}^2(\mathbb{R})$ have complex dimension two. (These agree with the dimensions assigned by the general definition in Chapter IV.) Therefore for curves C_1 and C_2 in $\mathbb{P}^2(\mathbb{C})$, our dimension statement becomes $\operatorname{cod}(C_1 \cap C_2) \leqslant 2$. Since by definition, $\dim(\emptyset) = -1$, in $\mathbb{C}^2$ or $\mathbb{P}^2(\mathbb{C})$ $\operatorname{cod}(\emptyset)$ is 3, so in this case we may rephrase our result this way:

Theorem 6.1. *Let C_1 and C_2 be algebraic curves in $\mathbb{P}^2(\mathbb{C})$. Then*

$$C_1 \cap C_2 \neq \emptyset.$$

This is the result we prove in this section.

Our main tool will be Theorem 6.2. Recall that two polynomials $p(X, Y)$, $q(X, Y) \in \mathbb{C}[X, Y]$ which are nonconstant in Y and which satisfy the conditions in Theorem 4.4 with $D = \mathbb{C}[X]$, have intersecting zero-sets ($\mathsf{V}(p) \cap \mathsf{V}(q) \neq \varnothing$) iff the resultant polynomial $\mathscr{R}_Y(p, q) \in \mathbb{C}[X]$ has a zero—that is, if $\mathscr{R}_Y$ is not a nonzero constant. But of course there are many cases in which two curves in $\mathbb{C}^2$ don't intersect, having instead all intersection points on the line at infinity. Two parallel lines in $\mathbb{C}^2$, given, say, by $X + Y = 0$ and $X + Y - 1 = 0$, or the two parabolas given by $Y^2 - X = 0$ and $Y^2 - X - 1 = 0$ are examples. This behavior is reflected in the resultants

$$\mathscr{R}_Y(X + Y, X + Y - 1) = \begin{vmatrix} 1 & X \\ 1 & X - 1 \end{vmatrix} = -1, \text{ a nonzero constant;}$$

and

$$\mathscr{R}_Y(Y^2 - X, Y^2 - X - 1) = \begin{vmatrix} 1 & 0 & -X & 0 \\ 0 & 1 & 0 & -X \\ 1 & 0 & -X - 1 & 0 \\ 0 & 1 & 0 & -X - 1 \end{vmatrix} = 1.$$

However the corresponding pairs of *homogeneous* surfaces (given by $X + Y = 0$, $X + Y - Z = 0$, and by $Y^2 - XZ = 0$, $Y^2 - XZ - Z^2 = 0$) each intersect in a 1-subspace of $\mathbb{C}_{XYZ}$ (that is, in a point of $\mathbb{P}^2(\mathbb{C})$); the first pair intersects in the 1-subspace through $(1, -1, 0)$, and the second pair, in the 1-subspace through $(1, 0, 0)$. Compare this with these resultants of each pair of homogenized polynomials:

$$\mathscr{R}_Y(X + Y, X + Y - Z) = \begin{vmatrix} 1 & X \\ 1 & X - Z \end{vmatrix} = -Z,$$

and

$$\mathscr{R}_Y(Y^2 - XZ, Y^2 - XZ - Z^2)$$

$$= \begin{vmatrix} 1 & 0 & -XZ & 0 \\ 0 & 1 & 0 & -XZ \\ 1 & 0 & -XZ - Z^2 & 0 \\ 0 & 1 & 0 & -XZ - Z^2 \end{vmatrix} = Z^4.$$

In each case the resultant does indeed have a zero. One might note in passing that the degree of the first resultant is $1 \cdot 1 = 1$, the degree of the second one is $2 \cdot 2 = 4$, which corresponds precisely to the "multiplication of degrees" property stated in Chapter I. See Bézout's theorem (Theorem 7.1 of Chapter IV).

We shall prove Theorem 6.1 by looking at resultants from the homogeneous viewpoint; the main fact we use is this:

Theorem 6.2. *Let p and q be homogeneous polynomials in $\mathbb{C}[X_0, \ldots, X_r]$ of positive degree m and n, respectively. Write*

$$p(X_0, \ldots, X_r) = p_0 X_r{}^m + \ldots + p_m \qquad (m > 0, p_0 \in \mathbb{C}\backslash\{0\})$$

$$q(X_0, \ldots, X_r) = q_0 X_r{}^n + \ldots + q_n \qquad (n > 0, q_0 \in \mathbb{C}\backslash\{0\})$$

where for $i \geqslant 1$, each $p_i, q_i \in \mathbb{C}[X_0, \ldots, X_{r-1}]$ is either the zero polynomial or is homogeneous of degree i. Then the resultant $\mathscr{R}_{X_r}$ is either the zero polynomial or is homogeneous of degree mn.

Remark 6.3. If p_0 and q_0 are not both different from 0, the resultant may be nonzero but of degree different from mn. For example, in the parabolas above, $p = Y^2 - XZ$ and $q = Y^2 - XZ - Z^2$, so $mn = 4$; but the coefficient of X^2 in both polynomials is zero, and

$$\mathscr{R}_X(p, q) = \begin{vmatrix} -Z & Y^2 \\ -Z & Y^2 - Z^2 \end{vmatrix} = Z^3.$$

A nice proof of Theorem 6.2 can be given by looking at homogeneity in a slightly different way. This alternate point of view is expressed in the following criterion.

Lemma 6.4. *A polynomial $p \in \mathbb{C}[X_0, \ldots, X_r]$ is homogeneous of degree m (or else is the zero polynomial) iff for a new indeterminate T,*

$$p(TX_0, \ldots, TX_r) = T^m p(X_0, \ldots, X_r) \tag{23}$$

holds in $\mathbb{C}[T, X_0, \ldots, X_r]$.

PROOF

$\Rightarrow$: Obvious.

$\Leftarrow$: Assume p is not the zero polynomial; suppose p has degree k and that p satisfies (23). Write

$$p = p_0 + p_1 + \ldots + p_k,$$

where $p_k \neq 0$ and $p_i = p_i(X_0, \ldots, X_r)$ is either 0 or is homogeneous of degree i. Then from (23), we have

$$T^m p = T^m p_0 + \ldots + T^m p_k = T^0 p_0 + \ldots + T^k p_k. \tag{24}$$

Since two polynomials are equal iff their coefficients agree, (24) implies that $m = k$, and that there is only one p_i; it has degree m—that is, p is homogeneous of degree m. □

We now use the above lemma in the

PROOF OF THEOREM 6.2. Denote $(X_0, \ldots, X_r)$ by X, and $(TX_0, \ldots, TX_r)$ by TX. Then

$$\mathcal{R}_{TX_r}(p(TX), q(TX)) = \begin{vmatrix} p_0 & T^1p_1 & \cdot & \cdot & \cdot & T^mp_m & & & \\ & p_0 & T^1p_1 & \cdot & \cdot & \cdot & T^mp_m & & \\ & & \cdot & & & & & \cdot & \\ & & & \cdot & & & & & \cdot \\ & & & & p_0 & T^1p_1 & \cdot & \cdot & \cdot\ T^mp_m \\ q_0 & T^1q_1 & \cdot & \cdot & \cdot & \cdot & \cdot & T^nq_n & \\ & q_0 & T^1q_1 & \cdot & \cdot & \cdot & \cdot & \cdot & T^nq_n \\ & & \cdot & \cdot & \cdot & \cdot & \cdot & \cdot & \cdot \\ & & & q_0 & T^1q_1 & \cdot & \cdot & \cdot & \cdot\ T^nq_n \end{vmatrix}$$

For each column, we may make all entries in that column appear with the same power of T by multiplying the ith row of p entries by T^{i-1}, and the jth row of q entries by T^{j-1}. The effect of this is to multiply the determinant by a total of T to the power

$$(0 + 1 + \ldots + (n-1)) + (0 + 1 + \ldots + (m-1)) = \frac{n(n-1)}{2} + \frac{m(m-1)}{2};$$

we then have:

$$T^{[n(n-1)+m(m-1)]/2}\mathcal{R}_{TX_r}(p(TX), q(TX))$$

$$= \begin{vmatrix} p_0 & T^1p_1 & \cdot & \cdot & \cdot & T^mp_m & & & \\ & T^1p_0 & \cdot & \cdot & \cdot & \cdot & T^{m+1}p_m & & \\ & & \cdot & & & & & \cdot & \\ & & & \cdot & & & & & \cdot \\ & & & & T^{n-1}p_0 & \cdot & \cdot & \cdot & \cdot\ T^{m+n-1}p_m \\ q_0 & T^1q_1 & \cdot & \cdot & \cdot & \cdot & \cdot & T^nq_n & \\ & T^1q_0 & \cdot & \cdot & \cdot & \cdot & \cdot & \cdot & T^{n+1}q_n \\ & & \cdot & \cdot & \cdot & \cdot & \cdot & \cdot & \cdot \\ & & T^{m-1}q_0 & \cdot & \cdot & \cdot & \cdot & \cdot & \cdot\ T^{m+n-1}q_n \end{vmatrix}$$

This determinant is the same as $\mathscr{R}_{X_r}(p(X), q(X))$ with the i^{th}column multiplied by T^{i-1}. In all, the power of T thus introduced into $\mathscr{R}_{X_r}$ is

$$1 + 2 + \ldots + (m + n - 1) = \frac{(m + n)(m + n - 1)}{2}.$$

Hence this last determinant expression may be written as

$$\mathscr{R}_{TX_r}(p(TX), q(TX)) = T^{[(m+n)(m+n-1)]/2-[n(n-1)+m(m-1)]/2}\mathscr{R}_{X_r}(p(X), q(X)).$$

The exponent of T simplifies to mn; Lemma 6.4 then implies that $\mathscr{R}_{X_r}$ is either the zero polynomial or is homogeneous of degree mn. □

Our dimension theorem for curves in $\mathbb{P}^2(\mathbb{C})$, namely Theorem 6.1, follows at once from

Lemma 6.5. *Let $p, q \in \mathbb{C}[X_0, \ldots, X_r]$ $(r \geqslant 2)$ be nonconstant homogeneous polynomials. Then p and q have a common zero other than* $(0, \ldots, 0)$.

PROOF. With notation as before, let

$$p = \sum_{i=0}^{m} p_i X_r^{m-i}, \qquad q = \sum_{i=0}^{n} q_i X_r^{n-i}.$$

By performing a linear change of coordinates if necessary, we may assume that $p_0 \neq 0$ and $q_0 \neq 0$. (The argument is essentially the same given for Assumption 4.2.) Since p and q are nonconstant, they are both of positive degree (m and n) in X_r, hence by Theorem 6.2 either $\mathscr{R}_{X_r} \in \mathbb{C}[X_0, \ldots, X_{r-1}]$ is the zero polynomial, or

$$\deg(\mathscr{R}_{X_r}) = mn > 0.$$

Now $r \geqslant 2$. Suppose without loss of generality that $\mathscr{R}_{X_r} = 0$ or the degree of $\mathscr{R}_{X_r}$ in X_{r-1} is positive; choose $a_0, \ldots, a_{r-2}$ where $a_i \in \mathbb{C}$ and not all $a_i = 0$, in such a way that the polynomial $\mathscr{R}_{X_r}$ evaluated at $X_0 = a_0, \ldots, X_{r-2} = a_{r-2}$ is the zero polynomial or a nonconstant polynomial in X_{r-1}; it then has a zero, say a_{r-1}. Then since $\mathbb{C}$ is algebraically closed, there is a common zero a_r of $p(a_0, \ldots, a_{r-1}, X_r)$ and $q(a_0, \ldots, a_{r-1}, X_r)$, since they have a common nonconstant factor. Since not every $a_i = 0$, Lemma 6.5 is proved. □

We can now at once prove our dimension theorem:

PROOF OF THEOREM 6.1. We have $r = 2$. Let C and C' be defined in $\mathbb{C}_{XYZ}$ by nonconstant homogeneous polynomials p and q. By homogeneity, a common nonzero solution to p and q implies that $\mathsf{V}(p) \cap \mathsf{V}(q)$ consists of at least a 1-subspace of $\mathbb{C}_{XYZ}$, i.e., that C and C' have at least one point in common. □

Exercises

6.1 Suppose, in Theorem 6.2, that p_0 and q_0 are both constants, but not necessarily nonzero. Can we still conclude that $\mathscr{R}_{X_r}$ is homogeneous?

6.2 Redo Exercise 3.3 of Chapter I using the ideas of this section.

6.3 Prove that a nonempty proper subvariety of an irreducible curve in $\mathbb{C}^2$ or in $\mathbb{P}^2(\mathbb{C})$ consists of finitely many points.

6.4 It is tempting to try for a quick proof of Bézout's theorem for curves in $\mathbb{P}^2(\mathbb{C})$, this way:

Let C_1 and $C_2 \subset \mathbb{P}^2(\mathbb{C})$ be two curves defined by homogeneous polynomials $p_1, p_2 \in \mathbb{C}[X, Y, Z]$, where p_1 and p_2 have no repeated factors, and no common factors. Assume without loss of generality that $\mathbb{P}^2(\mathbb{C})$ is the completion of $\mathbb{C}_{XY}$, that C_1 and C_2 have no points of intersection on the line at infinity, and that no two distinct points of $C_1 \cap C_2$ lie on the same translate of $\mathbb{C}_Y$ in $\mathbb{C}_{XY}$. Then define the *multiplicity of intersection of* C_1 *and* C_2 *at* $(a, b) \in \mathbb{C}_{XY}$ to be the multiplicity of the root a in $\mathscr{R}_Y(p_1, p_2) = 0$. Theorem 6.2 easily implies that with this choice of coordinates, $\deg(\mathscr{R}_Y(p_1, p_2)) = \deg p_1 \cdot \deg p_2$, thus proving Bézout's theorem.

What nontrivial geometric fact would one have to prove to get a proof using this idea?

7 A Jacobian criterion for nonsingularity

In Section 4 we saw that the points P of a curve $C \subset \mathbb{P}^2(\mathbb{C})$ fall into two classes. In one class, the points of C about P form a topological disk (or are *topologically nonsingular*); a neighborhood of C about any other point consists of the 1-point union of finitely many disks. (Such points are *topologically singular*.)

A little inspection of the topologically nonsingular points of C will reveal that they themselves fall into two quite different classes—points at which C is "smooth," and points where C is not. For instance, the parabola $\mathsf{V}(Y - X^2) \subset \mathbb{C}_{XY}$, being the graph of the smooth function $Y = X^2$, is smooth at each of its points. But what about, for instance, "ramp points"? From the winding nature of ramps, one might suspect for a minute that these are examples *par excellence* of points where C is nonsmooth. As it turns out, there are both smooth and nonsmooth ramps. We first look at some examples, then we shall prove a simple criterion for a topologically nonsingular point to be smooth (Theorem 7.4).

Example 7.1. The point (0, 0) of the parabola $V = \mathsf{V}(Y^2 - X) \subset \mathbb{C}_{XY}$ is a ramp point relative to the covering $(V, \mathbb{C}_X, \pi_Y)$ (π_X, π_Y as in Notation 5.3). Yet relative to the cover $(V, \mathbb{C}_Y, \pi_X)$, V becomes the graph of the smooth function $X = f(Y) = Y^2$. Hence *the "ramp" nature of the point changes as we change the direction of projection.*

Other ramp points are, however, essentially "ramplike;" these turn out to be the nonsmooth topologically nonsingular points.

EXAMPLE 7.2. Consider the curve $V = \mathsf{V}(Y^2 - X^3) \subset \mathbb{C}_{XY}$. If we start from a point of V and travel in a circle about $0 \in \mathbb{C}_X$, the Y-coordinate describes a circle in $\mathbb{C}_Y$. The argument of the point in $\mathbb{C}_Y$ increases $3/2$ as fast as the argument of the point in $\mathbb{C}_X$, since for increasing θ and fixed r, $X = re^{i\theta}$ describes a circle in $\mathbb{C}_X$, and $Y = r^{3/2}e^{(3/2)i\theta}$ describes the corresponding circle in $\mathbb{C}_Y$. Thus after one revolution in $\mathbb{C}_X$, $(1, 1) \in V$ is led into $(1, -1) \in V$—that is, one revolution leads to a new point of V. Since $(V, \mathbb{C}_X, \pi_Y)$ is a near 2-cover of $\mathbb{C}_X$, V forms a 2-ramp about $(0, 0)$ relative to π_Y.

Now let us look at V as a cover of $\mathbb{C}_Y$ relative to π_X. The argument of the point in $\mathbb{C}_X$ now increases just $2/3$ as fast as the argument of a point in $\mathbb{C}_Y$; hence after one revolution in C_Y, the point $(1, 1) \in V$ is led to $(e^{(2/3)i\pi}, 1)$; after two revolutions, to $(e^{(4/3)i\pi}, 1)$; and after three times, to $(e^{(6/3)i\pi}, 1) = (1, 1)$. Hence $(V, \mathbb{C}_Y, \pi_X)$ is a near 3-covering, and V forms a 3-ramp about $(0, 0)$ relative to this covering. Hence although the "order" of the ramp changes, it is still a ramp.

One can even choose arbitrary linear coordinates about $(0, 0) \in \mathbb{C}_{XY}$, given say by

$$X = aX' + bY' \quad \text{and} \quad Y = cX' + dY';$$

substituting these into $Y^2 - X^3 = 0$ yields a polynomial $p(X', Y')$ of order 2 at $(0, 0)$, having (up to a nonzero constant factor) either $(Y')^2$ or $(Y')^3$ as lowest-degree term of the form $(Y')^m$, so there are at least two distinct small Y'-values satisfying $p(X', Y') = 0$ for any sufficiently small $X' \neq 0$. Thus it is surely not locally a graph at $(0, 0)$, much less the graph of a smooth function. (We formalize this argument in the proof of Theorem 7.4.) The point $(0, 0)$ of $\mathsf{V}(Y^2 - X^3)$ is therefore in an essential way less well behaved than the point $(0, 0)$ of $\mathsf{V}(Y^2 - X)$.

In this section we look at such behavior more carefully, viewing the points of C less through the eyeglasses of a topologist, and more through those of an analyst (who takes differentiability and smoothness into account). Throughout this section, the ideas of Section 3 play an important role. We prove here one main theorem, a criterion which relates a geometric notion of nonsingularity ("smoothness") with an algebraic notion of nonsingularity. We, of course, state this theorem for curves, but our argument happens to generalize easily to varieties of any dimension.

First, recall Definition 3.1 of smoothness of a function $f: U \to \mathbb{R}^m$ (U open in $\mathbb{R}^n$).

Definition 7.3. A set $M \subset \mathbb{R}^n$ is **smooth at** $P \in M$ if in some neighborhood about P, for some choice of linear coordinates, M is the graph of a smooth function. The set M is **smooth** if it is smooth at each point.

Our main criterion is then

Theorem 7.4. *Let $p(X, Y) \in \mathbb{C}[X, Y]$ have no repeated factors. Then $\mathsf{V}(p) \subset \mathbb{C}_{XY}$ is smooth at $P \in \mathsf{V}(p)$ iff at least one of the following holds:*

$$\frac{\partial p}{\partial X}(P) \neq 0, \qquad \frac{\partial p}{\partial Y}(P) \neq 0.$$

In view of this theorem we make the

Definition 7.5. Let C be any curve in $\mathbb{C}_{XY}$, and let p be the polynomial of $\mathbb{C}[X, Y]$ having no repeated factors, for which $C = \mathsf{V}(p)$. Then with notation as in Theorem 7.4, $\mathsf{V}(p)$ is **singular** at P if $(\partial p/\partial X)(P) = (\partial p/\partial Y)(P) = 0$, and is **nonsingular** at P otherwise. We then say P is a **singular** (or **nonsingular**) **point of** C.

Before giving the proof of Theorem 7.4, recall that in Definition 3.1 of differentiability of a function, the tangent plane at $(a, f(a))$ to the graph of a smooth function $f : U \to \mathbb{R}_Y (U \subset \mathbb{R}_{X_1, \ldots, X_n})$ is given by

$$Y = f(a) + \sum_{j=1}^{n} \left(\frac{\partial f}{\partial X_j}(a)\right)(X_j - a_j).$$

If $f = (f_1, \ldots, f_m): U \to \mathbb{R}_{Y_1, \ldots, Y_m}$ of Definition 3.1 is differentiable at a, we have in $\mathbb{R}^{n+m}$ m hyperplanes through $(a, f(a))$, namely

$$Y_i = f_i(a) + \sum_{j=1}^{n} \frac{\partial f_i}{\partial X_j}(a)(X_j - a_j) \qquad (i = 1, \ldots, m). \tag{25}$$

Since these equations are linearly independent, the planes intersect in a real n-plane through $(a, f(a))$, which is the tangent plane to $V = \mathsf{V}(Y_1 - f_1, \ldots, Y_m - f_m)$ at $(a, f(a))$; it coincides with the set of limits of secant lines through $(a, f(a))$ and nearby points of V.

PROOF OF THEOREM 7.4

$\Leftarrow$: This is just Corollary 3.9.

$\Rightarrow$: We prove this half by contradiction. The strategy is this: Assume that $\mathsf{V}(p)$ is smooth at P and that $(\partial p/\partial x)(P) = (\partial p/\partial y)(P) = 0$. Then we shall find a neighborhood and coordinate system about P relative to which:

(a) $\mathsf{V}(p)$ smooth at P implies $\mathsf{V}(p)$ is locally at P a graph of some function.
(b) $(\partial p/\partial x)(P) = (\partial p/\partial y)(P) = 0$ implies $\mathsf{V}(p)$ is locally at P not a graph of any function.

First, write $\mathbb{C}_{XY} = \mathbb{R}_{X_1X_2Y_1Y_2} = \mathbb{R}^4$. Without loss of generality, let $P = (0) \in \mathbb{R}^4$ and let the part of $\mathsf{V}(p)$ near (0) be the graph of some smooth function $f = (f_1, f_2)$, i.e.

$$(Y_1, Y_2) = (f_1(X_1, X_2), f_2(X_1, X_2)): U \to U'$$

for some open sets U and U' containing 0 in $\mathbb{R}_{X_1X_2}$ and $\mathbb{R}_{Y_1Y_2}$ respectively. Since f_1 and f_2 are smooth, there are in $\mathbb{R}^4$ real 3-spaces

$$Y_1 = c_{11}X_1 + c_{12}X_2 \quad \text{and} \quad Y_2 = c_{21}X_1 + c_{22}X_2$$

which locally approximate f_1 and f_2, these 3-spaces intersecting in a real tangent plane T. (T will be our new "(X_1, X_2)-space" in a moment.)

We note two things about the plane T:

(7.6) This real 2-space is actually a complex 1-space;
(7.7) Let $F_i = Y_i - f_i(X_1, X_2)$, $i = 1, 2$. For each real line in T through (0), the corresponding directional derivative at (0) of each F_i is zero.

PROOF OF (7.6). Since $\mathsf{V}(p)$ is smooth at $(0) \in \mathbb{R}^4$, the tangent plane to $\mathsf{V}(p)$ at (0, 0) is given by

$$Y_i = \frac{\partial f_i}{\partial X_1}(0)(X_1 - 0) + \frac{\partial f_i}{\partial X_2}(0)(X_2 - 0) \qquad (i = 1, 2);$$

this is the limit as $(a) = (a_1, a_2) \to 0$ of

$$Y_i - f_i(a) = \frac{\partial f_i}{\partial X_1}(a)(X_1 - a_1) + \frac{\partial f_i}{\partial X_2}(a)(X_2 - a_2) \qquad (i = 1, 2). \quad (26)$$

Hence in this sense, the tangent plane T at (0) is the limit of tangent planes $T(P)$ to the graph at points P of $\mathsf{V}(p)$ near (0).

Since surely the limiting position of a sequence of complex lines in $\mathbb{C}^2$ is a complex line, it suffices to show that for $P \neq (0)$ near (0), each such $T(P)$ is a complex line. Now if $(\partial p/\partial X)(0) = (\partial p/\partial Y)(0) = 0$, then by Lemma 3.9, in some neighborhood of any $P \neq (0)$, (P in $\mathsf{V}(p)$ and P sufficiently close to (0)), the part of $\mathsf{V}(p)$ near P is the graph of the analytic function

$$Y = f(X) \qquad (f = f_1 + if_2).$$

The complex line through $(a, b) \in \mathbb{C}_{XY}$ in the corresponding complex definition of differentiability is

$$Y = f(a) + f'(a)(X - a).$$

By equating real and imaginary parts of this equation and making use of the Cauchy–Riemann equations, one may now verify that this real plane in $\mathbb{C}_{XY}$ is the same as that defined by the corresponding real equations in (26), so each such tangent plane is a complex line.

PROOF OF (7.7). Since f_i is differentiable there are planes

$$Y_i = c_{i1}X_1 + c_{i2}X_2 \qquad (i = 1, 2)$$

so that

$$\lim_{(x_1, x_2) \to 0} \frac{f_i(x_1, x_2) - (c_{i1}x_1 + c_{i2}x_2)}{|x_1| + |x_2|} = 0 \qquad (i = 1, 2).$$

This means

$$\lim_{(x_1, x_2, y_1, y_2) \to 0} \frac{(y_i - f_i(x_1, x_2)) - (y_i - (c_{i1}x_1 + c_{i2}x_2))}{|x_1| + |x_2|} \to 0 \qquad (i = 1, 2).$$

Now $y_i - (c_{i1}x_1 + c_{i2}x_2)$ is zero for $(x_1, x_2, y_1, y_2) \in T$, so if $(x_1, x_2, y_1, y_2) \to 0$ along points in T, the above limit becomes

$$\lim_{(x_1, x_2, y_1, y_2) \to 0} \frac{y_i - f_i(x_1, x_2)}{|x_1| + |x_2|} \to 0;$$

hence approaching along points in T, we have

$$\lim_{(x_1, x_2, y_1, y_2) \to 0} \frac{F_i(x_1, x_2, y_1, y_2) - F_i(0, 0, 0, 0)}{|x_1| + |x_2| + |y_1| + |y_2|} = 0.$$

Therefore (7.7) is proved. □

To continue with the proof of "⇒" in Theorem 7.4, note that the rank at (0) of the Jacobian matrix

$$J = \begin{pmatrix} \dfrac{\partial F_1}{\partial X_1} & \dfrac{\partial F_1}{\partial X_2} & \dfrac{\partial F_1}{\partial Y_1} & \dfrac{\partial F_1}{\partial Y_2} \\[2ex] \dfrac{\partial F_2}{\partial X_1} & \dfrac{\partial F_2}{\partial X_2} & \dfrac{\partial F_2}{\partial Y_1} & \dfrac{\partial F_2}{\partial Y_2} \end{pmatrix}$$

is two (the last two columns form an identity matrix). We now choose new coordinates in $\mathbb{C}^2$ about (0) as follows: Let $\mathbb{C}_{X'} = T$ be the tangent space to $\mathbf{V}(p)$ at (0, 0), and let $\mathbb{C}_{Y'}$ be any other complex 1-subspace of $\mathbb{C}^2$. If J' denotes J after changing to a new real basis with vectors in $\mathbb{C}_{X'}$ and $\mathbb{C}_{Y'}$, then rank $(J') = 2$, since the rank of a matrix is unchanged by a change of basis. Now derivatives at (0) in any direction in $\mathbb{R}_{X_1'X_2'}$ are all zero (by (7.7)), so the Y'-columns of J' are linearly independent; hence by the implicit function theorem (Theorem 3.6), the part of $\mathsf{V}(p)$ near (0) is also a graph of a function relative to these new coordinates.

Now, if $(\partial p/\partial X)(0) = (\partial p/\partial Y)(0) = 0$, then p has zero derivative in any direction, so if $X = \alpha X' + \beta Y'$ and $Y = \gamma X' + \delta Y'$, then $p(\alpha X' + \beta Y', \gamma X' + \delta Y') = p^*(X', Y')$ satisfies $(\partial p^*/\partial X') = (\partial p^*/\partial Y') = 0$. Hence $p^*(X', Y')$ expanded about $(0, 0)$ has order $s \geqslant 2$.

Now let us, in particular, choose $\mathbb{C}_{Y'}$ so that $(Y')^s$ is a term of $p^*(X', Y')$ when expanded about $(0, 0)$. (A proof of this is similar to that of Assumption 4.2; for "almost all" choices of ε, $X' = X'' + \varepsilon Y''$ and $Y' = Y''$ will bring this about.) We continue to denote these new coordinates by X' and Y'. We note that if a polynomial has no repeated factors it will continue to have no repeated factors after a linear change of coordinates. Therefore $\mathscr{D}_{Y'}(p^*(X', Y')) \neq 0$—that is, at all but finitely many values of x', $p^*(x', Y')$ has exactly n (= degree p) zeros. But we know $p^*(0, Y')$ has $Y' = 0$ as a zero of order $s \geqslant 2$ since $p^*(0, Y')$ is a polynomial in Y' of order s.

Now one can apply the argument principle of Theorem 3.8 (much as in the proof of Theorem 3.6, or as in Section 4 where we determined the structure of C at the finitely many exceptional points) to conclude that there are neighborhoods $\Delta_{X'}$ and $\Delta_{Y'}$ about 0 in $\mathbb{C}_{X'}$ and $\mathbb{C}_{Y'}$ respectively, such that for each $x' \in \Delta_{X'}$, there are within $\Delta_{Y'}$ exactly s zeros of $p^*(x', Y')$; these zeros are distinct for x' different from 0. Hence the part of $\mathsf{V}(p)$ in $\Delta_{X'} \times \Delta_{Y'}$ cannot be a graph of a function with respect to axes $\mathbb{C}_{X'}$ and $\mathbb{C}_{Y'}$, a contradiction. Hence "$\Rightarrow$" of Theorem 7.4 is proved. □

Remark 7.8. A statement analogous to Theorem 7.4 does not hold for real curves. That is, a real curve $\mathsf{V}(p) \subset \mathbb{R}_{XY}$ may be singular at P in the algebraic sense that $(\partial p/\partial X)(P) = (\partial p/\partial Y)(P) = 0$, yet it may be smooth there. For instance, $p(X, Y) = (X^2 + Y^2)(Y - X^2)$ has no repeated factors, and describes the union of the sets $\mathsf{V}(X^2 + Y^2)$ and $\mathsf{V}(Y - X^2)$ in $\mathbb{R}^2$. This set is the real parabola $\mathsf{V}(Y - X^2)$, since $\mathsf{V}(X^2 + Y^2)$ is only the origin of $\mathbb{R}^2$. The parabola is of course perfectly smooth everywhere, yet $(\partial p/\partial X)(0, 0) = (\partial p/\partial Y)(0, 0) = 0$. But if we look at all this in $\mathbb{C}_{XY}$, the lack of smoothness guaranteed by the vanishing of the partials becomes strongly manifested—$\mathsf{V}(p) \subset \mathbb{C}_{XY}$ is not even *topologically* a manifold, but is instead the union of a complex parabola with two distinct complex lines, $X = iY$ and $X = -iY$. Topologically the part near $(0, 0)$ is the one-point union of three disks.

We conclude this section with the important global analogue of Definition 7.5.

Definition 7.9

(7.9.1) Let C be a curve in $\mathbb{C}_{XY}$. Then C is **nonsingular** if each point $P \in C$ is a nonsingular point of C.

(7.9.2) Let C be a curve in $\mathbb{P}^2(C)$. Then C is **nonsingular** if each point $P \in C$ is a nonsingular point in some affine representative of C containing P.

Remark 7.10. It is easy to check that in (7.9.2), if $P \in C$ is nonsingular in *some* affine representative of C containing P, then it is nonsingular in *all* affine representatives of C containing P. (Cf. Exercise 7.1; also, cf. Section IV,4 where we prove this in a more general setting.)

Exercises

7.1 It is evident from Definition 7.3 that *smoothness at* P of a given curve $C \subset \mathbb{C}_{XY}$ is a geometric notion—that is, it is independent of the choice of coordinates X, Y in $\mathbb{C}^2$. From Theorem 7.4 we then conclude that singularity and nonsingularity at P are also geometric notions. Prove directly from Definition 7.5 that singularity and nonsingularity at points of C are properties independent of the linear coordinates chosen in $\mathbb{C}^2$.

7.2 Suppose curves $C_1, C_2 \subset \mathbb{C}_{XY}$ have no common components. Show that any point of intersection of C_1 and C_2 is singular in $C_1 \cup C_2$.

7.3 Show that if the curve $C \subset \mathbb{P}^2(\mathbb{C})$ is nonsingular, then C can be defined by an irreducible homogeneous polynomial in $\mathbb{C}[X, Y, Z]$.

7.4 By squeezing to points appropriate topological circles in the projective completion of $C_g = \mathsf{V}(Y^2 - X(X - 1^2)(X - 2^2) \cdot \ldots \cdot (X - g^2)) \subset \mathbb{C}_{XY}$, find the equation of an irreducible curve having $n > 0$ singular points which are "topologically nonsingular," and $m > 0$ other singular points which are "topologically singular." Note that for $g > 1$, C_g has a singular point.

7.5 Find an irreducible polynomial $p(X, Y) \in \mathbb{C}[X, Y]$ such that $\mathsf{V}(p)$ is singular at $(0, 0)$, but such that the part of $\mathsf{V}(p) \cap \mathbb{R}_{XY}$ near $(0, 0)$ is the graph $Y = f(X)$ of a smooth function f.

8 Curves in $\mathbb{P}^2(\mathbb{C})$ are connected[1]

In this and the next two sections, we look at the global topology of curves in $\mathbb{P}^2(\mathbb{C})$. In this section we answer this global question: Is it possible for an algebraic curve in $\mathbb{P}^2(\mathbb{C})$ to consist of separate parts, like two disjoint tori, or must "all parts touch," as in Figure 1.20? Theorem 8.4, the main result of this section, tells us that all parts must touch, i.e., that the curve is connected.

First, recall the following definition and basic facts:

Definition 8.1. A subset A of a topological space S is **connected** if A cannot be decomposed into a union $A = B \cup C$ satisfying

$$B \neq \varnothing, \qquad C \neq \varnothing,$$
$$\bar{B} \cap C = B \cap \bar{C} = \varnothing$$

(the bar means topological closure in S).

[1] This section requires more background in complex analysis than we assume elsewhere. Therefore on first reading, the student may read up to Theorem 8.4 and skip the remainder of the section.

Lemma 8.2. *If connected subsets A and B of S intersect, then $A \cup B$ is connected.*

Lemma 8.3. *If A is a connected subset of S, so is its topological closure $\overline{A}$.*

Now at this stage, we do know this much: If $C \subset \mathbb{P}^2(\mathbb{C})$ is defined by the homogeneous variety $\mathsf{V}(q(X, Y, Z)) \subset \mathbb{C}_{XYZ}$, and if $q = q_1^{r_1} \cdot \ldots \cdot q_k^{r_k}$ is the factorization of q into irreducibles (so that $\mathsf{V}(q) = \mathsf{V}(q_1) \cup \ldots \cup \mathsf{V}(q_k)$), then if we know each projective curve C_i defined by $\mathsf{V}(q_i)$ is connected, we know that their union is too. (Any two of the $\mathsf{V}(q_i)$ intersect by the dimension theorem, Theorem 6.1, so Lemma 8.2 applies.) But it is still quite conceivable that C_i itself is not connected. In fact this can happen for real varieties. For example, $\mathsf{V}(ZY^2 - X(X^2 - Z^2)) \subset \mathbb{P}^2(\mathbb{R})$ consists topologically of two disjoint real circles, but in $\mathbb{P}^2(\mathbb{C})$ it forms one connected piece, a torus. (See Example 2.4 of Chapter I.) This is always true in the complex setting, and our main result of this section is

Theorem 8.4. *Any complex algebraic curve $C \subset \mathbb{P}^2(\mathbb{C})$ is connected.*

Besides assuming $q(X, Y, Z)$ is irreducible, we may further reduce the problem to considering only affine varieties: If we dehomogenize with respect to a projective line containing a point of $\mathbb{P}^2(\mathbb{C})$ not in C (say, without loss of generality, at Z), then C intersects this line in a proper algebraic variety (that is, in a finite number of points—see Exercise 6.3). But from Lemma 4.1, C has no isolated points, so

$$C = \overline{\mathsf{V}(q(X, Y, 1))};$$

hence by Lemma 8.3, if $\mathsf{V}(q(X, Y, 1)) \subset \mathbb{C}_{XY}$ is connected, so is C.

By a linear change of coordinates in $\mathbb{C}_{XY}$ we may also assume q is monic in Y. Therefore, to prove Theorem 8.4 it suffices to prove

Theorem 8.5. *Let*

$$p(X, Y) = Y^n + a_1(X)Y^{n-1} + \ldots + a_n(X) \in \mathbb{C}[X, Y] \qquad (n \geqslant 1)$$

be irreducible. Then $\mathsf{V}(p) \subset \mathbb{C}_{XY}$ is connected.

Our general strategy in proving Theorem 8.5 is this: We prove that for a particular finite set of points $\{P_i\}$, $\mathsf{V}(p)\backslash\{P_i\}$ is connected (which implies, by Lemma 8.3, that the closure $\mathsf{V}(p) \subset \mathbb{C}_{XY}$ is also connected). In fact, we show $\mathsf{V}(p)\backslash\{P_i\}$ is connected in an even stronger sense—that it is *chainwise connected* (Definition 8.7). This will be done by contradiction: The assumption that $\mathsf{V}(p)\backslash\{P_i\}$ is not chainwise connected will imply the existence of a polynomial $\phi \in \mathbb{C}[X, Y]$ of Y-degree less than n, having a nonconstant factor in common with p; but this is impossible since p is assumed irreducible.

Let π_Y denote, as usual, the natural projection along the Y-axis: $\mathbb{C}_X \times \mathbb{C}_Y \to \mathbb{C}_X$, and let $\mathscr{D}$ be the discriminant variety

$$\mathscr{D} = \mathsf{V}(\mathscr{D}_Y(p(X, Y)) \subset \mathbb{C}_X.$$

Because $p(X, Y)$ is irreducible, $\mathscr{D}$ consists of only finitely many points. In our proof we shall be looking at $\mathsf{V}(p)\backslash\{P_i\}$ as the cover

$$(\mathsf{V}(p)\backslash\pi_Y{}^{-1}(\mathscr{D}), \mathbb{C}\backslash\mathscr{D}, \pi_Y).$$

Note that since above any point of $\mathbb{C}_X$ there are only finitely many points of $\mathsf{V}(p)$, $\pi_Y{}^{-1}(\mathscr{D})$ removes only finitely many points from $\mathsf{V}(p)$.

We now give some definitions and facts used in the proof of Theorem 8.5.

Definition 8.6. Let S be a topological space; a finite sequence of connected open subsets $(\mathcal{O}_1, \mathcal{O}_2, \ldots, \mathcal{O}_m)$ such that $\mathcal{O}_i \cap \mathcal{O}_{i+1}$ is connected and nonempty ($i = 1, \ldots, m - 1$), is called a **chain in** S, a **chain from** $\mathcal{O}_1$ **to** $\mathcal{O}_m$, or a **chain**. If P_1 and P_m are points in $\mathcal{O}_1$ and $\mathcal{O}_m$, respectively, we say $(\mathcal{O}_1, \ldots, \mathcal{O}_m)$ is a **chain from** P_1 **to** P_m.

From Lemma 8.2 we see that $\mathcal{O}_1 \cup \ldots \cup \mathcal{O}_m$ is a connected subset of S.

Definition 8.7. Two connected open sets $\mathcal{O}$, $\mathcal{O}'$ in S are **chain connectible** if there is a chain from $\mathcal{O}$ to $\mathcal{O}'$. Two points $P, P' \in S$ are **chain connectible** if there is a chain from P to P'. If every two points $P, P' \in S$ are chain connectible, then S is **chainwise connected**.

Lemma 8.8. *Any chainwise connected topological space S is connected.*

PROOF. If S is not connected, then for two nonempty subsets B and C we have $S = B \cup C$, where $\bar{B} \cap C = B \cap \bar{C} = \varnothing$. Let $b \in B, c \in C$, and let $(\mathcal{O}_1, \ldots, \mathcal{O}_m)$ be a chain from b to c. Then $\mathcal{O} = \mathcal{O}_1 \cup \ldots \cup \mathcal{O}_m$ is a connected set containing b and c. But $\mathcal{O} = \mathcal{O} \cap S = (\mathcal{O} \cap B) \cup (\mathcal{O} \cap C)$; since $\mathcal{O} \cap B \subset B$ and $\mathcal{O} \cap C \subset C$, it follows that

$$(\overline{\mathcal{O} \cap B}) \cap (\mathcal{O} \cap C) = (\mathcal{O} \cap B) \cap (\overline{\mathcal{O} \cap C}) = \varnothing.$$

Since $\mathcal{O} \cap B$ and $\mathcal{O} \cap C$ are nonempty, $\mathcal{O}$ is not connected, a contradiction. □

Definition 8.9. Let (A, M, π) be a near cover (Definition 5.2). A connected open set $\mathcal{O} \subset M$ is said to be **liftable to** A if there is an open set $\underset{\sim}{\mathcal{O}} \subset A$ such that $\pi|\underset{\sim}{\mathcal{O}}$ is a homeomorphism from $\underset{\sim}{\mathcal{O}}$ to $\mathcal{O}$; $\underset{\sim}{\mathcal{O}}$ is then a **lifting of** $\mathcal{O}$. If $P \in \underset{\sim}{\mathcal{O}}$ we say $\underset{\sim}{\mathcal{O}}$ is a **lifting through** P, and that $\underset{\sim}{\mathcal{O}}$ **lifts** $\mathcal{O}$ **through** P.

A chain $(\mathcal{O}_1, \ldots, \mathcal{O}_m)$ in M is **liftable to** A if there is a chain $(\underset{\sim}{\mathcal{O}}_1, \ldots, \underset{\sim}{\mathcal{O}}_m)$ in A such that each $\underset{\sim}{\mathcal{O}}_i$ is a lifting of $\mathcal{O}_i$. Then $(\underset{\sim}{\mathcal{O}}_1, \ldots, \underset{\sim}{\mathcal{O}}_m)$ is called a **lifting of** $(\mathcal{O}_1, \ldots, \mathcal{O}_m)$, and a **lifting through** P if $P \in \underset{\sim}{\mathcal{O}}_1 \cup \ldots \cup \underset{\sim}{\mathcal{O}}_m$.

Definition 8.10. Let $\mathcal{O}$ be a connected open subset of $\mathbb{C} = \mathbb{C}_X$. The graph in $\mathcal{O} \times \mathbb{C}_Y$ of a function single-valued and complex-analytic on $\mathcal{O}$, is called an **analytic function element**. Note that an analytic function element describes in a natural way a lifting of $\mathcal{O}$; we therefore write $\underset{\sim}{\mathcal{O}}$ to denote such a function element. If $P \in \underset{\sim}{\mathcal{O}}$, then $\underset{\sim}{\mathcal{O}}$ is an **analytic function element through** P. A chain $(\underset{\sim}{\mathcal{O}}_1, \ldots, \underset{\sim}{\mathcal{O}}_m)$ of analytic function elements lifting a chain $(\mathcal{O}_1, \ldots, \mathcal{O}_m)$ of $\mathbb{C}$ is called the **analytic continuation from** $\underset{\sim}{\mathcal{O}}_1$ **to** $\underset{\sim}{\mathcal{O}}_m$ **along** $(\mathcal{O}_1, \ldots, \mathcal{O}_m)$, or the **analytic continuation of** $\underset{\sim}{\mathcal{O}}_1$ **along** $(\mathcal{O}_1, \ldots, \mathcal{O}_m)$; if $P \in \underset{\sim}{\mathcal{O}}_1$ and $P' \in \underset{\sim}{\mathcal{O}}_m$, $(\underset{\sim}{\mathcal{O}}_1, \ldots, \underset{\sim}{\mathcal{O}}_m)$ is the **analytic continuation from** P **to** P' **along** $(\mathcal{O}_1, \ldots, \mathcal{O}_m)$.

Relative to the cover of special interest to us, namely $(\mathsf{V}(p)\backslash\pi_Y{}^{-1}(\mathcal{D}), \mathbb{C}\backslash\mathcal{D}, \pi_Y)$, there is about each point of $\mathbb{C}\backslash\mathcal{D}$ a connected open neighborhood $\mathcal{O}$ which has a lifting $\underset{\sim}{\mathcal{O}}$. Any such lifting is the graph of a function analytic on $\mathcal{O}$ (from Theorem 3.6)—that is, *any such* $\underset{\sim}{\mathcal{O}}$ *is an analytic function element.*

When considering chains in our proof of Theorem 8.5, it will be of technical convenience to restrict our attention to connected open sets $\mathcal{O}_i$ of $\mathbb{C}\backslash\mathcal{D}$ which are liftable through each point of $\pi_Y{}^{-1}(\mathcal{O}_i)$ (which means that $\pi_Y{}^{-1}(\mathcal{O}_i)$ consists of n $(=\deg_Y p)$ functional elements. Note that there is such an $\mathcal{O}$ about each point of $\mathbb{C}\backslash\mathcal{D}$.

Definition 8.11. Relative to $(\mathsf{V}(p)\backslash\pi_Y{}^{-1}(\mathcal{D}), \mathbb{C}\backslash\mathcal{D}, \pi_Y)$, any connected open set $\mathcal{O}$ of $\mathbb{C}\backslash\mathcal{D}$ which lifts through each point of $\pi_Y{}^{-1}(\mathcal{O})$ is an **allowable set.** Any chain of allowable open sets is an **allowable chain.**

Lemma 8.13 below is used in our proof of Theorem 8.5 and gives an important class of allowable open sets.

Definition 8.12. An open set $\Omega \subset \mathbb{C}$ is **simply connected** if it is homeomorphic to an open disk.

Examples are: $\mathbb{C}$ itself; $\mathbb{C}\backslash$(nonnegative real axis); $\mathbb{C}\backslash\Phi$, where Φ is any closed, non-self-intersecting polygonal path that goes out to the infinite point of $\mathbb{P}^1(\mathbb{C})$ (see Figure 20).

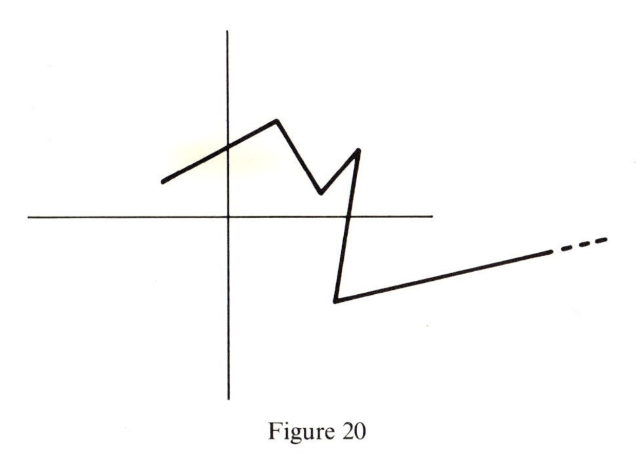

Figure 20

Lemma 8.13. *Relative to* $(\mathsf{V}(p)\backslash\pi_Y{}^{-1}(\mathscr{D}), \mathbb{C}\backslash\mathscr{D}, \pi_Y)$, *any simply connected open subset of* $\mathbb{C}\backslash\mathscr{D}$ *is allowable.*

This is an immediate consequence of the familiar "monodromy theorem" (proved in most standard texts on elementary complex analysis). To state it, we use the following ideas: First, let U be a nonempty open subset of $\mathbb{C}$. A **polygonal path** in U is the union of closed line segments $\overline{P_i, P_{i+1}} \subset U$ $(i = 0, \ldots, r-1)$ connecting finitely many ordered points $(P_0, \ldots, P_r)$ $(P_i \neq P_{i+1})$ in U. Now suppose $\underset{\sim}{Q}$ is an analytic function element which is a lifting of a connected open set $\mathcal{O} \subset U$. We say $\underset{\sim}{Q}$ can be **continued along a polygonal path** $\overline{P_0, P_1} \cup \ldots \cup \overline{P_{r-1}, P_r}$ in U if $P_0 \in \mathcal{O}$ and if there is a chain $\mathcal{O}, \mathcal{O}_1, \ldots, \mathcal{O}_r$ in U such that $\overline{P_i, P_{i+1}} \subseteq \mathcal{O}_{i+1}(i = 0, \ldots, r-1)$, and such that there is an analytic continuation of $\underset{\sim}{Q}$ along $\mathcal{O}, \ldots, \mathcal{O}_r$.

Theorem 8.14 (Monodromy theorem). *Let* Ω *be a simply connected open set in* $\mathbb{C}$, *and suppose an analytic function element* $\underset{\sim}{Q}$ *is a lifting of a connected open set* $\mathcal{O} \subset \Omega$. *If* $\underset{\sim}{Q}$ *can be analytically continued along any polygonal path in* Ω, *then* $\underset{\sim}{Q}$ *has a unique extension to a* (*single-valued*) *function which is analytic at each point of* Ω.

For a proof of Theorem 8.14, see, e.g., [Ahlfors, Chapter VI, Theorem 2].

To prove Lemma 8.13, we need only verify that in our case, the hypothesis of Theorem 8.14 is satisfied, i.e., that for any simply connected open subset Ω of $\mathbb{C}\backslash\mathscr{D}$, we can analytically continue any analytic function element along any polygonal path in Ω. The argument is easy, and may be left to the exercises (Exercise 8.1). □

We now prove that $\mathsf{V}(p)\backslash\pi_Y{}^{-1}(\mathscr{D})$ is chainwise connected by contradiction. Suppose P and Q are two points of $\mathsf{V}(p)\backslash\pi_Y{}^{-1}(\mathscr{D})$ such that there is no analytic continuation from P to Q along any allowable chain in $\mathbb{C}\backslash\mathscr{D}$.

Choose a non-self-intersecting polygonal path Φ in $\mathbb{C}$ connecting the finitely many points of $\mathscr{D}$, and the infinite point of $\mathbb{P}^1(\mathbb{C}) = \mathbb{C}_X \cup \{\infty\}$, as suggested by Figure 20. We can obviously choose Φ so it does not go through $\pi_Y(P)$ or $\pi_Y(Q)$. The "slit sphere" $\mathbb{P}^1(\mathbb{C})\backslash\Phi$ is then topologically an open disk of $\mathbb{C}$, and is therefore simply connected. Now each point of C above any point of $\mathbb{C}\backslash\Phi$ is contained in an analytic function element, and by Lemma 8.13 each such function element extends to an analytic function on $\mathbb{C}\backslash\Phi$. Since there are n points of C above each point of $\mathbb{C}\backslash\Phi$, there are just n such functions f_i on $\mathbb{C}\backslash\Phi$. Call their graphs $F_1, \ldots, F_n$. Suppose, to be specific, that $P \in F_1$ and $Q \in F_n$.

Now let P and P' be two points of C lying over $\mathbb{C}\backslash\Phi$, and suppose that we can analytically continue from P to P' along some allowable chain $(\mathcal{O}_1, \ldots, \mathcal{O}_r)$ in $C\backslash\mathscr{D}$. Choose open sets such that $\mathcal{O}_0, \mathcal{O}_{n+1} \subset \mathbb{C}\backslash\Phi, P \in \mathcal{O}_0 \subset \mathcal{O}_1$ and $P' \in \mathcal{O}_{r+1} \supset \mathcal{O}_r$. Since $\mathbb{C}\backslash\Phi$ is simply connected, it is allowable by Lemma 8.13. Hence its subsets $\mathcal{O}_0, \mathcal{O}_{r+1}$ are also allowable, and therefore $(\mathbb{C}\backslash\Phi, \mathcal{O}_0,$

$\mathcal{O}_1, \ldots, \mathcal{O}_r, \mathcal{O}_{r+1}, \mathbb{C}\backslash\Phi)$ is an allowable chain. Thus we may assume without loss of generality that any such analytic continuation in $\mathbb{C}\backslash\mathcal{D}$ from a point $P \in C$ to any other point $P' \in C$, where $\pi_Y(P)$ and $\pi_Y(P') \in \mathbb{C}\backslash\Phi$, is the lifting of some allowable chain from $\mathbb{C}\backslash\Phi$ to $\mathbb{C}\backslash\Phi$. If $P \in F_i$ and $P' \in F_j$, then this same chain also defines a continuation from any point in F_i to any point in F_j. Now consider any F_k. This chain also lifts to an analytic continuation from F_k to some one of $F_1, \ldots, F_n$—that is, it defines a mapping ρ from the set $\{F_1, \ldots, F_n\}$ into itself. Clearly if this chain defines a continuation from F_i to F_j, then the "reverse chain" $(\mathbb{C}\backslash\Phi, \mathcal{O}_{r+1}, \mathcal{O}_r, \ldots, \mathcal{O}_1, \mathcal{O}_0, \mathbb{C}\backslash\Phi)$ defines a continuation from F_j to F_i. Hence ρ has an inverse and is therefore 1:1 and onto—that is, ρ is a permutation. The set of all allowable chains in $\mathbb{C}\backslash\mathcal{D}$ from $\mathbb{C}\backslash\Phi$ to $\mathbb{C}\backslash\Phi$ then defines a set of permutations (actually a group of permutations, as the reader may check for himself).

Let us consider again the points P and Q; we are assuming they are not connectible by an allowable chain. If $P \in F_1$ and $Q \in F_n$, then no point of F_1 can be so analytically continued to any point of F_n; hence the permutations permute F_1 together with possibly some other elements of $\{F_1, F_2, \ldots, F_n\}$, say $\{F_1, \ldots, F_m\}$, to form a closed cycle. The function F_n is not in this cycle (but is of course permuted within its own disjoint cycle).

We can now get a contradiction as follows: Let ϕ be any polynomial in $\mathbb{C}[X, Y]$ which vanishes on all of $F_1, \ldots, F_m$, $m < n$. Then the resultant polynomial $\mathcal{R}_Y(p, \phi) \in \mathbb{C}[X]$ is zero at each point of $\mathbb{C}\backslash\Phi$, so must itself be the zero polynomial. Hence p and ϕ have a common (nonconstant) factor (Theorem 4.4). But p is irreducible, so ϕ must have p as a factor. Hence ϕ is of degree $\geqslant n$ in Y. *We shall get our contradiction by constructing such a ϕ of degree less than n in Y.*

For this, first consider the ($\mathbb{C}$-valued) function

$$(Y - f_1) \cdot \ldots \cdot (Y - f_m) \qquad (m < n),$$

defined on $(\mathbb{C}_X\backslash\Phi) \times \mathbb{C}_Y$. Its coefficients are, of course, symmetric functions of $f_1, \ldots, f_m$, namely,

$$\begin{aligned}
\sigma_0(X) &= 1 \\
\sigma_1(X) &= -(f_1(X) + \ldots + f_m(X)), \\
\sigma_2(X) &= f_1(X) \cdot f_2(X) + \ldots + f_{m-1}(X) \cdot f_m(X), \\
&\vdots \\
\sigma_m(X) &= (-1)^m f_1(X) \cdot \ldots \cdot f_m(X).
\end{aligned}$$

Hence the functions σ_i are all analytic at each point of $\mathbb{C}\backslash\Phi$. Let Q be any point of $\Phi\backslash\mathcal{D}$. There are then m distinct points of $\bar{F}_1, \ldots, \bar{F}_m$ above Q (the bar means topological closure in $\mathbb{C}_{XY}$), and m analytic function elements through these m points. Obviously the symmetric functions of these function elements agree with the σ_i near Q. In this way the functions σ_i may be extended to functions analytic on $\mathbb{C}\backslash\mathcal{D}$. (We still denote the extension of σ_i by σ_i.)

But σ_i even extends analytically to all of $\mathbb{C}$. First note that if we analytically continue any σ_i from $P \in \mathbb{C}\backslash\mathscr{D}$ back to P along an allowable chain in $\mathbb{C}\backslash\mathscr{D}$, we must arrive at the same value for σ_i. For the worst that can happen to the points above P of the $\bar{F}_j$'s under such a continuation, is that they are permuted among themselves; each symmetric function $\sigma_i(P)$ of these values is left unchanged by any such permutation. Hence each σ_i is single valued and analytic in $\mathbb{C}\backslash\mathscr{D}$. Now we may apply the Riemann extension theorem (Theorem 4.12), for since $p(X, Y)$ is monic in Y, the n zeros of p near any $P \in \mathscr{D}$ are all bounded, so the same is true of the functions σ_i. Hence they are analytic at each $P \in \mathscr{D}$, and therefore single valued and analytic in all of $\mathbb{C}$.

All that is left to check is that each $\sigma_i(X)$ is actually a polynomial. For this, write

$$\sigma_i(X) = \sum_j c_{ij} X^j. \tag{27}$$

Set $X = 1/X'$. Then for X' small but nonzero,

$$\sigma_i\left(\frac{1}{X'}\right) = \sum_j c_{ij} \frac{1}{(X')^j} \tag{28}$$

represents σ_i at points close to $X = P_\infty = \mathbb{P}^1(\mathbb{C})\backslash\mathbb{C}_X$. Now the zero-set of

$$Y^n + \sigma_1\left(\frac{1}{X'}\right)Y^{n-1} + \ldots + \sigma_n\left(\frac{1}{X'}\right) = 0 \tag{29}$$

for small $X' \neq 0$ gives that part of $\mathsf{V}(p)$ near the line $\mathbb{P}^2(\mathbb{C})\backslash\mathbb{C}_{XY}$. Let $\max_i(\deg(\sigma_i)) = M$. Multiplying each side of the equation in (29) by $(X')^{Mn}$ gives

$$(X'^M Y)^n + b_1(X')(X'^M Y)^{n-1} + \ldots + b_n(X') = 0,$$

where each b_i is a polynomial in X'. Since this is a monic polynomial in the variable $X'^M Y$, from Theorem 4.13 the solutions about $(X' = 0,\ X'^M Y = 0)$ are given by finitely many fractional-power series

$$(X'^M Y)_j = \text{a fractional-power series in } X'.$$

Then each Y_j is a fractional-power series with at most finitely many negative-power terms. Hence any symmetric function σ_i, being a sum of products of the functions Y, likewise has only finite many negative-power terms. Hence all but finitely many coefficients c_{ij} in (28) are zero—that is, the expansion in (27) is finite. Therefore each $\sigma_i(X)$ is in $\mathbb{C}[X]$, and we have found a polynomial

$$\phi(X, Y) = Y^m + \sigma_1(X)Y^{m-1} + \ldots + \sigma_m(X)$$

vanishing on $F_1 \cup \ldots \cup F_m$, with $\deg_Y \phi = m < n$. We have thus obtained the promised contradiction; hence Theorems 8.4 and 8.5 are proved. □

EXERCISES

8.1 Prove Lemma 8.13.

8.2 Although we have shown that any irreducible curve in $\mathbb{P}^2(\mathbb{C})$ is connected, it is still conceivable that, for instance, the one-point union of two topological spheres is the underlying space of an irreducible curve, so that the irreducible curve would in this sense split up. Prove this can never happen by showing that the underlying space of each irreducible component of a curve $C \subset \mathbb{P}^2(\mathbb{C})$ is the topological closure of a connected component of $C \backslash S_C$, where S_C is the set of those points of C which are not smooth.

9 Algebraic curves are orientable

We stated in Theorem 2.7 of Chapter I that any curve in $\mathbb{P}^2(\mathbb{C})$ is obtainable from a compact connected orientable real 2-manifold by identifying finitely many points to finitely many points. We will at last have a proof of this after considering orientability, which we do in this section.

First, what is an *orientable 2-manifold*? Intuitively, it is a 2-manifold on which one can specify in a consistent way a direction of spinning in the manifold at each of its points, much as suggested by Figure 21; the division

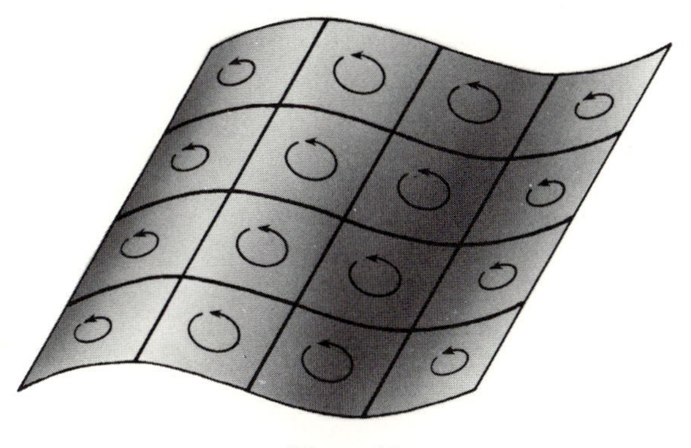

Figure 21

of the manifold into parallelograms (triangles can be used just as well) may be made as fine as desired. If a 2-manifold is orientable, it has two possible orientations, which we may call *positive* and *negative*.

Not all 2-manifolds are orientable; one example is the Möbius strip (without boundary points), a model being obtained by taking a long rectangular strip of paper and pasting its ends together after giving one end a 180° twist. This is indicated in Figure 22. Note that if one starts from the point P with a given orientation and travels once along the dotted line, one arrives at P with the opposite orientation. Intuitively, we see one cannot make the orientations at the various points fit together in a compatible way.

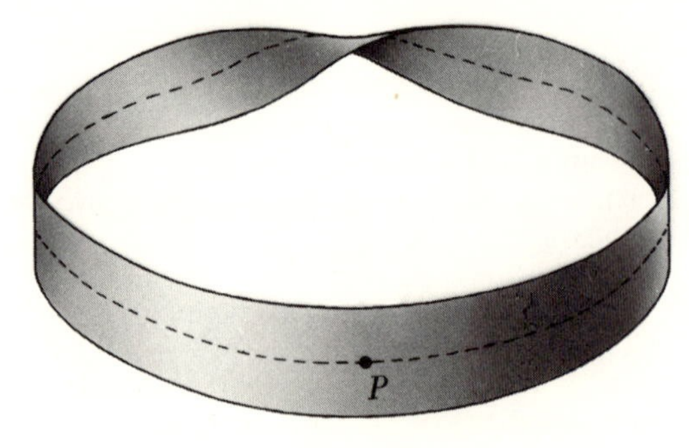

Figure 22

We can, however, orient $\mathbb{R}^2$ in our intuitive sense. We can arrive at a precise definition of orientation for $\mathbb{R}^2$ very easily: If $\{v_1, v_2\}$ is a basis of $\mathbb{R}^2$, we can order this basis in two different ways, as the ordered pair (v_1, v_2) and as (v_2, v_1). Each ordering specifies an orientation at the origin; for (v_1, v_2) this is done as follows (the case of (v_2, v_1) is treated similarly): Within $\mathbb{R}^2$, we can rotate (or spin) the first vector v_1 into the second one v_2 in less than $180°$ in exactly one of two ways—counterclockwise or clockwise; if counterclockwise we say $\mathbb{R}^2$ together with (v_1, v_2) is *positively oriented*, otherwise *negatively oriented*. By parallel translation of the basis vectors there is induced an orientation at each point of $\mathbb{R}^2$. Thus for $\mathbb{R}^2$ it suffices to work only at the origin. As an example, if in $\mathbb{R}^2 = \mathbb{R}_{XY}$, $v_1 = (1, 0)$, $v_2 = (a_1, a_2)\,(a_2 \neq 0)$, then (v_1, v_2) defines a positive orientation if $a_2 > 0$, and a negative one if $a_2 < 0$. One can of course associate to this basis the matrix

$$\begin{pmatrix} 1 & 0 \\ a_1 & a_2 \end{pmatrix}.$$

Since its determinant is a_2,

$$\det\begin{pmatrix} 1 & 0 \\ a_1 & a_2 \end{pmatrix}$$

is positive or negative according to whether the orientation is positive or negative. Furthermore any basis can be rotated so that one vector lies along $\mathbb{R}_X$, this rotation being given by a proper orthogonal matrix

$$\begin{pmatrix} \cos\theta & \sin\theta \\ -\sin\theta & \cos\theta \end{pmatrix}.$$

Since this has determinant 1, rotating any ordered basis leaves the associated determinant unchanged. The above definition of orientation of $\mathbb{R}^2$ relative to a basis (v_1, v_2) can therefore be recast as follows:

Definition 9.1. Let (v_1, v_2) be an ordered basis of $\mathbb{R}^2$, where $v_1 = (a_{11}, a_{12})$, $v_2 = (a_{21}, a_{22})$, and $A = (a_{ij})$. Then (v_1, v_2) defines a **positive orientation** on $\mathbb{R}^2$ if det A is positive, and a **negative orientation** if det A is negative.[1]

Now evidently any nonsingular linear transformation T of an oriented $\mathbb{R}^2$ induces an orientation (Tv_1, Tv_2) on its image space. If T maps $\mathbb{R}^2$ into itself, the new orientation may well be different. For instance if $\mathbb{R}_{XY}$ is flipped about its X-axis by $T:(x, y) \to (x, -y)$, then the canonical basis $v_1 = (1, 0)$, $v_2 = (0, 1)$ is mapped into $T(v_1) = (1, 0)$, $T(v_2) = (0, -1)$; $(T(v_1), T(v_2))$ now defines the opposite orientation on $\mathbb{R}_{XY}$. If B is any nonsingular matrix representing such a linear automorphism, then the new basis is given by AB, and the sign of det AB = det $A \cdot$ det B is preserved or reversed according to whether det B is positive or negative.

We can easily extend these ideas to smooth manifolds. Let U, U' be open neighborhoods of an arbitrary point $(0, 0) \in R_{X_1X_2}$, and suppose $\phi: U \to U'$ is a smooth map (Definition 3.1) given by real-valued functions $X_i' = \phi_i(X_1, X_2)$, where ϕ_i is smooth for $i = 1, 2$, and where $\phi_i(0, 0) = 0$. For U sufficiently small, this map is well approximated by the linear map

$$(X_1', X_2') = (X_1, X_2)\begin{pmatrix} \dfrac{\partial \phi_1}{\partial X_1} & \dfrac{\partial \phi_2}{\partial X_1} \\ \dfrac{\partial \phi_1}{\partial X_2} & \dfrac{\partial \phi_2}{\partial X_2} \end{pmatrix}_{(X_1, X_2) = (0, 0)} = (X_1, X_2) J_\phi(0, 0).$$

We now make the

Definition 9.2. Let U, U' be open sets in $\mathbb{R}_{X_1X_2} = \mathbb{R}_X$. A smooth map $\phi = (\phi_1, \phi_2): U \to U$ is **orientation preserving** *at* $(x) = (x_1, x_2) \in U$ if

$$\det\left(\frac{\partial \phi_i}{\partial X_j}\right)_{X = x} = \det J_\phi(x) > 0;$$

ϕ is **orientation reversing** *at* x if det $J_\phi(x) < 0$. The map $\phi: U \to U$ is **orientation preserving** if it is orientation preserving at each point of U (that is, if det $J_\phi(x) > 0$ for each $x \in U$).

We can now give the following

Definition 9.3. A **smooth real** 2-**manifold** M is a real 2-manifold together with an open cover $\{U_\alpha\}$ of M by open neighborhoods U_α, such that intersecting neighborhoods U_α attach to each other smoothly—that is, there

[1] We can in an analogous way define an orientation on any $\mathbb{R}^n$. To any ordering of the vectors in a basis $\{v_i\} = \{(a_{i1}, \ldots, a_{in})\}$ of $\mathbb{R}^n$, one may associate det $A = \det(a_{ij}) \neq 0$. Then the ordered basis orients $\mathbb{R}^n$ positively if det $A > 0$, and negatively otherwise. Note that switching two vectors in any ordering of the basis switches two rows of A, thus changing the sign of the determinant and so also the orientation of $\mathbb{R}^n$.

are homeomorphisms ϕ_α from open sets of $\mathbb{R}^2$ to U_α such that each homeomorphism $\phi_\beta^{-1} \circ \phi_\alpha : \phi_\alpha^{-1}(U_\alpha \cap U_\beta) \to \phi_\beta^{-1}(U_\alpha \cap U_\beta)$ is smooth. The manifold M is **orientable** if each such $\phi_\beta^{-1} \circ \phi_\alpha$ is orientation preserving—that is, if

$$\det(J_{\phi_\beta^{-1} \circ \phi_\alpha}(x)) > 0 \quad \text{for each } x \in \phi_\alpha^{-1}(U_\alpha \cap U_\beta).$$

Remark 9.4. This definition generalizes in the obvious way to real n-manifolds (that is, Hausdorff spaces in which each point has an open neighborhood homeomorphic to an open ball in $\mathbb{R}^n$). Also, the hypothesis of smoothness allows us the convenience of an "analysis-type" definition of orientability; this will be useful, as we see in the proof of Lemma 9.5. In a certain sense this definition is like specifying an orientation on "infinitely small" parallelograms or triangles of M. The reader familiar with combinatorial topology will recognize that if M is any (not necessarily smooth) topological 2-manifold, one can more generally define orientability using triangles of finite size (or n-simplices in the case of topological n-manifolds).

Lemma 9.5. *Let $C \subset \mathbb{P}^2(\mathbb{C})$, and let $\{P_i\}$ be the (finite) set of singular points of C. Then $C \backslash \{P_i\}$ is an orientable real 2-manifold.*

Corollary 9.6. *If $C \subset \mathbb{P}^2(\mathbb{C})$ is nonsingular, it is orientable.*

Proof of Lemma 9.5. By Theorem 3.6, for each $P \in C\backslash\{P_i\}$, the part $U(P)$ of C within a small $\mathbb{C}^2$-neighborhood of P, forms the graph of a function f_P analytic in $U(P)$. The set of $U(P)$ where $p \in C\backslash\{P_i\}$, covers $C\backslash\{P_i\}$ and will serve as the open cover of Definition 9.3. If the domain of f_P is an open set of, say $\mathbb{R}^2 = \mathbb{C}_X$, then the functions ϕ_P mapping a neighborhood of $\mathbb{R}^2$ to $U(P)$ are defined by

$$\phi_P(x) = (x, f(x)) \qquad (x = (x_1, x_2) \in \mathbb{C}_X). \tag{30}$$

Now let P be a fixed point of $C\backslash\{P_i\}$, and let $U(Q)$ and $U(Q')$ be any two of the above neighborhoods which contain P ($Q, Q' \in C\backslash\{P_i\}$). We want to show that

$$\phi_{Q'}^{-1} \circ \phi_Q : \phi_Q^{-1}(U(Q) \cap U(Q')) \to \phi_{Q'}^{-1}(U(Q) \cap U(Q'))$$

is orientation-preserving. Any point in $U(Q) \cap U(Q')$ has two sets of real coordinates—one relative to the system about Q (say (X_1, X_2)) transferred from a neighborhood of $\mathbb{C}_X = \mathbb{R}_{X_1X_2}$ by the homeomorphism ϕ_Q; and the other system relative to the coordinates about Q' (say (X_1', X_2') induced by $\phi_{Q'}$. Then, from (30) we see that $\phi_{Q'}^{-1} \circ \phi_Q$ becomes a map from an open set in $\mathbb{C}_X = \mathbb{R}_{X_1X_2}$ to one in $\mathbb{C}_{X'} = \mathbb{R}_{X_1'X_2'}$. We are to show this map is orientation preserving, that is, writing

$$\phi_{Q'}^{-1} \circ \phi_Q(X) = \phi(X_1, X_2) = \phi_1(X_1, X_2) + i\phi_2(X_1, X_2)$$

(where ϕ_1 and ϕ_2 are real valued), we want to show $\det J = \det(\partial\phi_i/\partial X_j)$ is positive at each point of $\phi_Q^{-1}(U(Q) \cap U(Q')$. We first show that ϕ is analytic.

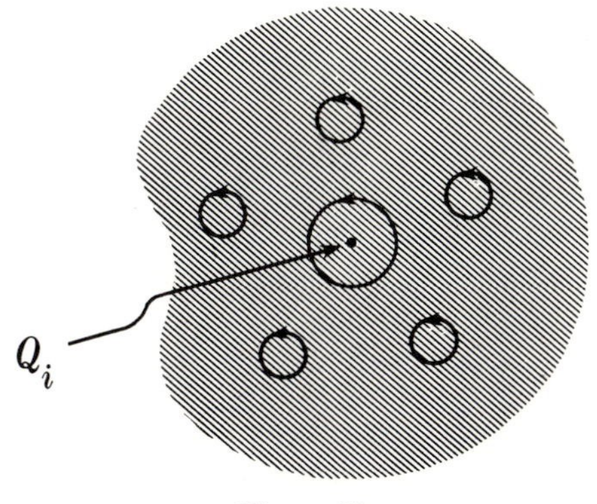

Figure 23

For this, let (X, Y) be coordinates in $\mathbb{C}^2$ about Q with respect to which $U(Q)$ is the graph of an analytic function $Y = f_Q(X)$; similarly, let (X', Y') be coordinates about Q' with respect to which $U(Q')$ is the graph of an analytic function $Y' = f_{Q'}(X')$. Now the projection $\pi_{Y'}$ along Y' to $\mathbb{C}_{X'}$ is analytic; since $\phi_{Q'}^{-1} \circ \phi_Q = \pi_{Y'} \circ \phi_Q$, $\phi = \phi_{Q'}^{-1} \circ \phi_Q$ is analytic in $\phi_Q^{-1}(U(Q) \cap U(Q'))$. Similarly, ϕ^{-1} is analytic in $\phi_{Q'}^{-1}(U(Q) \cap U(Q'))$. Now $\det(\partial\phi_i/\partial X_j)$ is nonzero at each point of $\phi_Q^{-1}(U(Q) \cap U(Q'))$, since ϕ is invertible there. By the proof of Lemma 3.11, this determinant is just $|d\phi/dX|^2$, so it is positive. We have therefore established Lemma 9.5. □

Now let us turn to the whole topological space C. Suppose P is a topologically singular point—that is, suppose there is some neighborhood $U(P)$ about P such that $U(P) \cap C$ is the one-point union of at least two disks. Then $U(P) \cap (C\backslash\{P\})$ consists of the union of a finite number of punctured disks, $\Delta_i\backslash P$. To each such punctured disk let us add a new point Q_i and let a typical small neighborhood about Q_i consist of Q_i together with $\Delta_i\backslash P$ intersected with a small open set (in $\mathbb{P}^2(\mathbb{C})$) containing P. This in effect "separates" the one-point union of disks into disjoint disks; this new topological space is now a manifold M. We have shown this manifold is orientable except possibly at finitely many points $\{Q_1, Q_2, \ldots\}$. But we can extend an orientation of $M\backslash\{Q_1, Q_2, \ldots\}$ to all of M in a way suggested by Figure 23, which shows a neighborhood of Q_i. Since C is obtained from M by identifying finitely many points to finitely many points, we see that C is obtained from a compact connected orientable 2-manifold by identifying finitely many points to finitely many points.

10 The genus formula for nonsingular curves

We have seen that topologically, any irreducible curve $C \subset \mathbb{P}^2(\mathbb{C})$ may be obtained from a compact connected orientable 2-manifold by identifying finitely many points to finitely many points. If C is in addition nonsingular in the sense of Definition 7.9, then it is just a compact connected orientable 2-manifold. Now there is a basic classification theorem for such manifolds. It says that any such manifold is a sphere with a finite number g of "handles," g being the genus of the manifold, and that any two such surfaces are homeomorphic iff they have the same genus. For a proof of this classification theorem, see, for instance [Massey, Chapter I] or [Cairns, Chapter 2].

Now suppose that $C \subset \mathbb{P}^2(\mathbb{C})$ is nonsingular, defined by the polynomial $p(X, Y) \subset \mathbb{C}[X, Y]$ or by its homogenization $H_Z(p) = q(X, Y, Z)$ (p, q, of degree n). Certainly the topology of C is determined once p or q is specified. It is therefore reasonable to ask if there is any way of finding the genus g directly from p or q, without recourse to a careful geometric investigation of the curve. There indeed is—it is given directly by a "genus formula"; the main object of this section is to prove this formula. Even at this stage in our study of algebraic geometry, we can give a fairly complete proof of this formula. In this section we prove this formula, except that the treatment of some purely topological facts (proved in many standard texts) is more informal here, and a few details at the end of the proof will be left to exercises later in the book (Exercises 6.9 and 7.4 of Chapter IV), when establishing them will be both easy and natural.

Before stating the formula, first note that if $C \subset \mathbb{P}^2(\mathbb{C})$ is nonsingular, then C can be defined by an irreducible homogeneous polynomial in $\mathbb{C}[X, Y, Z]$ (Exercise 7.3).

Theorem 10.1 (Genus formula). *Let $C \subset \mathbb{P}^2(\mathbb{C})$ be a nonsingular projective curve defined by the irreducible polynomial $p(X, Y)$. If $\deg p = n$, then the genus g of C is*

$$g = \frac{(n-1)(n-2)}{2}.$$

The basic outline of the proof is this:

First, we note that any compact connected orientable 2-manifold M may be looked at, topologically, as a polyhedron having g handles.

Second, we recall the basic fact that one can compute g from M looked at as any polyhedron having V vertices, E edges, and F faces. (Specifically,

$$g = 1 - \frac{V - E + F}{2},$$

sometimes called Euler's formula.)

Finally, we look at C as a near n-sheeted covering of the sphere $\mathbb{P}^1(\mathbb{C}) = \mathbb{C}_X \cup \{\infty\}$; we in turn regard $\mathbb{P}^1\mathbb{C})$ as a polyhedron, its set of vertices containing the set of discriminant points of the covering. Then above each face, edge and vertex of $\mathbb{P}^1(\mathbb{C})$ lie n faces, n edges, and n vertices, except that over any discriminant point there are *fewer* than n vertices. Using some facts about discriminants, we will find out just how many fewer, and from this information we will be able to compute g grom Euler's formula.

The first two parts above are purely topological in nature, and are included in a number of standard texts on topology. We therefore consider these two points more informally.

To begin with the first part of the proof, a **(topological triangulated) polyhedron** will mean, for us, a compact connected orientable real 2-manifold M together with a covering of M by finitely many closed subsets $\{S_1, \ldots, S_n\}$, and a corresponding set of homeomorphisms $S_i \leftrightarrow T_i$, where each T_i is a triangle in $\mathbb{R}^2$ (that is, a compact subset of $\mathbb{R}^2$ whose boundary consists of three line segments). The subsets of S_i homeomorphic to the interiors of the triangles T_i are called the **faces** of the polyhedron; the subsets of S_i homeomorphic to the edges (=open line segments) of T_i, and to the vertices of T_i, are still called the **edges** and **vertices** of the polyhedron, respectively. Finally, we require that any two different S_i and S_j be disjoint, or have exactly one vertex in common, or have exactly one edge (and its two vertices) in common. It turns out that the set consisting of all edges and vertices of a polyhedron is connected.

We now give an outline of the proof of the second topological assertion, Euler's formula.

Lemma 10.2. *Let M be any polyhedron in the above sense, having V vertices, E edges, and F faces. Suppose that M has genus g. Then*

$$V - E + F = 2 - 2g.$$

(Or equivalently, $g = 1 - (\frac{1}{2})(V - E + F)$.)

PROOF. We first consider the case $g = 0$. Therefore, assume M is a sphere. Let e be any edge of M; e belongs to some closed polygonal curve C consisting of edges and vertices of M. The union of e and the two faces on either side of e is a single connected open set, which we continue to call a *face*; the subset e is *no longer an edge*, or, loosely, e has been "removed." In this operation, E and F have each decreased by 1, and $V - E + F$ remains unchanged. Also, the system of edges and vertices is still connected, for one can travel from one vertex of e to the other by going around the remaining edges and vertices of C.

Now remove in the same way another edge e' belonging to a closed polygonal curve C' selected from the edges and vertices remaining after e was removed; as before, the new $V - E + F$ remains unchanged and the system of edges and vertices is still connected. Continue this process until no remaining edge is an edge of a closed polygonal curve. At this stage, (a) the system of edges and vertices is connected; (b) there must be some vertex which is the vertex of only one edge, since if every vertex were the vertex of at least two edges, one could continue traveling, eventually traversing a closed path. Remove this vertex and edge; $V - E + F$ remains unchanged. This process yields no new closed polygonal curves, and the system of edges and vertices still remains connected. Continue this process until there are no

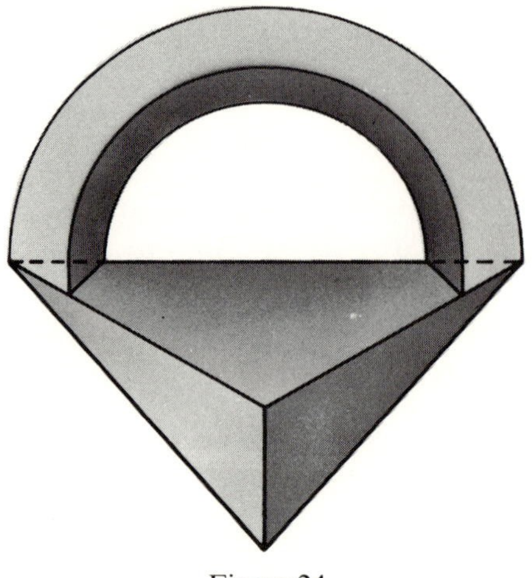

Figure 24

more edges; we are left with one vertex and obviously only one face, whence $V - E + F = 2$. Thus Lemma 10.2 is proved when $g = 0$.

To prove it for any g, note that any surface of our type of genus g may be transformed to a polyhedral sphere with polyhedral handles. Figure 24 shows an example for $g = 1$. In transforming the surface to such a polyhedron, it may be necessary to add some extra vertices, edges, and faces, but it is easy to check that this can be done leaving $V - E + F$ unchanged. It may also be assumed that the handles and sphere are joined along edges of the polyhedron. Now cut each handle at one of the two places where it joins the sphere (see Figure 25). If there are g handles, then the new figure is

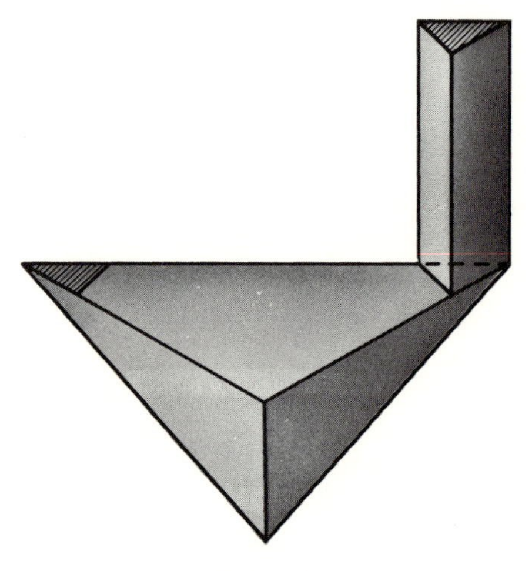

Figure 25

homeomorphic to a spherical polyhedron with $2g$ faces missing. These missing faces then make our formula read

$$V - E + F = 2 - 2g. \qquad \square$$

With these topological preliminaries taken care of, we now turn to the third part of our argument.

We will use the following definition and lemma.

Definition 10.3. Let f be a function complex analytic at $(a_1, \ldots, a_n) \in \mathbb{C}_{X_1, \ldots, X_n}$. Then f **has order** s **in** X_i **at** $(a_1, \ldots, a_n)$ if $f(a_1, \ldots, a_{i-1}, X_i, a_{i+1}, \ldots a_n)$ has order s at a_i.

In $\mathbb{C}_{X_1, \ldots, X_n}$, any product $\Delta_1 \times \ldots \times \Delta_n$ of disks $\Delta_i \subset \mathbb{C}_{X_i}$ is called a **polydisk**.

Lemma 10.4. *Let* $f : \mathbb{C}_{X_1, \ldots, X_n} \to \mathbb{C}$ *be complex analytic at* $a = (a_1, \ldots, a_n)$, *and suppose it has order* s *in* X_i *at* a. *Then there is an open polydisk about* a *in* $\mathbb{C}^n$, $\Delta(a) = \Delta = \Delta_1 \times \ldots \times \Delta_n$ (Δ_i, *an open disk in* $\mathbb{C}_{X_i}$ *centered at* a_i, *and* f *analytic in* Δ) *such that for each point*

$$(a'_1, \ldots, a'_{i-1}, a'_{i+1}, \ldots, a'_n) \in \Delta_1 \times \ldots \times \Delta_{i-1} \times \Delta_{i+1} \times \ldots \times \Delta_n,$$

$f(a'_1, \ldots, a'_{i-1}, X_i, a'_{i+1}, \ldots, a'_n)\colon \mathbb{C}_{X_i} \to \mathbb{C}$ *has exactly* s *zeros in* Δ_i, *these zeros being counted with multiplicity.*

Proof. The proof exactly parallels part of the proof of the implicit function theorem, Theorem 3.6: The hypothesis that f has order s in X_i at a just says that $f(a'_1, \ldots, a'_{i-1}, X_i, a'_{i+1}, \ldots, a'_n)$ has a as a zero of order s. Hence if Δ_i is sufficiently small with boundary $\partial\Delta_i$, then by the argument principle (Theorem 3.8.2) we have

$$\frac{1}{2\pi i} \int_{\partial\Delta_i} \frac{f_{X_i}(a'_1, \ldots, a'_{i-1}, X_i, a'_{i+1}, \ldots, a'_n)}{f(a'_1, \ldots, a'_{i-1}, X_i, a'_{i+1}, \ldots, a'_n)} \, dX_i = s$$

Now f and f_{X_i} are continuous in X_i; since $f(a'_1, \ldots, a'_{i-1}, X_i, a'_{i+1}, \ldots, a'_n)$ is bounded away from zero on the compact set $\partial\Delta_i$, the values on $\partial\Delta_i$ of the above integrand vary continuously in X_i. Hence as we vary each a'_j within sufficiently small disks Δ_j, the integral still has value nearly s. But this integral is always an integer, so it is *exactly* s—i.e., the argument principle says that for Δ_i sufficiently small, there are exactly s zeros of f in Δ above any point of a sufficiently small neighborhood $\Delta_1 \times \ldots \times \Delta_{i-1} \times \Delta_{i+1} \times \ldots \times \Delta_n$. $\square$

We may now indicate how the proof of the genus formula can be carried out. First, we look at C as the near n-cover $\{C, \mathbb{C}_X \cup \{\infty\}, \pi_Y\}$. Next, look at $\mathbb{P}^1(\mathbb{C}) = \mathbb{C}_X \cup \{\infty\}$ as a polyhedron. We may assume that all the points where $\mathscr{D}_Y(p)$ vanishes are included in the finite set of vertices of the polyhedron. (If $\{v_1, v_2, \ldots\}$ is the set of discriminant points in a given face, we may draw edges, starting from v_1, to the vertices of the face, these edges not touching $v_2, v_3 \ldots$; we may continue this way until all discriminant points are made vertices of a polyhedron.) For any face f and any edge e of $\mathbb{P}^1(\mathbb{C})$, $\pi_Y{}^{-1}(f)$ and $\pi_Y{}^{-1}(e)$ consist of n faces and edges, respectively. And for a vertex v of $\mathbb{P}^1(\mathbb{C})$, $\pi^{-1}(v)$ consists of n vertices if $\mathscr{D}_Y(p)$ does not vanish at v, and fewer than n vertices if it does vanish at v. If the number of distinct points of C above v is $n - m$, it is natural to say that we have "lost" m points. If we knew how many were lost in this way, we could compute the genus of C as follows: If V, E, F are the number of vertices, edges and faces of $\mathbb{P}^1(\mathbb{C})$, then $V - E + F = 2$. Also, the number of edges and faces of C are nE and nF. If we can show that the number of vertices of C is $nV - n(n - 1)$—i.e., that $n(n - 1)$ vertices were lost, then the genus of C would be

$$g = 1 - \frac{(nV - n(n-1)) - nE + nF}{2}$$

$$= 1 - \frac{2n - (n)(n-1)}{2} = \frac{(n-1)(n-2)}{2},$$

which would establish Theorem 10.1.

Now in Exercises 6.9 and 7.4 of Chapter IV, we will show that coordinates in $\mathbb{C}^2$ can be selected so that relative to these coordinates, the order with respect to Y of P at any point $P \in C$, is either one or two. We see from Lemma 10.4 that we lose one point in our covering at precisely those points $P \in C$ where the order is two. It follows directly from Definition 10.3 that the points P of order two are just the intersection points of $\mathsf{V}(p)$ with $\mathsf{V}(p_Y)$. Since p and p_Y have (total) degree n and $n - 1$, respectively, by Bézout's Theorem (Theorem IV, 7.1), they intersect in $n(n - 1)$ points, counting multiplicity. But an easy argument (Exercise 6.1 of Chapter IV) will show that each such intersection is of multiplicity one; hence the total number of points lost in our covering is exactly $n(n - 1)$, thus proving Theorem 10.1. $\square$

CHAPTER III

Commutative ring theory and algebraic geometry

1 Introduction

In Chapter II, all our results were proved for plane algebraic curves. It is natural to try to extend these results to arbitrary complex-algebraic varieties in affine or projective n-space. What do arbitrary varieties look like in the neighborhood of a point? How does one prove a general dimension theorem? What can be said about the connectedness or orientability of arbitrary varieties? Can we get topological invariants (like the genus) directly from a defining set of polynomials instead of analyzing each variety separately?

In answering these questions for curves, we made constant use of $p(X, Y)$ (or the homogeneous polynomial $q(X, Y, Z)$) defining the curve. We factored p and q into irreducibles; we used facts like

$$\begin{aligned} \mathsf{V}(p) &= \mathsf{V}(p_1{}^{r_1} \cdot \ldots \cdot p_k{}^{r_k}) = \mathsf{V}(p_1{}^{r_1}) \cup \ldots \cup \mathsf{V}(p_k{}^{r_k}) \\ &= \mathsf{V}(p_1) \cup \ldots \cup \mathsf{V}(p_k) = \mathsf{V}(p_1 \cdot \ldots \cdot p_k), \end{aligned}$$

so that p could be assumed to have no repeated factors; we used the degree of a polynomial, the discriminant, the argument principle, and the implicit function theorem for p, to name just a few of our tools.

Now in any attempt to generalize our plane-curve results to arbitrary varieties in $\mathbb{P}^n(\mathbb{C})$, we immediately run into a difficulty: An arbitrary variety in $\mathbb{P}^n(\mathbb{C})$ cannot in general be defined by only one polynomial. In fact it can be easily shown, using the implicit mapping theorem (Theorem 3.5 of Chapter II), that one (nonconstant) polynomial $p(X_1, \ldots, X_n)$ always defines a subset of complex codimension one in $\mathbb{C}^n$, or in $\mathbb{P}^n(\mathbb{C})$. The reader may well ask: "What's so bad about not being able to define a variety by just one polynomial?" Let us see what happens when we try to prove something like "Any irreducible variety is connected." In Chapter II, we took for an irreducible variety a curve $\mathsf{V}(p)$, where p is nonconstant and irreducible, and found it to be connected. Suppose a variety V requires at least $s \geqslant 2$ polynomials to define it, say $V = \mathsf{V}(p_1, \ldots, p_s)$. We are now faced with deciding whether the *collection* of polynomials $\{p_1, \ldots, p_s\}$ is in some sense irreducible. The reader may reply: "Why not say the collection is irreducible if each member p_i is irreducible?" Let us try this. For example, let $p_1(X, Y) = X - 1$ and $p_2(X, Y) = Y^2 - X$; both p_1 and p_2 are irreducible. Now consider $V = \mathsf{V}(p_1, p_2) \subset \mathbb{C}^2$. This variety cannot be defined by just one polynomial, for it consists of two distinct points $(1, 1)$ and $(1, -1)$. It is thus the union of two point-varieties $\mathsf{V}(X - 1, Y - 1) \cup \mathsf{V}(X - 1, Y + 1)$, and is therefore not connected. Geometrically, there is certainly nothing irreducible about $\mathsf{V}(p_1, p_2)$. Obviously, one meets even greater difficulties in trying to generalize the notion of factorization to a collection of polynomials. Other examples show that defining the degree, or the discriminant of a collection of polynomials, cannot be handled by looking individually at each polynomial.

What is needed is a whole new approach, a new language and new machinery to handle what is algebraically a quite different type of question from that of the "one-polynomial" theory. That is the object of this chapter—to introduce this new approach.

To begin, let us consider the specific problem of defining a circle in 3-space. For ease of illustration, let us look at the case of a real circle in the $\mathbb{R}_{XY}$-plane of $\mathbb{R}_{XYZ}$. This circle C might be looked at as the intersection of the plane $\mathbb{R}_{XY}$ $(= \mathsf{V}(Z))$ and a cylinder through C—algebraically as, say, $\mathsf{V}(Z, X^2 + Y^2 - 1)$. But C could just as well be defined by the plane and a sphere $\mathsf{V}(Z, X^2 + Y^2 + Z^2 - 1)$, or by a cylinder and sphere, $\mathsf{V}(X^2 + Y^2 - 1, X^2 + Y^2 + Z^2 - 1)$. One could also intersect ellipsoids, paraboloids, hyperboloids, and so on, to get C.

There are many pairs which have equal claim to being the "most natural" pair of defining equations. A more symmetric approach is simply to consider the collection of *all* the polynomials whose zero-set contains C! That is, we consider this subset of $\mathbb{R}[X, Y, Z]$:

$$\mathfrak{a} = \{p(X, Y, Z) \mid p(a, b, c) = 0 \text{ for every point } (a, b, c) \in C\}.$$

It is an all-important fact that $\mathfrak{a}$ is an *ideal* in $\mathbb{R}[X, Y, Z]$. (Recall that a subset $\mathfrak{a}$ of a commutative ring R is an **ideal** if it is closed under subtraction and has the "absorption property"—that is, $r_1 \in \mathfrak{a}$ and $r_2 \in R$ implies $r_1 r_2 \in \mathfrak{a}_1$). For if $p \in \mathfrak{a}$ and $q \in \mathfrak{a}$, then $p(a, b, c) = q(a, b, c) = 0$ implies that $(p - q)(a, b, c) = p(a, b, c) - q(a, b, c) = 0$—that is, $p - q \in \mathfrak{a}$. Similarly, for any polynomial $r \in \mathbb{R}[X, Y, Z]$, $(p \cdot r)(a, b, c) = p(a, b, c) \cdot r(a, b, c) = 0 \cdot r(a, b, c) = 0$—that is, $p \cdot r \in \mathfrak{a}$. Hence $\mathfrak{a}$ is an ideal. Of course conversely, any ideal of $\mathbb{R}[X, Y, Z]$, being a set of polynomials, also defines a variety in $\mathbb{R}_{XYZ}$. *We have thus associated an ideal to an algebraic variety, and an algebraic variety to an ideal.*

But this can be extended much further. For instance, we will see that "irreducible polynomial" generalizes to "prime ideal," and that factorization of a polynomial generalizes to a "decomposition of an ideal." One can also generalize to the ideal-theoretic setting the important assumption that a polynomial has no repeated factors (i.e., that the ideal is its own "radical," as in Definition 1.1).

We now make these ideas a little more precise. First, essentially everything we did in Chapters I and II involved the polynomial ring in one, two, or three variables over $\mathbb{R}$ or $\mathbb{C}$. However, most of the generalizations we consider here make sense in the commutative ring $k[X_1, \ldots, X_n] = k[X]$ (k any field), so we shall state them at this level. We now look at translations into ideal theory of a few concepts from the "one-polynomial" theory of Chapter II.

Let $p(X_1, \ldots, X_n) = p(X)$ be any polynomial of $k[X]$. Then the set $(p) = \{pq \mid q \in k[X]\} \subset k[X]$ is an ideal of $k[X]$, the **principal ideal of** $k[X]$ **generated by** p.

Thus to each polynomial p corresponds the ideal (p). More generally, if $\{p_\alpha\}$ is any collection of polynomials in $k[X]$, the set $(\{p_\alpha\})$ of all finite sums $\{r_1 p_1 + \ldots + r_s p_s \mid r_i \in k[X],\ p_i \in \{p_\alpha\}\}$ is an ideal in $k[X]$, **the ideal generated by** $\{p_\alpha\}$. Hence to each collection of polynomials we associate an ideal.

How is the product $p \cdot q$ of two polynomials $p, q \in k[X]$ related to their ideals (p) and (q)? We may define the product of principal ideals by

$$(p) \cdot (q) = (pq).$$

(This definition is independent of the choice of generators of (p) and of (q).) More generally, if $\mathfrak{a}$ and $\mathfrak{b}$ are ideals of any commutative ring R, we may define their **product** by $\mathfrak{a} \cdot \mathfrak{b} = \{$all finite sums of products $a \cdot b$, where $a \in \mathfrak{a}, b \in \mathfrak{b}\}$. It is easy to check that $\mathfrak{a} \cdot \mathfrak{b}$ is an ideal.

We note that if $p, q \in k[X]$ have no factors in common, then

$$(p) \cdot (q) = (p) \cap (q).$$

PROOF. "$\subset$" follows directly from the definitions of ideal and product ideal. For "$\supset$", note that $r \in (p) \cap (q)$ implies both p and q divide r. Now $k[X]$ is a unique factorization domain, so since p and q are relatively prime, $p \cdot q$ must be a factor of r. □

Next, suppose $p \in k[X]$ is irreducible. Then (p) is a *prime ideal* in the usual sense. Thus if $r_1 \notin (p)$ and $r_2 \notin (p)$, then $r_1 \cdot r_2 \notin (p)$. For otherwise, $r_1 \cdot r_2 = pr_3$, meaning that either r_1 or r_2 has p as a factor, which is impossible.

Finally, let us look at the assumption we made so often in Chapter II, that p has no repeated nonconstant factors. Let its decomposition into irreducibles be

$$p = p_1^{n_1} \cdot \ldots \cdot p_s^{n_s}; \quad \text{then}$$
$$p^* = p_1 \cdot \ldots \cdot p_s$$

is the corresponding polynomial having no repeated factors. Ideal-theoretically, how do we go from (p) to (p^*)? Certainly $(p) \subset (p^*)$. Now let a be any element of (p^*). It is clear that $a^m \in (p)$ for some sufficiently high power a^m of

a. Conversely, suppose $a \notin (p^*)$. Then at least one polynomial p_i is not a factor of a, hence no power of a is divisible by p_i—that is, $a^m \notin (p)$, all m. Thus (p^*) consists precisely of those elements of $k[X]$ some power of which is in (p). A generalization of this idea is

Definition 1.1. For any ideal $\mathfrak{a}$ in a commutative ring R, the **radical of** $\mathfrak{a}$ (written $\sqrt{\mathfrak{a}}$) is

$$\sqrt{\mathfrak{a}} = \{a \in R \mid a^m \in \mathfrak{a}, \text{ for some positive integer } m\}.$$

It is easily checked that $\sqrt{\mathfrak{a}}$ is an ideal.

As we develop the theory we shall see that these translations into ideal theory turn out to be the "correct" ones. Later on we consider translations of other concepts. For instance, the fundamental theorem of algebra may be generalized to the so-called "Hilbert zero theorem," or "Nullstellensatz" (Theorem 5.1). And in working with higher dimensional varieties, it is important to be able to deal satisfactorily with subvarieties and projections of these subvarieties. We never brought this facet out explicitly for the algebraic curves of Chapter II, because the only proper subvarieties of irreducible algebraic curves are collections of finitely many points, and these are essentially of a trivial nature.

Our new language, then, will be that of commutative ring and ideal theory. This whole area is a large one, and we will not be able to generalize to ideal theory all the ideas used in Chapter II. However, we will make a start.

2 Some basic lattice-theoretic properties of varieties and ideals

In this section we establish some lattice-theoretic properties of varieties and ideals. We begin with a brief review of a few basic notions from lattice theory. We will ultimately use the language of lattices to put our "dictionary" into a quite compact form (cf. Diagrams 3.3 and 3.4). Also, just as point-set topology helps to unify various notions of convergence, and group theory unifies various different geometries, in algebraic geometry lattice theory unifies various "dictionaries" (for instance at the "local" and "analytic" levels). We begin with partially ordered sets.

Definition 2.1. A **partially ordered set** or **p.o. set** $(T, \leqslant)$ is a set T together with a binary relation $\leqslant$ (called the **partial order**) so that any three elements a, b, c of T satisfy: $a \leqslant a$ (identity); if $a \leqslant b$ and $b \leqslant a$, then $a = b$ (anti-symmetry; if $a \leqslant b$ and $b \leqslant c$, then $a \leqslant c$ (transitivity). (If, in addition, $a \leqslant b$ or $b \leqslant a$ for any two elements of T, then T is **totally ordered** (**t.o.**).)

A **sub-partially ordered set** $(U, \leqslant)$ *of* $(T, \leqslant)$ is any subset U of T with the partial order induced from $\leqslant$. Finally, if $(T, \leqslant)$ and $(T', \leqslant')$ are any two p.o. sets, then a **p.o. homomorphism from** $(T, \leqslant)$ **to** $(T', \leqslant')$ is a mapping ϕ from T to T'' such that if $x \leqslant y$ $(x, y \in T)$, then $\phi(x) \leqslant' \phi(y)$. The mapping ϕ is a **p.o. isomorphism** if it is 1 : 1-onto, and $x \leqslant y$ iff $\phi(x) \leqslant' \phi(y)$.

A particular kind of homomorphism of a p.o. set into itself (which will be of importance to us a little later) is the closure map. A **closure map** on a p.o. set $(T, \leqslant)$ is a function $a \to \bar{a}$ from T to T satisfying: For any two elements a, b of T, we have $a \leqslant \bar{a}$, $\bar{a} = \bar{\bar{a}}$, and $a \leqslant b$ implies $\bar{a} \leqslant \bar{b}$. (Note that the last condition says that a closure map is a p.o. homomorphism of T into itself.)

In a number of areas of mathematics, and in algebraic geometry at all levels in particular, there is a general notion of "decomposition into irreducibles." For instance at the one-polynomial level, there is unique factorization into irreducible polynomials of any $p \in k[X_1, \ldots, X_n]$, where k is any field. (This follows from the well-known Gauss lemma.) As mentioned earlier, we want as much as possible to be able to manipulate the ideals occurring in algebraic geometry like polynomials; as an example, we would like to "decompose any ideal into irreducible ideals." We now lead up to a general necessary condition for this.

First, any unique factorization domain D must satisfy a "divisor condition"—that is, for any element $a \in D$, any "strict chain"

$$a_1 > a_2 > a_3 > \ldots$$

must be of finite length, where $a_i > a_{i+1}$ means that a_{i+1} is a proper divisor of a_i (that is, $a_i = na_{i+1}$ for some non-invertible element $n \in D$). Now for polynomials $p, q \in k[X_1, \ldots, X_n]$, $p > q$ iff $(p) \subsetneqq (q)$; then by unique factorization, any "strict chain"

$$(p_1) \subsetneqq (p_2) \subsetneqq (p_3) \subsetneqq \ldots$$

must be of finite length. This condition may be translated into a chain condition on general ideals in a ring R—that is, every chain of ideals $\mathfrak{a}_1 \subsetneqq \mathfrak{a}_2 \subsetneqq \ldots$ (where $\mathfrak{a}_i \subset R$) is of finite length. We will be able to unify a number of decomposition theorems by generalizing this still further to the following definition:

Definition 2.2. Let $(T, \leqslant)$ be a p.o. set.

(2.2.1) $(T, \leqslant)$ satisfies the **ascending chain condition (a.c.c.)** iff there is no infinite strictly ascending chain of elements from T—that is, any chain $a_1 < a_2 < \ldots$ (where $a_i \in T$) must terminate after finitely many steps; dually,

(2.2.2) $(T, \leqslant)$ satisfies the **descending chain condition (d.c.c)** iff any strictly descending chain $b_1 > b_2 > \cdots$ must terminate after finitely many steps.

The reader may easily find examples showing that the a.c.c. and d.c.c. are independent conditions—that is, the a.c.c. may hold or not, independent of whether the d.c.c. holds.

We now give an alternate form of the a.c.c. (and of the d.c.c.) which we will need later on. Recall that if $(T, \leqslant)$ is a p.o. set, and if U is a subset of T, then an element $a \in U$ is **maximal in** U if no other element of U is larger than a—that is, if for $t \in U$, $a \leqslant t$ implies $a = t$; similarly, $b \in U$ is **minimal in** U if for $t \in U$, $t \leqslant b$ implies $b = t$. Then an easy contrapositive argument shows that $(T, \leqslant)$ *satisfies the* a.c.c. (d.c.c.) *iff in each nonempty subset* U *of* T *there is some maximal* (*minimal*) *element*. (We then say $(T, \leqslant)$ **satisfies the maximal (minimal) condition**.)

We shall also use the notions of upper and lower bounds: An element $x \in T$ is called an **upper bound** of two elements $a, b \in T$ provided that $a \leqslant x$ and $b \leqslant x$. If in addition, for any $z \in T$, $a \leqslant z$ and $b \leqslant z$ implies $x \leqslant z$, then x is called the **least upper bound** of a, b. An element $y \in T$ is a **lower bound** or the **greatest lower bound** of a, b if analogous conditions hold with $\geqslant$ in place of $\leqslant$. We write $x = \text{l.u.b.}(a, b)$, or $x = a \vee b$, and $y = \text{g.l.b.}(a, b)$, or $y = a \wedge b$. The element $a \vee b$ is sometimes called the **join of** a **and** b, and $a \wedge b$, the **meet of** a **and** b.

It is clear that if there is a least upper bound of two elements, then it must be unique, for if x and y are both l.u.b.'s of a and b, then we obviously have $x \leqslant y$ and $y \leqslant x$; hence $x = y$. Similarly, the greatest lower bound must be unique when it exists.

Definition 2.3. A **lattice** is a p.o. set $(L, \leqslant)$ in which any two elements of L have a least upper bound and a greatest lower bound. Depending on context, lattices will be denoted by L, M, $(L, \leqslant)$, $(L, \vee, \wedge)$, $(L, \leqslant, \vee, \wedge)$, etc. A **sublattice** $(M, \leqslant)$ **of** $(L, \leqslant)$ is a sub-p.o. set M of L, such that any $a, b \in M$ have a g.l.b. and l.u.b. in M, and these coincide with the g.l.b. and l.u.b. of a, b in L. Finally, a mapping ϕ from lattices $(L, \leqslant, \wedge, \vee)$ to $(L', \leqslant', \wedge', \vee')$ is a **lattice homomorphism** provided:

For any two elements $a, b \in L$,

(2.3.1) $\phi(a \wedge b) = \phi(a) \wedge' \phi(b)$

(2.3.2) $\phi(a \vee b) = \phi(a) \vee' \phi(b)$.

If in addition ϕ is 1 : 1 onto and ϕ^{-1} is a lattice homomorphism, then ϕ is a **lattice isomorphism** and L and L' are **lattice-isomorphic**, or simply **isomorphic**. A mapping ζ from lattices $(L, \leqslant, \wedge, \vee)$ to $(L', \leqslant', \wedge', \vee')$ is said to be a **lattice-reversing homomorphism** provided:

(2.3.3) $\zeta(a \wedge b) = \zeta(a) \vee' \zeta(b)$

(2.3.4) $\zeta(a \vee b) = \zeta(a) \wedge' \zeta(b)$.

If in addition ζ is 1 : 1 onto and ζ^{-1} is a lattice-reversing homomorphism, then ζ is a **lattice-reversing isomorphism**, and L and L' are **reverse isomorphic**.

Remark 2.4. The requirement in (2.3.1) implies that ϕ is a p.o. homomorphism. This follows at once from the easily-established relation $a \leqslant b$ iff $a \wedge b = a$. Similarly, the requirement in (2.3.2) implies that ϕ is a p.o. homomorphism, and requirements (2.3.3) and (2.3.4) each imply that ζ is, in the obvious sense, a p.o.-reversing homomorphism.

EXAMPLE 2.5. The set $\mathscr{I}$ of all ideals of any commutative ring forms a lattice $(\mathscr{I}, \subset, \cap, +)$. We see this as follows: $(\mathscr{I}, \subset)$ is surely a p.o. set. Since the intersection of any two ideals $\mathfrak{a}_1, \mathfrak{a}_2$ is an ideal, $\mathfrak{a}_1 \cap \mathfrak{a}_2 = \text{g.l.b.}(\mathfrak{a}_1, \mathfrak{a}_2)$. Now $\text{l.u.b.}(\mathfrak{a}_1, \mathfrak{a}_2)$ is the smallest ideal containing $\mathfrak{a}_1$ and $\mathfrak{a}_2$; since any ideal is closed under addition, the l.u.b. contains the set

$$\mathfrak{a}_1 + \mathfrak{a}_2 = \{a_1 + a_2 \mid a_1 \in \mathfrak{a}_1 \quad \text{and} \quad a_2 \in \mathfrak{a}_2\};$$

but this is already an ideal, so $\text{l.u.b.}(\mathfrak{a}_1, \mathfrak{a}_2) = \mathfrak{a}_1 + \mathfrak{a}_2$.

There is an important way of obtaining new lattices from old ones:

Lemma 2.6. *Any lattice $(L, \leqslant, \vee, \wedge)$ together with a closure map $a \to \bar{a}$ on the underlying p.o. set determines a new lattice $(L', \leqslant, \vee', \wedge')$; L' is a sub-p.o. set of L (but not a sublattice in general). For the elements of L', we take the* **closed** *elements of L—that is, the elements a such that $a = \bar{a}$. The new l.u.b. and g.l.b. on L' are then the following: For $a, b \in L'$,*

$$a \vee' b = \overline{a \vee b}$$
$$a \wedge' b = a \wedge b.$$

PROOF. $\vee'$: Let c be any closed upper bound of a and b. Then $a \vee b \leqslant c$, therefore $\overline{a \vee b} \leqslant \bar{c} = c$. But $\overline{a \vee b}$ is itself an upper bound of a and b, so it is the least closed one—that is, $a \vee' b = \overline{a \vee b}$.

$\wedge'$: $a \wedge b \leqslant a$, so $\overline{a \wedge b} \leqslant \bar{a} = a$; similarly $\overline{a \wedge b} \leqslant b$. Hence $\overline{a \wedge b} \leqslant a \wedge b$. But of course $a \wedge b \leqslant \overline{a \wedge b}$, so $a \wedge b = \overline{a \wedge b} \in L'$—that is, $a \wedge' b = a \wedge b$. □

Definition 2.7. A lattice $(L, \vee, \wedge)$ is **distributive** provided: For every $a, b, c \in L$,

(2.7.1) $a \wedge (b \vee c) = (a \wedge b) \vee (a \wedge c)$, and
(2.7.2) $a \vee (b \wedge c) = (a \vee b) \wedge (a \vee c)$.

Remark 2.8. We show in Exercise 2.7 that (2.7.1) holds iff (2.7.2) does. Hence in checking a lattice for distributivity, it is enough to check just one of these properties. Also note that any sublattice of a distributive lattice is distributive. (One may say distributivity is "inherited" by sublattices.)

As indicated earlier, lattice theory allows us to state decomposition theorems in a very general form, thus unifying the statements of a number of other decomposition theorems in mathematics. Statements of such theorems in algebraic geometry are special cases of the general lattice theorems. The proofs of the *general* theorems are simple; but showing that the hypotheses are satisfied at more concrete levels may of course be much more difficult.

Definition 2.9. An element x of a lattice $L = (L, \vee, \wedge)$ is said to be **$\vee$-irreducible** if $x = a \vee b$ implies $x = a$ or $x = b$ $(a, b \in L)$; and $y \in L$ is **$\wedge$-irreducible** if $y = c \wedge d$ implies $y = c$ or $y = d$ $(c, d \in L)$.

Definition 2.10. A representation $a = a_1 \vee \ldots \vee a_m$ of an element a in a lattice $(L, \vee, \wedge)$ is **irredundant** if no proper subset of $\{a_1, \ldots, a_m\}$ has join equal to a. A representation $b = b_1 \wedge \ldots \wedge b_n$ is **irredundant** if no proper subset of $\{b_1, \ldots, b_n\}$ has meet equal to b. In either case, the representation is **redundant** if it is not irredundant.

Theorem 2.11 (Basic decomposition theorem for lattices). *Let* $(L, \vee, \wedge)$ *be a lattice.*

(2.11.1) *If L satisfies the a.c.c., then there exists an irredundant representation of any* $a \in L$ *as the meet* $a = a_1 \wedge \ldots \wedge a_m$ *of* $\wedge$*-irreducible elements* $a_i \in L$. *Dually, if L satisfies the d.c.c., then any* $b \in L$ *is an irredundant join* $b = b_1 \vee \ldots \vee b_n$ *of* $\vee$*-irreducibles* $b_i \in L$.

(2.11.2) *If L is distributive, then if either of the above representations exists, it is unique* (*up to order of the irreducibles*).

A number of examples of this theorem are indicated in Exercise 2.1; two particularly important examples for us are the lattices of ideals of those rings occurring in algebraic geometry, and the lattice of algebraic varieties (under $\cup$ and $\cap$). Although the lattice of ideals of an arbitrary ring generally satisfies neither chain condition (and isn't distributive either), it will turn out that many (but not all) of the rings arising in algebraic geometry satisfy the a.c.c.; and, as we will see just after Definition 4.5, the "closed ideals" of many of these rings form distributive lattices as well as do the zero-sets of their ideals. There are thus a number of basic decomposition theorems at each level of our dictionary—affine, projective, and local.

We now give the

Proof of the basic decomposition theorem (Theorem 2.11)

(2.11.1): By symmetry it suffices to prove only one of the statements in (2.11.1). Therefore assume that L satisfies the a.c.c. If a is $\wedge$-irreducible we are done; if not we may write $a = a_1 \wedge a_2$, where $a < a_1$, $a < a_2$. We may similarly split up any reducible a_i into $a_{i1} \wedge a_{i2}$. Continuing in this way, we get a strict sequence $a_1 < a_i < a_{ij} < \ldots$. This process must end after finitely many steps, otherwise L wouldn't satisfy the a.c.c. One can make the resulting representation into irreducibles irredundant by simply erasing as many of the irreducible elements in the decomposition as possible, until the representation becomes irredundant.

(2.11.2): Again, it is enough to prove (2.11.2) for only the "$\wedge$" part. Suppose $a = a_1 \wedge \ldots a_m = a'_1 \wedge \ldots \wedge a'_n$ are two irredundant decompositions into irreducibles. Then

$$a_i \geqslant a'_1 \wedge \ldots \wedge a'_n \quad (i = 1, \ldots m),$$

so

$$a_i = a_i \vee (a'_1 \wedge \ldots \wedge a'_n).$$

Now use distributivity:

$$a_i = (a_i \vee a'_1) \wedge \ldots \wedge (a_i \vee a'_n).$$

Therefore, since a_i is $\wedge$-irreducible, $a_i = a_i \vee a'_j$ for some j. Hence $a'_j \leqslant a_i$. By symmetry, also $a_k \leqslant a'_j$, some k, so $a_k \leqslant a_i$. If $a_k < a_i$, the first decomposition would not be irredundant. Therefore $a_k = a_i$. Since a'_j is sandwiched between a_k and a_i, $a_i = a'_j$. Hence each a_i equals some a'_j, and no two different a_i equal the same a'_j, by irredundancy of $a_1 \wedge \ldots \wedge a_n$. Similarly, each a'_j equals some a_i, and no two different a'_j's equal the same a_i. We thus obtain uniqueness up to order. □

We now begin our exploration of some lattice-theoretic properties of varieties and ideals. First, we make the

Convention: ALL RINGS IN THIS BOOK ARE COMMUTATIVE AND HAVE AN IDENTITY ELEMENT.

We start with this general setting: Let:

(a) S be any set;
(b) k be any field;
(c) R be any ring of k-valued functions defined on S. We assume that the identity element of R is the function identically 1 on S.

The operations of R are pointwise—that is, for any f, $g \in R$ and any $P \in S$, we define $f + g$ and $f \cdot g$ by

$$(f + g)(P) = f(P) + g(P),$$
$$(f \cdot g)(P) = f(P) \cdot g(P).$$

Now let A be any collection of functions $f \in R$; A defines a subset $\mathsf{V}(A)$ of S as follows:

$$\mathsf{V}(A) = \{s \in S \mid f(s) = 0 \text{ for all } f \in A\};$$

we call $\mathsf{V}(A)$ the **variety defined by the collection** A, we say A **defines** $\mathsf{V}(A)$, and write $A \to \mathsf{V}(A)$. Conversely, given any subset B of S, B defines an ideal $\mathsf{J}(B)$ of R as follows:

$$\mathsf{J}(B) = \{f \in R \mid f(s) = 0 \text{ for each } s \in B\};$$

we call $\mathsf{J}(B)$ the **ideal defined by the subset** B, we say B **defines** $\mathsf{J}(B)$, and write $B \to \mathsf{J}(B)$.

Throughout this book, sans-serif print (V, J, etc.) *will always connote an operation.*

One can now ask:

(a) Given A, what does $\mathsf{V}(A)$ look like?
(b) Given B, what can we say about $\mathsf{J}(B)$?

We shall usually take S to be $\mathbb{C}^n$ or $\mathbb{R}^n$, k to be $\mathbb{C}$ or $\mathbb{R}$, and R to be a finitely generated extension of $\mathbb{C}$ or $\mathbb{R}$. However, much of what we do in this section holds in the general setting.

We denote the set of all ideals of R by $\mathscr{I}(R)$, or by $\mathscr{I}$ if reference to R is clear, and we denote the set of all varieties $\mathsf{V}(A) \subset S$ by $\mathscr{V}(R)$ or by $\mathscr{V}$. Note that $(\mathscr{J}, \subset)$ and $(\mathscr{V}, \subset)$ are p.o. sets. The functions J and V are *order-reversing* homomorphisms in the sense that

$$\mathfrak{a}_1 \subset \mathfrak{a}_2 \quad \text{implies} \quad \mathsf{V}(\mathfrak{a}_1) \supset \mathsf{V}(\mathfrak{a}_2), \quad \text{and}$$
$$V_1 \subset V_2 \quad \text{implies} \quad \mathsf{J}(V_1) \supset \mathsf{J}(V_2);$$

these facts follow directly from the definitions.

Do the functions J and V have other nice properties? We show now that J is 1:1 on $\mathscr{V}(R)$, but not in general onto.

J is 1:1 on $\mathscr{V}(R)$: Let V, W be two different varieties. We show $\mathsf{J}(V) \neq \mathsf{J}(W)$. Since $V \neq W$, there is a point s in one of the varieties not in the other, say $s \in V$ and $s \notin W$. Now W is the zero set of *some* set of polynomials, so it is the zero set of $\mathsf{J}(W)$. Then for some $f \in \mathsf{J}(W)$, $f(s) \neq 0$; yet for each $g \in \mathsf{J}(V)$, $g(s) = 0$. Hence $\mathsf{J}(W) \neq \mathsf{J}(V)$.

J is not always onto $\mathscr{I}(R)$: For instance, taking $S = \mathbb{C}$, $k = \mathbb{C}$, and $R = \mathbb{C}[X]$, the origin in $\mathbb{C}$ defines (X) in $\mathbb{C}[X]$. But no subset of $\mathbb{C}$ maps onto any of (X^2), (X^3), ...

Similarly, the map V is obviously onto $\mathscr{V}(R)$; the above example shows it is not 1 : 1 on $\mathscr{I}(R)$.

The next lemma says that by looking at a subset of $\mathscr{I}$, we may make both V and J 1:1-onto.

Lemma 2.12. *Let* $\mathfrak{a}, \mathfrak{b} \in \mathscr{I}$, *and suppose that* $\mathsf{V}(\mathfrak{a}) \to \mathfrak{b}$. *Then* $\mathfrak{a} \subset \mathfrak{b}$, *and* $\mathfrak{b} \to \mathsf{V}(\mathfrak{a}) \to \mathfrak{b} \to \mathsf{V}(\mathfrak{a}) \to \ldots$.

PROOF. That $\mathfrak{a} \subset \mathfrak{b}$ is obvious. This in turn implies $\mathsf{V}(\mathfrak{a}) \supset \mathsf{V}(\mathfrak{b})$. But from $\mathsf{V}(\mathfrak{a}) \to \mathfrak{b}$, $s \in \mathsf{V}(\mathfrak{a})$ implies $f(s) = 0$ for each $f \in \mathfrak{b}$; hence $\mathsf{V}(\mathfrak{a}) \subset \mathsf{V}(\mathfrak{b})$. Thus $\mathsf{V}(\mathfrak{a}) = \mathsf{V}(\mathfrak{b})$, which proves the lemma. □

Of all those ideals defining $\mathsf{V}(\mathfrak{a})$ above, $\mathfrak{b}$ is clearly the unique largest; Lemma 2.12 then tells us there is a *1 : 1-onto correspondence between varieties and their largest defining ideals*. The subset of $\mathscr{I}$ of *largest* defining ideals is denoted by $\mathscr{J}(R)$ or by $\mathscr{J}$; note that J sends sets to elements of $\mathscr{J}$. We continue to denote the restrictions $\mathsf{V}|\mathcal{J}$ and $\mathsf{J}|\mathcal{V}$ by V and J, resp. (The meaning of V and of J will be clear from context.) These restricted functions are both 1:1-onto and are inverses of each other. One may easily verify

Lemma 2.13. J *and* V *are both p.o.-reversing isomorphisms between* $(\mathcal{V}, \subset)$ *and* $(\mathcal{J}, \subset)$*; furthermore* $\mathsf{J} = \mathsf{V}^{-1}$.

If $V \in \mathcal{V}$ and $\mathfrak{b} \in \mathcal{J}$ correspond under the above map, we say V and $\mathfrak{b}$ are **associated**.

From Example 2.5, we see that $(\mathcal{I}, \subset, \cap, +)$ is a lattice. It is natural to ask if $(\mathcal{V}, \subset)$ also has a lattice structure, and if so, to what extent we can get a lattice generalization of Lemma 2.13. The next lemma gives information about this.

Lemma 2.14. *For any* $\mathfrak{a}_1, \mathfrak{a}_2 \subset R$, *we have*

(2.14.1) $\mathsf{V}(\mathfrak{a}_1 \cap \mathfrak{a}_2) = \mathsf{V}(\mathfrak{a}_1 \cdot \mathfrak{a}_2) = \mathsf{V}(\mathfrak{a}_1) \cup \mathsf{V}(\mathfrak{a}_2)$,
(2.14.2) $\mathsf{V}(\mathfrak{a}_1 + \mathfrak{a}_2) = \mathsf{V}(\mathfrak{a}_1) \cap \mathsf{V}(\mathfrak{a}_2)$.

Remark 2.15. We will not use the $\mathsf{V}(\mathfrak{a}_1 \cdot \mathfrak{a}_2)$ part of (2.14.1) right away. but it will be important later, and it is efficient to include its proof here.

PROOF OF (2.14.1). $\subset$: We show

$$\mathsf{V}(\mathfrak{a}_1 \cap \mathfrak{a}_2) \subset \mathsf{V}(\mathfrak{a}_1) \cup \mathsf{V}(\mathfrak{a}_2),$$
$$\mathsf{V}(\mathfrak{a}_1 \cdot \mathfrak{a}_2) \subset \mathsf{V}(\mathfrak{a}_1) \cup \mathsf{V}(\mathfrak{a}_2).$$

Since $\mathfrak{a}_1 \cdot \mathfrak{a}_2 \subset \mathfrak{a}_1 \cap \mathfrak{a}_2$,

$$\mathsf{V}(\mathfrak{a}_1 \cap \mathfrak{a}_2) \subset \mathsf{V}(\mathfrak{a}_1 \cdot \mathfrak{a}_2).$$

Now suppose $s \notin \mathsf{V}(\mathfrak{a}_1) \cup \mathsf{V}(\mathfrak{a}_2)$, i.e., $s \notin \mathsf{V}(\mathfrak{a}_1)$ and $s \notin \mathsf{V}(\mathfrak{a}_2)$. Then there are functions $f_i \in \mathfrak{a}_i$ such that $f_i(s) \neq 0$, for $i = 1, 2$. Then $f_1(s) \cdot f_2(s) \neq 0$. But $f_1 \cdot f_2 \in \mathfrak{a}_1 \cdot \mathfrak{a}_2$, so $s \notin \mathsf{V}(\mathfrak{a}_1 \cdot \mathfrak{a}_2)$. This proves the second of our inclusions,

$$\mathsf{V}(\mathfrak{a}_1 \cdot \mathfrak{a}_2) \subset \mathsf{V}(\mathfrak{a}_1) \cup \mathsf{V}(\mathfrak{a}_2).$$

From this the first inclusion follows, because $\mathsf{V}(\mathfrak{a}_1 \cap \mathfrak{a}_2) \subset \mathsf{V}(\mathfrak{a}_1 \cdot \mathfrak{a}_2)$.

$\supset$: We show

$$\mathsf{V}(\mathfrak{a}_1) \cup \mathsf{V}(\mathfrak{a}_2) \subset \mathsf{V}(\mathfrak{a}_1 \cap \mathfrak{a}_2),$$
$$\mathsf{V}(\mathfrak{a}_1) \cup \mathsf{V}(\mathfrak{a}_2) \subset \mathsf{V}(\mathfrak{a}_1 \cdot \mathfrak{a}_2).$$

Suppose $s \in \mathsf{V}(\mathfrak{a}_1)$. Now $\mathfrak{a}_1 \supset \mathfrak{a}_1 \cap \mathfrak{a}_2$ and $\mathfrak{a}_1 \supset \mathfrak{a}_1 \cdot \mathfrak{a}_2$. Thus $\mathsf{V}(\mathfrak{a}_1) \subset \mathsf{V}(\mathfrak{a}_1 \cap \mathfrak{a}_2)$ and $\mathsf{V}(\mathfrak{a}_1) \subset \mathsf{V}(\mathfrak{a}_1 \cdot \mathfrak{a}_2)$. Similarly, $\mathsf{V}(\mathfrak{a}_2) \subset \mathsf{V}(\mathfrak{a}_1 \cap \mathfrak{a}_2)$ and $\mathsf{V}(\mathfrak{a}_2) \subset \mathsf{V}(\mathfrak{a}_1 \cdot \mathfrak{a}_2)$. Thus $\mathsf{V}(\mathfrak{a}_1) \cup \mathsf{V}(\mathfrak{a}_2) \subset \mathsf{V}(\mathfrak{a}_1 \cap \mathfrak{a}_2)$, and $\mathsf{V}(\mathfrak{a}_1) \cup \mathsf{V}(\mathfrak{a}_2) \subset \mathsf{V}(\mathfrak{a}_1 \cdot \mathfrak{a}_2)$.

Proof of (2.14.2)

$\subset$: Let $s \in \mathsf{V}(\mathfrak{a}_1 + \mathfrak{a}_2)$. Then $f_1(s) + f_2(s) = 0$ for any $f_i \in \mathfrak{a}_i$. Choosing $f_2 = 0$ shows $s \in \mathsf{V}(\mathfrak{a}_1)$; choosing $f_1 = 0$ shows $s \in \mathsf{V}(\mathfrak{a}_2)$, hence

$$s \in \mathsf{V}(\mathfrak{a}_1) \cap \mathsf{V}(\mathfrak{a}_2).$$

$\supset$: Let $s \in \mathsf{V}(\mathfrak{a}_1) \cap \mathsf{V}(\mathfrak{a}_2)$. Then $f_i(s) = 0$ for any $f_i \in \mathfrak{a}_i$. Hence $f_1(s) + f_2(s) = 0$, i.e., $s \in \mathsf{V}(\mathfrak{a}_1 + \mathfrak{a}_2)$. Thus Lemma 2.14 is proved. □

Lemma 2.14 *implies that* $(\mathscr{V}, \subset, \cap, \cup)$ *is a lattice*, for it shows that $\mathsf{V}(\mathfrak{a}_1) \cup \mathsf{V}(\mathfrak{a}_2)$ and $\mathsf{V}(\mathfrak{a}_1) \cap \mathsf{V}(\mathfrak{a}_2)$ are the varieties defined by $\mathfrak{a}_1 \cap \mathfrak{a}_2$ and $\mathfrak{a}_1 + \mathfrak{a}_2$ respectively; hence $(\mathscr{V}, \subset)$ is closed under $\cap$ and $\cup$. Since $\subset$ is our p.o., the l.u.b. $V_1 \vee V_2$ of V_1 and V_2 by definition contains at least $V_1 \cup V_2$; but since $V_1 \cup V_2$ is already in $\mathscr{V}$, $V_1 \vee V_2$ is exactly $V_1 \cup V_2$. Similarly, $V_1 \wedge V_2 = V_1 \cap V_2$.

Lemma 2.14 together with the fact that V is a p.o.-reversing homomorphism from $(\mathscr{I}, \subset)$ to $(\mathscr{V}, \subset)$ implies

Lemma 2.16. V *is a lattice-reversing homomorphism from* $(\mathscr{I}, \subset, \cap, +)$ *to* $(\mathscr{V}, \subset, \cap, \cup)$.

Since V is not in general 1 : 1, it is natural to ask if we can try the same trick as in Lemma 2.13, using the subset $\mathscr{J}$ of $\mathscr{I}$ to get a 1 : 1-onto lattice-reversing map between $\mathscr{J}$ and $\mathscr{V}$. We can do this, but not directly, since $\mathscr{J}$ is not in general a sublattice of $(\mathscr{I}, \subset, \cap, +)$.

Example 2.17. Let $S = \mathbb{R}^2$, $k = \mathbb{R}$, and $R = \mathbb{R}[X, Y]$. We show that there are ideals $\mathfrak{a}_1, \mathfrak{a}_2$ in $\mathscr{J}$ such that $\mathfrak{a}_1 + \mathfrak{a}_2 \notin \mathscr{J}$. Let $\mathfrak{a}_1 = (Y)$, $\mathfrak{a}_2 = (Y - X^2)$. Now Y and $Y - X^2$ are both irreducible in $\mathbb{R}[X, Y]$; this implies that (Y) and $(Y - X^2)$ are the *largest* defining ideals of their varieties, for if a polynomial p vanishing on all of, say, $\mathsf{V}(Y)$ were not in (Y), then p would be relatively prime to Y. Then $\mathscr{R}_Y(p, Y) \neq 0$, and one sees that $\mathsf{V}(p) \cap \mathsf{V}(Y)$ would consist of only finitely many points. Hence p could not vanish on all of $\mathsf{V}(Y)$. Therefore $(Y) \in \mathscr{J}$. Likewise $(Y - X^2) \in \mathscr{J}$.

We now show that $(Y) + (Y - X^2) \notin \mathscr{J}$. First, $\mathsf{V}((Y) + (Y - X^2)) = \mathsf{V}((Y)) \cap \mathsf{V}((Y - X^2)) = \{(0, 0)\} \subset \mathbb{R}^2$. Now a typical element of $(Y) + (Y + X^2)$ is of the form $q_1(X, Y)Y + q_2(X, Y)(Y - X^2)$; it is easily checked that the polynomial X is not of this form. Yet $\mathsf{V}(X)$ surely contains $(0, 0)$. Hence the ideal $((Y) + (Y - X^2)) + (X)$ which is strictly larger than $(Y) + (Y - X^2)$, also defines $\{(0, 0)\}$. (Incidentally, note that $((Y) + (Y - X^2)) + (X)$ is just (X, Y).)

Figure I, 25 gives a justification of this kind of behavior. A small upward translate of $\mathsf{V}(Y)$ intersects $\mathsf{V}(Y - X^2)$ in two points; correspondingly, $\mathsf{V}(Y)$ and $\mathsf{V}(Y - X^2)$ intersect in a "double point." In a sense, the chain $2\{(0, 0)\}$ is "bigger" than $\{(0, 0)\}$, and ought to correspond to a *smaller* ideal than the one defined by $\{(0, 0)\}$. This is exactly what happens. For instance the axes $\mathsf{V}(X)$ and $\mathsf{V}(Y)$ intersect with multiplicity one, and the ideal $(X) + (Y)$ $(=(X, Y))$ turns out to be the largest defining ideal of $\{(0, 0)\}$. Ideals like $(Y) + (Y - X^3)$ or $(Y) + (Y - X^4)$ differ even further from (X, Y); they correspond to $3\{(0, 0)\}$ and $4\{(0, 0)\}$, respectively.

We return to the question of getting a 1 : 1-onto lattice-reversing map between $\mathscr{V}$ and $\mathscr{J}$. The way out of the above difficulty is actually very simple. We first note

Lemma 2.18. *The map* $\mathfrak{a} \to \bar{\mathfrak{a}} = \mathsf{J}(\mathsf{V}(\mathfrak{a}))$ *is a closure map on the p.o. set* $(\mathscr{I}, \subset)$.

PROOF. It follows immediately from Lemma 2.12 that $\mathfrak{a} \subset \bar{\mathfrak{a}}$ and $\bar{\mathfrak{a}} = \bar{\bar{\mathfrak{a}}}$. If $\mathfrak{a}_1 \subset \mathfrak{a}_2$, then $\mathsf{V}(\mathfrak{a}_2) \subset \mathsf{V}(\mathfrak{a}_1)$, which means $\mathsf{J}(\mathsf{V}(\mathfrak{a}_1)) \subset \mathsf{J}(\mathsf{V}(\mathfrak{a}_2))$—that is, $\bar{\mathfrak{a}}_1 \subset \bar{\mathfrak{a}}_2$. ☐

In view of Lemmas 2.12 and 2.18 we shall often refer to elements of $\mathscr{J}$ as *defined ideals* or *closed ideals*.

Although $\mathscr{J}$ with $\subset, \cap$, and $+$ is not in general a sublattice of $(\mathscr{I}, \subset, \cap, +)$, we see from Lemma 2.6 that $\mathscr{J}$ equipped with $\mathfrak{c}_1 \wedge \mathfrak{c}_2 = \mathfrak{c}_1 \cap \mathfrak{c}_2$ and $\mathfrak{c}_1 \vee \mathfrak{c}_2 = \overline{\mathfrak{c}_1 + \mathfrak{c}_2}$ ("closed sum") is a lattice. We denote this new lattice by $(\mathscr{J}, \subset, \cap, \boldsymbol{+})$. Of course the maps J and V are still p.o.-reversing isomorphisms between $(\mathscr{V}, \subset)$ and $(\mathscr{J}, \subset)$. We now prove

Theorem 2.19. J *and* V *are lattice-reversing isomorphisms between* $(\mathscr{V}, \subset, \cap, \cup)$ *and* $(\mathscr{J}, \subset, \cap, \boldsymbol{+})$.

PROOF. We need only show that for $\mathfrak{c}_1, \mathfrak{c}_2 \in \mathscr{J}$,

(2.19.1) $\mathsf{V}(\mathfrak{c}_1 \cap \mathfrak{c}_2) = \mathsf{V}(\mathfrak{c}_1) \cup \mathsf{V}(\mathfrak{c}_2)$

(2.19.2) $\mathsf{V}(\mathfrak{c}_1 \boldsymbol{+} \mathfrak{c}_2) = \mathsf{V}(\mathfrak{c}_1) \cap (\mathfrak{c}_2)$.

(2.19.1): This follows from Lemma 2.14.1.

(2.19.2): Lemma (2.14.2) implies $\mathsf{V}(\mathfrak{c}_1 + \mathfrak{c}_2) = \mathsf{V}(\mathfrak{c}_1) \cap \mathsf{V}(\mathfrak{c}_2)$. This, with Lemma 2.12, yields

$$\mathfrak{c}_1 + \mathfrak{c}_2 \to \mathsf{V}(\mathfrak{c}_1) \cap \mathsf{V}(\mathfrak{c}_2) \to \overline{(\mathfrak{c}_1 + \mathfrak{c}_2)} = \mathfrak{c}_1 \boldsymbol{+} \mathfrak{c}_2 \to \mathsf{V}(\mathfrak{c}_1) \cap \mathsf{V}(\mathfrak{c}_2) \to \ldots$$

Hence (2.19.2) follows. ☐

EXERCISES

2.1 In each case, show that the set with the indicated $\wedge$ and $\vee$ forms a lattice. Then find what specific decomposition theorems Theorem 2.11 yields for each of the lattices.

(a) All subsets of a finite set, using $\cap$ and $\cup$.
(b) All complements of finite subsets of any given set S, using $\cap$ and $\cup$.

(c) All subspaces of a finite-dimensional vector space, using meet and join of subspaces.
(d) Any totally-ordered set, using "min" and "max."
(e) The natural numbers, using "least common multiple" and "greatest common divisor."

2.2 Let the set S consist of three points, and let $\mathscr{F}(S)$ denote the set of all real-valued functions on S; $\mathscr{F}(S)$ becomes a commutative ring with identity when supplied with pointwise addition and multiplication. (Geometrically, $\mathscr{F}(S)$ is $\mathbb{R}^3$ with componentwise addition and multiplication.)

(a) Describe, geometrically, the ideals of $\mathscr{F}(S)$, and show that they form a lattice in a natural way.
(b) Establish a lattice-reversing isomorphism between the lattice of ideals of $\mathscr{F}(S)$ (with $\vee = +$ and $\wedge = \cap$) and the lattice of subsets of S (with $\vee = \cup$ and $\wedge = \cap$).

2.3 Let $R = \mathbb{C}[X]$, and let $\mathfrak{a} = (p_1{}^{n_1} \cdot \ldots \cdot p_r{}^{n_r})$ be an arbitrary ideal of R (where the polynomials p_i are distinct irreducibles and the integers n_i are positive). Show that $\sqrt{\mathfrak{a}} = (p_1 \cdot \ldots \cdot p_r)$.

2.4 Let $S = \mathbb{C}$, $k = \mathbb{C}$, and $R = \mathbb{C}[X]$. Explicitly describe $\mathscr{I}(R)$ and $\mathscr{V}(R)$. Show that there are always infinitely many different ideals defining any subvariety (other than $\varnothing$ and $\mathbb{C}$) of $\mathbb{C}$. What additional information are these ideals trying to give us? We supply an answer as follows:

First, define a **positive point chain** in $\mathbb{C}$ to be any finite formal sum $n_1\{c_1\} + \ldots + n_s\{c_s\}$, n_i nonnegative integers and $\{c_i\} \in \mathbb{C}$. Show how to define equality on these point chains so that

$$n_1\{c_1\} + \ldots + n_s\{c_s\} \leftrightarrow ((X - c_1)^{n_1} \cdot \ldots \cdot (X - c_s)^{n_s})$$

describes a natural 1:1-onto correspondence between the set of all positive point chains in $\mathbb{C}$, and the set of all nonzero ideals in $\mathbb{C}[X]$.

Prove that for arbitrary ideals (p) and (q) of $\mathbb{C}[X]$, we have:

$$(p) \cap (q) = (p \text{ l.c.m. } q)$$
$$(p) + (q) = (p \text{ g.c.d. } q),$$

where p l.c.m. q is the least common multiple of p and q, and p g.c.d. q is the greatest common divisor of p and q.

Next, denote by $\mathsf{C}(p)$ the chain corresponding to (p); then the 1:1-onto correspondence between ideals and chains translates to the following: If $\mathsf{C}(p) = \sum_i m_i\{c_i\}$ and $\mathsf{C}(q) = \sum_i n_i\{c_i\}$, then

$$\mathsf{C}((p) \cap (q)) = \sum_i (m_i \max n_i)\{c_i\} \tag{1}$$

and

$$\mathsf{C}((p) + (q)) = \sum_i (m_i \min n_i)\{c_i\}. \tag{2}$$

Also, we may define a partial order on $\mathscr{C}$, the set of all chains of $\mathbb{C}$, by

$$\sum_i m_i\{c_i\} \leqslant \sum_i n_i\{c_i\} \quad \text{iff} \quad m_i \leqslant n_i \quad \text{all } i.$$

We may define the maximum of two chains by the right-hand side of equation (1), and the minimum by the right-hand side of equation (2). Show that the set $\mathscr{I}$ of all nonzero ideals of $\mathbb{C}[X]$ forms a lattice using $\subset, \cap, +$, and that the set $\mathscr{C}$ forms a lattice with the above partial order, maximum, and minimum. Show that these lattices are reverse isomorphic.

2.5 One may ask about the operations of "product" and "taking the radical" for ideals. These, too, have geometric translations.

(a) Show that the product of two ideals corresponds to the sum of their chains in the sense that

$$\mathsf{C}(p \cdot q) = \mathsf{C}(p) + \mathsf{C}(q) = \sum_i (m_i + n_i)\{c_i\}.$$

(b) Show that taking the radical of an ideal corresponds to taking the "support" of the chain, or the "variety of the chain," namely, if $\mathsf{C}(p) = \sum_i m_i\{c_i\}$, where each $m_i > 0$, then

$$\mathsf{C}(\sqrt{(p)}) = \sum_i 1\{c_i\};$$

this last sum may be identified with $\bigcup_i \{c_i\}$; that is, with $\mathsf{V}(p)$.

2.6 An ideal $\mathfrak{a} \subset \mathbb{C}[X_1, \ldots, X_n]$ is called **homogeneous** iff it is generated by a set of homogeneous polynomials. (We agree that the empty set of polynomials defines the 0 ideal.) If $\mathfrak{a} = (\{p_\alpha\})$ is a homogeneous ideal, its **dehomogenization** $D_{X_i}(\mathfrak{a})$ **at** X_i is defined to be $(\{D_{X_i}(p_\alpha)\})$. Show that the set of all homogeneous ideals of $\mathbb{C}[X_1, \ldots, X_n]$ forms a lattice under $\cap$ and $+$; show that the set of homogeneous varieties of $\mathbb{C}_{X_1, \ldots, X_n}$ forms a lattice under $\cup$ and $\cap$; show that the set of defined homogeneous ideals forms a lattice under $\cap$ and closed sum. State and prove homogeneous (or projective) analogues of Lemma 2.16 and Theorem 2.19.

2.7 Let a, b, c be arbitrary elements of a lattice $(L, \vee, \wedge)$. Prove that

$$a \vee (b \wedge c) = (a \vee b) \wedge (a \vee c) \tag{3}$$

implies

$$a \wedge (b \vee c) = (a \wedge b) \vee (a \wedge c) \tag{4}$$

by justifying each of the following steps:

$$\begin{aligned} a \wedge (b \vee c) &= a \wedge [(a \vee c) \wedge (b \vee c)] \\ &= [(a \wedge b) \vee a] \wedge [(a \wedge b) \vee c] \\ &= (a \wedge b) \vee (a \wedge c). \end{aligned}$$

Then prove that (3) holds iff (4) does.

3 The Hilbert basis theorem

In Theorem 2.11 we saw that if a lattice satisfies the ascending or descending chain condition, then each element decomposes into irreducibles, and the decomposition is unique if the lattice is distributive. Let S, k, R be as in the last section; since $(\mathscr{I}, \subset, \cap, +)$, $(\mathscr{J}, \subset, \cap, +)$ and $(\mathscr{V}, \subset, \cap, \cup)$ are all lattices, it is natural to ask in specific cases if these lattices satisfy any chain conditions or are distributive. We can then formulate possible decomposition theorems. In this and the next section we do this for an important class of rings occurring in algebraic geometry.

Definition 3.1. Let R be a ring and suppose the associated p.o. set $(\mathscr{I}, \subset)$ satisfies the a.c.c.—that is, each strictly ascending chain of ideals of R, $\mathfrak{a}_1 \subsetneqq \mathfrak{a}_2 \subsetneqq \ldots$, terminates after finitely many steps. Then by abuse of language, we say that R **satisfies the** a.c.c. Similarly, R **satisfies the** d.c.c. if $(\mathscr{I}, \subset)$ does.

We now turn our attention to proving that the a.c.c. holds for polynomial rings over a field. We begin by giving an equivalent formulation of the a.c.c. on R (Lemma 3.3).

Definition 3.2. A **basis** (or **base**) **for an ideal** $\mathfrak{a}$ in R is any collection $\{a_\gamma\}$ of elements $a_\gamma \in \mathfrak{a}$ (γ in some indexing set Γ) such that

$$\mathfrak{a} = \{r_{\gamma_1}a_{\gamma_1} + \ldots + r_{\gamma_k}a_{\gamma_k} | r_{\gamma_i} \in R \text{ and } \gamma_i \in \Gamma\}.$$

We write $\mathfrak{a} = (\{a_\gamma\})$, or $\mathfrak{a} = (a_1, a_2, \ldots)$ if Γ is countable, and $\mathfrak{a} = (a_1, \ldots, a_n)$ if Γ is finite. If we can write $\mathfrak{a} = (a_1, \ldots, a_n)$, we say $\mathfrak{a}$ *has a finite basis.*

Lemma 3.3. *R satisfies the* a.c.c. *iff every ideal of R has a finite basis.*

PROOF. $\Rightarrow$: Suppose some ideal $\mathfrak{a}$ did not have a finite basis. Then one could find a sequence of elements $a_1, a_2, \ldots$ ($a_k \in \mathfrak{a}$) such that

$$(a_1) \subsetneqq (a_1, a_2) \subsetneqq \ldots,$$

and R would not satisfy the a.c.c.

$\Leftarrow$: Suppose R did not satisfy the a.c.c.; let $\mathfrak{a}_1 \subsetneqq \mathfrak{a}_2 \subsetneqq \ldots$ be an infinite strict sequence. Then $\mathfrak{a} = \bigcup_j \mathfrak{a}_j$ is an ideal. The ideal $\mathfrak{a}$ cannot have a finite basis $a_1, \ldots, a_n$, since surely $a_1 \in \mathfrak{a}_{j_1}$ for some j_1, $a_2 \in \mathfrak{a}_{j_2}$ for some $j_2, \ldots,$ and so on. This would mean $\bigcup_{k=1}^n \mathfrak{a}_{jk} = \mathfrak{a}$, so the ideals $\mathfrak{a}_j$ could strictly increase at most up to $\mathfrak{a}_{j_n}$. □

This explains the commonly-used alternate

Definition 3.4. A ring satisfying the a.c.c. is said to satisfy the **finite basis condition**; such a ring is further called **Noetherian**. (This term is named after the German mathematician Emmy Noether (1882–1935), the daughter of Max Noether (1844–1921). M. Noether was the "father of algebraic geometry." E. Noether was a central figure in the development of modern ideal theory.)

If R is any ring, then $R[X]$ as usual denotes the ring of all polynomials in X with coefficients in R.

Our main result of this section is

Theorem 3.5 (Hilbert basis theorem). *If R is Noetherian, so is $R[X]$.*

Before proving it, let us note

Corollary 3.6. *If k is a field, then $k[X_1, \ldots, X_n]$ is Noetherian.*

PROOF. Certainly k satisfies the a.c.c. since it has only two ideals. Then by repeated application of Theorem 3.5, $k[X_1]$, $k[X_1][X_2] = k[X_1, X_2], \ldots$, $k[X_1, \ldots, X_{n-1}][X_n] = k[X_1, \ldots, X_n]$ must all be Noetherian. □

Remark 3.7. In the next section we apply the Hilbert basis theorem to get at once decomposition into irreducibles in $\mathscr{I}$, and unique decomposition in $\mathscr{J}$ and in $\mathscr{V}$.

Remark 3.8. The Basis Theorem does not have a dual—that is, no polynomial ring $R[X_1, \ldots, X_n]$ where $n \geqslant 1$ ever satisfies the d.c.c.; one strictly descending sequence is always

$$(X_1) \supsetneqq (X_1{}^2) \supsetneqq (X_1{}^3) \supsetneqq \ldots.$$

Note on the Hilbert basis theorem

The basis theorem lies at the very foundations of algebraic geometry; it shows there are "fundamental building blocks," in the sense that each variety is uniquely the finite union of irreducible varieties (Theorem 4.4). This is very much akin to the fundamental theorem of arithmetic, which lies at the foundations of number theory; it says that every integer is a product of primes (the "building blocks"), and that this representation is unique (up to order and units.) The essential idea of the basis theorem, though couched in older language, led at once to a solution of one of the outstanding unsolved problems of mathematics in the period 1868–1888, known as "Gordan's problem" (in honor of Paul Gordan).

Gordan's computational abilities were recognized as a youth, and he became the world's leading expert in unbelievably extended algorithms in a field of mathematics called *invariant theory*. In 1868 he found a long, computational proof of the basis theorem for two variables which showed, in essence, how to construct a specific base for a given ideal. Proving the generalization to n variables defied the attempts of some of the world's most distinguished mathematicians. All their attempts were along the same basic path that Gordan followed and, one by one, they became trapped in a dense jungle of complicated algebraic computations.

Now it was Hilbert's belief that the trick in doing mathematics is to start at the right end, and there can hardly be a more beautiful example of this than Hilbert's own solution to Gordan's problem. He looked at it as an *existence* problem rather than as a *construction* problem (wherein a basis is actually produced). In a short notice submitted in 1888 in the *Nachrichten* he showed in the n-variable case the existence of a finite basis for any ideal. Many in the mathematical community reacted by doubting that this was even mathematics; the philosophy of their day was that if you want to prove that something exists, you must explicitly *find* it. Thus Gordan saw the proof as akin to those of theologians for the existence of God, and his comment has become forever famous: "Das ist nicht Mathematik. Das ist Theologie." However, later Hilbert was able to build upon his existence proof, and he actually found a general constructive proof. This served as a monumental vindication of Hilbert's outlook and began a revolution in mathematical thinking. Even Gordan had to admit that theology had its merits. Hilbert's philosophy, so simple, yet so important, may perhaps be looked at this way: If we see a fly in an airtight room and then it hides from us, we still know there is a fly in the room even though we cannot specify its coordinates. Acceptance of this broader viewpoint has made possible some of the most elegant and important contributions to mathematics, and mathematicians of today would find themselves hopelessly straitjacketed by a reversion to the attitude that you must *find* it to show it exists. (For an absorbing account of Hilbert's life and times, see [Reid].)

The following proof is essentially Hilbert's—his language was a bit different, and he took R to be the integers, but the basic ideas are all the same.

PROOF OF THE BASIS THEOREM. We show that if R satisfies the finite basis condition, then so does $R[X]$. First, if $r_0 X^n + \ldots + r_n$ ($r_0 \neq 0$) is any nonzero polynomial of $R[X]$, we call r_0 the **leading coefficient** of the polynomial. Now let $\mathfrak{A}$ be any ideal of $R[X]$. Then $\mathfrak{A}$ induces an ideal $\mathfrak{a}$ in R, as well as smaller ideals $\mathfrak{a}_k$ in R, as follows:

Let $\mathfrak{a}$ consist of 0 together with all leading coefficients of all polynomials in $\mathfrak{A}$. (We show that this is an ideal in a moment.) Since R is Noetherian, for some N, $\mathfrak{a} = (a_1, \ldots, a_N)$, where $a_i \in R$. Let $p_i(X) \in \mathfrak{A}$ have a_i as leading coefficient and let $m^* = \max(\deg p_1, \ldots, \deg p_N)$. Then for each $k < m^*$, let $\mathfrak{a}_k$ consist of 0 together with all leading coefficients of all polynomials in $\mathfrak{A}$ whose degree is equal to or less than k.

We now show $\mathfrak{a}$ is an ideal. (The proof for $\mathfrak{a}_k$ is similar.) First, $\mathfrak{a}$ is closed under subtraction, for $a, b \in \mathfrak{a}$ implies that there are polynomials $p(X) = aX^m + \sum_{i=1}^m c_i X^{m-i}$ and $q(X) = bX^n + \sum_{i=1}^n d_i X^{n-i}$ in $\mathfrak{A}$. Then $m \geqslant n$ implies that $p(X) - (X^{m-n}q(X)) \in \mathfrak{A}$; if $a = b$, then $a - b = 0 \in \mathfrak{a}$, and if $a \neq b$, then $a - b \in \mathfrak{a}$ since $a - b$ is then the leading coefficient of $p(X) - (X^{m-n}q(X))$.

Second, $\mathfrak{a}$ has the absorption property, for if $r \in R$, then $r \neq 0$ implies that the leading coefficient of $rp(X)$ is $ra \in \mathfrak{a}$, and $r = 0$ implies that $ra = 0 \in \mathfrak{a}$.

Now write $\mathfrak{a}_k = (a'_{k1}, \ldots, a'_{kn_k})$, and let $q_1(X), \ldots, q_M(X)$ be polynomials of $\mathfrak{A}$ whose leading coefficients are the basis elements a'_{kj} of the ideals $\mathfrak{a}_1, \ldots, \mathfrak{a}_{m^*-1}$. We claim that

$$\mathfrak{A} = (p_1, \ldots, p_N, q_1, \ldots, q_M). \tag{5}$$

Let us denote $(p_1, \ldots, p_N, q_1, \ldots, q_M)$ by $\mathfrak{A}^\dagger$; we show $\mathfrak{A} = \mathfrak{A}^\dagger$. Since all polynomials p_i and q_j were chosen from $\mathfrak{A}$, obviously $\mathfrak{A}^\dagger \subset \mathfrak{A}$. We show $\mathfrak{A} = \mathfrak{A}^\dagger$ by assuming $\mathfrak{A}^\dagger \subsetneqq \mathfrak{A}$ and deriving a contradiction. Thus if $\mathfrak{A}^\dagger \subsetneqq \mathfrak{A}$, let p be any polynomial of lowest degree which is in $\mathfrak{A}$ but not in $\mathfrak{A}^\dagger$. We may write p's leading coefficient as $a = \sum_{i=1}^N r_i a_i$. Now surely either $\deg p \geqslant m^*$ or $\deg p < m^*$. Suppose first that $\deg p \geqslant m^*$. This would imply there are monomials $m_i(X) \in R[X]$ such that $\sum_i m_i p_i$ has the same leading term as p. (Specifically, if we take m_i to be $r_i(X^{\deg p - \deg p_i})$, then

$$\sum r_i(X^{\deg p - \deg p_i})p_i \tag{6}$$

has leading term $aX^{\deg p}$. The effect of $X^{\deg p - \deg p_i}$ is to "jack up" the degree of each p_i so that all the N summands in (6) have the same degree. This is possible since $\deg p - \deg p_i \geqslant 0$ for each $i = 1, \ldots, N$.) We thus get

$$\deg((\textstyle\sum m_i p_i) - p) < \deg p.$$

But p is a polynomial of lowest degree which is in $\mathfrak{A}$ and not in $\mathfrak{A}^\dagger$. Thus $(\sum m_i p_i) - p \in \mathfrak{A}^\dagger$. But surely also $\sum m_i p_i \in \mathfrak{A}^\dagger$, so $p \in \mathfrak{A}^\dagger$, a contradiction.

Now suppose $\deg p < m^*$. Now we may use the q_i! For some monomials $v_i(X) \in R[X]$, we have

$$\deg((\textstyle\sum v_i q_i) - p) < \deg p,$$

so as before, $p \in \mathfrak{A}^\dagger$.

Thus p cannot exist, (5) holds, and the basis theorem is proved. □

EXERCISES

3.1 Follow through the proof of the basis theorem for the ideal $\mathfrak{A} \subset \mathbb{Z}[X]$, where $\mathfrak{A}$ is generated by the set $\{2nX + 3m \mid n \text{ and } m \text{ positive integers}\}$ to arrive at $\mathfrak{A} = (2X, 3)$.

3.2 Let the ideal $\mathfrak{A} \subset \mathbb{C}[X]$ be generated by $\{n + X^n \mid n \in \mathbb{Z}^+\}$. Use the proof of the basis theorem to find a single generator of $\mathfrak{A}$.

4 Some basic decomposition theorems on ideals and varieties

Now that we have proved the Hilbert basis theorem we may apply it, together with the basic decomposition theorems of lattice theory, to reap some of the important decomposition results of algebraic geometry.

In any lattice of ideals $(\mathscr{I}, \subset, \cap, +)$ one may make the following

Definition 4.1. An ideal $\mathfrak{a}$ in any ring R is **irreducible** if $\mathfrak{a} = \mathfrak{a}_1 \cap \mathfrak{a}_2$ implies $\mathfrak{a} = \mathfrak{a}_1$ or $\mathfrak{a} = \mathfrak{a}_2$.

For the rest of this chapter we shall state our results over the field $\mathbb{C}$; in this section we shall always use the ring $\mathbb{C}[X_1, \ldots, X_n]$. Later on homomorphic images and "localizations" of $\mathbb{C}[X_1, \ldots, X_n]$ will become important, but generalizations of our decomposition results to these cases will be very easy.

Applying the basis theorem and Theorem 2.11 gives us at once this important

Theorem 4.2. *Every ideal* $\mathfrak{a} \subset \mathbb{C}[X_1, \ldots, X_n]$ *is a finite irredundant intersection* $\mathfrak{a} = \mathfrak{a}_1 \cap \ldots \cap \mathfrak{a}_s$ *of irreducible ideals* $\mathfrak{a}_i$.

The above representation need not be unique (see Exercise 4.4).

Let us now look at decompositions from a geometric viewpoint. In $(\mathscr{V}, \subset, \cap, \cup)$, irreducibility becomes

Definition 4.3. A variety V is **irreducible** if $V = V_1 \cup V_2$ implies $V = V_1$ or $V = V_2$.

Since for $\mathbb{C}[X_1, \ldots, X_n]$ the corresponding $(\mathscr{I}, \subset, \cap, +)$ satisfies the a.c.c. and since for closed ideals $\mathfrak{a}_1$ and $\mathfrak{a}_2$ we have $\mathfrak{a}_1 \subsetneqq \mathfrak{a}_2$ iff $\mathsf{V}(\mathfrak{a}_1) \supsetneqq \mathsf{V}(\mathfrak{a}_2)$, $(\mathscr{V}, \subset, \cap, \cup)$ must satisfy the d.c.c. But in this case we know even more: Because varieties are subsets of a set, $(\mathscr{V}, \subset, \cap, \cup)$ is *distributive*. Hence we have this basic result:

Theorem 4.4. *Each variety* $V \subset \mathbb{C}^n$ *is a finite irredundant union* $V = V_1 \cup \ldots \cup V_s$ *of irreducible varieties* V_i; *this decomposition is unique up to order of the* V_i.

Definition 4.5. We call the V_i in Theorem 4.4. the **irreducible components of** V, or simply the **components of** V.

Now $(\mathscr{V}, \subset, \cap, \cup)$ is reverse isomorphic to the lattice of closed ideals $(\mathscr{J}, \subset, \cap, +)$, so (in great contrast to $\mathscr{I}$) $\mathscr{J}$ is distributive, and we have

Theorem 4.6. *Every closed ideal* $\mathfrak{c} \subset \mathbb{C}[X_1, \ldots, X_n]$ *is the irredundant intersection* $\mathfrak{c} = \mathfrak{c}_1 \cap \ldots \cap \mathfrak{c}_s$ *of finitely many irreducible closed ideals; any irredundant representation of* $\mathfrak{c}$ *by irreducible closed ideals is unique up to order.*

Of course, under the isomorphism between $\mathscr{V}$ and $\mathscr{J}$, irreducible varieties correspond to irreducible closed ideals.

At this point it is fair to ask if there is some direct way of deciding when an ideal of $\mathbb{C}[X_1, \ldots, X_n]$ is irreducible and closed. It turns out there is a simple characterization of these ideals: *They are precisely the prime ideals of* $\mathbb{C}[X_1, \ldots, X_n]$.

Remark 4.7. The behavior of ideals in $\mathscr{I}$ is quite different; not every irreducible ideal is prime in $\mathscr{I}$! For example if $n > 1$, then $(X^n) \subset \mathbb{C}[X]$ is nonprime; however, it is irreducible, because the only ideals larger than (X^n) are (X^m) where $m < n$, and these can never intersect to give (X^n).

It is easy to show that in $\mathbb{C}[X_1, \ldots, X_n]$ every irreducible closed ideal is prime. (We do this next in Lemma 4.8.) But it is conceivable that *other* ideals besides irreducible closed ones are prime, too. Showing that this is not so is a somewhat longer story; we answer this question in the next section (Corollary 5.9 and Theorem 5.11). The following lemma is actually valid in the general "S, k, R" setting of Section 2.

Lemma 4.8. *Every irreducible closed ideal in* $\mathbb{C}[X_1, \ldots, X_n]$ *is prime.*

PROOF. Let $\mathfrak{c}$ be irreducible and closed in $\mathbb{C}[X_1, \ldots, X_n]$, and suppose it is not prime—say $a_1, a_2 \in \mathbb{C}[X_1, \ldots, X_n]\backslash\mathfrak{c}$ satisfy $a_1 \cdot a_2 \in \mathfrak{c}$. Now $\mathfrak{c}_1 = \mathfrak{c} + (a_1)$ and $\mathfrak{c}_2 = \mathfrak{c} + (a_2)$ are both strictly larger than $\mathfrak{c}$. If $\mathfrak{c} \to V$, then $\mathfrak{c}_1 \to V_1 \subsetneqq V$ and $\mathfrak{c}_2 \to V_2 \subsetneqq V$. Hence $V_1 \cup V_2 \subset V$. Since $a_1 \cdot a_2 \in \mathfrak{c}$, any product

$$(c_1 + r_1 a_1) \cdot (c_2 + r_2 a_2) = c_1 c_2 + c_1 r_2 a_2 + c_2 r_1 a_1 + r_1 r_2 a_1 a_2 \qquad (c_i \in \mathfrak{c}_i)$$

is in $\mathfrak{c}$, hence $\mathfrak{c}_1 \cdot \mathfrak{c}_2 \subset \mathfrak{c}$. But by Lemma 2.14 it follows that $\mathfrak{c}_1 \cdot \mathfrak{c}_2 \to V_1 \cup V_2$, so $\mathfrak{c}_1 \cdot \mathfrak{c}_2 \to V_1 \cup V_2 \supset V$. Hence $V = V_1 \cup V_2$, $V_1 \neq V$, and $V_2 \neq V$ which contradicts the fact that $\mathfrak{c}$ is an irreducible closed ideal (i.e., that V is irreducible). □

As soon as we prove that every prime ideal is irreducible and closed (in Theorem 5.11 and Corollary 5.9), we will have

Theorem 4.9. *Each closed ideal of* $\mathbb{C}[X_1, \ldots, X_n]$ *is a finite irredundant intersection of prime ideals, this representation being unique up to order. Conversely, every finite intersection of prime ideals is closed. Hence distinct irredundant intersections of prime ideals define distinct closed ideals.*

EXERCISES

4.1 "Each ideal $(X - i)$ $(i \in \mathbb{Z})$ is irreducible in $\mathbb{C}[X]$, and defines the point $i \in \mathbb{C}$. Therefore $\bigcap_{i=-\infty}^{+\infty} (X - i)$ defines $\mathbb{Z} \subset \mathbb{C}$." What is wrong with this argument?

4.2 Show that no proper algebraic subvariety V of $\mathbb{C}^n$ is intersection-irreducible.

4.3 Let $p \in \mathbb{C}[X_1, \ldots, X_n]$ be irreducible. Show that $\mathsf{V}(p) \subset \mathbb{C}^n$ is irreducible.

4.4 Show that decompositions in Theorem 4.2 may not be unique by showing that (X^2, XY) has two different irredundant decompositions into irreducibles. [*Hint*: Show that $(X^2, XY) = (X) \cap (X^2, Y) = (X) \cap (X^2, X + Y)$. Then show that (X), (X^2, Y) and $(X^2, X + Y)$ are irreducible, and that there are no containment relations between any of these irreducibles.

4.5 Although Remark 4.7 shows that in a Noetherian ring not every irreducible ideal need be prime, it *is* true that *in a Noetherian ring, every irreducible ideal is primary* in the following sense: An ideal $\mathfrak{a}$ in any ring R is **primary** provided $(x \notin \mathfrak{a}$ and $xy \in \mathfrak{a}) \Rightarrow (y^m \in \mathfrak{a}$ for some $m \in \mathbb{Z}^+)$. Prove the italicized statement as follows: First, if $\mathfrak{a}, \mathfrak{b}$ are ideals of any ring R, define the "quotient" $\mathfrak{a} : \mathfrak{b}$ to be $\{x \in R \mid x\mathfrak{b} \subset \mathfrak{a}\}$.

(a) Show that $\mathfrak{a} : \mathfrak{b}$ is an ideal of R, and that $\mathfrak{a} \subset \mathfrak{a} : \mathfrak{b}$.
In (b)–(d) assume that R is Noetherian, and that $\mathfrak{a}$ is *not* primary.
(b) Show that there are elements $x, y \in \mathfrak{a}$ with $xy \in \mathfrak{a}$, $y^m \notin \mathfrak{a}$, for all $m \in Z^+$.
(c) Show that there is an $m_0 \in Z^+$ such that $\mathfrak{a} : (y^{m_0}) = \mathfrak{a} : (y^{m_0 + 1})$.
(d) Show that $\mathfrak{a} = (\mathfrak{a}, x) \cap (\mathfrak{a}, y^{m_0})$.
(e) Show how (b)–(d) imply that in a Noetherian ring, every irreducible ideal is primary.

4.6 State and prove the homogeneous (i.e. projective) analogues of the results in this section.

4.7 Prove the statement in Remark 2.10 of Chapter II.

4.8 (a) Prove that a variety $V \subset \mathbb{C}^n$ is irreducible iff its projective completion is irreducible.
(b) Is a projective variety W in $\mathbb{P}^n(\mathbb{C})$ irreducible iff a given affine part of it is irreducible? Is W irreducible iff *all* affine parts of it are irreducible?

5 The Nullstellensatz: Statement and consequences

As mentioned before, in order to generalize the results of Chapter II to arbitrary varieties in n-space we want, as much as possible, to generalize to ideals various concepts and operations on polynomials. One of the most central facts about polynomials in $\mathbb{C}[X]$ is the fundamental theorem of algebra. In this section we generalize this to ideals, getting the Nullstellensatz (Theorem 5.1) We then look at some consequences. Its proof is given in Section 6.

Geometrically, the fundamental theorem of algebra says that any nonconstant polynomial $p(X) \in \mathbb{C}[X]$ has a zero in $\mathbb{C}_X$. Since $\mathbb{C}[X]$ is a principal ideal domain, every ideal of $\mathbb{C}[X]$ is of the form (p), for some p. Hence in ideal language, the fundamental theorem of algebra becomes: "The variety in $\mathbb{C}_X$ defined by any proper ideal of $\mathbb{C}[X]$ contains at least one point." We may also phrase this as: "Every proper ideal in $\mathbb{C}[X]$ has a zero in $\mathbb{C}_X$." We would thus like to prove more generally that the variety defined by any proper ideal of $\mathbb{C}[X_1, \ldots, X_n]$ contains at least one point. We may also generalize the notion of "zero of a polynomial": A **zero of an ideal** $\mathfrak{a}$ is any point $(a) = (a_1, \ldots, a_n)$ such that $p(a) = 0$ for each $p \in \mathfrak{a}$. With this terminology our ideal-theoretic generalization of the fundamental theorem of algebra becomes

Theorem 5.1 (Hilbert Nullstellensatz, or zero theorem). *Every proper ideal of* $\mathbb{C}[X_1, \ldots, X_n]$ *has a zero in* $\mathbb{C}_{X_1, \ldots, X_n}$.

Remark 5.2. Our proof will show that Theorem 5.1 holds if $\mathbb{C}$ is replaced by any algebraically closed field. The hypothesis of algebraic closure is necessary, for when $n = 1$, the theorem reduces to the fundamental theorem of algebra, and we know $X^2 + 1 \in \mathbb{R}[X]$ has no zero in $\mathbb{R}$.

Just as the fundamental theorem of algebra has a number of equivalent forms and many important implications, so does the Nullstellensatz. We devote the remainder of this section to looking at some of them.

First, since $\mathbb{C}[X_1, \ldots, X_n]$ is Noetherian, its set of ideals satisfies the maximal condition (see the discussion after Definition 2.2); hence every proper ideal of $\mathbb{C}[X_1, \ldots, X_n]$ is contained in a maximal ideal. This yields this equivalent form of the Nullstellensatz:

Theorem 5.3. *Every maximal ideal of* $\mathbb{C}[X_1, \ldots, X_n]$ *has a zero in* $\mathbb{C}^n$.

Corollary 5.4. *There is a* 1:1-*onto correspondence between points of* $\mathbb{C}^n$ *and maximal ideals of* $\mathbb{C}[X_1, \ldots, X_n]$. *The maximal ideal corresponding to* $(c_1, \ldots, c_n) \in \mathbb{C}^n$ *is* $((X_1 - c_1), \ldots, (X_n - c_n))$.

We shall denote the maximal ideal corresponding to $P \in \mathbb{C}^n$ by $\mathfrak{m}_P$.

Proof of Corollary 5.4. First note that in $\mathbb{C}^n$ any single point is irreducible, but any finite union of two or more points is reducible. Then Theorem 2.19 implies that J restricted to points defines a 1:1 onto correspondence between points of $\mathbb{C}^n$ and closed maximal ideals. It follows from Theorem 5.3 that every maximal ideal is closed, so the first assertion of the corollary is proved. To show that for any $(c_1, \ldots, c_n) \in \mathbb{C}^n$, $\mathsf{J}((c_1, \ldots, c_n)) = ((X - c_1), \ldots, (X - c_n))$, we need only show that $\mathfrak{a} = ((X - c_1), \ldots, (X - c_n))$ is maximal; this follows from $\mathbb{C}[X_1, \ldots, X_n]/\mathfrak{a} = \mathbb{C}$.

The above corollary leads to some other important facts. If P and Q are two points of $\mathbb{C}^n$, then the closed ideal of the variety $\{P\} \cup \{Q\}$ is of course $\mathfrak{m}_P \cap \mathfrak{m}_Q$. Since we require in a lattice only that the l.u.b. and g.l.b. of finitely many elements exist, the isomorphism J in Theorem 2.19 says nothing about preserving operations under, for example, infinite intersection. However if V is a variety in $\mathbb{C}^n$, it is natural to ask whether $\mathsf{J}(V)$ is the same as $\bigcap_{P \in V} \mathfrak{m}_P$. Now since any $p \in \bigcap_{P \in V} \mathfrak{m}_P$ lies in each $\mathfrak{m}_P$, by Corollary 5.4

p vanishes at each point of V; hence $\bigcap_{P \in V} \mathfrak{m}_P \subset \mathsf{J}(V)$. Conversely, any $q \notin \bigcap_{P \in V} \mathfrak{m}_P$ fails to be in some $\mathfrak{m}_P$, say $\mathfrak{m}_{P_0}$, where $P_0 \in V$. Then $q(P_0) \neq 0$, so q also could not be in $\mathsf{J}(V)$, so the reverse inclusion holds. Hence we indeed have

$$\mathsf{J}(V) = \bigcap_{P \in V} \mathfrak{m}_P.$$

Now suppose $\mathfrak{a} \to V$. Then the point P defined by an arbitrary maximal ideal $\mathfrak{m}$ lies in V iff $\mathfrak{m}$ contains $\mathfrak{a}$. ($\mathfrak{m} \supset \mathfrak{a}$ implies $\{P\} \subset V$; conversely, if $\mathfrak{m} \not\supset \mathfrak{a}$, then there is a polynomial $p \in \mathfrak{a}$, $p \notin \mathfrak{m}$, meaning that p vanishes on all of V, but not at P.)

We summarize these facts in

Corollary 5.5. *If $\mathfrak{a}$ is any ideal in $\mathbb{C}[X_1, \ldots, X_n]$, then the closure $\bar{\mathfrak{a}}$ of $\mathfrak{a}$ is the intersection of all maximal ideals containing it—that is,*

$$\bar{\mathfrak{a}} = \bigcap_{\mathfrak{m} \supset \mathfrak{a}} \mathfrak{m}.$$

The next theorem gives another equivalent form of the Nullstellensatz. The German letter $\mathfrak{p}$ always denotes a prime ideal.

Theorem 5.6. *Let $\mathfrak{a}$ be any ideal of $\mathbb{C}[X_1, \ldots, X_n]$. Then*

$$\bigcap_{\mathfrak{p} \supset \mathfrak{a}} \mathfrak{p} = \bigcap_{\mathfrak{m} \supset \mathfrak{a}} \mathfrak{m}.$$

This is a very useful form of the Nullstellensatz, and is called the "strong form" of the theorem, though, of course, it is no stronger than Theorem 5.1 (which is often referred to as the "weak form"); the strong form easily gives, for instance, a proof of Theorem 4.9. In a moment we prove Theorem 5.6, assuming Theorem 5.1. That Theorem 5.6 implies Theorem 5.1 follows immediately from Theorem 5.8.

The above ideal $\bigcap_{\mathfrak{p} \supset \mathfrak{a}} \mathfrak{p}$ turns out to be the radical $\sqrt{\mathfrak{a}}$ of $\mathfrak{a}$ (Definition 1.1). We prove this next (Lemma 5.7); we will then not be far from a proof of Theorem 5.6.

Lemma 5.7. *Let R be any ring and let $\mathfrak{a}$ be any ideal of R. Then*

$$\sqrt{\mathfrak{a}} = \bigcap_{\mathfrak{p} \supset \mathfrak{a}} \mathfrak{p}.$$

PROOF

$\sqrt{\mathfrak{a}} \subset \bigcap_{\mathfrak{p} \supset \mathfrak{a}} \mathfrak{p}$: If $r \in \sqrt{\mathfrak{a}}$, then $r^n \in \mathfrak{a}$ for some n, so r^n is in each $\mathfrak{p} \supset \mathfrak{a}$. Since $\mathfrak{p}$ is prime, r itself is in each $\mathfrak{p} \supset \mathfrak{a}$, hence $r \in \bigcap_{\mathfrak{p} \supset \mathfrak{a}} \mathfrak{p}$.

$\sqrt{\mathfrak{a}} \supset \bigcap_{\mathfrak{p} \supset \mathfrak{a}} \mathfrak{p}$: Suppose $r \notin \sqrt{\mathfrak{a}}$; we show $r \notin \bigcap_{\mathfrak{p} \supset \mathfrak{a}} \mathfrak{p}$ by showing that there is a prime ideal $\mathfrak{p}^*$ containing $\mathfrak{a}$ but such that $r \notin \mathfrak{p}^*$. Now $r \notin \sqrt{\mathfrak{a}}$

means that for each positive integer n, $r^n \notin \mathfrak{a}$; we may write this as $\mathfrak{a} \cap \{r, r^2, r^3, \ldots\} = \varnothing$. By Zorn's lemma there is an ideal $\mathfrak{p}^*$ such that

1. $\mathfrak{p}^* \supset \mathfrak{a}$
2. $\mathfrak{p}^* \cap \{r, r^2, r^3, \ldots\} = \varnothing$,
3. there is no ideal strictly larger than $\mathfrak{p}^*$ satisfying (1) and (2).

(If one assumes that R is Noetherian, one may replace the reference to Zorn's lemma by the maximal condition, given just after Definition 2.2.) We now show $\mathfrak{p}^*$ is prime. Suppose $s_1 \notin \mathfrak{p}^*$ and $s_2 \notin \mathfrak{p}^*$. We show $s_1 \cdot s_2 \notin \mathfrak{p}^*$. First, since the ideals $\mathfrak{p}^* + (s_1)$ and $\mathfrak{p}^* + (s_2)$ are strictly larger than $\mathfrak{p}^*$, they both must intersect the set $\{r, r^2, r^3, \ldots\}$—that is, there are elements r^{n_1}, r^{n_2} such that

$$r^{n_1} = a_1 + r_1 s_1 \in \mathfrak{p}^* + (s_1) \quad \text{and} \quad r^{n_2} = a_2 + r_2 s_2 \in \mathfrak{p}^* + (s_2).$$

Now $r^{n_1 + n_2} \in \{r, r^2, r^3, \ldots\}$; however

$$r^{n_1 + n_2} = a_1 a_2 + a_1 r_2 s_2 + a_2 r_1 s_1 + r_1 r_2 s_1 s_2.$$

If $s_1 \cdot s_2$ were in $\mathfrak{p}^*$ then clearly $r^{n_1 + n_2} \in \mathfrak{p}^*$ too, meaning that $r^{n_1 + n_2} \in \mathfrak{p}^* \cap \{r, r^2, r^3, \ldots\}$, which is impossible. Thus $s_1 \cdot s_2 \notin \mathfrak{p}^*$, as promised. □

Proof of Theorem 5.6, *assuming Theorem* 5.1

That $\bigcap_{\mathfrak{p} \supset \mathfrak{a}} \mathfrak{p} \subset \bigcap_{\mathfrak{m} \supset \mathfrak{a}} \mathfrak{m}$ is trivial since any maximal ideal is prime.

To show the reverse inclusion, let p be a nonzero element of $\bigcap_{\mathfrak{m} \supset \mathfrak{a}} \mathfrak{m}$. In view of Lemma 5.7 it suffices to show that $p^m \in \mathfrak{a}$, for some $m > 0$. For this, consider, in $\mathbb{C}[X_1, \ldots, X_n, X_{n+1}]$, the ideal $(\mathfrak{a}, 1 - X_{n+1} \cdot p)$. Now $\mathfrak{a}$ defines a variety $V = \mathsf{V}(\mathfrak{a})$ in $\mathbb{C}_{X_1, \ldots, X_{n+1}}$, and $\mathsf{V}(\mathfrak{a}, 1 - X_{n+1} \cdot p)$ is a subset of this V; also, $p \equiv 0$ on V, so $1 - X_{n+1} \cdot p$ is nonzero at each point of V. Hence $\mathsf{V}(\mathfrak{a}, 1 - X_{n+1} \cdot p) = \varnothing$. Therefore by Theorem 5.1, $1 \in (\mathfrak{a}, 1 - X_{n+1} \cdot p)$—that is,

$$1 = \sum r_j q_j + s(1 - X_{n+1} \cdot p) \qquad (q_j \in \mathfrak{a}, r_j, s \in \mathbb{C}[X_1, \ldots, X_{n+1}]). \tag{7}$$

Now set $X_{n+1} = 1/p$ in (7). Then $s(1 - (1/p) \cdot p) = 0$, and (7) becomes

$$1 = \sum \bar{r}_j q_j, \quad \text{where } \bar{r}_j \in \mathbb{C}\left[X_1, \ldots, X_n, \frac{1}{p}\right].$$

Clearing this equation of denominators yields an equation

$$p^m = \sum_j (p^m \bar{r}_j) q_j \qquad (p^m \bar{r}_j \in \mathbb{C}[X_1, \ldots, X_n]).$$

But each $q_j \in \mathfrak{a}$, so this equation implies $p^m \in \mathfrak{a}$, as desired. □

Corollary 5.5, Theorem 5.6, and Lemma 5.7 together yield the Nullstellensatz in a more explicit form, and is essentially the way Hilbert stated it:

Theorem 5.8. *Suppose $\mathfrak{a} \to V$, where $\mathfrak{a} \subset \mathbb{C}[X_1 \ldots X_n]$ and $V \subset \mathbb{C}^n$. If p is any polynomial vanishing on V, then $p^n \in \mathfrak{a}$ for some $n > 0$.*

We next note that Corollary 5.5 and Theorem 5.6 at once imply

Corollary 5.9. *Every prime ideal of* $\mathbb{C}[X_1, \ldots, X_n]$ *is closed.*

Remark 5.10. This important corollary of the Nullstellensatz will establish Theorem 4.9 (as soon, of course, as we prove the Nullstellensatz itself, in the next section). All that remains to prove in Theorem 4.9 is the assertion, "Every prime ideal is an irreducible closed ideal." Now since from Corollary 5.9 every prime ideal is closed, we need only show:

Theorem 5.11. *In any ring, every prime ideal is irreducible.*

PROOF. Suppose $\mathfrak{a} = \mathfrak{a}_1 \cap \mathfrak{a}_2$, $\mathfrak{a}_1 \supsetneqq \mathfrak{a}$, and $\mathfrak{a}_2 \supsetneqq \mathfrak{a}$. Surely we cannot have $\mathfrak{a}_1 \subset \mathfrak{a}_2$ or $\mathfrak{a}_2 \subset \mathfrak{a}_1$. Hence let $a \in \mathfrak{a}_1$, $a \notin \mathfrak{a}_2$, and $b \in \mathfrak{a}_2$, $b \notin \mathfrak{a}_1$. Then $a \notin \mathfrak{a}$ and $b \notin \mathfrak{a}$. But $a \in \mathfrak{a}_1$ implies $ab \in \mathfrak{a}_1$ while $b \in \mathfrak{a}_2$ implies $ab \in \mathfrak{a}_2$, so $ab \in \mathfrak{a}_1 \cap \mathfrak{a}_2 = \mathfrak{a}$. Hence $\mathfrak{a}$ could not be prime. □

EXERCISES

5.1 Prove that the strong form of the Nullstellensatz implies the weak form.

5.2 Show that if $\mathfrak{q}$ is primary in any ring R (cf. Exercise 4.5), then $\sqrt{\mathfrak{q}}$ is prime.

5.3 Show that if $\mathfrak{q}$ is any irreducible ideal in a Noetherian ring R, then $\sqrt{\mathfrak{q}}$ is prime. [*Hint*: Use Exercise 4.5.]

5.4 Show that taking the closure $\sqrt{\mathfrak{a}}$ of an ideal $\mathfrak{a}$ is actually a closure operation (as defined after Definition 2.1).

5.5 Show that in any ring R, for any two ideals $\mathfrak{a}$, $\mathfrak{b}$ we have $\sqrt{\mathfrak{a} \cap \mathfrak{b}} = \sqrt{\mathfrak{a}} \cap \sqrt{\mathfrak{b}}$, and $\sqrt{\mathfrak{a} + \mathfrak{b}} = \sqrt{\sqrt{\mathfrak{a}} + \sqrt{\mathfrak{b}}}$.

6 Proof of the Nullstellensatz

We prove the Nullstellensatz in this form:

$$\text{Every maximal ideal } \mathfrak{m} \text{ of } \mathbb{C}[X_1, \ldots, X_n] \text{ has a zero in } \mathbb{C}^n.$$

Although we state it using $\mathbb{C}$, any algebraically closed field can be used in place of $\mathbb{C}$. Our proof will use the concepts of transcendental and algebraic extension of a field.

First, since $\mathfrak{m}$ is maximal, $\mathbb{C}[X_1, \ldots, X_n]/\mathfrak{m}$ is a field. Suppose we were able to prove that $\mathbb{C}[X_1, \ldots, X_n]/\mathfrak{m}$ is isomorphic to $\mathbb{C}$, i.e., that the images $X_i + \mathfrak{m}$ may be identified in the natural way with elements of $\mathbb{C}$, say $X_i + \mathfrak{m} = a_i \in \mathbb{C}$. Then under the evaluation map $X_i \to a_i$ mapping $\mathbb{C}[X_1, \ldots, X_n]$ to $\mathbb{C}$, each polynomial $p \in \mathfrak{m}$ maps to $0 \in \mathbb{C}$—that is,

$p(a_1, \ldots, a_n) = 0$. Thus $(a_1, \ldots, a_n)$ would be a zero of $\mathfrak{m}$. Now suppose we could establish the following result:

Lemma 6.1. *Let k be any field. If an extension $k[y_1, \ldots, y_n]$ of k is a field, then each y_i must be algebraic over k.*

Then using $k = \mathbb{C}$ and $y_i = X_i + \mathfrak{m}$ we see that each y_i is algebraic over $\mathbb{C}$, hence $y_i \in \mathbb{C}$, thus proving the Nullstellensatz. We therefore next give the

PROOF OF LEMMA 6.1. We prove this by induction on n. For $n = 1$ the lemma is obvious—if y_1 were not algebraic over k, $k[y_1]$ would be a polynomial ring, which surely is not a field.

Now assume the lemma is true for $n - 1$; we prove it for n. Suppose, then, that $k[y_1, \ldots, y_n]$ is a field. We may look at this as $k(y_n)[y_1, \ldots, y_{n-1}]$, since the field $k(y_n)$ is a subfield of $k[y_1, \ldots, y_n]$. (But we cannot assume $k(y_n) = k[y_n]$ since, *a priori*, y_n might be transcendental over k.) By our induction hypothesis, each of $y_1, \ldots, y_{n-1}$ is algebraic over $k(y_n)$. If we prove y_n is algebraic over k then, by transitivity, it follows that each of $y_1, \ldots, y_n$ is algebraic over k. Our proof that y_n is algebraic is by contradiction; the remainder of this section is devoted to proving this.

Suppose, therefore, that in (6.1.1)–(6.1.2), each $y_1, \ldots, y_{n-1}$ is algebraic over $k(y_n)$ and that y_n is transcendental over k.

(6.1.1) If an element z in $k(y_n)$ satisfies

$$z^m + p_1(y_n)z^{m-1} + \ldots + p_m(y_n) = 0 \qquad (p_i(y_n) \in k[y_n]), \tag{8}$$

then $z \in k[y_n]$—that is, z is a polynomial.

PROOF. Write $z = z_1/z_2$, where $z_1, z_2 \in k[y_n]$ and where z_1, z_2 are relatively prime. Then (8) becomes

$$\left(\frac{z_1}{z_2}\right)^m = -p_1(y_n)\left(\frac{z_1}{z_2}\right)^{m-1} - \ldots - p_m(y_n), \quad \text{or}$$

$$z_1{}^m = -z_2[p_1(y_n)z_1{}^{m-1} + \ldots + p_m(y_n)z_2{}^{m-1}].$$

Hence z_2 is a divisor of z_1. But since the z_i are relatively prime, z_2 must be a unit in $k[y_n]$; therefore $z = z_1/z_2 \in k[y_n]$. □

In (8), z is, of course, algebraic over $k[y_n]$. But in addition the leading coefficient is 1. (Recall that for an integral domain D, any element w satisfying an equation

$$w^n + a_1w^{n-1} + \ldots + a_n = 0 \qquad (a_i \in D)$$

is called **integral over** D.)

Continuing towards our goal of a contradiction, we show next

(6.1.2) There is a polynomial $p(y_n) \in k[y_n]$ so that each of $p(y_n) \cdot y_1, \ldots, p(y_n) \cdot y_{n-1}$ is integral over $k[y_n]$.

PROOF. Let the equations satisfied by y_i $(i = 1, \ldots, n - 1)$ be

$$p_{i0}(y_n) \cdot y_i^{m_i} + \ldots + p_{im_i}(y_n) = 0. \tag{9}$$

Then let $p(y_n) = \prod_{i=1}^{n-1} p_{i0}$; each $p(y_n) \cdot y_i$ is integral over $k[y_n]$, as can be seen at once by multiplying each side of the equation in (9) by the polynomial $p^{m_i}(y_n)/p_{i0}(y_n)$; thus (6.12) is proved. □

Now let $f(y_n)$ be any element of $k(y_n)$; since $k(y_n) \subset k[y_1, \ldots, y_n]$, $f(y_n)$ is actually a polynomial $q(y_1, \ldots, y_n)$ (involving some or all of $y_1, \ldots, y_n$). Let d be its total degree in $y_1, \ldots, y_{n-1}$, and let $p(y_n)$ be as in (9). Then $(p(y_n))^d \cdot q(y_1, \ldots, y_n)$ can be looked at as a polynomial in $p(y_n) \cdot y_1, \ldots, p(y_n) \cdot y_{n-1}$ with coefficients in $k[y_n]$, for multiplying $q(y_1, \ldots, y_n)$ by $p(y_n)^d$ transforms a monomial $y_1^{\alpha_1} \cdot \ldots \cdot y_n^{\alpha_n}$ of $q(y_1, \ldots, y_n)$ to

$$p(y_n)^{d-(\alpha_1 + \ldots + \alpha_{n-1})}[p(y_n) \cdot y_1]^{\alpha_1} \cdot \ldots \cdot [p(y_n) \cdot y_{n-1}]^{\alpha_{n-1}}.$$

Now if we knew that the integral elements formed a ring, we could complete the proof of the Nullstellensatz as follows: By (6.1.2), each $p(y_n)y_1, \ldots, p(y_n)y_{n-1}$ is integral over $k[y_n]$; by what we have just said, $p^d q$ is a polynomial in $p \cdot y_1, \ldots, p \cdot y_{n-1}$. Thus if $f(y_n) = q(y_1, \ldots, y_n)$ is any element of $k(y_n)$, then $p^d q$ (which is a sum of products of $p \cdot y_1, \ldots, p \cdot y_{n-1}$, and elements of $k[y_n]$) would also be integral over $k[y_n]$. Hence by (6.1.1), $p^d q$ would be an element of $k[y_n]$—that is,

$$p(y_n)^d \cdot f(y_n) = P(y_n) \in k[y_n].$$

But surely if y_n is transcendental, not every element of $k(y_n)$ is of the form $P(y_n)/p(y_n)^d$, where $p(y_n)$ is a *fixed* polynomial. Hence y_n cannot be transcendental over k; hence it is algebraic over k, and the induction is complete.

Our remaining task, then, is to establish the following

Lemma 6.2. *Let $D \subset D^*$ be Noetherian integral domains. The elements of D^* which are integral over D form an integral domain.*

(In the application above, $D^* = k[y_1, \ldots, y_n]$, $D = k[y_n]$.)

PROOF OF LEMMA 6.2. It is clearly enough to show that if elements $a, b \in D^*$ are integral over D, then so are $a - b$ and ab. For this, note that w is integral over D if

$$w^n = a_1 w^{n-1} + \ldots + a_n;$$

that is, w is integral over D if some (positive) power of w is a linear combination of lower powers with coefficients in D. Hence if u, v are integral over D, we want to show:

For some integer $N > 0$ and $b_i \in D$,

$$(u - v)^N = \sum_{i=0}^{N-1} b_i(u - v)^i; \tag{10}$$

similarly for uv.

Now suppose the equations for u and v are

$$u^n = \sum_{i=0}^{n-1} c_i u^i \quad \text{and} \quad v^m = \sum_{j=0}^{m-1} d_j v^j.$$

Then all powers $u^n, u^{n+1}, \ldots$ are expressible as linear combinations of $1, u, \ldots, u^{n-1}$; likewise for the analogous powers of v. Therefore all positive powers of $u - v$ and uv are expressible as linear combinations of $u^i v^j$, where $0 \leqslant i \leqslant n - 1$ and $0 \leqslant j \leqslant m - 1$. But to show that, e.g., $u - v$ is integral, we want all higher powers of $u - v$ to be expressed in terms of lower powers of $u - v$, not the $u^i v^j$! For this, consider the set of all D-linear combinations of the $u^i v^j$. This is an example of the important notion of *module* over D (which generalizes the notion of vector space).

Definition 6.3. A **module** M **over a ring** R is a commutative group $(M, +)$ together with a map from $R \times M$ to M satisfying:

For any $m, n \in M$ and $r, s \in R$,

(6.3.1) $(r + s)m = rm + sm$
(6.3.2) $r(m + n) = rm + rn$
(6.3.3) $r(sm) = (rs)m$
(6.3.4) for $1 \in R$, $1 \cdot m = m$

A **submodule** M' is any subgroup of M closed under the above multiplication by elements of R. The submodule **generated by a subset** $T \subset M$ is the set

$$(T) = \left\{\sum_{\alpha} r_\alpha t_\alpha \,|\, r_\alpha \in R \quad \text{and} \quad t_\alpha \in T\right\},$$

each sum of course being a finite sum.

In our case, M is the additive group of $k[y_1, \ldots, y_n]$, and $R = D$, the ring $k[y_n]$.

Now the submodules $M_1 = (u - v), M_2 = ((u - v), (u - v)^2), \ldots$ obviously form an increasing sequence of submodules of M. If we had an a.c.c. on submodules of M so that all the M_i were equal after some stage, then we could write $(u - v)^N = \sum_{i=0}^{N-1} b_i(u - v)^i$, which is just the equation in (10), with a similar equation for $u \cdot v$. This would prove Lemma 6.1 and with it, the Nullstellensatz.

Now in our case, $R = k[y_n]$ is Noetherian, and $M = (u^0v^0, \ldots, u^{n-1}v^{m-1})$ is finitely generated. It turns out that any finitely generated module over a Noetherian ring does indeed satisfy the a.c.c. The proof of this is essentially the same as that of the Hilbert Basis Theorem. We outline its analogous proof in Exercises 6.1 and 6.2.

EXERCISES

6.1 Prove that a module M over any ring R satisfies the "a.c.c. for submodules" iff every submodule of M is finitely generated.

6.2 Using the above exercise, prove that any finitely generated module M over a Noetherian ring R satisfies the a.c.c. as follows:

Let $\{a_1, \ldots, a_n\}$ be a basis of a finitely generated module M, and let M' be any submodule of M. The leading coefficients r_i of all those elements of M' of the form $r_ia_i + r_{i-1}a_{i-1} + \ldots + r_1a_1$ form an ideal I_i of R; say $I_i = (s_{i1}, \ldots, s_{ij})$. Let $m_{i1}, \ldots, m_{ij}$ be elements of M' whose a_i-coefficients are $s_{i1}, \ldots, s_{ij}$, respectively. Show that $\{m_{ij}\}$ is a basis of M', where i runs from 1 to n.

7 Quotient rings and subvarieties

We have established a lattice-reversing isomorphism between the closed ideals of $\mathbb{C}[X_1, \ldots, X_n]$ and the varieties of $\mathbb{C}^n$. From the standpoint of dictionary building, closed ideals thus translate into varieties, intersection and closed sum translate into union and intersection, and prime ideals correspond to irreducible varieties. There are many other operations on $\mathbb{C}[X_1, \ldots, X_n]$ and its ideals, and on $\mathbb{C}^n$ and its subvarieties; it is natural to seek the geometric meaning of operations on algebraic objects, and the algebraic translation of operations on geometric objects. For example, one can ask for the effect on the subvarieties of $\mathbb{C}^n$ of taking a homomorphism of $\mathbb{C}[X_1, \ldots, X_n]$. Or one can look for the geometric meaning of the direct sum of various polynomial rings, or of their tensor product. One could instead start with a geometric operation—for instance, one might restrict attention to the subvarieties of a fixed subvariety V of $\mathbb{C}^n$, and ask what this means in algebraic terms. Again, one might take the Cartesian product $V \times W \subset \mathbb{C}^{n+m}$ of varieties $V \subset \mathbb{C}^n$ and $W \subset \mathbb{C}^m$. Is this product again a variety? If so, how does this multiplication translate into algebraic language? And so on. We investigate some of these questions in the remaining sections of this chapter. In this section and the next, we consider the algebraic effect of restricting attention to an irreducible subvariety of $\mathbb{C}^n$.

First, the ring $\mathbb{C}[X_1, \ldots, X_n]$ is called the *affine ring* of $\mathbb{C}_{X_1,\ldots X_n}$; it consists of all polynomial functions on $\mathbb{C}_{X_1,\ldots,X_n}$. Since $\mathbb{C}[X_1, \ldots, X_n]$ is determined by the canonical affine coordinate functions $X_1, \ldots, X_n$, it is also called the affine *coordinate* ring of C^n. Now if we restrict our attention from $\mathbb{C}^n$ to a fixed irreducible variety V in $\mathbb{C}^n$, one would also like a corresponding notion of affine or coordinate ring on V.

We begin with an example—the parabola $\mathsf{V}(Y - X^2) \subset \mathbb{C}^2$. We want something corresponding to the notion of "all distinct polynomial functions on $\mathsf{V}(Y - X^2)$." Now from the viewpoint of an observer restricting his attention to only V, he cannot distinguish, e.g., the function $0 \in \mathbb{C}[X, Y]$ from $Y - X^2 \in \mathbb{C}[X, Y]$, since they are both identically zero on V. More generally, any polynomial in the ideal $\mathsf{J}(V)$ is identically zero for him, while any polynomial outside $\mathsf{J}(V)$ is not identically zero for him. Two polynomials p, q are the "same" for him iff $p - q \in \mathsf{J}(V)$; he thus lumps together polynomials in a coset of $\mathsf{J}(V)$ in $\mathbb{C}[X, Y]$. We therefore have some justification in calling the quotient ring $\mathbb{C}[X, Y]/(Y - X^2)$ the "coordinate ring of $V(Y - X^2)$."

We now make the

Definition 7.1. Let $\mathfrak{p}$ be any prime ideal of $\mathbb{C}[X_1, \ldots, X_n]$, and let $V = \mathsf{V}(\mathfrak{p}) \subset \mathbb{C}_{X_1, \ldots, X_n}$. Then $\mathbb{C}[X_1, \ldots, X_n]/\mathfrak{p}$ is the **affine coordinate ring of** V (or, commonly, the **affine ring of** V or the **coordinate ring of** V); we denote it by R_V. More generally, any domain having the form $\mathbb{C}[x_1, \ldots, x_n]$ is called an **affine coordinate ring over** $\mathbb{C}$, a **coordinate ring over** $\mathbb{C}$, etc.

Remark 7.2. Since any coordinate ring $\mathbb{C}[x_1, \ldots, x_n]$ may be written as $\mathbb{C}[X_1, \ldots, X_n]/\mathfrak{p}$ for some prime ideal $\mathfrak{p}$, any coordinate ring is the coordinate ring of some irreducible variety.

Remark 7.3. Definition 7.1 generalizes to any coefficient field k.

We may now ask the very same questions concerning correspondences between V and $\mathbb{C}[X_1, \ldots, X_n]/\mathsf{J}(V) = R_V$ as we did between $\mathbb{C}^n$ and $\mathbb{C}[X_1, \ldots, X_n]$. For example, instead of asking for a lattice-reversing isomorphism between subvarieties of $\mathbb{C}^n$ and closed ideals of $\mathbb{C}[X_1, \ldots, X_n]$, we might ask for an appropriate definition of closed ideal in R_V so that there is a corresponding lattice-reversing isomorphism between the subvarieties of V and the closed ideals of R_V. Also, since it is easily verified that R_V satisfies the a.c.c., one might ask about expressing its closed ideals as the intersection of prime ideals. We look at these questions next. The remainder of this section is devoted to proving Lemmas 7.4–7.8 below.

Lemma 7.4. *If $h_\mathfrak{p}$ is the natural homomorphism*

$$h_\mathfrak{p}: \mathbb{C}[X_1, \ldots, X_n] \to \mathbb{C}[X_1, \ldots, X_n]/\mathfrak{p},$$

then $h_\mathfrak{p}^{-1}$ induces a natural lattice-embedding $\mathfrak{a} \to h_\mathfrak{p}^{-1}(\mathfrak{a})$ of $(\mathscr{I}(\mathbb{C}[X_1, \ldots, X_n]/\mathfrak{p}), \subset, \cap, +)$ into $(\mathscr{I}(\mathbb{C}[X_1, \ldots, X_n]), \subset, \cap, +)$.

Proof. Let $\mathfrak{a}_1 \neq \mathfrak{a}_2$ be distinct ideals of $\mathbb{C}[X_1, \ldots, X_n]/\mathfrak{p}$; say $p \in \mathfrak{a}_1$ but $p \notin \mathfrak{a}_2$. Then for any $q \in \{h_\mathfrak{p}^{-1}(p)\}$, we have that $q \in h_\mathfrak{p}^{-1}(\mathfrak{a}_1)$ and $q \notin h_\mathfrak{p}^{-1}(\mathfrak{a}_2)$—that is, $h_\mathfrak{p}^{-1}(\mathfrak{a}_1) \neq h_\mathfrak{p}^{-1}(\mathfrak{a}_2)$. Hence the mapping is 1:1 on the set of ideals.

Next, for any two ideals $\mathfrak{a}_1, \mathfrak{a}_2$ of $\mathbb{C}[X_1, \ldots, X_n])/\mathfrak{p}$, we have

$$h_{\mathfrak{p}}^{-1}(\mathfrak{a}_1 \cap \mathfrak{a}_2) = h_{\mathfrak{p}}^{-1}(\mathfrak{a}_1) \cap h_{\mathfrak{p}}^{-1}(\mathfrak{a}_2), \tag{11}$$

$$h_{\mathfrak{p}}^{-1}(\mathfrak{a}_1 + \mathfrak{a}_2) = h_{\mathfrak{p}}^{-1}(\mathfrak{a}_1) + h_{\mathfrak{p}}^{-1}(\mathfrak{a}_2); \tag{12}$$

these follow at once from the definitions of $h_{\mathfrak{p}}^{-1}$, $\cap$, and $+$. Hence the embedding preserves the lattice structure. □

For an analogous $\mathscr{J}$-result, we need to define closed ideal of

$$\mathbb{C}[X_1, \ldots, X_n]/\mathfrak{p} = R_V.$$

Since $\mathscr{J}(\mathbb{C}[X_1, \ldots, X_n])$ is a subset of

$$\mathscr{I}(\mathbb{C}[X_1, \ldots, X_n]),$$

it would seem natural to define the closed ideals of R_V by means of Lemma 7.4—that is, to be simply the ideals corresponding under $h_{\mathfrak{p}}^{-1}$ to the closed ideals of $\mathbb{C}[X_1, \ldots, X_n]$. But there is a problem: $\mathbb{C}[X_1, \ldots, X_n]/\mathfrak{p}$ can be represented in many different ways as a quotient ring of $\mathbb{C}[X_1, \ldots, X_m]$, for some m. We would have to show this definition of closed ideal is independent of this ring's representation as a quotient ring. However, since the closed ideals of $\mathbb{C}[X_1, \ldots, X_n]$ are precisely intersections of prime ideals (Theorem 4.9), one possible definition is

Definition 7.5. An ideal in $\mathbb{C}[X_1, \ldots, X_n]/\mathfrak{p}$ is **closed** if it is the intersection of some set of prime ideals in $\mathbb{C}[X_1, \ldots, X_n]/\mathfrak{p}$.

For any ideal $\mathfrak{a} \subset \mathbb{C}[X_1, \ldots, X_n]/\mathfrak{p}$, the map $\mathfrak{a} \to \sqrt{\mathfrak{a}} = \bigcap_{\mathfrak{P} \supset \mathfrak{a}} \mathfrak{P}$ is a closure map. Thus we see (from Lemma 2.6) that for closed ideals $\mathfrak{c}_1, \mathfrak{c}_2 \subset \mathbb{C}[X_1, \ldots, X_n]/\mathfrak{p}$, defining $\boldsymbol{+}$ by $\mathfrak{c}_1 \boldsymbol{+} \mathfrak{c}_2 = \sqrt{\mathfrak{c}_1 + \mathfrak{c}_2}$, makes the set $\mathscr{J}$ of closed ideals into a lattice $(\mathscr{J}(\mathbb{C}[X_1, \ldots, X_n]/\mathfrak{p}), \cap, \boldsymbol{+})$.

That these closed ideals do indeed correspond under $h_{\mathfrak{p}}^{-1}$ to the closed ideals of $\mathbb{C}[X_1, \ldots, X_n]$ containing $\mathfrak{p}$ is shown by

Lemma 7.6. $h_{\mathfrak{p}}^{-1}$ *defines a natural lattice-embedding of*

$$(\mathscr{J}(\mathbb{C}[X_1, \ldots, X_n]/\mathfrak{p}), \cap, \boldsymbol{+})$$

into $(\mathscr{J}(\mathbb{C}[X_1, \ldots, X_n]), \cap, \boldsymbol{+})$.

Proof. Clearly $h_{\mathfrak{p}}^{-1}$ defines a set-injection of $\mathscr{J}(\mathbb{C}[X_1, \ldots, X_n]/\mathfrak{p})$ into $\mathscr{I}(\mathbb{C}[X_1, \ldots, X_n])$. That this injection is actually into $\mathscr{J}(\mathbb{C}[X_1, \ldots, X_n])$, i.e., that $h_{\mathfrak{p}}^{-1}$ embeds $\mathscr{J}(\mathbb{C}[X_1, \ldots, X_n]/\mathfrak{p})$ into $\mathscr{J}(\mathbb{C}[X_1, \ldots, X_n])$, may be seen as follows: First note that for any ideal $\mathfrak{a} \supset \mathfrak{p}$ in $\mathbb{C}[X_1, \ldots, X_n]$, $h_{\mathfrak{p}}^{-1}$ induces a 1:1-onto map $\mathfrak{P} \to h_{\mathfrak{p}}^{-1}(\mathfrak{P})$ from the set of prime ideals $\mathfrak{P}$ of $\mathbb{C}[X_1, \ldots, X_n]/\mathfrak{p}$ containing $\mathfrak{a}/\mathfrak{p}$, to the set of prime ideals of $\mathbb{C}[X_1, \ldots, X_n]$ containing $\mathfrak{a}$. (Note that $h_{\mathfrak{p}}^{-1}$ preserves primality of ideals in

$$\mathbb{C}[X_1, \ldots, X_n]/\mathfrak{p},$$

as does $h_{\mathfrak{p}}$ for those ideals of $\mathbb{C}[X_1, \ldots, X_n]$ containing $\mathfrak{p}$.) Now let $\mathfrak{c}$ be any ideal in $\mathscr{J}(\mathbb{C}[X_1, \ldots, X_n]/\mathfrak{p})$; then

$$h_{\mathfrak{p}}^{-1}(\mathfrak{c}) = h_{\mathfrak{p}}^{-1}\left(\bigcap_{\mathfrak{P} \supset \mathfrak{c}} \mathfrak{P}\right) = \bigcap_{\mathfrak{P} \supset \mathfrak{c}} (h_{\mathfrak{p}}^{-1}(\mathfrak{P})) = \bigcap_{\mathfrak{P}' \supset h_{\mathfrak{p}}^{-1}(\mathfrak{c})} \mathfrak{P}'$$

$$= \sqrt{h_{\mathfrak{p}}^{-1}(\mathfrak{c})} \in \mathscr{J}(\mathbb{C}[X_1, \ldots, X_n]).$$

It is easy to see that this embedding preserves lattice structure, for $h_{\mathfrak{p}}^{-1}$ preserves intersection (from (11)); it also preserves $+$, that is, for closed ideals $\mathfrak{c}_1, \mathfrak{c}_2$,

$$h_{\mathfrak{p}}^{-1}(\sqrt{\mathfrak{c}_1 + \mathfrak{c}_2}) = \sqrt{h_{\mathfrak{p}}^{-1}(\mathfrak{c}_1) + h_{\mathfrak{p}}^{-1}(\mathfrak{c}_2)}.$$

This follows since $h_{\mathfrak{p}}^{-1}$ preserves sum (from (12)) and radical. □

Now that we have shown that $h_{\mathfrak{p}}^{-1}$ induces lattice embeddings, it is natural to ask if it likewise preserves decomposition of ideals into irreducibles. It does indeed. Since any homomorphic image of a Noetherian ring is Noetherian, the p.o. set $\mathscr{I}(\mathbb{C}[X_1, \ldots, X_n]/\mathfrak{p})$ satisfies the a.c.c., so *a fortiori* $\mathscr{J}(\mathbb{C}[X_1, \ldots, X_n]/\mathfrak{p})$ does too. Hence any element in either of these sets has an irredundant decomposition into irreducibles. This decomposition is unique in the case of $\mathscr{J}(\mathbb{C}[X_1, \ldots, X_n]/\mathfrak{p})$ since it is distributive. It is isomorphic to a sublattice of $\mathscr{J}(\mathbb{C}[X_1, \ldots, X_n])$, which itself is isomorphic to the distributive lattice of subvarieties $\mathscr{V}(\mathbb{C}[X_1, \ldots, X_n])$.

Now if $\mathfrak{a} \subset \mathbb{C}[X_1, \ldots, X_n]/\mathfrak{p}$ is irreducible, then $h_{\mathfrak{p}}^{-1}(\mathfrak{a}) \subset \mathbb{C}[X_1, \ldots, X_n]$ is too; this is obvious from the definition of irreducibility. Therefore if $\mathfrak{a} = \mathfrak{a}_1 \cap \ldots \cap \mathfrak{a}_r$ is a decomposition of $\mathfrak{a} \in \mathscr{I}(\mathbb{C}[X_1, \ldots, X_n]/\mathfrak{p})$ into irreducibles, then

$$h_{\mathfrak{p}}^{-1}(\mathfrak{a}) = h_{\mathfrak{p}}^{-1}(\mathfrak{a}_1) \cap \ldots \cap h_{\mathfrak{p}}^{-1}(\mathfrak{a}_r)$$

is a decomposition of $h_{\mathfrak{p}}^{-1}(\mathfrak{a})$ into irreducibles in $\mathbb{C}[X_1, \ldots, X_n]$. And if $\mathfrak{a} \in \mathscr{J}(\mathbb{C}[X_1, \ldots, X_n]/\mathfrak{p})$, then since $\mathscr{J}(\mathbb{C}[X_1, \ldots, X_n])$ is distributive, this decomposition is unique.

Notice that in the proof of Lemmas 7.4 and 7.6, no use was made that one of the rings was of the specific form $\mathbb{C}[X_1, \ldots, X_n]$; as the reader can easily verify, the proofs go through verbatim using any coordinate ring R and natural homomorphism $h_{\mathfrak{p}}$ to $R/\mathfrak{p}$. The same comments apply to unique decomposition of $h_{\mathfrak{p}}^{-1}(\mathfrak{a})$, for any ideal $\mathfrak{a}$ in $R/\mathfrak{p}$. Since we refer later to the more general forms of these lemmas, we state them explicitly:

Lemma 7.7. *Let R be any coordinate ring, $\mathfrak{p}$ any prime ideal of R, and $h_{\mathfrak{p}}$, the natural homomorphism $h_{\mathfrak{p}}: R \to R/\mathfrak{p}$. Then $h_{\mathfrak{p}}^{-1}$ induces a natural lattice-embedding $\mathfrak{a} \to h_{\mathfrak{p}}^{-1}(\mathfrak{a})$ of $(\mathscr{I}(R/\mathfrak{p}), \cap, +)$ into $(\mathscr{I}(R), \cap, +)$.*

Lemma 7.8. *$h_{\mathfrak{p}}^{-1}$ above defines a natural lattice-embedding of $(\mathscr{J}(R/\mathfrak{p}), \cap, +)$ into $(\mathscr{J}(R), \cap, +)$.*

In this section we have so far made no mention of a lattice

$$\text{“}\mathscr{V}(\mathbb{C}[X_1, \ldots, X_n]/\mathfrak{p}, \cap, \cup).\text{”}$$

Since $\mathscr{I}(\mathbb{C}[X_1, \ldots, X_n]/\mathfrak{p})$ is embedded in $\mathscr{I}(\mathbb{C}[X_1, \ldots, X_n])$, it is natural to ask if in some sense we can make a statement like "$\mathscr{V}(\mathbb{C}[X_1, \ldots, X_n]/\mathfrak{p})$ is embedded in $\mathscr{V}(\mathbb{C}[X_1, \ldots, X_n])$." For this we need an appropriate definition of $\mathscr{V}(\mathbb{C}[X_1, \ldots, X_n]/\mathfrak{p})$. As we saw at the beginning of this section, $\mathbb{C}[X_1, \ldots, X_n]/\mathfrak{p}$ may be looked at as a ring of functions on the subvariety $\mathsf{V}(\mathfrak{p})$ in $\mathbb{C}_{X_1, \ldots, X_n}$; $\mathscr{V}(\mathbb{C}[X_1, \ldots, X_n]/\mathfrak{p})$ then becomes the lattice of subvarieties of $\mathsf{V}(\mathfrak{p})$. In this way the embedding statement would indeed hold.

But there arises a question: We were able to produce the above $\mathsf{V}(\mathfrak{p})$ because the coordinate ring was presented in the particular form $\mathbb{C}[X_1, \ldots, X_n]/\mathfrak{p}$. However there are many ways of writing a given coordinate ring R as a quotient ring of some $\mathbb{C}[X_1, \ldots, X_m]$, and for each such representation one would in general end up with a different variety, hence *a priori* a different lattice of subvarieties, too. (See Example 8.1.) To make the notion "$\mathscr{V}(R)$" well-defined we shall in the next section develop a notion of isomorphism of varieties, so that varieties with isomorphic coordinate rings are isomorphic, and so that their associated lattices of subvarieties are also isomorphic.

Exercises

7.1 Prove that any homomorphic image of any Noetherian ring is Noetherian.

7.2 Let R_1 and R_2 be coordinate rings over $\mathbb{C}$. Supply the cartesian product $R_1 \times R_2$ with componentwise addition and multiplication, thus obtaining a new commutative ring with identity. (This new ring is never an integral domain.) What algebraic variety (or varieties) can be naturally associated with $R_1 \times R_2$? Interpret the existence of zero-divisors geometrically.

7.3 The cartesian product $\mathbb{C}_{X_1, \ldots, X_n} \times \mathbb{C}_{Y_1, \ldots, Y_m}$ has as coordinate ring the *tensor product* $\mathbb{C}[X_1, \ldots, X_n, Y_1, \ldots, Y_m]$ of the coordinate rings $\mathbb{C}[X_1, \ldots, X_n]$ and $\mathbb{C}[Y_1, \ldots, Y_m]$. Generalize this to more arbitrary affine varieties and their coordinate rings. (Cf. Theorem 2.24 of Chapter IV.)

8 Isomorphic coordinate rings and varieties

In Section 7 we attached to each irreducible affine variety V a coordinate ring R_V. For a good dictionary, we want our correspondences to be as faithful as possible. Hence we ask if isomorphic coordinate rings determine in some sense the "same" variety. In this book, "isomorphic coordinate rings" will always mean "$\mathbb{C}$-isomorphic coordinate rings"—that is, our isomorphisms are the identity on $\mathbb{C}$.

EXAMPLE 8.1. These coordinate rings are all isomorphic:

$$\mathbb{C}[X] \simeq \frac{\mathbb{C}[X, Y]}{(Y)} \simeq \frac{\mathbb{C}[X, Y]}{(Y - X^2)} \simeq \frac{\mathbb{C}[X, Y, Z]}{(Y - X^2, X + Z)}.$$

In Example 8.1 the first coordinate ring is that of $\mathbb{C}_X$ in $\mathbb{C}_X$; the second is that of $\mathbb{C}_X$ in $\mathbb{C}_{XY}$; the third, that of the parabola $\mathsf{V}(Y - X^2)$ in $\mathbb{C}_{XY}$; and the last is the coordinate ring of the space curve $\mathsf{V}(Y - X^2, X + Z)$ in $\mathbb{C}_{XYZ}$. Hence isomorphic rings surely need not determine identical varieties! Of course one could simply say that two varieties are isomorphic iff their coordinate rings are. But there is then the obvious question of how isomorphic varieties are related from a *geometric* viewpoint. We would like a corresponding *geometric* definition of isomorphism of varieties, so that isomorphic coordinate rings determine geometrically isomorphic varieties, and conversely.

As it turns out, such a definition can be looked at as an analogue of corresponding differentiable or analytic notions. Write $X = (X_1, \ldots, X_n)$ and $Y = (Y_1, \ldots, Y_n)$. Then recall, for instance, that if ϕ is a 1:1-onto map from $\mathbb{R}^n$ to $\mathbb{R}^n$, say

$$\phi:(X) \to (\phi_1(X), \ldots, \phi_n(X)) = \phi(X) = Y,$$

with inverse

$$\phi^{-1} = \psi:(Y) \to (\psi_1(Y), \ldots, \psi_n(Y)) = X,$$

then ϕ is a topological, differentiable, analytic, or linear isomorphism if the ϕ_i and ψ_j are continuous, differentiable, analytic, or linear, respectively. Since we are studying varieties defined by polynomials, it is natural to ask for a "polynomial isomorphism." We say that such a 1:1-onto map $\phi:\mathbb{R}^n \to \mathbb{R}^n$ is a *polynomial isomorphism* if the ϕ_i and ψ_j are all polynomials over $\mathbb{R}$. We analogously define polynomial isomorphisms from $\mathbb{C}^n$ to $\mathbb{C}^n$. For polynomial isomorphisms, we will use the more suggestive letters p and q in place of ϕ and ψ. (A more general, formal definition of polynomial isomorphism will be given in Definition 8.6.)

EXAMPLE 8.2. Let

$$p:(X_1, X_2) \to (X_1, X_1{}^2 + X_2) = (Y_1, Y_2)$$

map $\mathbb{R}_{X_1X_2}$ to $\mathbb{R}_{Y_1Y_2}$. Then

$$p^{-1}:(Y_1, Y_2) \to (Y_1, -Y_1{}^2 + Y_2) = (X_1, -X_1{}^2 + (X_1{}^2 + X_2)) = (X_1, X_2)$$

is the inverse of p, so p is a polynomial isomorphism. Under p, the horizontal lines X_2 = constant map into parabolas.

As for subvarieties, given a polynomial isomorphism p and subvariety V of $\mathbb{C}^n$ (or $\mathbb{R}^n$), it is natural to define V to be polynomially isomorphic with its image $p(V)$. But if, as in Example 8.1, $\mathbb{C}_X$ in $\mathbb{C}_{XY}$ is to be isomorphic to the space curve $\mathsf{V}(Y - X^2, Y + Z)$ in $\mathbb{C}_{XYZ}$, we furthermore want a notion of when varieties in spaces of different dimensions are isomorphic.

One may start by recalling various kinds of embeddings (say of $\mathbb{R}^n$ into $\mathbb{R}^m$, where $m \geqslant n$). For instance a 1:1 map $\phi:\mathbb{R}^n \to \mathbb{R}^m$ is a differentiable (analytic, linear, etc.) embedding iff the inverse $\phi^{-1} = \psi:\phi(\mathbb{R}^n) \to \mathbb{R}^n$ has at each point $P \in \phi(\mathbb{R}^n)$ a differentiable (analytic, linear, etc.) extension—that is, if for some open neighborhood $U_P \subset \mathbb{R}^m$ of P, there is such a map $\Psi_P:U_P \to \mathbb{R}^n$ agreeing with ψ on $\phi(\mathbb{R}^n) \cap U_P$. Since the values of a polynomial on any open subset of $\mathbb{R}^m$ or $\mathbb{C}^m$ determine its values in all of $\mathbb{R}^m$ or $\mathbb{C}^m$, in the polynomial case we may just as well take U_P to be all of m-space. We thus say that a 1:1 map $p:\mathbb{R}^n \to \mathbb{R}^m$ is a *polynomial embedding* if the components of both p and some $q^*:\mathbb{R}^m \to \mathbb{R}^n$ are polynomials such that $(q^* | p(\mathbb{R}^n) \circ p$ is the identity map on $\mathbb{R}^n$, with a corresponding definition for $\mathbb{C}$.

EXAMPLE 8.3. $p:(X) \to (X, X^2) = (Y_1, Y_2)$ is a polynomial embedding of $\mathbb{R}_X$ in $\mathbb{R}_{Y_1Y_2}$. We have $p_1(X) = X = Y_1$, and $p_2(X) = X^2 = Y_2$; hence the image of $\mathbb{R}_X$ is the parabola $V = \mathsf{V}(Y_2 - Y_1{}^2) \subset \mathbb{R}_{Y_1Y_2}$. Let $q^*:\mathbb{R}_{Y_1Y_2} \to \mathbb{R}_X$ be the projection map $(Y_1, Y_2) \to Y_1 = X \in \mathbb{R}_X$. Clearly $q^*|V$ is the inverse of p, so p is a polynomial embedding of $\mathbb{R}_X$ in $\mathbb{R}_{Y_1Y_2}$. We will say that $\mathbb{R}_X$ and the parabola are isomorphic (see Definition 8.6); this fits in nicely with the fact that $\mathbb{R}[X] \simeq \mathbb{R}[Y_1, Y_2]/(Y_2 - Y_1{}^2)$, noted (in the complex setting) in Example 8.1.

EXAMPLE 8.4. Let $p:(X) \to (X, X^2, -X) = (Y_1, Y_2, Y_3)$ map $\mathbb{R}_X$ to $\mathbb{R}_{Y_1Y_2Y_3}$. Eliminating the parameter X from $Y_1 = X$, $Y_2 = X^2$, $Y_3 = -X$ yields

$$Y_2 - Y_1{}^2 = 0 \qquad Y_1 + Y_3 = 0;$$

this defines the real part of the space curve of Example 8.1. Define q^* by $q^*(Y_1, Y_2, Y_3) = Y_1 = X$. Then p is a polynomial embedding, and our space curve is isomorphic with $\mathbb{R}_X$ and the parabola above.

EXAMPLE 8.5. Here is an example of a 1:1 map of $\mathbb{R}_X$ into $\mathbb{R}_{Y_1Y_2}$ defined by polynomials, which is not a polynomial embedding. Let $\phi:(X) \to (X^2, X^3) = (Y_1, Y_2)$. The image $\phi(\mathbb{R}_X) \subset \mathbb{R}_{Y_1Y_2}$ of ϕ is the cusp curve $Y_1 = X^2$, $Y_2 = X^3$—that is, $Y_1{}^3 = Y_2{}^2$. The graph in $\mathbb{R}_{XY_1Y_2}$ of ϕ consists of more than finitely many points, and is the real part of a complex curve; we call it a real curve. It is easy to check, by taking secant lines through (0, 0, 0) and points of the graph near (0, 0, 0), that $\mathbb{R}_X$ is the unique line tangent to the curve at (0, 0, 0). Now the graph of any function $\Psi(Y_1, Y_2)$ from $\mathbb{R}_{Y_1Y_2}$ to $\mathbb{R}_X$ whose restriction to $\phi(\mathbb{R}_X)$ defines the inverse of ϕ, must surely contain the graph of ϕ. Since ϕ's graph is tangent to $\mathbb{R}_X$, the derivative at (0, 0) of $\Psi(Y_1, Y_2)$ approaching (0, 0) along the cusp in $\mathbb{R}_{Y_1Y_2}$, is infinite. Hence Ψ cannot be a polynomial function.

It is natural to call a space isomorphic to any polynomially embedded image of it, and different embedded images of the same space should themselves be called "isomorphic." The following definition expresses this idea, and includes our previous informal definitions. We state it for varieties over $\mathbb{C}$, but it extends to varieties over an arbitrary field; it also yields in $\mathbb{R}^n$ and $\mathbb{C}^n$ the corresponding differentiable or analytic notions, using the arbitrary neighborhood U_P in place of all $\mathbb{R}^n$ or $\mathbb{C}^n$.

Definition 8.6. Let $V \subset \mathbb{C}^n$ and $W \subset \mathbb{C}^m$ be irreducible affine varieties. A mapping $p: V \to W$ which is 1:1 from V onto W is a **polynomial isomorphism** if there are polynomial maps

$$p^*: \mathbb{C}^n \to \mathbb{C}^m \quad \text{and} \quad q^*: \mathbb{C}^m \to \mathbb{C}^n$$

such that

$$p^* | V = p \quad \text{and} \quad q^* | W = p^{-1}. \tag{13}$$

We then say V and W are *isomorphic*.

We now come to the result which is the main rationale of this section, namely

Theorem 8.7. *Two irreducible affine varieties $V \subset \mathbb{C}_{X_1, \ldots, X_n}$ and $W \subset \mathbb{C}_{Y_1, \ldots, Y_m}$ are isomorphic iff their coordinate rings are $\mathbb{C}$-isomorphic.*

Before giving the proof, let us first note that a representation of any coordinate ring R as $R = \mathbb{C}[z_1, \ldots, z_k] = \mathbb{C}[Z_1, \ldots, Z_k]/\mathfrak{p}$, where $\mathfrak{p}$ is a prime ideal and $z_i = Z_i + \mathfrak{p}$, determines the variety $\mathsf{V}(\mathfrak{p})$ in $\mathbb{C}_{Z_1, \ldots, Z_n}$. We regard $(z_1, \ldots, z_k)$ as an *ordered* k-tuple; the points of the variety are precisely those $(c_1, \ldots, c_k) \in \mathbb{C}^k$ such that $(Z_1 - c_1, \ldots, Z_k - c_k)$ is a maximal ideal of $\mathbb{C}[Z_1, \ldots, Z_k]$ containing $\mathfrak{p}$—that is, those points $(c_1, \ldots, c_k)$ such that $(z_1 - c_1, \ldots, z_k - c_k)$ is a maximal ideal of $\mathbb{C}[z_1, \ldots, z_k]$, or, what is the same, those points $(c_1, \ldots, c_k)$ such that when we substitute c_i for $z_i (i = 1, \ldots, k)$, we obtain a single-valued mapping from $\mathbb{C}[z_1, \ldots, z_k]$ to $\mathbb{C}$. For any such $c = (c_1, \ldots, c_k)$, we may denote this "evaluation mapping" by ϕ_c; for any $q \in \mathbb{C}[z_1, \ldots, z_k]$, $\phi_c(q) = q(c)$. It is clear that ϕ_c is a $\mathbb{C}$-ring homomorphism—that is, a ring homomorphism on $\mathbb{C}[z_1, \ldots, z_k]$ which is the identity map on $\mathbb{C}$. Because of the importance of this ring homomorphism, we make the following

Definition 8.8. A mapping of k-tuples $(z_1, \ldots, z_k) \to (c_1, \ldots, c_k)$ is a $\mathbb{C}$-**specialization**, or a $\mathbb{C}$-**specialization of** $(z_1, \ldots, z_k)$, if it defines a single-valued mapping from $\mathbb{C}[z_1, \ldots, z_k]$ to $\mathbb{C}$, via substitution. This mapping is then a $\mathbb{C}$-ring homomorphism. If reference to $\mathbb{C}$ is clear, we abbreviate the term to **specialization**. The point $(c_1, \ldots, c_k)$ is called a **specialization point (of** $(z_1, \ldots, z_k)$). The set of all specialization points of $(z_1, \ldots, z_k)$ forms a variety in $\mathbb{C}^k$; $(z_1, \ldots, z_k)$ is called a **generic point** of this variety.

PROOF OF THEOREM 8.7

$\Leftarrow$: Let $\mathbb{C}[x] = \mathbb{C}[x_1, \ldots, x_n] \simeq \mathbb{C}[y_1, \ldots, y_m] = \mathbb{C}[y]$ be a given isomorphism. Then of course each $y_i \in \mathbb{C}[y_1, \ldots, y_m]$ corresponds under this isomorphism to some element of $\mathbb{C}[x_1, \ldots, x_n]$—that is, there is a polynomial $p_i^*(X)$ such that $y_i = p_i^*(x_1, \ldots, x_n)$. Similarly, for each x_j there is a polynomial $q_j^*(Y)$ such that $x_j = q_j^*(y_1, \ldots, y_m) \in \mathbb{C}[y_1, \ldots, y_m]$. (Note that p_i^* and q_j^* are not necessarily uniquely determined.) Write $(X_1, \ldots, X_n) = (X)$ and $(Y_1, \ldots, Y_m) = (Y)$, and define the polynomial maps $p^*: \mathbb{C}_X \to \mathbb{C}_Y$ and $q^*: \mathbb{C}_Y \to \mathbb{C}_X$ by

$$p^*(X) = (p_1^*(X), \ldots, p_m^*(X)) \quad \text{and} \quad q^*(Y) = (q_1^*(Y), \ldots, q_n^*(Y)). \tag{14}$$

We now show that $p^*|V = (q^*|W)^{-1}$ (which just says that there is a 1:1-onto map from V to W such that it and its inverse have polynomial extensions to $\mathbb{C}_X$ and $\mathbb{C}_Y$, respectively). For this, let $(a) = (a_1, \ldots, a_n)$ be any point of V. Then (a) has the associated maximal ideal $\mathfrak{m} = (x_1 - a_1, \ldots, x_n - a_n)$ in $\mathbb{C}[x_1, \ldots, x_n]$. As noted above, the points of V are in 1:1-onto correspondence with the maximal ideals of $\mathbb{C}[x]$, and the points of W are in 1:1-onto correspondence with the maximal ideals of $\mathbb{C}[y]$. But the isomorphism $\mathbb{C}[x] \simeq \mathbb{C}[y]$ links the maximal ideals $\mathfrak{m}$ of $\mathbb{C}[x]$ with those in $\mathbb{C}[y]$ in a 1:1-onto way—say $\mathfrak{m}$ $(\subset \mathbb{C}[x])$ maps under the isomorphism to $\mathfrak{m}'$ $(\subset \mathbb{C}[y])$—so there is defined a natural 1:1-onto correspondence between the points of V and those of W, which we may write as

$$(x_1 + \mathfrak{m}, \ldots, x_n + \mathfrak{m}) = (a) \quad \longleftrightarrow \quad (y_1 + \mathfrak{m}', \ldots, y_m + \mathfrak{m}') = (b).$$

Since each y_j corresponds to $p_j^*(x)$, then $(a) \in V$ corresponds to $(p_1^*(a), \ldots, p_m^*(a)) \in W$—that is, $(a) \leftrightarrow p^*(a)$. Similarly for $(b) \in W$, $(b) \leftrightarrow q^*(b)$. Hence $p^*|V$ defines our 1:1-onto map from V to W and $q^*|W$ is its inverse, so "$\Leftarrow$" is proved.

$\Rightarrow$: Assume that V (which is a subvariety of $\mathbb{C}_{X_1, \ldots, X_n} = \mathbb{C}_X$) and W (a subvariety of $\mathbb{C}_{Y_1, \ldots, Y_m} = \mathbb{C}_Y$) are isomorphic—that is, that $Y = p^*(X)$ and $X = q^*(Y)$ are polynomial maps from $\mathbb{C}_X$ to $\mathbb{C}_Y$ and $\mathbb{C}_Y$ to $\mathbb{C}_X$, respectively, and that $p^*|V$ and $q^*|W$ are 1:1-onto and mutual inverses. We show that $R_V = \mathbb{C}[x]$ is isomorphic to $R_W = \mathbb{C}[y]$. For this, define z_i by $z_i = p_i^*(x)$, and let W' be the variety in $\mathbb{C}_Y$ having coordinate ring $R_{W'} = \mathbb{C}[z_1, \ldots, z_m] = \mathbb{C}[z]$. We show $R_V = R_W$ by showing $R_V = R_{W'}$ (that is, that $\mathbb{C}[x] = \mathbb{C}[z]$), and then that $W' = W$.

By hypothesis, for any $(a) \in V$, we have $(a) = q^*(p^*(a))$. Hence each X_i agrees with $q_i^*(p^*(x))$ on V. Therefore X_i and $q_i^*(p^*(X))$ differ only by an element of $\mathsf{J}(V)$; thus $x_i = X_i + \mathsf{J}(V) = q_i^*(p(X)) + \mathsf{J}(V) = q_i^*(p^*(x))$. Hence with $(z) = p^*(x)$, we have $(x) = q^*(z)$. Therefore $\mathbb{C}[x] \subset \mathbb{C}[z] \subset \mathbb{C}[x]$, so $\mathbb{C}[x] = \mathbb{C}[z]$.

We now show $W' = W$. For $W \subset W'$, let (b) be a typical point of W. Then $(b) = p^*(a)$ for a unique $(a) \in V$. Now we know $(x) \to (a)$ is a specialization of $\mathbb{C}[x]$; thus $z = p^*(x) \to p^*(a)$ is a specialization of $\mathbb{C}[z]$ and this in turn means that $(b) = (p^*(a)) \in W'$. Hence $W \subset W'$.

For $W' \subset W$, let (b') be a typical point of W'; then $z \to (b')$ is a specialization of $\mathbb{C}[z]$. Now we have just seen that $x = q^*(z)$, so $x = q^*(z) \to q^*(b')$ is a specialization of x, meaning that $q^*(b') \in V$; therefore $p^*(q^*(b')) = (b') \in W$. Hence $W' \subset W$. □

As we have seen, whenever a coordinate ring R is represented as $R = \mathbb{C}[x_1, \ldots, x_n]$, there is induced a corresponding variety of specializations in $\mathbb{C}^n$ of $(x_1, \ldots, x_n)$; each way of so expressing R yields a variety in some $\mathbb{C}^n$, and all such varieties are isomorphic. For example, $\mathbb{C}[X]$ may be rewritten in the redundant form $\mathbb{C}[X, X]$, which may be thought of as $\mathbb{C}[X, Y]/(Y - X)$. The ring $\mathbb{C}[X]$ defines the whole space $\mathbb{C}_X$ of $\mathbb{C}_X$, and $\mathbb{C}[X, X]$ yields the set $\mathsf{V}(Y - X) = \{(c, c) \mid c \in \mathbb{C}\} \subset \mathbb{C}_{XY}$; $\mathbb{C}_X$ is isomorphic to $\mathsf{V}(Y - X)$ via the polynomial map $X \to (X, X)$. Likewise, $\mathbb{C}[X]$ may be represented in any of the following forms; the associated varieties are all isomorphic: $\mathbb{C}[X, X, X]$ (a complex 1-space in $\mathbb{C}^3$); $\mathbb{C}[X, X^2]$ (the parabola $\mathsf{V}(Y - X^2) \subset \mathbb{C}_{XY}$); $\mathbb{C}[X, X^2, X^3]$ (a space curve in $\mathbb{C}^3$). Note that interchanging the order of the x_i in $\mathbb{C}[x_1, \ldots, x_n]$ yields in general a different, but isomorphic, variety.

So far we have not answered the question of what is *the* variety associated with a given coordinate ring—we have only constructed a bunch of mutually isomorphic varieties, each embedded in some surrounding space. Toward this end, first notice that for any $\mathbb{C}$-ring homomorphism $\phi: \mathbb{C}[x_1, \ldots, x_n] \to \mathbb{C}$ defined by $\phi(x_i) = c_i$ $(i = 1, \ldots, n)$, ϕ's kernel $\mathfrak{m} = (x_1 - c_1, \ldots, x_n - c_n)$ completely determines ϕ. Let $V \subseteq \mathbb{C}^n$ be the variety of specializations of $(x_1, \ldots, x_n)$, and let $W \subseteq \mathbb{C}^m$ be the variety of specializations of $p^*(x_1, \ldots, x_n)$. Now $\phi(x_i) = c_i$ implies that $\phi(p_j^*(x_1, \ldots, x_n)) = p_j^*(c_1, \ldots, c_n)$, so if $\mathfrak{m}$ defines the point $(c_1, \ldots, c_n)$ in V, then it defines the point $p^*(c_1, \ldots, c_n)$ in W. *The ideal* $\mathfrak{m}$ *thus connects corresponding points of varieties isomorphic under the equations in* (14). Looked at this way, $\mathfrak{m}$ takes on the role of "the essential notion of point." Let us now formalize this idea.

Definition 8.9. Let R be a coordinate ring. We call the set of all maximal ideals of R the **abstract variety of** R, and denote it by $\underset{\sim}{\mathsf{V}}_R$; any maximal ideal $\mathfrak{m}$ of R will be called an **abstract point**, or a **point of** $\underset{\sim}{\mathsf{V}}_R$. If $\mathfrak{a}$ is any ideal of R, we call the set of maximal ideals of R containing $\mathfrak{a}$ the **abstract subvariety of** $\underset{\sim}{\mathsf{V}}_R$ **defined by** $\mathfrak{a}$, and denote it by $\underset{\sim}{\mathsf{V}}(\mathfrak{a})$. If $\underset{\sim}{\mathsf{W}}$ is any such abstract subvariety of $\underset{\sim}{\mathsf{V}}_R$, the ideal $\bigcap_{\mathfrak{m} \in \underset{\sim}{\mathsf{W}}} \mathfrak{m}$ of R is called the **ideal defined by** $\underset{\sim}{\mathsf{W}}$ and is denoted by $\underset{\sim}{\mathsf{J}}(\underset{\sim}{\mathsf{W}})$.

Remark 8.10. A point of $\mathbb{C}^n$ is in a variety $\mathsf{V}(\mathfrak{a}) \subset \mathbb{C}^n$ if and only if the corresponding maximal ideal contains $\mathfrak{a}$; this is the justification of the last parts of the above definition.

Definition 8.11. Let $\mathbb{C}[x_1, \ldots, x_n]$ be any representation of a coordinate ring R. The variety V in $\mathbb{C}_{X_1, \ldots, X_n}$ of all $\mathbb{C}$-specializations of $(x_1, \ldots, x_n)$

will be called a **concrete model** in $\mathbb{C}_{X_1,\ldots,X_n}$ of the abstract variety $\underset{\sim}{V}_R$; we denote this concerete model by $V_{\mathbb{C}[x_1,\ldots,x_n]}$.

From Theorem 8.7, we see that any two concrete models are isomorphic.

Now let $\mathscr{V}(\mathbb{C}[x_1, \ldots, x_n])$ be our lattice of subvarieties of $V_{\mathbb{C}[x_1,\ldots,x_n]}$, and let J be the isomorphism from $\mathscr{V}(\mathbb{C}[x_1, \ldots, x_n])$ to $\mathscr{J}(\mathbb{C}[x_1, \ldots, x_n])$. The restriction of J to the point varieties of $V_{\mathbb{C}[x_1,\ldots,x_n]}$ maps these points to abstract points of $\underset{\sim}{V}_{\mathbb{C}[x_1,\ldots,x_n]}$—that is, to maximal ideals of $\mathbb{C}[x_1, \ldots, x_n]$. This map is 1:1 and onto $\underset{\sim}{V}_{\mathbb{C}[x_1,\ldots,x_n]}$. (First, J is a lattice isomorphism, hence 1:1 on the points; second, the map is onto since every maximal ideal $\mathfrak{m}$ in $\mathbb{C}[x_1, \ldots, x_n]$ defines a specialization point in $V_{\mathbb{C}[x_1,\ldots,x_n]}$.) We may extend this restricted map to arbitrary subsets S of $V_{\mathbb{C}[x_1,\ldots,x_n]}$ in the usual manner: $\underset{\sim}{S} = \{\mathfrak{m} = \mathsf{J}(\{P\}) \mid P \in S\}$. As with any 1:1 map from one set to another, this map preserves unions and intersections—that is, for any subsets S, T of $V_{\mathbb{C}[x_1,\ldots,x_n]}$ we have

$$S \cup T \to \underset{\sim}{S} \cup \underset{\sim}{T} \quad \text{and} \quad S \cap T \to \underset{\sim}{S} \cap \underset{\sim}{T}.$$

Hence such concepts in $V_{\mathbb{C}[x_1,\ldots,x_n]}$ as subvariety, irreducibility, decomposition, or topology, may all be transferred to the set $\underset{\sim}{V}_{\mathbb{C}[x_1,\ldots,x_n]}$ of abstract points. One can easily see that all these notions are well defined on $\underset{\sim}{V}_R$, since any two concrete models of a given coordinate ring are isomorphic. In particular, one may define $(\mathscr{V}(R), \subset, \cap, \cup)$ to be the lattice of subvarieties of the abstract variety $\underset{\sim}{V}_R$.

Now that we've introduced abstract varieties, the reader may well ask, "Since the abstract definition is an invariant notion, independent of any surrounding space or any particular representation of R, shouldn't we now simply abandon the notion of concrete variety and just work in the abstract setting?" Not at all. Both forms are useful. In stating results, it is frequently neater and more natural to use an abstract formulation, but in proving these results it is often convenient to use a concrete model.

Exercises

8.1 Find the coordinate ring of each of the three canonical dehomogenizations of the variety in $\mathbb{P}^2(\mathbb{C})$ defined by $\mathsf{V}(aX + bY) \subset \mathbb{C}_{XY}$ (a, $b \in \mathbb{C}$). Show that all three coordinate rings are isomorphic iff $ab \neq 0$. Interpret this fact geometrically.

8.2 Show that the transcendence degree over $\mathbb{C}$ of the coordinate ring of any irreducible curve in $\mathbb{C}^2$ is one.

8.3 Let R_C and $R_{C'}$ be the coordinate rings of irreducible plane curves C and C'. Show that the quotient fields of R_C and $R_{C'}$ may be isomorphic without C and C' being polynomially isomorphic.

8.4 If ϕ is a polynomial isomorphism from an irreducible plane curve C to another irreducible plane curve C', show that $P \in C$ is nonsingular iff $\phi(\mathrm{P}) \in C'$ is.

8.5 Find an irreducible curve $C \subset \mathbb{C}_{XY}$ and an element x of C's coordinate ring which tends to infinity at some, but not all, of C's points at infinity. Is it possible to choose an irreducible $C \subset \mathbb{C}_{XY}$ and a nonconstant x in C's coordinate ring so that x tends to infinity at *none* of C's points at infinity?

9 Induced lattice properties of coordinate ring surjections; examples

In this section we consolidate some of the earlier results of this chapter; this will lead to some further questions whose answers will extend our algebra–geometry dictionary in an important way.

First, recall that to any variety $V \subset \mathbb{C}^n$ there is associated a coordinate ring $R_V = R$ and three lattices—$(\mathscr{V}(R), \subset, \cap, \cup)$, $(\mathscr{J}(R), \subset, \cap, +)$, and $(\mathscr{I}(R), \subset, \cap, +)$; these lattices may be put into a sequence in a natural way (Diagram 1),

$$(\mathscr{I}(R), \subset, \cap, +) \underset{\mathrm{i}}{\overset{\sqrt{\ }}{\rightleftarrows}} (\mathscr{J}(R), \subset, \cap, +) \underset{\mathsf{J}}{\overset{\mathsf{V}}{\rightleftarrows}} (\mathscr{V}(R), \subset, \cap, \cup)$$

Diagram 1

where $\sqrt{\ }$ and i are the radical and embedding maps, respectively, and V, J are the lattice-isomorphisms of Theorem 2.19. Conversely, any coordinate ring R determines a variety, hence also a sequence of the form in Diagram 1. Therefore such a sequence may be generated by either a variety or by a coordinate ring.

Now if two varieties (or two coordinate rings) are related in some way, it is natural to ask if there are corresponding relations between the associated sequences. Since so many of the important relations between rings may be defined via ring homomorphisms, we make the following definition, which will be used throughout the book:

Definition 9.1. Let R, R^* be commutative rings with identity. Let $h: R \to R^*$ be a ring homomorphism. For any ideal $\mathfrak{a} \subset R$, the ideal $(h(\mathfrak{a})) \subset R^*$ generated by the set $h(\mathfrak{a}) = \{h(a) \mid a \in \mathfrak{a}\}$ is called the **extension of** $\mathfrak{a}$ **in** R^* (or just the **extension of** $\mathfrak{a}$ if no confusion can arise) and is denoted by $\mathfrak{a}^e$. For any ideal $\mathfrak{b} \subset R^*$, the inverse image $h^{-1}(\mathfrak{b})$ is an ideal in R and is called the **contraction of** $\mathfrak{b}$ (**in** R) and is denoted by $\mathfrak{b}^c$.

We saw a special case of an induced relation on two sequences in Section 1—there we looked at a coordinate ring $\mathbb{C}[x_1, \ldots, x_n] = R_V = R$ related to $R_W = R^* = R/\mathfrak{p}$ by the natural homomorphism $h_\mathfrak{p}$, and this implied that $(\mathscr{V}(R/\mathfrak{p}), \subset, \cap, \cup) = \mathscr{V}(R^*)$ is lattice-embedded in $(\mathscr{V}(R), \subset, \cap, \cup) = \mathscr{V}(R)$.

The homomorphism $h_\mathfrak{p}$ induces maps from the lattices in one sequence to those in the other as follows:

(a): $h_\mathfrak{p}$ induces the extension map $(\ )^e : \mathfrak{a} \to \mathfrak{a} + \mathfrak{p} = \mathfrak{a}^e$ from $\mathscr{I}(R)$ to $\mathscr{I}(R/\mathfrak{p}) = \mathscr{I}(R^*)$; $h_\mathfrak{p}^{-1}$ induces the contraction map $(\ )^c : \mathfrak{b} \to h^{-1}(\mathfrak{b}) = \mathfrak{b}^c$ from $\mathscr{I}(R^*)$ to $\mathscr{I}(R)$.

(b): $h_\mathfrak{p}$ induces the "closed extension"

$$\mathfrak{a} \to \sqrt{\mathfrak{a} + \mathfrak{p}} = \sqrt{\mathfrak{a}^e}$$

from $\mathscr{J}(R)$ to $\mathscr{J}(R^*)$; $h_\mathfrak{p}^{-1}$ defines the contraction

$$\mathfrak{b} \to h_\mathfrak{p}^{-1}(\mathfrak{b}) = \mathfrak{b}^c$$

from $\mathscr{J}(R^*)$ to $\mathscr{J}(R)$. We continue to denote this restriction of $(\ )^c$ in (a) to $\mathscr{J}(R^*)$ by $(\ )^c$.

Note that in the closed extension, the radical is necessary (Example 2.17); in the contraction, $h_\mathfrak{p}^{-1}(\mathfrak{b})$ is already closed (Lemma 7.6).

(c): Let V and J be the lattice-reversing isomorphisms between $\mathscr{V}(R)$ and $\mathscr{J}(R)$, and let V^* and J^* be the lattice-reversing isomorphisms between $\mathscr{V}(R^*)$ and $\mathscr{J}(R^*)$. Then $h_\mathfrak{p}$ and $h_\mathfrak{p}^{-1}$ induce maps

$$V \to \mathsf{V}^* \circ (\ )^e \circ \mathsf{J}(V) = \mathsf{V}^*((\mathsf{J}(V))^e)$$

from $\mathscr{V}(R)$ to $\mathscr{V}(R^*)$, and

$$V^* \to \mathsf{V} \circ (\ )^c \circ \mathsf{J}^*(V^*) = \mathsf{V}((\mathsf{J}^*(V^*))^c)$$

from $\mathscr{V}(R^*)$ to $\mathscr{V}(R)$.

Under these two maps, objects in $\mathscr{J}(R)$ and $\mathscr{V}(R)$ which correspond under V or J map to objects in $\mathscr{J}(R^*)$ and $\mathscr{V}(R^*)$ which correspond under V^* or J^*.

Let $\mathfrak{a} \subset R = \mathbb{C}[x_1, \ldots, x_n]$. Then $\mathsf{V}(\mathfrak{a})$ is a subvariety of the variety V_R of specializations in $\mathbb{C}^n$ of $(x_1, \ldots, x_n)$. Since $\mathfrak{a}$ maps under closed extension to $\sqrt{\mathfrak{a} + \mathfrak{p}}$, $\mathsf{V}^* \circ \sqrt{(\ )^e} \circ \mathsf{J}$ maps $\mathsf{V}(\mathfrak{a})$ to $\mathsf{V}(\mathfrak{a}) \cap \mathsf{V}(\mathfrak{p})$—that is, to $\mathsf{V}(\mathfrak{a})$'s intersection with $\mathsf{V}(\mathfrak{p})$. Similarly, since $\mathfrak{b}$ $(\subset R/\mathfrak{p})$ maps under contraction to $h^{-1}(\mathfrak{b})$, $\mathsf{V} \circ (\ )^e \circ \mathsf{J}^*$ embeds $\mathsf{V}(\mathfrak{b})$ $(\subset V_{R/\mathfrak{p}})$ in V_R; we denote this embedding by $\imath$. We then get a double sequence of lattices as in Diagram 2:

$$\begin{array}{ccccc}
(\mathscr{I}(R), \subset, \cap, +) & \underset{\mathrm{i}}{\overset{\sqrt{\ }}{\rightleftarrows}} & (\mathscr{J}(R), \subset, \cap, +) & \underset{\mathsf{J}}{\overset{\mathsf{V}}{\rightleftarrows}} & (\mathscr{V}(R), \subset, \cap, \cup) \\
(\ )^e \downarrow \uparrow (\ )^c & & \sqrt{(\ )^e} \downarrow \uparrow (\ )^c & & \text{Intersection with } V(\mathfrak{p}) \downarrow \uparrow \imath \\
(\mathscr{I}(R^*), \subset, \cap, +) & \underset{\mathrm{i}^*}{\overset{\sqrt{\ }}{\rightleftarrows}} & (\mathscr{J}(R^*), \subset, \cap, +) & \underset{\mathsf{J}^*}{\overset{\mathsf{V}^*}{\rightleftarrows}} & (\mathscr{V}(R^*), \subset, \cap, \cup)
\end{array}$$

Diagram 2

A number of previous and future results are incorporated in this diagram. Certain results tell us that various maps are well defined; others essentially make comments on how much lattice structure or decomposition into irreducibles is preserved. For instance, various results earlier in this chapter (for instance, in Section 2) make comments about the horizontal maps. Lemmas 7.4 and 7.6 tell us that the three upward maps (contraction of ideals, contraction of closed ideals, and embedding ι are all lattice-embeddings. The downward maps preserve $+$ and $\boldsymbol{+}$ of ideals and intersection of varieties in the respective lattices, but need not preserve lattice structure. For more about this, and the question of preserving decomposition into irreducibles, see Exercise 9.1.

So far we have looked only at *onto* homomorphisms. Since the double sequence in Diagram 2 contains so much information, one may wonder if considering more general ring homomorphisms might represent a new and important way of extending our geometric knowledge. This is indeed the case.

Besides onto homomorphisms, other important homomorphisms are the 1 : 1 homomorphisms, or embeddings.

Note that for a fixed coordinate ring R, any coordinate ring R^* which is an onto image of R is simply R modulo a prime ideal of R; but in the case of 1 : 1 homomorphisms, R^* can be any ring containing R. In a sense, then, there are more possible choices for R^* in the 1 : 1 case. In this and the next three sections we look at some of these. We devote the remainder of this section to examples; these will give us an idea of the direction our development will take.

EXAMPLE 9.2. The natural embedding $h: \mathbb{C}[X] \subset \mathbb{C}[X, Y]$. First note that for any embedding $R \subset R^*$, contraction becomes just intersection with R—that is, $\mathfrak{a}^* \subset R^*$ implies $h^{-1}(\mathfrak{a}^*) = (\mathfrak{a}^*)^c = \mathfrak{a} \cap R$.

Now in $\mathbb{C}[X, Y]$, any maximal ideal is of the form $(X - c, Y - d)$ where $c, d \in \mathbb{C}$, so the contraction $(X - c, Y - d)^c$ is $(X - c, Y - d) \cap \mathbb{C}[X] = (X - c)$. Of course the isomorphisms V and J between maximal ideals in $\mathscr{J}(\mathbb{C}[X])$ and points in $\mathscr{V}(\mathbb{C}[X])$ are given by

$$(X - c) \underset{\mathsf{J}}{\overset{\mathsf{V}}{\rightleftarrows}} c;$$

the analogous correspondences at the V^*-J^* level are

$$(X - c, Y - d) \underset{\mathsf{J}^*}{\overset{\mathsf{V}^*}{\rightleftarrows}} (c, d).$$

Hence $(X - c, Y - d) \cap \mathbb{C}[X] = (X - c)$ defines the projection $\pi_Y(c, d) = c \in \mathbb{C}_X$ of the point $(c, d) \in \mathbb{C}_{XY}$—that is, $\mathsf{V} \circ (\ \ )^c \circ \mathsf{J}^*$ π_Y-projects $\mathbb{C}_{XY}$ to $\mathbb{C}_X$.

Now let us look at h, which induces the map $\mathfrak{a} \to (h(\mathfrak{a})) = \mathfrak{a}^e$ for $\mathfrak{a} \subset R$. In $\mathbb{C}[X]$, any maximal ideal is of the form $(X - c)$, so $\mathfrak{a}^e$ is just the principal ideal $(X - c)$ in $\mathbb{C}[X, Y]$. But in $\mathbb{C}_{XY}$, $\mathsf{V}(X - c)$ is the line $X - c$, so $\mathsf{V}(X - c)$

consists of the set of all points of $\mathbb{C}_{XY}$ π_Y-projecting to $c \in \mathbb{C}_X$. It is thus the *largest* variety in $\mathbb{C}_{XY}$ projecting onto c. Note that the principal ideal $(X - c)$ in $\mathbb{C}[X, Y]$ is the *smallest* ideal of $\mathbb{C}[X, Y]$ whose intersection with $\mathbb{C}[X]$ is the ideal $(X - c)$ in $\mathbb{C}[X]$! Theorems 10.1 and 10.8 easily imply a basic generalization of this phenomenon.

EXAMPLE 9.3. In the preceding example we looked at the effect of h and h^{-1} on only maximal ideals of $\mathbb{C}[X]$ and $\mathbb{C}[X, Y]$ (or, geometrically, the effect of $\mathsf{V}^* \circ \sqrt{(\ )^e} \circ \mathsf{J}$ and of $\mathsf{V} \circ (\ )^c \circ \mathsf{J}^*$ on point varieties in $\mathbb{C}_X$ and $\mathbb{C}_{XY}$). It is natural to ask about the corresponding effects on more general ideals and subvarieties.

For instance, consider the ideal

$$\mathfrak{a}^* = (X - c_1, Y - d_1) \cap (X - c_2, Y - d_2)$$

in $\mathbb{C}[X, Y]$. This defines the variety in $\mathbb{C}_{XY}$ consisting of the points (c_1, c_2) and (d_1, d_2). Then

$$\begin{aligned} h^{-1}(\mathfrak{a}^*) &= (\mathfrak{a}^*)^c = \mathfrak{a} \cap \mathbb{C}[X] \\ &= ((X - c_1, Y - d_1) \cap \mathbb{C}[X]) \cap ((X - c_2, Y - d_2) \cap \mathbb{C}[X]) \\ &= (X - c_1) \cap (X - c_2) \subset \mathbb{C}[X], \end{aligned}$$

so geometrically, $\mathsf{V} \circ (\ )^c \circ \mathsf{J}^*$ projects $\{(c_1, d_1), (c_2, d_2)\}$ to $\{c_1, c_2\}$.

Similarly, if $\mathfrak{b}^* = (Y - X^2)$ in $\mathbb{C}[X, Y]$, then

$$(\mathfrak{b}^*)^c = (Y - X^2) \cap \mathbb{C}[X] = (0),$$

so $\mathsf{V} \circ (\ )^c \circ \mathsf{J}^*$ projects the parabola $\mathsf{V}(Y - X^2)$ onto $\mathbb{C}_X$—that is, contraction in this instance again has the geometric effect of projecting onto $\mathbb{C}_X$.

In both cases extension corresponds geometrically to sending the variety into the largest variety lying above it: The ideal $(X - c_1) \cap (X - c_2)$ in $\mathbb{C}[X]$ extends to $(X - c_1) \cap (X - c_2)$ in $\mathbb{C}[X, Y]$, and this defines in $\mathbb{C}_{XY}$ the union of the lines $X = c_1$ and $X = c_2$. In the parabola example, $(Y - X^2)^c = (0)$; $(0)^e = (0)$ in $\mathbb{C}[X, Y]$, which defines $\mathbb{C}_{XY}$, this being the set of all points of $\mathbb{C}_{XY}$ lying above $\mathbb{C}_X$.

The reader can also easily verify that the ideal

$$\mathfrak{c}^* = (X^2 + Y^2 - 1, X^2 + Y^2 + Z^2 - 4)$$

in $\mathbb{C}[X, Y, Z]$ defines in $\mathbb{C}_{XYZ}$ two circles (the intersection of a cylinder and sphere); clearly $\mathfrak{c}^* \cap \mathbb{C}[X, Y] = (X^2 + Y^2 - 1)$; this defines the circle in $\mathbb{C}_{XY}$ which is the projection of the two circles in $\mathbb{C}_{XYZ}$. And $(X^2 + Y^2 - 1)^e$ in $\mathbb{C}[X, Y, Z]$ defines the largest variety of $\mathbb{C}_{XYZ}$ lying above the circle, namely, the cylinder above the circle.

Let us look again at contraction. First, we have the following

Lemma 9.4. *For any two embedded coordinate rings $R \subset R^*$, if $\mathfrak{m}^*$ is maximal in R^*, then $\mathfrak{m}^* \cap R = \mathfrak{m}$ is maximal in R.*

PROOF. Write $R = \mathbb{C}[x_1, \ldots, x_n]$ and $R^* = \mathbb{C}[x_1, \ldots, x_m]$ where $m \geqslant n$. Then $\mathfrak{m}^*$ is of the form

$$(x_1 - a_1, \ldots, x_m - a_m),$$

so $R^*/\mathfrak{m}^*$ is isomorphic to $\mathbb{C}[a_1, \ldots, a_m] = \mathbb{C}$. Since $R/(\mathfrak{m}^* \cap R)$ is in a natural way isomorphic to a subring of $R^*/\mathfrak{m}^*$, we see that $R/(\mathfrak{m}^* \cap R) \subset \mathbb{C}$. But also clearly $\mathbb{C} \subset R/(\mathfrak{m}^* \cap R)$, so $R/(\mathfrak{m}^* \cap R)$ is isomorphic to the field $\mathbb{C}$, hence $\mathfrak{m}^* \cap R$ is maximal in R. $\square$

From this lemma and the behavior of $\mathbf{V} \circ (\ )^c \circ \mathbf{J}^*$ in our examples so far, one might guess that the "projected variety" $\mathbf{V} \circ (\ )^c \circ \mathbf{J}^*(V^*)$ of any variety V^* in $\mathbb{C}_{XY}$, is just the union of the projections on $\mathbb{C}_X$ of each of the points in V^*, or equivalently, that the function $\mathbf{V} \circ (\ )^c \circ \mathbf{J}^*$ sending varieties to varieties is induced by its restriction to the point varieties. Recall that any function $f: D \to D^*$ automatically induces a function $\mathscr{S}(D) \to \mathscr{S}(D^*)$ from the set of all subsets $\mathscr{S}(D)$ of D to the set of all subsets $\mathscr{S}(D^*)$ of D^*. We just define the map so it preserves union, i.e., for $S = \cup \{P\} \subset D$, let $f(\cup \{P\}) = \cup \{f(P)\}$ (that is, $f(S) = \{f(P) | P \in S\}$). Now obviously an arbitrary function defined on the set of subsets of a set D is not in general induced by its restriction to the singleton subsets. However, since for any $\mathfrak{a}^*, \mathfrak{b}^* \subset R^*$,

$$(\mathfrak{a}^* \cap \mathfrak{b}^*) \cap R = (\mathfrak{a}^* \cap R) \cap (\mathfrak{b}^* \cap R)$$

(that is, contraction preserves intersections), and since the $\mathscr{J}$- and $\mathscr{V}$-lattices are reverse-isomorphic, we see that $\mathbf{V} \circ (\ )^c \circ \mathbf{J}^*$ preserves unions, so it would seem that $\mathbf{V} \circ (\ )^c \circ \mathbf{J}^*$ is induced by its restriction to the point varieties—that is, by a function on $\mathbb{C}_{XY}$.

But let us consider $V = \mathbf{V}(XY - 1)$; if our guess were correct, then

$$\mathbf{V} \circ (\ )^c \circ \mathbf{J}^*(V) \text{ would equal } \bigcup_{P \in V} \{\mathbf{V} \circ (\ )^c \circ \mathbf{J}^*(P)\}. \tag{5}$$

The first expression we know is $\mathbb{C}_X$, since $(XY - 1) \cap \mathbb{C}[X] = (0)$. The second is just the set-theoretic projection of V on C_X, and this not $\mathbb{C}_X$, but $\mathbb{C}_X \backslash \{0\}$! (This is not even a variety in $\mathbb{C}_X$.)

What went wrong? Since $V = \bigcup_{P \in V} \{P\}$, we see that to have equality in (5), $\mathbf{V} \circ (\ )^c \circ \mathbf{J}^*$ would have to preserve infinite union. But we have proved only that $\mathbf{V} \circ (\ )^c \circ \mathbf{J}^*$ is a *lattice* homomorphism, meaning that it preserves union and intersection of two (and hence, by induction, finitely many) varieties. In general, the g.l.b. and l.u.b. of an arbitrary collection of elements in a lattice need not exist; even when they do, one can find many examples showing that a lattice homomorphism may not preserve g.l.b. and l.u.b. of infinite sets. In extending such finite operations to infinite ones, one often meets topological notions. In our case, the set-theoretic projection $\mathbb{C}_X \backslash \{0\}$ is in a topological sense close to the variety-theoretic projection

$\mathsf{V} \circ (\ )^c \circ \mathsf{J}^*(V) = \mathbb{C}_X$, for $C_X \backslash \{0\}$ is dense in $\mathbb{C}_X$. More generally, *the set-theoretic projection always turns out to be dense in the variety-theoretic projection*; this follows at once from Theorem 10.8.1.

One thing we have learned from this is that the *variety projection* $\mathsf{V} \circ (\ )^c \circ \mathsf{J}^*$ is essentially a map from varieties to varieties, not from points to points.

EXAMPLE 9.6. In Examples 9.2 and 9.3 we took both R and R^* to be coordinate rings of the very simple form $\mathbb{C}[X_1, \ldots, X_n]$. Now let us consider arbitrary embedded coordinate rings $R \subset R^*$. For R and R^* as in Example 9.3, for any maximal ideal $\mathfrak{m}^* \subset R^*$, $\mathfrak{m}^* \cap R$ is maximal in R; hence a point $\mathfrak{m}^*$ of $\underset{\sim}{V}_{R^*}$ maps to the point $\mathfrak{m} = \mathfrak{m}^* \cap R$ of $\underset{\sim}{V}_R$. Then the question arises: Can this map be looked at as a *projection*, as in Examples 9.2 and 9.3? To answer this, let us write $R \subset R^*$ as

$$\mathbb{C}[x_1, \ldots, x_n] \subset \mathbb{C}[x_1, \ldots, x_m] \qquad (m \geqslant n).$$

By so writing R and R^* we have, of course, selected affine models $V_R \subset \mathbb{C}^n$ and $V_{R^*} \subset \mathbb{C}^m$ of $\underset{\sim}{V}_R$ and $\underset{\sim}{V}_{R^*}$. The model V_{R^*} consists of all specialization points $(c_1, \ldots, c_m)$ in $\mathbb{C}_{X_1, \ldots, X_m}$ of $(x_1, \ldots, x_m)$, the maximal ideal in $\mathbb{C}[x_1, \ldots, x_m]$ corresponding to $(c_1, \ldots, c_m)$ being $\mathfrak{m}^* = (x_1 - c_1, \ldots, x_m - c_m)$.

Let us now note the general

Lemma 9.7. *Let* $\mathbb{C}[x_1, \ldots, x_n] \subset \mathbb{C}[x_1, \ldots, x_m]$ *where* $m \geqslant n$. *For any maximal ideal* $\mathfrak{m}^* = (x_1 - c_1, \ldots, x_m - c_m) \subset \mathbb{C}[x_1, \ldots, x_m]$, *we have*

$$(x_1 - c_1, \ldots, x_m - c_m) \cap \mathbb{C}[x_1, \ldots, x_n] = (x_1 - c_1, \ldots, x_n - c_n) = \mathfrak{m};$$

$\mathfrak{m}$ *is maximal in* $\mathbb{C}[x_1, \ldots, x_n]$.

PROOF. Since $\mathbb{C}[x_1, \ldots, x_n]/(x - c_1, \ldots, x_n - c_n) \simeq \mathbb{C}[c_1, \ldots, c_n] = \mathbb{C}$, $(x_1 - c_1, \ldots, x_n - c_n)$ is clearly maximal in $\mathbb{C}[x_1, \ldots, x_n]$. From Lemma 9.4 we see that $\mathfrak{m}^* \cap \mathbb{C}[x_1, \ldots, x_n]$ is maximal in $\mathbb{C}[x_1, \ldots, x_n]$. Obviously $(x_1 - c_1, \ldots, x_n - c_n)$ is contained in the maximal ideal $\mathfrak{m}^* \cap \mathbb{C}[x_1, \ldots, x_n]$, so these two maximal ideals of $\mathbb{C}[x_1, \ldots, x_n]$ must coincide. $\square$

Since $(x_1 - c_1, \ldots, x_n - c_n)$ corresponds to $(c_1, \ldots, c_n) \in V_{\mathbb{C}[x_1, \ldots, x_n]}$, Lemma 9.7 shows that $(\mathfrak{m}^*)^c = \mathfrak{m}^* \cap \mathbb{C}[x_1, \ldots, x_n]$ does indeed define a projection of the points in $V_{\mathbb{C}[x_1, \ldots, x_m]}$ to those in $V_{\mathbb{C}[x_1, \ldots, x_n]}$. We therefore make the following

Definition 9.8. Let $\mathfrak{m}^*$ and $\mathfrak{m}$ be points in the abstract varieties $\underset{\sim}{V}_{R^*}$ and $\underset{\sim}{V}_R$ of embedded coordinate rings $R \subset R^*$; then $\mathfrak{m}^*$ is said to **lie above** $\mathfrak{m}$ if $\mathfrak{m} = \mathfrak{m}^* \cap R$; we write $\mathfrak{m} = \underset{\sim}{\pi}(\mathfrak{m}^*)$ and call $\underset{\sim}{\pi}$ the **natural projection** from

$\underset{\sim}{V}_{R^*}$ to $\underset{\sim}{V}_R$. For $\underset{\sim}{V}(\mathfrak{a}^*) = \{\mathfrak{m}^* | \mathfrak{m}^* \supset \mathfrak{a}^*\}$, the **set-theoretic projection** $\underset{\sim}{\pi}(\underset{\sim}{V}(\mathfrak{a}^*))$ is $\{\underset{\sim}{\pi}(\mathfrak{m}^*) | \mathfrak{m}^* \in \underset{\sim}{V}(\mathfrak{a}^*)\}$.

Thus, for example, $R = \mathbb{C}[X]$ is embedded in $R^* = \mathbb{C}[X, 1/X]$, $\mathbb{C}[X]$ determines the affine model $\mathbb{C}_X$, while $\mathbb{C}[X, 1/X]$ defines the hyperbola consisting of all points $(c, 1/c)$ in $\mathbb{C}_{XY}$. Each maximal ideal $\mathfrak{m}^*$ of $\mathbb{C}[X, 1/X]$ is of the form $(X - c, (1/X) - (1/c))$ where $c \neq 0$, and $\mathfrak{m}^* \cap R$ is $(X - c) \subset \mathbb{C}[X]$; hence any point $(c, 1/c) \in \mathbb{C}_{XY}$ projects to $c \in \mathbb{C}_X$.

What happens if we use a trivial embedding, for instance $\mathbb{C}[X] \subset \mathbb{C}[X, X^2]$, in which the two rings are actually the same? Then $\mathbb{C}[X]$ has $\mathbb{C}_X$ for an affine model, and $\mathbb{C}[X, X^2]$ defines the parabola $\mathsf{V}(Y - X^2) \subset \mathbb{C}_{XY}$. Each maximal ideal $(X - c, X^2 - c^2)$ intersects $\mathbb{C}[X]$ in $(X - c)$; this defines the projection $(c, c^2) \to c$. Hence the parabola set-theoretically projects onto $\mathbb{C}_X$. Note that since $\mathbb{C}[X] = \mathbb{C}[X, X^2]$, $\mathbb{C}_X$ and the parabola are just different models of one and the same abstract affine variety $\underset{\sim}{V}_{\mathbb{C}[X]} = \underset{\sim}{V}_{\mathbb{C}[X, X^2]}$. Our projection defines an isomorphism between the two models; the points c and (c, c^2) correspond via the one maximal ideal $(X - c) = (X - c, X^2 - c^2) \subset \mathbb{C}[X]$.

Likewise, $\mathbb{C}[X] \subset \mathbb{C}[X, X^3]$ defines a projection of the cubic

$$\mathsf{V}(Y - X^3) \subset \mathbb{C}_{XY} \quad \text{into} \quad \mathbb{C}_X,$$

this projection being an isomorphism. And $\mathbb{C}[X] \subset \mathbb{C}[X, X^2, X^3]$ defines a projection (which is also an isomorphism) from a curve in $\mathbb{C}_{XYZ}$ onto $\mathbb{C}_X$. Note that in the parabola above, the projection of it into $\mathbb{C}_Y$ is all of $\mathbb{C}_Y$, which is isomorphic to the parabola; but the projection is not $1:1$ and does not define an isomorphism.

In each of Examples 9.2 and 9.3, $(h(\mathfrak{a})) = \mathfrak{a}^e \subset R^*$ defines the largest subvariety of V_R lying above $\mathsf{V}(\mathfrak{a}) \subset V_R$. Even for $\mathbb{C}[X] \subset \mathbb{C}[X, 1/X]$ having as a model the hyperbola $\mathsf{V}(Y - (1/X)) \subset \mathbb{C}_{XY}$, we see that above $0 \in \mathbb{C}_X$ there lies no point of $\mathsf{V}(Y - (1/X))$; correspondingly, the set of all points above $0 \in \mathbb{C}_X$ is the empty set in $\mathsf{V}(Y - (1/X))$. This is reflected in the fact that $(X)^e$ in $\mathbb{C}[X, 1/X]$ is actually all of $\mathbb{C}[X, 1/X]$ (since $X \cdot (1/X) = 1 \in (X)^e$). Of course the ideal $(1) = \mathbb{C}[X, 1/X]$ defines the empty set in $\mathsf{V}(Y - (1/X))$. In the next section we prove that the set of all points of $\underset{\sim}{V}_{R^*}$ projecting into a given subvariety $\underset{\sim}{V}(\mathfrak{a})$ of $\underset{\sim}{V}_R$ is a subvariety of $\underset{\sim}{V}_{R^*}$ (Theorem 10.1) and that this subvariety is defined by $\mathfrak{a}^e \subset R^*$.

Exercises

9.1 In Diagram 2, how much lattice structure is preserved by each of the three downward maps? Show that for none of the three downward maps is irreducibility preserved.

9.2 Let W be a subvariety of $\mathbb{C}_{X_1, \ldots, X_n}$. Find, for any $m > 0$, a variety V in $\mathbb{C}_{X_1, \ldots, X_{n+m}}$ whose natural set-theoretic projection on $\mathbb{C}_{X_1, \ldots, X_n}$ is $\mathbb{C}_{X_1, \ldots, X_n} \backslash W$.

10 Induced lattice properties of coordinate ring injections

In the last section, an onto homomorphism $h_{\mathfrak{p}} : R \to R^*$ generated the double sequence of Diagram 2. More generally, any coordinate ring homomorphism $h : R \to R^*$ generates a double sequence as shown in Diagram 3. In this and the next section, we will consider the important case when h is one-to-one.

To begin with, any homomorphism $h : R \to R^*$ yields

$$
\begin{array}{ccccc}
(\mathscr{I}(R), \subset, \cap, +) & \underset{\mathrm{i}}{\overset{\sqrt{\ }}{\rightleftarrows}} & (\mathscr{J}(R), \subset, \cap, +) & \underset{\underset{\sim}{\mathrm{J}}}{\overset{\underset{\sim}{\mathrm{V}}}{\rightleftarrows}} & (\mathscr{V}(R), \subset, \cap, \cup) \\
(\ )^e \downarrow \uparrow (\ )^c & & \sqrt{(\ )^e} \downarrow \uparrow (\ )^c & & \underset{\sim}{\mathrm{V}}^* \circ \sqrt{(\ )^e} \circ \underset{\sim}{\mathrm{J}} \downarrow \uparrow \underset{\sim}{\mathrm{V}} \circ (\ )^c \circ \underset{\sim}{\mathrm{J}}^* \\
(\mathscr{I}(R^*), \subset, \cap, +) & \underset{\mathrm{i}^*}{\overset{\sqrt{\ }}{\rightleftarrows}} & (\mathscr{J}(R^*), \subset, \cap, +) & \underset{\underset{\sim}{\mathrm{J}}^*}{\overset{\underset{\sim}{\mathrm{V}}^*}{\rightleftarrows}} & (\mathscr{V}(R^*), \subset, \cap, \cup)
\end{array}
$$

Diagram 3

where extension $(\ )^e$ and contraction $(\ )^c$ are as in Definition 9.1. The lattice $\mathscr{V}(R)$ may be looked at as the lattice of subvarieties of either the abstract variety $\underset{\sim}{\mathrm{V}}_R$, or of some concrete model of it; likewise for $\mathscr{V}(R^*)$. The maps in Diagram 3 are pretty much self-explanatory, except that in two of the upward maps, we must show that $(\ )^c$ actually does map a closed ideal in R^* to a closed ideal in R—that is, that $(\ )^c = \sqrt{(\ )^c}$. This is easy: If $\mathfrak{c}^*$ is a closed ideal in R^*, it suffices to show that if $a \in R \backslash h^{-1}(\mathfrak{c}^*)$, then $a^n \in R \backslash h^{-1}(\mathfrak{c}^*)$, for all positive integers n. For this, note that $a \in R \backslash h^{-1}(\mathfrak{c}^*)$ implies that $h(a) \in R^* \backslash \mathfrak{c}^*$. Therefore $(h(a))^n \in R^* \backslash \mathfrak{c}^*$ for all $n > 0$, which implies that $h(a^n) \in R^* \backslash \mathfrak{c}^*$ for all $n > 0$. Since distinct elements of R^* map under h^{-1} to distinct cosets of $h^{-1}(0)$, $h(a^n) \in R^* \backslash \mathfrak{c}^*$ implies that any element in $h^{-1}(h(a^n))$ must be in $R \backslash h^{-1}(\mathfrak{c}^*)$. Obviously a^n is such an element.

Now if h is $1:1$, then contraction is just *intersection with* R. Extension is $\mathfrak{a}^e = \mathfrak{a}R^*$ for $\mathfrak{a} \subset R$, and $\mathfrak{c}^e = \sqrt{\mathfrak{c}R^*}$ for $\mathfrak{c}$ closed in R. Note that this radical is actually necessary—that is, if $\mathfrak{c}$ is closed in R, $\mathfrak{c}R^*$ may not be closed in R^*. From a geometric standpoint, we might expect this to happen if, for example, above a point of multiplicity one there lies exactly one point of multiplicity greater than one. For instance there are two distinct points of the parabola $\mathsf{V}(Y^2 - X) \subset \mathbb{C}_{XY}$ π_Y-lying above any point $X \neq 0$ in $\mathbb{C}_X$. At $X = 0$ these two points coalesce to one *double point*. The coordinate rings of $\mathbb{C}_X$ and $\mathsf{V}(Y^2 - X)$ are $\mathbb{C}[X]$ and $\mathbb{C}[X, X^{1/2}] = \mathbb{C}[X^{1/2}]$. The origin of $\mathbb{C}_X$ is defined by $\mathfrak{m} = (X) \subset \mathbb{C}[X]$, and $\mathfrak{m}^e$ is (X) in $\mathbb{C}[X^{1/2}]$. The extended ideal $\mathfrak{m}^e$ is not closed; the only maximal ideal containing (X) in $\mathbb{C}[X^{1/2}]$ is $(X^{1/2})$, so $(X^{1/2})$ is the closure of (X).

We next prove the important facts that when h is $1:1$, $\underset{\sim}{\mathrm{V}}^* \circ \sqrt{(\ \ )^e} \circ \underset{\sim}{\mathrm{J}}$ and $\underset{\sim}{\mathrm{V}} \circ (\ \ )^c \circ \underset{\sim}{\mathrm{J}}^*$ become "inverse projection" $\underset{\sim}{\pi}^{-1}$ and "closed projection" $\bar{\underset{\sim}{\pi}}$, respectively. These will fit in with our examples in Section 9.

For the statement of the next theorem, recall Definition 9.8.

Theorem 10.1. *Let $R \subset R^*$ be coordinate rings, and let $\mathfrak{a}$ be any ideal of R. The set of all points in* $\underset{\sim}{\mathrm{V}}_{R^*}$ lying above $\underset{\sim}{\mathrm{V}}(\mathfrak{a}) \subset \underset{\sim}{\mathrm{V}}_R$ *is the variety* $\underset{\sim}{\mathrm{V}}(\mathfrak{a}^e)$. *(Hence* $\underset{\sim}{\mathrm{V}}(\mathfrak{a}^e) = \underset{\sim}{\pi}^{-1}(\underset{\sim}{\mathrm{V}}(\mathfrak{a}))$, *where* $\underset{\sim}{\pi}$ *is the natural projection of Definition* 9.8.)

PROOF. First recall from Definition 8.9 that for any ideal $\mathfrak{b}$ in any coordinate ring R, $\underset{\sim}{\mathrm{V}}(\mathfrak{b})$ is the subset of $\underset{\sim}{\mathrm{V}}_R$ consisting of all maximal ideals in R containing $\mathfrak{b}$. In particular,

(10.2) $\underset{\sim}{\mathrm{V}}(\mathfrak{a}^e)$ is the set $\{\mathfrak{m}^*\}$ of all maximal ideals of R^* containing $\mathfrak{a}^e$.

Now from Example 9.6 we know that any maximal ideal $\mathfrak{m}^\dagger$ in $\underset{\sim}{\mathrm{V}}(\mathfrak{a}^e)$ lies above the point $\mathfrak{m}^\dagger \cap R$ in $\underset{\sim}{\mathrm{V}}_R$; thus from (10.2) we know that $\mathfrak{m}^\dagger$ lies above $\underset{\sim}{\mathrm{V}}(\mathfrak{a})$ iff $\mathfrak{m}^\dagger \cap R \supset \mathfrak{a}$. Therefore

> The set of points in $\underset{\sim}{\mathrm{V}}_{R^*}$ lying above $\underset{\sim}{\mathrm{V}}(\mathfrak{a})$ is just the set $\{\mathfrak{m}^\dagger\}$ of maximal ideals $\mathfrak{m}^\dagger$ in R^* satisfying $\mathfrak{m}^\dagger \cap R \supset \mathfrak{a}$.

Hence we will have proved Theorem 10.1 once we show that $\{\mathfrak{m}^*\} = \{\mathfrak{m}^\dagger\}$. But this is easy; both $\{\mathfrak{m}^*\} \subset \{\mathfrak{m}^\dagger\}$ and $\{\mathfrak{m}^\dagger\} \subset \{\mathfrak{m}^*\}$ follow at once from the fact that $\mathfrak{a}^e$ is the smallest ideal in R^* containing $\mathfrak{a}$. □

Before stating the next theorem we make the following definitions:

Definition 10.3. The **natural topology** on any affine abstract variety $\underset{\sim}{\mathrm{V}}_R$ is the topology induced on $\underset{\sim}{\mathrm{V}}_R$ by any of its concrete models $V_{\mathbb{C}[x_1,\ldots,x_n]} \subset \mathbb{C}_{X_1,\ldots,X_n}$ (via the map $\mathfrak{m} = (x_1 - c_1, \ldots, x_n - c_n) \to (c_1, \ldots, c_n)$).

Remark 10.4. The above topology is well defined since any two concrete models are isomorphic; hence in particular they are homeomorphic relative to the topologies induced on them from their surrounding spaces.

The next definition is an extension of Definition 9.8.

Definition 10.5. Let $\mathfrak{a} \subset R$ and $\mathfrak{b}^* \subset R^*$ be ideals in coordinate rings $R \subset R^*$, let $\underset{\sim}{\mathrm{V}}(\mathfrak{a}) \subset \underset{\sim}{\mathrm{V}}_R$ and $\underset{\sim}{\mathrm{V}}(\mathfrak{b}^*) \subset \underset{\sim}{\mathrm{V}}_{R^*}$ be the associated abstract varieties, and let $\underset{\sim}{\pi}$ be the natural projection $\mathfrak{m}^* \to \mathfrak{m}^* \cap R$ from $\underset{\sim}{\mathrm{V}}_{R^*}$ onto $\underset{\sim}{\mathrm{V}}_R$. If the set $\underset{\sim}{\pi}(\underset{\sim}{\mathrm{V}}(\mathfrak{b}^*))$ is dense in $\underset{\sim}{\mathrm{V}}(\mathfrak{a})$ relative to $\underset{\sim}{\mathrm{V}}(\mathfrak{a})$'s natural topology, we say that the variety $\underset{\sim}{\mathrm{V}}(\mathfrak{b}^*)$ **lies over** the variety $\underset{\sim}{\mathrm{V}}(\mathfrak{a})$.

The corresponding concrete-model form of this is:

Definition 10.6. Let π be the natural projection of $\mathbb{C}^n \times \mathbb{C}^m$ onto the first factor. A variety $V^* \subset \mathbb{C}^n \times \mathbb{C}^m$ **lies over** a variety $V \subset \mathbb{C}^n$ if the set $\pi(V^*)$ is dense in V.

Definition 10.7. Let $R \subset R^*$ be coordinate rings. An ideal $\mathfrak{a}^* \subset R^*$ is said to **lie over** an ideal $\mathfrak{a} \subset R$ if $\mathfrak{a} = \mathfrak{a}^* \cap R$.

Our next theorem connects the above geometric and algebraic notions of lying over as follows:

Theorem 10.8. *Let $R \subset R^*$ be coordinate rings, and let $\underset{\sim}{\pi}$ be the natural projection $\mathfrak{m}^* \to \mathfrak{m}^* \cap R$ from $\underset{\sim}{V}_{R^*}$ to $\underset{\sim}{V}_R$.*

(10.8.1) *If $\mathfrak{a}$ and $\mathfrak{a}^*$ are ideals in R and R^* respectively, if $\mathfrak{a}$ defines $\underset{\sim}{W}$ ($\subset \underset{\sim}{V}_R$), and if $\mathfrak{a}^*$ defines $\underset{\sim}{W}^*$ ($\subset \underset{\sim}{V}_{R^*}$) then*

$$(\mathfrak{a}^* \textit{ lies over } \mathfrak{a}) \Rightarrow (\underset{\sim}{W}^* \textit{ lies over } \underset{\sim}{W}).$$

(10.8.2) *If $\underset{\sim}{W}$ and $\underset{\sim}{W}^*$ are varieties in $\underset{\sim}{V}_R$ and $\underset{\sim}{V}_{R^*}$ respectively, if $\underset{\sim}{W}$ defines $\mathfrak{c}$ ($\subset R$), and if $\underset{\sim}{W}^*$ defines $\mathfrak{c}^*$ ($\subset R^*$), then*

$$(\underset{\sim}{W}^* \textit{ lies over } \underset{\sim}{W}) \Rightarrow (\mathfrak{c}^* \textit{ lies over } \mathfrak{c}).$$

In proving Theorem 10.8 we shall assume the following fact:

Lemma 10.9. *Let V be any irreducible variety in $\mathbb{C}^n$, and let V' be a proper subvariety of V. Then $V \setminus V'$ is dense in V.*

This says that any proper subvariety of V is in a sense much smaller than V. Proving this lemma here would somewhat disrupt the continuity of our development; it will fit in easily and naturally in Chapter IV (Exercise 2.6 in Chapter IV).

Remark 10.10. Lemma 10.9 need not hold for reducible varieties, e.g., $(\mathbb{C}_X \cup \mathbb{C}_Y)\setminus\mathbb{C}_X$ is not dense in $\mathbb{C}_X \cup \mathbb{C}_Y \subset \mathbb{C}_{XY}$. Also note that it need not hold for *real* irreducible varieties. For example, let $V = V(Y^2 - X^2(X-1)) \subset \mathbb{R}_{XY}$, and let $V' =$ origin of $\mathbb{R}_{XY}$. (The origin is an isolated point of V.)

Proof of Theorem 10.8. We prove (10.8.1) first for $\mathfrak{a} = (0) \subset R$ and $\mathfrak{a}^* = (0) \subset R^*$. That is, we show that $\underset{\sim}{\pi}(\underset{\sim}{V}^*)$ is dense in $\underset{\sim}{V}$. First, clearly $\underset{\sim}{\pi}(\underset{\sim}{V}^*) \subset \underset{\sim}{V}$, for as noted above, $\mathfrak{m}^*$ maximal in R^* implies that $\mathfrak{m}^* \cap R$ is maximal in R. To prove density, let us choose without loss of generality arbitrary concrete models of $\underset{\sim}{V}$ and $\underset{\sim}{V}^*$ by letting, for instance, $R = \mathbb{C}[x_1, \ldots, x_n]$ and $R^* = \mathbb{C}[x_1, \ldots, x_m]$ where $m \geqslant n$; let $V \subset \mathbb{C}^n$ and $V^* \subset \mathbb{C}^m$ be the associated varieties of specializations, with $\pi: \mathbb{C}_{X_1,\ldots,X_m} \to \mathbb{C}_{X_1,\ldots,X_n}$ the natural projection. We shall show that $\pi(V^*)$ is dense in V.

For this, let $(c_1, \ldots, c_n)$ be any point of V. Now there is a point $(c_1, \ldots, c_m)$ of V^* above $(c_1, \ldots, c_n)$ iff mapping each x_i to c_i $(i = 1, \ldots, m)$ defines a homomorphism from $\mathbb{C}[x_1, \ldots, x_m]$ to $\mathbb{C}$. We suggestively write $\mathbb{C}[x_1, \ldots, x_m] \to \mathbb{C}[c_1, \ldots, c_m]$ for this homomorphism; of course $\mathbb{C}[c_1, \ldots, c_m] = \mathbb{C}$. Using analogous notation, we see that there is a homomorphism $\mathbb{C}[x_1, \ldots, x_m] \to \mathbb{C}[c_1, \ldots, c_m]$ iff we can successively extend the homomorphism

$$\mathbb{C}[x_1, \ldots, x_n] \to \mathbb{C}[c_1, \ldots, c_n]$$

to

$$\mathbb{C}[x_1, \ldots, x_{n+1}] \to \mathbb{C}[c_1, \ldots, c_{n+1}],$$

then to

$$\mathbb{C}[x_1, \ldots, x_{n+1}, x_{n+2}] \to \mathbb{C}[c_1, \ldots, c_{n+1}, c_{n+2}],$$

and so on, up to the full homomorphism

$$\mathbb{C}[x_1, \ldots, x_m] \to \mathbb{C}[c_1, \ldots, c_m].$$

Let us denote the variety of specializations of $\mathbb{C}[x_1, \ldots, x_n, \ldots, x_{n+i}]$ in $\mathbb{C}_{X_1, \ldots, X_n, \ldots, X_{n+i}}$ by V_i. (Hence $V_0 = V$ and $V_{m-n} = V^*$.)

Now if x_{n+1} is transcendental over $\mathbb{C}[x_1, \ldots, x_n]$, then any choice of $c_{n+1} \in \mathbb{C}$ yields a homomorphism $\mathbb{C}[x_1, \ldots, x_{n+1}] \to \mathbb{C}[c_1, \ldots, c_{n+1}]$, so in this case the first extension can always be made. If x_{n+1} is algebraic over $\mathbb{C}(x_1, \ldots, x_n)$, let its minimal polynomial be

$$p_{10}(x_1, \ldots, x_n)X_{n+1}{}^{N_1} + \ldots + p_{1N_1}(x_1, \ldots, x_n) \qquad (p_{10} \neq 0). \tag{15}$$

Since $N_1 > 0$, if $p_{10}(c_1, \ldots, c_n) \neq 0$, then there is a root of the equation

$$p_{10}(c_1, \ldots, c_n)X_{n+1}{}^{N_1} + \ldots + p_{1N_1}(c_1, \ldots, c_n) = 0, \tag{16}$$

and any of its roots c_{n+1} yields a homomorphism $x_{n+1} \to c_{n+1}$. The only time anything can go wrong, therefore, is when the leading coefficient $p_{10}(c_1, \ldots, c_n)$ is zero, for then there may be no zero of the polynomial in (15). (For example if the polynomial is $XY - 1$, the leading coefficient is X; when $X = 0$ we get $0Y - 1 = -1$, which has no zero.) Hence we can make this first extension except possibly when $(c_1, \ldots, c_n)$ lies in the zero-set of the nonzero polynomial $p_{10}(x_1, \ldots, x_n)$, so the points of V not in $\pi_{X_{n+1}}(V_1)$ are contained in the zero-set of $p_{10}(x_1, \ldots, x_n)$. This zero-set is a proper subvariety W_0 of V; since V is irreducible, $V \setminus W_0$ is dense in V (Lemma 10.9), so $\pi_{X_{n+1}}(V_1)$ is dense in V.

We can similarly extend

$$\mathbb{C}[x_1, \ldots, x_{n+1}] \to \mathbb{C}[c_1, \ldots, c_{n+1}]$$

to a homomorphism

$$\mathbb{C}[x_1, \ldots, x_{n+2}] \to \mathbb{C}[c_1, \ldots, c_{n+2}]$$

for all $(c_1, \ldots, c_{n+1})$ except possibly off a proper subvariety W_1 of V_1; hence the set-theoretic projection $\pi_{X_{n+2}}(V_2)$ is dense in V_1. This means that $\pi_{X_{n+1}} \circ \pi_{X_{n+2}}(V_2) = \pi_{X_{n+1}X_{n+2}}(V_2)$ is also dense in V_0, because, more generally, if $f: S \to T$ and $g: T \to U$ are continuous maps, if $f(S)$ is dense in T, and if $g(T)$ is dense in U, then $g \circ f(S)$ is dense U. (PROOF: An arbitrary open neighborhood of any point P in U contains points of $g(T)$; since there are points of $f(S)$ arbitrarily close to any given point of T, continuity implies there are points of $g \circ f(S)$ arbitrarily close to any point of $g(T)$, hence arbitrarily close to P.)

Likewise, $\pi_{X_{n+1}X_{n+2}X_{n+3}}(V_3)$ is dense in V; continuing in this fashion, we are finally led to the result that $\pi_{X_{n+1}\ldots X_m}(V_{m-n}) = \pi_{X_{n+1}\ldots X_m}(V^*)$ is dense in V.

This argument generalizes at once to the case when $\mathfrak{a}^*$ is any prime ideal $\mathfrak{p}^*$, for $R/(\mathfrak{p}^* \cap R)$ may in a natural way be looked at as a subring of $R^*/\mathfrak{p}^*$. Then W^* and W are defined by the 0-ideals in $R^*/\mathfrak{p}^*$ and $R/(\mathfrak{p}^* \cap R)$, respectively. This brings us back to the case just considered.

Finally, suppose $\mathfrak{a}^*$ is an arbitrary ideal in R^*. Now if $\mathfrak{a}^*$ lies above $\mathfrak{a}$, then $\sqrt{\mathfrak{a}^*}$ lies above $\sqrt{\mathfrak{a}}$—that is,

$$\mathfrak{a}^* \cap R = \mathfrak{a} \text{ implies } \sqrt{\mathfrak{a}^*} \cap R = \sqrt{\mathfrak{a}}.$$

(The proof of this fact is easy: $a \in \sqrt{\mathfrak{a}}$ means that $a^n \in \mathfrak{a}$ for some n, so $a^n \in \mathfrak{a}^*$ and $a \in R$—that is, $a \in \sqrt{\mathfrak{a}^*} \cap R$. Conversely, $b \in \sqrt{\mathfrak{a}^*} \cap R$ implies $b^m \in \mathfrak{a}^*$ for some m, so also $b^m \in \mathfrak{a}^* \cap R = \mathfrak{a}$, hence $b \in \sqrt{\mathfrak{a}}$.) We may now factor $\sqrt{\mathfrak{a}^*}$ into prime ideals (by Theorem 4.9):

$$\sqrt{\mathfrak{a}^*} = \mathfrak{p}_1^* \cap \ldots \cap \mathfrak{p}_r^*.$$

The preceding argument then applies to each prime separately:

$$\sqrt{\mathfrak{a}^*} \quad \text{defines} \quad \mathsf{V}(\mathfrak{p}_1^*) \cup \ldots \cup \mathsf{V}(\mathfrak{p}_r^*)$$

and

$$\sqrt{\mathfrak{a}} \quad \text{defines} \quad \mathsf{V}(\mathfrak{p}_1^* \cap R) \cup \ldots \cup \mathsf{V}(\mathfrak{p}_r^* \cap R);$$

hence

$$\begin{aligned}\pi_{X_{n+1}\ldots X_m}(\mathsf{V}(\mathfrak{a}^*)) &= \pi_{X_{n+1}\ldots X_m}(\mathsf{V}(\mathfrak{p}_1^*) \cup \ldots \cup \mathsf{V}(\mathfrak{p}_r^*)) \\ &= \pi_{X_{n+1}\ldots X_m}(\mathsf{V}(\mathfrak{p}_1^*)) \cup \ldots \cup \pi_{X_{n+1}\ldots X_m}(\mathsf{V}(\mathfrak{p}_r^*)).\end{aligned}$$

Thus $\pi_{X_{n+1}\ldots X_m}(\mathsf{V}(\mathfrak{a}^*))$ is dense in $\mathsf{V}(\mathfrak{a}) = \mathsf{V}(\mathfrak{p}_1^* \cap R) \cup \ldots \cup \mathsf{V}(\mathfrak{p}_r^* \cap R)$ since each $\pi_{X_{n+1}\ldots X_m}(\mathsf{V}(\mathfrak{p}_i^*))$ is dense in $\mathsf{V}(\mathfrak{p}_i^* \cap R)$. Thus (10.8.1) is proved.

We now prove (10.8.2). By hypothesis, $\underset{\sim}{W}^*$ lies over $\underset{\sim}{W}$—that is, $\overline{\underset{\sim}{\pi}(\underset{\sim}{W}^*)} = \underset{\sim}{W}$; also $\underset{\sim}{W} = \underset{\sim}{V}(\mathfrak{c})$ since $\underset{\sim}{W}$ defines $\mathfrak{c}$ implies that $\mathfrak{c}$ defines $\underset{\sim}{W}$. Likewise $\mathfrak{c}^*$ defines $\underset{\sim}{W}^*$, so by (10.8.1), $\underset{\sim}{W}^*$ lies over $\underset{\sim}{V}(\mathfrak{c}^* \cap R)$—that is, $\overline{\underset{\sim}{\pi}(\underset{\sim}{W}^*)} = \underset{\sim}{V}(\mathfrak{c}^* \cap R)$. Therefore

$$\underset{\sim}{V}(\mathfrak{c}) = \underset{\sim}{V}(\mathfrak{c}^* \cap R).$$

Since $\mathfrak{c}$ and $\mathfrak{c}^*$ are both closed, so is $\mathfrak{c}^* \cap R$ (proved at the beginning of this section). We know $\mathscr{J}(R)$ is lattice reverse-isomorphic to $\mathscr{V}(R)$, so $\underset{\sim}{\mathsf{V}}(\mathfrak{c}) = \underset{\sim}{\mathsf{V}}(\mathfrak{c}^* \cap R)$ implies that $\mathfrak{c} = \mathfrak{c}^* \cap R$—that is, $\mathfrak{c}^*$ lies over $\mathfrak{c}$. This completes the proof of (10.8.2) and therefore of Theorem 10.8. □

It is natural to ask next just how much lattice structure in Diagram 3 is preserved under an embedding $h: R \subset R^*$. Some elementary results, valid for arbitrary coordinate rings $R \subset R^*$ are outlined in Exercise 10.1. We shall henceforth assume these very easily-established results. Other results hold only for specific types of extensions. We look at these in the next section.

Exercise

10.1 In this exercise we look at properties of Diagram 3 when R is embedded in R^*. The following letters denote typical ideals and varieties in the diagram: $\mathfrak{a}, \mathfrak{b} \in \mathscr{J}(R)$; $\mathfrak{a}^*, \mathfrak{b}^* \in \mathscr{J}(R^*)$; $\mathfrak{c}, \mathfrak{d} \in \mathscr{J}(R)$; $\mathfrak{c}^*, \mathfrak{d}^* \in \mathscr{J}(R^*)$; $W_1, W_2 \in \mathscr{V}(R)$; and $W_1^*, W_2^* \in \mathscr{V}(R^*)$.

(a) Show that all six vertical maps are p.o. homomorphisms.

(b) Prove that the three upward maps, from left to right, preserve $\cap$, $\cap$, and $\cup$, respectively, but yield only inclusions, e.g., $(\mathfrak{a}^* + \mathfrak{b}^*)^c \supset (\mathfrak{a}^*)^c + (\mathfrak{b}^*)^c$, for $+$, $+$ and $\cap$, respectively. Show by example that we cannot strengthen these inclusions to equalities. [*Hint*: Find simple varieties W_1, W_2 such that $\bar{\pi}(W_1 \cap W_2) \subsetneqq \bar{\pi}(W_1) \cap \bar{\pi}(W_2)$, then translate into ideal language.]

(c) Prove that the three downward maps, from left to right, preserve $+$, $+$, and $\cap$, respectively. Show that $(\mathfrak{a} \cap \mathfrak{b})^e \subset \mathfrak{a}^e \cap \mathfrak{b}^e$. Show that $(\mathfrak{c} \cap \mathfrak{d})^e = \mathfrak{c}^e \cap \mathfrak{d}^e$ and that $\pi^{-1}(\mathsf{V}(\mathfrak{c}) \cup \mathsf{V}(\mathfrak{d})) = \pi^{-1}(\mathsf{V}(\mathfrak{c})) \cup \pi^{-1}(\mathsf{V}(\mathfrak{d}))$. (*Hint*: Use the "$\mathscr{V} \leftrightarrow \mathscr{J}$" lattice isomorphism.)

(d) Prove that the three upward maps preserve primality of ideals and irreducibility of varieties; show by example that the downward maps need not.

(e) Do the three upward maps preserve decomposition into primes or irreducibles? Show that in general, irredundancy in the decomposition is not preserved.

11 Geometry of coordinate ring extensions

We now turn our attention to the geometry of specific kinds of coordinate ring extensions. From elementary algebra, we know that if a domain D^* is a finitely generated extension of a domain D, it may be looked at as a pure transcendental extension followed by a pure algebraic extension. The geometric effect of any coordinate ring extension can be determined once we know the effects of these pure extensions. In this section we study these two types of extensions.

Pure transcendental extensions

Let an arbitrary coordinate ring $R = \mathbb{C}[x_1, \ldots, x_n]$ define the variety $V \subset \mathbb{C}_{X_1, \ldots, X_n}$, and let Y be a transcendental element over R. Each maximal ideal of R is of the form $\mathfrak{m} = (x_1 - c_1, \ldots, x_n - c_n)$, where $x_1 \to c_1, \ldots, x_n \to c_n$

defines a homomorphism of R to $\mathbb{C}$; and the maximal ideals $\mathfrak{m}^*$ of $R^* = \mathbb{C}[x_1, \ldots, x_n, Y]$ are $(x_1 - c_1, \ldots, x_n - c_n, Y - d)$, where $d \in \mathbb{C}$ may be chosen arbitrarily. (Since Y is transcendental, for any $d \in \mathbb{C}$, $x_1 \to c_1, \ldots, x_n \to c_n$, $Y \to d$ defines a homomorphism of $R^* \to \mathbb{C}$.) Hence the set of all points in $\mathbb{C}_{X_1,\ldots,X_n,Y}$ lying above a given $(c_1, \ldots, c_n) \in V$ is $\{(c_1, \ldots, c_n, d) | d \in \mathbb{C}\}$. Thus $\pi^{-1}(V)$ is the "product variety" $V \times \mathbb{C}_Y \subset \mathbb{C}_{X_1,\ldots,X_n,Y}$. (We briefly introduce product varieties in Section IV,2.) Similarly, one sees that for a pure transcendental coordinate ring extension $R[Y_1, \ldots, Y_m]$ of R, above each point $(c_1, \ldots, c_n) \in V$ lies the set $\{(c_1, \ldots, c_n, d_1, \ldots, d_m) | d_i \in \mathbb{C}, i = 1, \ldots, m\}$; hence the extension defines the variety $V \times \mathbb{C}^m \subset \mathbb{C}^{n+m}$. Likewise, for any ideal $\mathfrak{a} \subset R$, $\mathfrak{a}^e \subset R[Y_1, \ldots, Y_m]$ defines in $\mathbb{C}^{n+m}$ the variety $\mathsf{V}(\mathfrak{a}^e)$ consisting of all points in $\mathbb{C}^{n+m}$ set-theoretically projecting into $\mathsf{V}(\mathfrak{a})$; thus $\mathsf{V}(\mathfrak{a}^e) = \mathsf{V}(\mathfrak{a}) \times \mathbb{C}^m \subset \mathbb{C}^{n+m}$.

Under the assumption that R^* is a pure transcendental extension $R[Y_1, \ldots, Y_m]$ of R, we can easily sharpen some of our results about the behavior of ideals under extension. (See Exercise 11.1.) For instance, the inclusion in Exercise 10.1(c) becomes equality, and unique irredundant decomposition is preserved. (Exercise 11.1(d).) However, under contraction from $R[Y_1, \ldots, Y_m]$ to R, the results of Exercise 10.1(b) cannot be improved, as is shown by the example of $\{(1, 1) \cup (1, -1)\} \subset C_{Y_1 Y_2}$, which projects to $\{1\} \subset C_{Y_1}$.

Algebraic extensions

We next turn to the geometric significance of finite algebraic extensions of coordinate rings, and of a particularly important kind of algebraic extension, the integral extensions. We shall assume that the reader is familiar with the basic definitions and properties of algebraic and integral extensions. In particular, recall that for integral domains R, S, T, if S is algebraic (or integral) over R, and T is algebraic (or integral) over S, then T is algebraic (or integral) over R. If $k \subset K \subset L$ are algebraic field extensions, then one has the degree relation

$$[L : k] = [L : K] \cdot [K : k]. \tag{17}$$

We shall also use the important *theorem of the primitive element*.

The basic geometric facts about algebraic and integral coordinate ring extensions which we prove in this section are contained in the following theorem. To make our arguments a bit more intuitive, we use concrete models, though of course everything can be translated into the abstract setting. If R is any integral domain, $\tilde{R}$ denotes its quotient field.

Theorem 11.1

(11.1.1) *Let $R \subset R^*$ be a finite algebraic extension, where R and R^* are coordinate rings, and let $V = V_R$ and $V^* = V_{R^*}$ be concrete models of R and R^*. Let $[\tilde{R}^* : \tilde{R}] = D$. Then there is a proper subvariety W of V such that above any point of $V \backslash W$, there are precisely D distinct points of V^*.*

(11.1.2) *If, furthermore, R^* is integral over R, then over each point of W there are k distinct points, where $1 \leqslant k \leqslant D$.*

Notation 11.2. In proving our theorem, it turns out that there are fewer notational difficulties if we stick to one basic letter for the various generic points. Therefore we let $(x_1, \ldots, x_n) = (x)$ be a generic point of $V \subset \mathbb{C}_{X_1,\ldots,X_n} = \mathbb{C}_X$, and let $(x_1, \ldots, x_{n+m}) = (x, \bar{x})$ be a generic point of $V^* \subset \mathbb{C}_{X,\bar{X}} = \mathbb{C}^{n+m}$. We shall further write $(x) = (x', x'')$, where $x' = (x_1, \ldots, x_t)$ is a transcendence base of $\mathbb{C}[x]$.

Proof of (11.1.1). Let us first consider the special case when R is a pure transcendental extension of $\mathbb{C}$—that is, when $R = \mathbb{C}[x] = \mathbb{C}[x']$. Now if m happens to be 1, meaning that V^* has generic point (x, x_{n+1}), then R^* is isomorphic to $\mathbb{C}[x, X_{n+1}]/(p(X, X_{n+1}))$, where $p(X, X_{n+1}) \in \mathbb{C}[x, X_{n+1}]$ is a minimal polynomial of x_{n+1} over $\mathbb{C}(x)$; hence V^* is simply the variety $\mathsf{V}(p(X, X_{n+1})) \subset \mathbb{C}^{n+1}$. If $[\mathbb{C}(x, x_{n+1}) : \mathbb{C}(x)] = D$, then $\deg p = D$, so there are exactly D points of $\mathsf{V}(p)$ over each point of $\mathbb{C}_X$ off the proper discriminant variety $W = \mathsf{V}(\mathscr{D}_{X_{n+1}}(p))$.

To extend this result to an arbitrary algebraic extension $R^* = \mathbb{C}[x, \bar{x}]$ of the same ring $R = \mathbb{C}[x]$, we may use the theorem of the primitive element to reduce the problem to the above. Extending our result to the full statement of (11.1.1) will then be trivial.

Let us therefore write $\mathbb{C}(x)[\bar{x}]$ as $\mathbb{C}(x)[y]$, where y is a single quantity (a primitive element) algebraic over $\mathbb{C}(x)$. Let $V^\dagger$ be the variety in $\mathbb{C}_{X, X_{n+1}}$ with generic point (x, y). Now there are two natural maps between our generic points of V^* and $V^\dagger$: the first map is from $(x, \bar{x})$ to (x, y), defined by sending (x) to itself and $(\bar{x})$ to y. Since y is a primitive element, y can be looked at as a polynomial in $x_{n+1}, \ldots, x_{n+m}$ with coefficients in $\mathbb{C}(x)$; the nonzero coefficients will be denoted by f_j. The second map is the inverse of the first—(x) maps to (x), and y maps to $(\bar{x})$. Each of $x_{n+1}, \ldots, x_{n+m}$ is a polynomial in y with coefficients $g_{ik} \neq 0$ in $\mathbb{C}(x)$.

Now V^* consists of the set of all $\mathbb{C}$-specializations of $(x, \bar{x})$, and $V^\dagger$ is the set of specializations of (x, y). The above mutually inverse maps of generic points may not induce (via specialization) 1:1-onto maps between all the points of V^* and $V^\dagger$, since the f_j and g_{ik} $(\in \mathbb{C}(x))$, after reducing to lowest terms, may still have denominators which are zero at certain points of $\mathbb{C}_X$. However at any point (a) of $\mathbb{C}_X$ not in the union of the zero sets of the denominators of all the finitely many f_j and g_{ik}, any point $(a, \bar{a}) \in V^*$ corresponds in a one-to-one manner to a point $(a, b) \in V^\dagger$ via our relation between the generic points. Hence over any point of $\mathbb{C}_X$ not in this union, there lie just as many points of $V^*(\subset \mathbb{C}_{X,\bar{X}})$ as there lie points of $V^\dagger$ $(\subset \mathbb{C}_{X, X_{n+1}})$. Now if q is the minimal polynomial over $\mathbb{R}(x)$ of y, then removing from $\mathbb{C}_X$ the discriminant variety $\mathsf{V}(\mathscr{D}_{X_{n+1}}(q))$, as well as the above union, leaves us with a subset of $\mathbb{C}_X$ over which there are exactly D $(=\deg q)$ distinct points. We have thus proved our theorem in this case if we let W be $\mathsf{V}(\mathscr{D}_{X_{n+1}}(q))$ together with the above union.

The full statement of (11.1.1) may now be easily established. Recall Notation 11.2. If $[\tilde{R}^* : \tilde{R}] = D$, and $[\tilde{R} : \mathbb{C}(x')] = d$, then $[\tilde{R}^* : \mathbb{C}(x')] = D \cdot d$, and there is a proper subvariety W' of $\mathbb{C}_{X_1,\ldots,X_t} = \mathbb{C}_{X'}$ so that over each point of $\mathbb{C}_{X'} \backslash W'$ there lie exactly d points of $V \subset \mathbb{C}_X$ and exactly $D \cdot d$ points of $V^* \subset \mathbb{C}_{X\bar{X}}$. Now there is a point $(a, \bar{a}) \in V^*$ lying over $(a) \in V$ iff we can extend the specialization $\mathbb{C}[x] \to \mathbb{C}[a]$ to $\mathbb{C}[x, \bar{x}] \to \mathbb{C}[a, \bar{a}]$. It is clear that since $[\tilde{R}^* : \tilde{R}] = D$, for a fixed point $(a) \in V$ there can be at most D such extensions. But since there are exactly $D \cdot d$ points of V^* over any $(a') = (a_1, \ldots, a_t) \in \mathbb{C}_{X'} \backslash W'$, the maximum number D of extensions must be attained above any $(a) \in V$ which lies over (a'). Thus, letting $W = (W' \times \mathbb{C}_{X_{t+1},\ldots,X_n}) \cap V$, we see that (11.1.1) is proved. □

Proof of (11.1.2). Let $(a) \in V$. We know that a point $(a, \bar{a})$ is in V^* iff the specialization $\mathbb{C}[x] \to \mathbb{C}[a]$ extends to a specialization $\mathbb{C}[x, \bar{x}] \to \mathbb{C}[a, \bar{a}]$; this holds, of course, iff we can successively extend the homomorphism

$$\mathbb{C}[x_1, \ldots, x_{n+i-1}] \to \mathbb{C}[a_1, \ldots, a_{n+i-1}]$$

to

$$\mathbb{C}[x_1, \ldots, x_{n+i}] \to \mathbb{C}[a_1, \ldots, a_{n+i}] \qquad (i = 1, \ldots, m).$$

Now if $R^* = \mathbb{C}[x, \bar{x}]$ is integral over $R = \mathbb{C}[x]$, then in particular x_{n+i} is integral over $\mathbb{C}[x_1, \ldots, x_{n+i-1}]$ for $i = 1, \ldots, m$; hence the leading coefficient of a polynomial p over $\mathbb{C}[x_1, \ldots, x_{n+i-1}]$ such that $p(x_{n+i-1}) = 0$, may be taken to be 1. Now p_i is a multiple of x_{n+i}'s minimal polynomial m_i over $\mathbb{C}(x_1, \ldots, x_{n+i-1})$. Say $\deg m_i = d_i$. Since every zero of m_i is a zero of p_i, and since at any $(a_1, \ldots, a_{n+i-1})$ there are $\deg p_i$ zeros (counted with multiplicity) of p_i, there are at $(a_1, \ldots, a_{n+i-1})$ d_i zeros of m_i. There is therefore at least one extension of our homomorphism, and no more than d_i of them. Hence for any fixed $(a) \in V$ there is at least one choice, but no more than d_i choices for the $(n + i)^{\text{th}}$ coordinate of V^*. Therefore there is at least one, but there are no more than $d_1 \cdot \ldots \cdot d_m$ points of V^* above $(a) \in V$; by (17), $d_1 \cdot \ldots \cdot d_m = D$. Thus (11.1.2) is proved, and therefore Theorem 11.1. □

Example 11.3. If R^* is algebraic but not integral over R, then the number of points of V^* over W may vary more wildly. For example, consider $V^* = \mathsf{V}(X_2 X_3 - X_1) \subset \mathbb{C}_{X_1 X_2 X_3}$. The coordinate ring of V^* is

$$\mathbb{C}[X_1, X_2, X_3]/(X_2 X_3 - X_1) \simeq \mathbb{C}[X_1, X_2, X_1/X_2].$$

$\mathbb{C}[X_1, X_2, X_1/X_2]$ is an algebraic extension of $\mathbb{C}[X_1, X_2]$; X_1/X_2 satisfies the minimal polynomial

$$q(Y) = X_2 Y - X_1.$$

The discriminant with respect to Y of q is

$$\mathscr{D}_Y(q) = X_2,$$

so W in Theorem 11.1 is $\mathbb{C}_{X_1}$ ($\subset \mathbb{C}_{X_1X_2} = V$). Above the point $X_1 = 0$ in $\mathbb{C}_{X_1}$ there are infinitely many points of V^*. Above any other point of $\mathbb{C}_{X_1}$ there are no points of V^*.

Now clearly any coordinate ring R is an algebraic extension of some polynomial ring $\mathbb{C}[X_1, \ldots, X_n]$. In view of Theorem 11.1 we see that this says, geometrically, that any irreducible variety may be looked at as a kind of near cover of $\mathbb{C}^n$. We call $X_1, \ldots, X_n$ a **transcendence base** of R. Although for a fixed co-ordinate ring R, any two transcendence bases have the same number of elements, R of course has many different transcendence bases; for instance in the ring $\mathbb{C}[X_1, X_2, X_1/X_2]$ of Example 11.3, we chose X_1 and X_2 to play this role. But we could just as well have selected X_1/X_2 and X_2. Then X_1 is algebraic over $\mathbb{C}[X_1/X_2, X_2]$ since $(X_1)^1 - X_2 \cdot X_1/X_2(X_1)^0 = 0$. In general, as we change the transcendence base we change the way in which the variety is a cover of $\mathbb{C}^n$. Thus with respect to the new base $Y_1 = X_1/X_2$, $Y_2 = X_2$, our affine model of the coordinate ring of Example 11.3 becomes $Z = Y_1Y_2$, a hyperboloid set-theoretically projecting onto $\mathbb{C}_{Y_1Y_2}$. For certain choices of base the associated cover may be simpler than for others; in view of (11.1.2) of Theorem 11.1, we might expect to get a particularly pleasant situation if R turns out to be integral over $\mathbb{C}[X_1, \ldots, X_n]$.

Our next result is the important *normalization lemma*, which tells us we can in fact always choose a transcendence base $X_1, \ldots, X_n$ of R so that R is integral over $\mathbb{C}[X_1, \ldots, X_n]$. We shall begin by looking at an example which will point the way to a proof.

EXAMPLE 11.4. Consider the hyperbola $V = \mathrm{V}(XY - 1) \subset \mathbb{C}_{XY}$. There is no point of V π_Y-lying over the origin of $\mathbb{C}_X$; this fits in with the fact that $XY - 1 = 0$ does not define an integral equation for Y over $\mathbb{C}[X]$. Now Figure 1 suggests that if we tilt the Y-axis a bit, this bad behavior disappears relative to the new coordinate system—every line parallel to the Y'-axis in

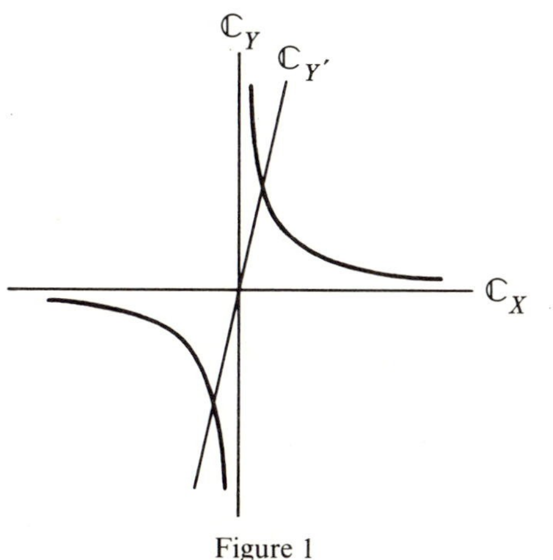

Figure 1

$\mathbb{R}^2$ intersects V in exactly two points. The new Y'-axis is the line $X' = X - aY = 0$, for some $a \neq 0$. The coordinate change from axes $\mathbb{C}_X$ and $\mathbb{C}_Y$ to $\mathbb{C}_X = \mathbb{C}_{X'}$ and $\mathbb{C}_{Y'}$ is given by

$$\begin{aligned} X' &= X - aY \qquad (a \neq 0), \\ Y' &= Y. \end{aligned}$$

Then the equation $XY - 1 = 0$ becomes

$$(X' + aY')Y' - 1 = 0, \quad \text{or } a(Y')^2 + X'Y' - 1 = 0. \tag{18}$$

Tilting the Y-axis in effect adds a leading Y'^2-term to $X'Y' - 1$, thus making Y' integral over $\mathbb{C}[X']$.

Essentially, our proof of the normalization lemma amounts to tilting enough of the axes to remove the bad behavior. Before stating this lemma, note that the coordinate axes $\mathbb{C}_{X_i}$ in $\mathbb{C}_{X_1,\ldots,X_n}$ are just the varieties $\mathsf{V}(X_1, \ldots, X_{i-1}, X_{i+1}, \ldots, X_n)$. If $X'_1, \ldots, X'_n$ are linearly independent $\mathbb{C}$-linear combinations of $X_1, \ldots, X_n$, then $\mathbb{C}[X_1, \ldots, X_n] = \mathbb{C}[X'_1, \ldots, X'_n]$, and X'_i are new coordinates in $\mathbb{C}^n$. Now in $\mathbb{C}_{X_1X_2}$, $X'_1 = X_1 - aX_2, X'_2 = X_2$ represents a tipping of $\mathbb{C}_{X_2}$. More generally, in $\mathbb{C}_{X_1,\ldots,X_n}$ the analogous coordinate change which tips one axis (say $\mathbb{C}_{X_n}$) and leaves the others fixed is

$$\begin{aligned} X'_1 &= X_1 - a_1X_n \\ &\ \vdots \\ X'_{n-1} &= X_{n-1} - a_{n-1}X_n \\ X'_n &= X_n \qquad\qquad (a_i \in \mathbb{C}). \end{aligned} \tag{19}$$

In an arbitrary coordinate ring

$$R = \mathbb{C}[x_1, \ldots, x_n] = \mathbb{C}[X_1, \ldots, X_n]/\mathfrak{p},$$

the relations

$$\begin{aligned} x'_1 &= x_1 - a_1x_n \\ &\ \vdots \\ x'_{n-1} &= x_{n-1} - a_{n-1}x_n \\ x'_n &= x_n \end{aligned}$$

are induced by the coordinate change given in (19) in the surrounding space $\mathbb{C}_{X_1,\ldots,X_n}$. Note that

$$\mathbb{C}[x_1, \ldots, x_n] = \mathbb{C}[x'_1, \ldots, x'_n].$$

We now state and prove the normalization lemma. Recall the notion of transcendence degree.

Lemma 11.5 (Normalization lemma). *If $R = \mathbb{C}[x_1, \ldots, x_n]$ has transcendence degree d over $\mathbb{C}$, then there are elements $y_1, \ldots, y_d$ in R such that R is integral over $\mathbb{C}[y_1, \ldots, y_d]$. The y_i may be chosen to be $\mathbb{C}$-linear combinations of the x_i.*

PROOF. If $d = n$, then we may take $y_i = x_i$, and the lemma is trivially true. Thus suppose without loss of generality that x_n is algebraic over $\mathbb{C}[x_1, \ldots, x_{n-1}]$. Let $q(X_1, \ldots, X_n)$ be a polynomial of lowest degree D (in $X_1, \ldots, X_n$) for which $q(x_1, \ldots, x_{n-1}, X_n)$ is a minimal polynomial of x_n over $\mathbb{C}(x_1, \ldots, x_{n-1})$. Now tilt $\mathbb{C}_{X_n}$. In the new coordinates X'_i given by (19), $q(X_1, \ldots, X_n)$ becomes

$$q(X'_1 + a_1X_n, \ldots, X'_{n-1} + a_{n-1}X_n, X_n), \tag{20}$$

which is still of degree D. Substituting x'_i for X'_i yields a polynomial for x_n over $\mathbb{C}[x'_1, \ldots, x'_{n-1}]$; *if we can make the coefficient of ${X_n}^D$ a nonzero constant, x_n will be integral over $\mathbb{C}[x'_1, \ldots, x'_{n-1}]$.* Now the coefficient of ${X_n}^D$ in (20) is just the coefficient of ${X_n}^D$ in (20) with each of $X'_1, \ldots, X'_{n-1}$ set equal to 0—that is, of ${X_n}^D$ in $q(a_1X_n, \ldots, a_{n-1}X_n, X_n)$. If $h(X_1, \ldots, X_n)$ is the homogeneous polynomial of all D-degree terms of $q(X_1, \ldots, X_n)$, then the D-degree term of $q(a_1X_n, \ldots, a_{n-1}X_n, X_n)$ is $h(a_1X_n, \ldots, a_{n-1}X_n, X_n) = {X_n}^D h(a_1, \ldots, a_{n-1}, 1)$; hence the coefficient of ${X_n}^D$ is $h(a_1, \ldots, a_{n-1}, 1)$ which may be made nonzero for an appropriate choice of the a_i, since h is not the zero polynomial. Hence if $d = n - 1$, we have proved the lemma with $y_1 = x'_1, \ldots, y_{n-1} = x'_{n-1}$.

Now if $n = d - 2$ choose $x'_1, \ldots, x'_{n-1}$ as above, and with no loss of generality assume that x'_{n-1} is algebraic over $x'_1, \ldots, x'_{n-2}$. The same argument as above shows that x'_{n-1} is integral over $\mathbb{C}[x''_1, \ldots, x''_{n-2}]$ for analogous $\mathbb{C}$-linear combinations $x''_1, \ldots, x''_{n-2}$ of $x'_1, \ldots, x'_{n-1}$. (Since the x'_i are linear combinations of $x_1, \ldots, x_n$, the x''_i are then also linear combinations of $x_1, \ldots, x_n$.) Now R is integral over $\mathbb{C}[x'_1, \ldots, x'_{n-1}]$, which in turn is integral over $\mathbb{C}[x''_1, \ldots, x''_{n-2}]$; hence R is integral over $\mathbb{C}[x''_1, \ldots, x''_{n-2}]$, so our lemma is proved for $d = n - 2$ with $y_1 = x''_1, \ldots, y_{n-2} = x''_{n-2}$. Proceeding by induction, we establish the lemma for any transcendence degree d, $0 \leqslant d \leqslant n$. □

EXERCISES

11.1 Suppose the coordinate ring R^* is a pure transcendental extension of a coordinate ring R. For these rings, establish the following:

(a) Extension on $\mathscr{I}(R)$ preserves intersections (cf. Exercise 10.1(c)).

(b) If $\mathfrak{p}$ is prime in R, then $\mathfrak{p}^e$ is prime in R^*.

(c) Let $\mathfrak{a} \subset R$ be closed, and write $\mathfrak{a}^e = \mathfrak{a}[Y_1, \ldots, Y_m] \subset R[Y_1, \ldots, Y_m] = R^*$, where the Y_i are independent transcendental elements over R. Show that arbitrary powers of elements in $R^* \backslash \mathfrak{a}^e$ are in $R^* \backslash \mathfrak{a}^e$ and thus conclude that $(\ \ )^e$ maps $\mathscr{I}(R)$ to $\mathscr{I}(R^*)$. (Thus $(\ \ )^e = \sqrt{(\ \ )^e}$ if R^* is pure transcendental over R.)

(d) If $\mathfrak{a} \in \mathscr{I}(R)$ has the unique irredundant decomposition $\mathfrak{a} = \mathfrak{p}_1 \cap \ldots \cap \mathfrak{p}_r$, show that $\mathfrak{a}^e$ has the unique irredundant decomposition ${\mathfrak{p}_1}^e \cap \ldots \cap {\mathfrak{p}_r}^e$.

11.2 With notation and assumptions as in (11.1.2) of Theorem 11.1, show that for any positive integer $k < D$, there is a subvariety $W_k \subsetneqq V$ of V so that there are $\leqslant k$ distinct points of V^* above each point in W_k, and $>k$ distinct points of V^* above each point of $V \backslash W_k$.

11.3 Let $R = \mathbb{C}[X, 1/X^2]$. Use the proof of the normalization lemma to find:
(a) a $\mathbb{C}$-linear combination y_1 of X and $1/X^2$ so that R is integral over $\mathbb{C}[y_1]$;
(b) an integral equation for $1/X^2$ over $\mathbb{C}[y_1]$.

CHAPTER IV

Varieties of arbitrary dimension

1 Introduction

In this chapter we generalize some of the results of Chapter II to varieties of arbitrary dimension. Let us begin by putting into perspective some of the results of Chapter II. A curve C in $\mathbb{C}^2$ or $\mathbb{P}^2(\mathbb{C})$ defines a topological space, its topology being induced from that of $\mathbb{C}^2$ or $\mathbb{P}^2(\mathbb{C})$. If, in particular, C is nonsingular, then it is a topological 2-manifold (cf. the discussion after Theorem 2.7 of Chapter I). In fact, it is actually an *analytic* manifold, in the sense that all the homeomorphisms $\phi_\beta{}^{-1} \circ \phi_\alpha$ of Definition 9.3 of Chapter II are analytic.

Now one can study curves locally or globally. In a *local* study, attention is focused on properties in the neighborhood of a point. An example of this is Theorem 4.13 of Chapter II giving the topological structure of any plane curve in the neighborhood of a point. In fact, in proving this theorem we actually obtained the *analytic* structure of a plane curve at any point.

In a *global* study one asks for results about the overall structure (in this case, topological or analytic). For example Theorem 8.4 of Chapter II, which says that any curve C in $\mathbb{P}^2(\mathbb{C})$ is connected, is a global theorem, in the sense that one cannot deduce that a topological space is connected by looking only at arbitrarily small neighborhoods of each point. Orientability is another global notion—small neighborhoods around any point of $\mathbb{P}^2(\mathbb{R})$ are topologically just like those of a sphere, yet the sphere is orientable and $\mathbb{P}^2(\mathbb{R})$ is not. (The reader may check that $\mathbb{P}^2(\mathbb{R})$ is not orientable by pushing an oriented circuit across the line at infinity.) Theorem 2.7 of Chapter I is another global result; it describes the overall topological structure of any curve in $\mathbb{P}^2(\mathbb{C})$. And for a nonsingular plane curve $C \subset \mathbb{P}^2(\mathbb{C})$ we have a particularly beautiful global result—we know that C is a compact connected orientable 2-manifold, its genus being very easily determined by the genus formula in

Theorem 10.1 of Chapter II; the genus then determines the overall topological structure of the curve in the sense that any two nonsingular curves of the same genus are homeomorphic. The manifold property is a local property, but compactness, connectedness, and orientability are all global. The genus (or, equally well, the *Euler characteristic*) is a "global invariant."

It is natural to seek generalizations of these results to an irreducible variety V of arbitrary dimension in $\mathbb{P}^n(\mathbb{C})$. Some of these generalizations are quite straightforward. For instance, showing that V is compact is trivial. Also, any irreducible variety V is connected (Theorem 5.1); this is easy to prove once connectedness is established for irreducible curves. See (5.2). The equivalence between smoothness and nonsingularity is established for arbitrary dimension in Theorem 4.1. And for a nonsingular variety, the definition and proof of orientability is straightforward (Theorem 5.3). The dimension theorem for curves (Theorem 6.1 of Chapter II) is another global result—two curves must intersect *somewhere in* $\mathbb{P}^2(\mathbb{C})$. The general dimension theorem for varieties $V, W \subset \mathbb{P}^n(\mathbb{C})$ says that $\operatorname{cod}(V \cap W) \leqslant \operatorname{cod} V + \operatorname{cod} W$. Hence if $\dim V + \dim W \geqslant n$, V and W must intersect somewhere in $\mathbb{P}^n(C)$; furthermore, at each point of intersection the dimension of $V \cap W$ is at least $\dim V + \dim W - n$. We prove these facts in Theorem 3.1, Corollary 3.2, and Theorem 3.8.

In Chapter I we briefly looked at Bézout's theorem for curves in $\mathbb{P}^2(\mathbb{C})$, and saw how it represents an important generalization of the fundamental theorem of algebra. In this chapter we prove Bézout's theorem for varieties in $\mathbb{P}^n(\mathbb{C})$ (Theorem 7.1); this both generalizes the fundamental theorem of algebra and extends the dimension theorem. In Section 6 we develop the fundamental notions of order and multiplicity used in stating and proving Bézout's theorem.

Now when we try to generalize to varieties in $\mathbb{P}^n(\mathbb{C})$ the structure results of curves in Chapter II, we meet a much more difficult problem, both at the local and global levels. For instance locally, one can have very bad singularities. (For example, if $V \subset \mathbb{C}^n$ is any affine variety, its homogenization in $\mathbb{C}^{n+1}$ has a singularity at the origin which has essentially the same complexity as V itself.) But even if one assumes a trivial local structure (for instance, if the variety is nonsingular) the corresponding global problem is no easy matter. The problem then becomes one of finding invariants that do for arbitrary nonsingular varieties what the genus does for nonsingular curves. Much work has been done towards finding the precise topological (and also analytic) structure; however, given a prime ideal $\mathfrak{p} \subset \mathbb{C}[X_1, \ldots, X_n]$ (corresponding to an irreducible polynomial $p(X) \in \mathbb{C}[X]$ in the case of a plane curve), there is no analogously simple formula yielding a complete set of invariants which fully determines the topological structure of the corresponding underlying manifold in $\mathbb{P}^n(\mathbb{C})$. In the case of curves, one invariant (for instance, the genus) suffices. But for higher dimensional manifolds, it in general requires more than one such number, as a reader familiar with Betti numbers will recognize. Even though Betti numbers, homology, cohomology, and homotopy all shed light on the topological structure, such information need not entirely determine the global topology. In fact, A. Markov ([Markov]) has proved that for closed, connected, compact manifolds of real dimension ≥ 4, it is impossible to get a general algorithm producing a complete set of topological invariants (so that two manifolds are homeomorphic iff they have the same set of invariants).

In this chapter, then, we generalize to varieties in $\mathbb{C}^n$ and in $\mathbb{P}^n(\mathbb{C})$ some of the results of Chapters I and II which can be proved in a chapter of reasonable length.

We remark that in Section I,4, we mentioned the usefulness of working over ground fields other than $\mathbb{C}$. In fact, one of the most productive directions recently taken in algebraic geometry is, even more generally, to replace the coordinate ring $k[x_1, \ldots, x_n]$ by an arbitrary Noetherian commutative ring R with identity; one can then let maximal ideals of R be "zero-dimensional points" of a kind of "variety," and can let the other prime ideals of R represent "higher-dimensional points," which are essentially "higher-dimensional irreducible subvarieties." It turns out that much of algebraic geometry can be treated in such a purely ring-theoretic fashion. Because of the growing importance of these ideas, we include in this and the next chapter some examples to give the reader some familiarity with them, and to let the reader see some translations from one level to another. In this chapter, the concept of *dimension* provides a nice example of how one can shift from the purely geometric, visual, notion to a purely algebraic notion; the algebraic form can then be used in settings far removed from the limited range of the "visual" definition, and the algebraic form can provide a more geometric way of looking at areas formerly thought to have little geometric content. As examples of the ring-theoretic viewpoint, in Section 2 of this chapter we include a ring-theoretic characterization of dimension; in Section V,4, we show how nonsingularity translates into ring-theoretic terms; and in Section V,5, the ring-theoretic forms of both dimension and nonsingularity are used to provide an example *par excellence* of a complete dictionary between ideals and geometric objects (chains).

2 Dimension of arbitrary varieties

In this section we look at the notion of dimension of varieties. We do this from three viewpoints. We begin by looking at dimension in a very geometric way, for complex varieties. (Definitions 2.3 and 2.7). For an irreducible variety, one can also look at dimension as the transcendence degree of its coordinate ring. For complex varieties, this is shown to be equivalent to the first definition (Theorem 2.14); it is perhaps less pictorial, but has the advantage of yielding a nice definition for varieties over any field. Finally, we show how to base the notion of dimension on sequences of prime ideals. This can be used as a definition of dimension for yet more general kinds of "variety" (cf. Section V,5) and is useful in purely ring-theoretic treatments of algebraic geometry.

We now turn to the geometric definition of dimension. First, we shall take for granted certain well-known facts about dimension: We assume that $\mathbb{C}^n$ has complex dimension n at each of its points, and that open subsets of $\mathbb{C}^n$, as well as all homeomorphic images of open subsets of $\mathbb{C}^n$, are of complex dimension n.

In defining dimension of a variety, we begin with a definition of dimension at a point of a variety, then extend this to a definition for the entire variety. Now in defining dimension at a point, we shall see that there are certain points having a V-open neighborhood which is indeed a homeomorphic image of an open subset of some $\mathbb{C}^r$—in fact, each sufficiently small such neighborhood is a subset of the graph of a complex-analytic function; this will follow from the general implicit mapping theorem (Theorem 3.5 of Chapter II), which we prove in this section. We then show that any point of V is a limit point of these "nice" ones, and then define the dimension of V at P to be the largest of all such limiting dimensions (Definitions 2.3 and 2.7). For example, the complex dimension which one would naturally assign at (0, 0, 0) to the subvariety $\mathbb{C}_X \cup \mathbb{C}_{YZ}$ of $\mathbb{C}_{XYZ}$ is 2; Definitions 2.3 and 2.7 are just generalizations of this idea. Because of the fundamental role the general implicit mapping theorem plays in this definition, we prove this theorem next.

Proof of Theorem 3.5 of Chapter II. We first prove the theorem for the case $q = r = 1$; then more generally, for any $q = r$. We then consider the full theorem $q \geqslant r$. (Of course, we never have $q < r$.)

Case $q = r = 1$. With notation as in the statement of the theorem, let $f = f_1 = f_q$; say $f = f(X_1, \ldots, X_{n-1}, Y)$ with $f_Y(0, \ldots, 0) \neq 0$. The proof for this case is, aside from minor modifications, the same as our proof of Theorem 3.6 of Chapter II. For the hypothesis

$$f(0, \ldots, 0) = 0, \qquad f_Y(0, \ldots, 0) \neq 0$$

expresses that the function $f(0, \ldots, 0, Y)$ in Y alone has $Y = 0$ as a zero of multiplicity one. Then the proof of Theorem 3.6 of Chapter II works essentially verbatim, with $f(0, \ldots, 0, Y)$, $(c_1, \ldots, c_{n-1})$, and $(X_1, \ldots, X_{n-1})$ in place of $p(0, Y)$, (c), and (X), respectively.

Case $q = r$. The theorem in this case gives us certain information about the zero-set $V = \mathsf{V}(f_1, \ldots, f_q) \subset \mathbb{C}^n$. In proving it we shall replace the functions f_i by other functions (which still define the same zero-set), so that the Jacobian becomes simplified; in this way we shall reduce the theorem's proof to a relatively simple induction.

We modify the functions f_i in two different ways:

First, if A is any nonsingular $(q \times q)$ matrix (entries in $\mathbb{C}$), then f_i may be replaced by g_i, where

$$(g_1, \ldots, g_q)^t = A(f_1, \ldots, f_q)^t;$$

here t denotes transpose. Each g_i is thus an invertible linear combination of the functions f_i, and $\mathsf{V}(f_1, \ldots, f_q) = \mathsf{V}(g_1, \ldots, g_q)$.

Second, if $\mathbb{C}^n$ has coordinates $X = (X_1, \ldots, X_n)$, then any nonsingular $(n \times n)$ matrix B over $\mathbb{C}$ induces a coordinate change to $(Z_1, \ldots, Z_n) = Z$ via

$$X = ZB;$$

thus

$$\mathsf{V}(g_1(X), \ldots, g_q(X)) = \mathsf{V}(g_1(ZB), \ldots, g_q(ZB)) = \mathsf{V}(g_1^*(Z), \ldots, g_q^*(Z)).$$

Hence we may replace the $f_i(X)$ by the $g_i^*(Z)$.

To get our desired simplification, we choose A and B as follows: Let $J(f)_{X=0}$ be as in the theorem's statement; write this as J_0. Then by elementary matrix theory, we may let A and B be nonsingular matrices of order q and n, respectively, so that in the $(q \times n)$ matrix AJ_0B, the last q columns of AJ_0B form the $(q \times q)$ identity matrix, and the first $n - q$ columns are all zero columns. If we let $(g_1, \ldots, g_q)^t = A(f_1, \ldots, f_q)^t$, and $X = ZB$, then

$$A\left(\frac{\partial f_i}{\partial X_j}\right) = \left(\frac{\partial g_i}{\partial X_j}\right),$$

and therefore

$$\begin{aligned} AJ(f)B &= \left(\frac{\partial g_i}{\partial X_j}\right)B \\ &= \left(\sum_k \frac{\partial g_i^*}{\partial Z_k} \cdot \frac{\partial Z_k}{\partial X_j}\right)B \qquad \text{(chain rule)} \\ &= \left(\frac{\partial g_i^*}{\partial Z_k}\right)\left(\frac{\partial Z_k}{\partial X_j}\right)B \\ &= \left(\frac{\partial g_i^*}{\partial Z_k}\right). \end{aligned}$$

(The last equality follows from the fact that $X = ZB \Rightarrow (\partial Z_k/\partial X_j)B = (\partial X_k/\partial X_j) = I$.) Therefore at $(X) = (Z) = (0)$, we have

$$AJ_0B = \left(\frac{\partial g_i^*}{\partial Z_j}\right)_{Z=(0)}$$

Since $\mathsf{V}(f_1, \ldots, f_q) = \mathsf{V}(g_1^*, \ldots, g_q^*)$, we may just as well have used the functions g_i^* to begin with; thus, writing f_i for g_i^*,

(2.1) *We may assume that the last q columns of $J(f)_{X=(0)}$ form the unit matrix.*

We now complete the proof of the case $q = r$ using induction on q. We have already established it for $q = 1$. Now suppose it has been established for $q - 1$; let $f_1, \ldots, f_{q-1}$ be $q - 1$ functions. To prove it for q, let f_q be a qth function. By (2.1), we may assume that $(\partial f_q/\partial X_n)(0, \ldots, 0) \neq 0$, so the theorem for a single function tells us that there are neighborhoods U^{n-1} and

U^1 of (0) in $\mathbb{C}_{X_1, \ldots, X_{n-1}}$ and $\mathbb{C}_{X_n}$ respectively, and a function $\phi_q(X_1, \ldots, X_{n-1})$ holomorphic in U^{n-1} such that in $U^{n-1} \times U^1$ the zero-set of f_q is the graph of the function $X_n = \phi_q(X_1, \ldots, X_{n-1})$.

Now consider

$$h_i(X_1, \ldots, X_{n-1}) = f_i(X_1, \ldots, X_{n-1}, \phi_q(X_1, \ldots, X_{n-1}))$$

where $i = 1, \ldots, q - 1$; each of these $q - 1$ functions is holomorphic in a neighborhood of $(0) \in \mathbb{C}^{n-1}$. The fact that $X_n = \phi_q(X_1, \ldots, X_{n-1})$ near (0) means that a point $(c) = (c_1, \ldots, c_n)$ near $(0) \in \mathbb{C}^n$ is in $\mathsf{V}(f_1, \ldots, f_q)$ iff $c_n = \phi_q(c_1, \ldots, c_{n-1})$ and $h_i(c_1, \ldots, c_{n-1}) = 0$ for $i = 1, \ldots, q - 1$ (that is, iff (c) is in $\mathsf{V}(f_q)$, and in $\mathsf{V}(f_1, \ldots, f_{q-1})$). So far, we have made the variable X_n "explicit." Our aim is to represent $\mathsf{V}(f_1, \ldots, f_q)$ locally as the graph of a function from U^{n-q} to U^q, so we want to simultaneously make all the variables $X_{n-q+1}, \ldots, X_n$ explicit.

To do this, we apply our induction hypothesis to the functions h_i; we show that our simplifying assumption (2.1) on $f_1, \ldots, f_q$ implies that a corresponding simplifying assumption holds for the new functions h_i. First, by the chain rule, we have

$$\begin{aligned} \frac{\partial h_i}{\partial X_j} &= \frac{\partial f_i}{\partial X_1} \cdot \frac{\partial X_1}{\partial X_j} + \ldots + \frac{\partial f_i}{\partial X_{n-1}} \cdot \frac{\partial X_{n-1}}{\partial X_j} + \frac{\partial f_i}{\partial X_n} \cdot \frac{\partial \phi_q}{\partial X_j} \\ &= \frac{\partial f_i}{\partial X_j} \cdot 1 + \frac{\partial f_i}{\partial X_n} \cdot \frac{\partial \phi_q}{\partial X_j}. \end{aligned}$$

This, together with the hypothesis that $(\partial f_i/\partial X_{n-q+j})(0) = \delta_{ij}$ (where $\delta_{ij} = 1$ if $i = j$ and 0 otherwise), implies that $(\partial f_i/\partial X_n)(0) = 0$ for $i = 1, \ldots, q - 1$. Hence

$$\frac{\partial h_i}{\partial X_{n-q+j}}(0) = \frac{\partial f_i}{\partial X_{n-q+j}}(0) = \delta_{ij},$$

so (2.1) is satisfied for $h_1, \ldots, h_{q-1}$. Hence there are functions $\phi_1, \ldots, \phi_{q-1}$ holomorphic in a neighborhood of $(0) \in \mathbb{C}_{X_1, \ldots, X_{n-q}}$ $(n - q = (n - 1) - (q - 1))$ so that in a neighborhood of $(0) \in \mathbb{C}_{X_1, \ldots, X_{n-1}}$, $\mathsf{V}(h_1, \ldots, h_{q-1})$ is the graph of the function $(\phi_1, \ldots, \phi_{q-1})$. Thus near $(0) \in \mathbb{C}^n$, $\mathsf{V}(f_1, \ldots, f_q)$ $= \mathsf{V}(h_1, \ldots, h_{q-1}, f_q)$ is the graph of

$$(\phi_1, \ldots, \phi_{q-1}, \phi_q(X_1, \ldots, X_{n-q}, \phi_1, \ldots, \phi_{q-1})).$$

We have thus proved the theorem when $q = r$.

Case $q > r$. Suppose that the first r rows of $(\partial f_i/\partial X_j)$ are linearly independent; denote $\mathsf{V}(f_1, \ldots, f_r)$ by V. We shall show that $\mathsf{V}(f_1, \ldots, f_q) = V$. Thus since near (0), V is the graph of an analytic function $(\phi_1, \ldots, \phi_r): U^{n-r} \to U^r$ (notation as before), near (0) $\mathsf{V}(f_1, \ldots, f_q)$ would be the graph of the same analytic function, which would prove this case.

We prove $\mathsf{V}(f_1, \ldots, f_q) = V$ by showing that for each $i > r$, $\mathsf{V}(f_i) \supset V$—that is, f_i vanishes on V for $i > r$. But it is easy to evaluate these f_i on V, since V is described near (0) by $(\phi_1, \ldots, \phi_r)$. The evaluated function is just

$$f_i(X_1, \ldots, X_{n-r}, \phi_1(X_1, \ldots, X_{n-r}), \ldots, \phi_r(X_1, \ldots, X_{n-r}));$$

we shall denote this by F_i. *We want to show that* $F_i \equiv 0$ *for each* $i > r$. (Note that trivially, $F_i \equiv 0$ for $i \leqslant r$).

Since $F_i(0) = 0$, it suffices to prove that for each $i > r$,

$$\frac{\partial F_i}{\partial X_1} = \ldots = \frac{\partial F_i}{\partial X_{n-r}} = 0$$

throughout a neighborhood of $(0) \in \mathbb{C}_{X_1, \ldots, X_{n-r}}$, since then each F_i would be the constant 0. For this, we need only use the chain rule together with the fact that the last $q - r$ rows of the Jacobian matrix at (0) are linear combinations of the first r, say

$$\frac{\partial f_i}{\partial X_k} = \sum_{l=1}^{r} a_{il} \frac{\partial f_l}{\partial X_k} \qquad (i = r + 1, \ldots, q).$$

We thus have, for $i > r$ and $1 \leqslant j \leqslant n - r$:

$$\begin{aligned}
\frac{\partial F_i}{\partial X_j} &= \sum_{k=1}^{n-r} \frac{\partial f_i}{\partial X_k} \frac{\partial X_k}{\partial X_j} + \sum_{k=n-r+1}^{n} \frac{\partial f_i}{\partial X_k} \frac{\partial \phi_{k+r-n}}{\partial X_j} \\
&= \left(\sum_{l=1}^{r} a_{il} \frac{\partial f_l}{\partial X_j}\right) \cdot 1 + \sum_{k=n-r+1}^{n} \left[\left(\sum_{l=1}^{r} a_{il} \frac{\partial f_l}{\partial X_k}\right) \cdot \frac{\partial \phi_{k+r-n}}{\partial X_j}\right] \\
&= \sum_{l=1}^{r} a_{il} \left(\frac{\partial f_l}{\partial X_j} + \sum_{k=n-r+1}^{n} \frac{\partial f_l}{\partial X_k} \cdot \frac{\partial \phi_{k+r-n}}{\partial X_j}\right) = \sum_{l=1}^{r} a_{il} \frac{\partial F_l}{\partial X_j},
\end{aligned}$$

where all the partials of f_l are evaluated at $(X_1, \ldots, X_{n-r}, \phi_1, \ldots, \phi_r)$. The last equality is simply the chain rule applied to $\partial F_l / \partial X_j$. Now for $l = 1, \ldots, r$, $F_l \equiv 0$, so $\partial F_l / \partial X_j \equiv 0$ for $l = 1, \ldots, r$. Hence for each $i > r$,

$$\frac{\partial F_i}{\partial X_1} = \ldots = \frac{\partial F_i}{\partial X_{n-r}} = 0$$

throughout a neighborhood of $(0) \in \mathbb{C}_{X_1, \ldots, X_{n-r}}$, as desired. □

Now that we have proved this theorem, we can easily show that any variety in $\mathbb{P}^n(\mathbb{C})$ or $\mathbb{C}^n$ has a dimension at each of its points. We begin with the following convenient

Notation 2.2. Let $V \subset \mathbb{C}_{X_1, \ldots, X_n}$ be a variety, and let P be any point of V. We denote by

$$\text{rank}(J(V)_P)$$

the number of linearly independent rows of the *Jacobian array* $(\partial p_\alpha/\partial X_1, \ldots, \partial p_\alpha/\partial X_n)_P$ of polynomials evaluated at P; there are infinitely many rows (corresponding to some indexing by α of the polynomials in $\mathsf{J}(V) \subset \mathbb{C}[X_1, \ldots, X_n]$), and n columns. Note that this rank is the same as the rank of any $(m \times n)$ submatrix, where $\{p_1, \ldots, p_m\}$ is a basis of $\mathsf{J}(V)$ (cf. Definition 3.2 of Chapter III). In the sequel it will be convenient to denote any such Jacobian array by $J(V)$, though, of course, the order in which the rows are written is not uniquely determined.

Our first definition of dimension will be based on the notion of "local analytic manifold points" of a variety. A **local analytic manifold point of a variety** V is any point Q of V near which V can be represented, as in Theorem 3.5 of Chapter II, as the graph of an analytic function. (Thus any point Q having a V-neighborhood throughout which rank $(J(V))$ is constant, is such a point.) Note that there are local analytic manifold points Q arbitrarily near each point P of a variety V in $\mathbb{C}^n$ or $\mathbb{P}^n(\mathbb{C})$. To see this, let r be the largest integer that $\text{rank}(J(V)_Q)$ attains, as Q runs over all points of V arbitrarily close to P. Let Q_0 be any point such that $\text{rank}(J(V)_{Q_0}) = r$. This rank cannot increase at points of V close to Q_0, for Q_0 sufficiently close to P; nor can it decrease, by continuity of the entries of any r linearly independent rows of the array. Hence the rank is constant throughout a neighborhood in V of Q_0, so the hypotheses of Theorem 3.5 of Chapter II are satisfied.

Definition 2.3. Let V be a nonempty variety in $\mathbb{C}^n$. The **(complex) dimension of V at a point** $P \in V$, written $\dim_P V$, is $\max_Q(\dim_Q V)$, or equivalently, $n - \min_Q(\text{rank}(J(V)_Q))$, where Q ranges over the local analytic manifold points of V arbitrarily near P, and where $\dim_Q V$ is the complex dimension of the part of V near Q. The **dimension** of V, written $\dim V$, is $\max_{P \in V}(\dim_P V)$. By convention, $\dim \varnothing = -1$.

In view of our comments above, we have

Theorem 2.4. *If $V \subset \mathbb{C}^n$ is any variety, then every point P of V has a dimension, as does V itself.*

We may also define dimension for projective varieties. Certainly the dimension at a point P of an affine variety should not change simply by taking the projective completion of the variety. Similarly, we would expect that the dimension at a point $P \in V \subset \mathbb{P}^n(\mathbb{C})$ should agree with the dimension of any dehomogenization of V containing P. We connect this idea with the above affine definition of dimension as follows. Let $\mathsf{H}(V)$ be the homogeneous variety in $\mathbb{C}^{n+1}$ corresponding to $V \subset \mathbb{P}^n(\mathbb{C})$, and let L_P be the 1-subspace of $\mathbb{C}^{n+1}$ corresponding to P. If H is any (complex) hyperplane of $\mathbb{C}^{n+1}$ whose intersection with L_P is a point other than $0 \in \mathbb{C}^{n+1}$, then H defines a dehomogenization of V containing P. The intersection $\mathsf{H}(V) \cap H$ is a variety in H

(that is, in a copy of $\mathbb{C}^n$) representing this dehomogenization of V, and $L_P \cap H$ is the point of $\mathsf{H}(V) \cap H$ corresponding to $P \in V$. The intersection $\mathsf{H}(V) \cap H$ is an **affine representative of V relative to H**, and $L_P \cap H$ is an **affine representative of P relative to H**. In a moment we shall prove:

> **(2.5)** The dimension of the affine representative $\mathsf{H}(V) \cap H$ at $L_P \cap H$ is independent of the choice of any H such that $0 \notin H$ and $\varnothing \neq L_P \cap H$.

This shows that we can base a definition of dimension of projective varieties on Definition 2.3. One can equally well define dimension using the homogenization $\mathsf{H}(V)$ instead of V. Since 1-subspaces of $\mathbb{C}^{n+1}$ correspond to projective points in V, we might guess that $\mathsf{H}(V)$, as an affine variety, has dimension one greater than that of the projective variety V. This is true; this and (2.5) easily follow from the following lemma:

Lemma 2.6. *With notation as immediately above and as in Notation 2.2, we have*

$$\operatorname{rank}(J(\mathsf{H}(V))_{L_P \cap H}) = \operatorname{rank}(J(\mathsf{H}(V) \cap H)_{L_P \cap H}),$$

where $\mathsf{H}(V) \cap H$ is regarded as a subvariety of H, where $0 \notin H$, and where $L_P \cap H \neq \varnothing$.

PROOF. Since these ranks are unaffected by any nonsingular change of coordinates in $\mathbb{C}^{n+1}$, we may assume that H is defined by $X_{n+1} = 1$, and that $L_P \cap H = (0, \ldots, 0, 1)$. Since relative to these coordinates, any Jacobian array for $\mathsf{H}(V) \cap H \subset \mathbb{C}^n$ consists of just the first n columns of the corresponding array for $\mathsf{H}(V) \subset \mathbb{C}^{n+1}$, to prove the lemma, it suffices to show that in these coordinates, the $(n+1)^{\text{st}}$ column of any Jacobian array for $\mathsf{H}(V)$, is zero at the point $(0, \ldots, 0, 1)$. For this, let $p = p(X_1, \ldots, X_{n+1})$ be any polynomial in $\mathsf{J}(\mathsf{H}(V))$; to show that $p_{X_{n+1}}(0, \ldots, 0, 1) = 0$, use the familiar Euler Theorem for homogeneous polynomials:

$$(\deg p)p = X_1 p_{X_1} + \ldots + X_{n+1} p_{X_{n+1}}.$$

Since $p(0, \ldots, 0, 1) = 0$, we have, at $(0, \ldots, 0, 1)$,

$$0 = 0p_{X_1} + \ldots + 0p_{X_1} + 1p_{X_{n+1}} = p_{X_{n+1}}. \qquad \square$$

PROOF OF (2.5). Proving (2.5) is equivalent to showing that

$$\operatorname{rank}(J(\mathsf{H}(V) \cap H)_{L_P \cap H})$$

is independent of the choice of any H satisfying $0 \notin H$ and $L_P \cap H \neq \varnothing$. For all such H, these ranks are equal to rank $(J(\mathsf{H}(V))_{L_P \cap H})$, and they are all the same since $\mathsf{H}(V)$ is homogeneous, which obviously implies that the rank of $J(\mathsf{H}(V))$ is constant on points (other than the origin) of any 1-subspace L_P of $\mathsf{H}(V)$. $\square$

From Lemma 2.6, it is now easily seen that for any H as above, and any $Q \in \mathsf{H}(V) \cap H$, $\dim_Q \mathsf{H}(V)$ is one greater than $\dim_Q(\mathsf{H}(V) \cap H)$.

Using Definition 2.3 for the dimension of affine varieties, we now define the dimension of projective varieties as follows:

Definition 2.7. Let $V \subset \mathbb{P}^n(\mathbb{C})$ be a nonempty variety and let P be any point of V. The **dimension of V at P**, written $\dim_P V$, is the dimension of any affine representative of V at the corresponding affine representative of P. Or, equivalently, $\dim_P V = (\dim_Q \mathsf{H}(V)) - 1$, where $\mathsf{H}(V)$ is the homogeneous variety in $\mathbb{C}^{n+1}$ corresponding to V, and Q is any nonzero point on the 1-subspace of $\mathbb{C}^{n+1}$ corresponding to P. The **dimension of V**, written $\dim V$, is $\max_{P \in V}(\dim_P V)$. (Again, we define $\dim \varnothing$ to be -1.)

We then obviously have

Theorem 2.8. *If $V \subset \mathbb{P}^n(\mathbb{C})$ is a projective variety, then V has a dimension at each point and a dimension.*

If V is irreducible, we can prove more:

Theorem 2.9. *If V is any irreducible variety of $\mathbb{P}^n(\mathbb{C})$ or $\mathbb{C}^n$, then every point of V has the same dimension.*

Corollary 2.10. *Let V be a variety of $\mathbb{P}^n(\mathbb{C})$ or $\mathbb{C}^n$, and let P be any point of V. Then $\dim_P V$ is the largest of the dimensions of those irreducible components of V which contain P; $\dim V$ is the largest of the dimensions of V's irreducible components.*

Proof of Theorem 2.9. Since for a projective variety $V \subset \mathbb{P}^n(\mathbb{C})$, any two points of V lie in some one affine representative of V, it clearly suffices to assume that V is affine. Thus let $V \subset \mathbb{C}^n$, and let $r = \max_{P \in V}(\operatorname{rank}(J(V)_P))$. Then our theorem says that the set of points of V where the rank is r, is dense in V. Now the set of points P where $\operatorname{rank}(J(V)_P)$ is strictly less than r forms a proper subvariety of V, since these points form the zero-set of the collection of $(r \times r)$ minors of our "$(\infty \times n)$" array, and each such minor is a polynomial in $X_1, \ldots, X_n$. Hence it suffices to show that for any subvariety W of an irreducible variety V, $V \backslash W$ is dense in V; or, equivalently, if a subvariety V' of an irreducible variety V contains an open set of V, then $V = V'$. This follows at once from Theorem 2.11.

Theorem 2.11 (Identity theorem for irreducible varieties). *Let V_1, V_2 be irreducible varieties (in $\mathbb{P}^n(\mathbb{C})$ or in $\mathbb{C}^n$), and let U be any open set (in $\mathbb{P}^n(\mathbb{C})$ or $\mathbb{C}^n$). If*

$$V_1 \cap U = V_2 \cap U \neq \varnothing,$$

then $V_1 = V_2$.

PROOF OF THEOREM 2.11. Since any variety in $\mathbb{P}^n(\mathbb{C})$ may be represented by an affine variety in $\mathbb{C}^{n+1}$, we may without loss of generality consider only the affine case. For this, it suffices to prove that any polynomial in $\mathbb{C}[X_1, \ldots, X_n]$ which is zero on an open subset of V_1 is zero on all of V_1, for then, likewise, it is zero on all of V_2, hence $V_2 \subset V_1$. Similarly, $V_1 \subset V_2$, so V_1 would equal V_2.

Since the value at any point in V_1 of an arbitrary polynomial in $\mathbb{C}[X_1, \ldots, X_n]$ coincides with the value of that polynomial mod $\mathsf{J}(V_1)$, it is enough to show this: Let p be any element in V_1's coordinate ring $\mathbb{C}[X_1, \ldots, X_n]/\mathsf{J}(V_1) = \mathbb{C}[x_1, \ldots, x_n]$ which vanishes on an open subset of V_1; then p vanishes on all of V_1. Now from Theorem 2.4, any open set of V_1 contains some point (0) of dimension d, such that after renumbering coordinates if necessary, the part of V_1 near (0) is the graph of a function analytic in a neighborhood of $(0) \in \mathbb{C}_{X_1, \ldots, X_d}$ $(\subset \mathbb{C}_{X_1, \ldots, X_n})$. Hence the natural projection on $\mathbb{C}_{X_1, \ldots, X_d}$ of the part of V_1 near (0) is an open set of $\mathbb{C}_{X_1, \ldots, X_d}$. We want to show that $p = p(x_1, \ldots, x_n)$ is the zero polynomial. A point $(a_1, \ldots, a_n)$ is in V_1 iff $(x_1, \ldots, x_n) \to (a_1, \ldots, a_n)$ defines a $\mathbb{C}$-homomorphism of $\mathbb{C}[x_1, \ldots, x_n]$; by hypothesis, for each $(a_1, \ldots, a_n) \in V_1$ near (0), $p(a_1, \ldots, a_n) = 0$—that is, p is in the kernel of each such specialization of $\mathbb{C}[x_1, \ldots, x_n]$. It is easily seen that we may assume $\{x_1, \ldots, x_d\}$ is a transcendence base of $\mathbb{C}[x_1, \ldots, x_n]$. (Note that within some neighborhood N of $(0) \in \mathbb{C}^n$, above each point of $N \cap \mathbb{C}_{X_1, \ldots, X_d}$ there $\pi_{X_{d+1}, \ldots, X_n}$-lies just one point of V_1; a higher transcendence degree would yield, for any N, infinitely many points above most points of $N \cap \mathbb{C}_{X_1, \ldots, X_d}$.) If p were *not* the zero polynomial, it would satisfy a minimal equation

$$q_0 p^m + \ldots + q_m = 0, \quad \text{where} \quad q_i \in \mathbb{C}[x_1, \ldots, x_d]; \tag{1}$$

note that by minimality,

(2.12) q_m cannot be the zero polynomial.

Since $p(a_1, \ldots, a_d) = 0$, (1) implies that $q_m(a_1, \ldots, a_d) = 0$. But since $\{x_1, \ldots, x_d\}$ is a transcendence base $(a_1, \ldots, a_d)$ may be arbitrarily chosen in this specialization, so $q_m = 0$ throughout some neighborhood of $(0) \in \mathbb{C}_{X_1, \ldots, X_d}$. It is then easily proved that q_m is the zero polynomial, and this is a contradiction to (2.12). Therefore p is the zero polynomial in $\mathbb{C}[x_1, \ldots, x_n]$, which is what we wanted to prove. Hence Theorem 2.11 is proved, and therefore also Theorem 2.9. □

We next translate dimension into purely algebraic terms, based on the transcendence degree of an affine irreducible variety's coordinate ring (Theorem 2.14). This characterization often yields simple proofs of dimensional properties, and extends naturally to a definition for varieties over an arbitrary field (where ideas like "smoothness" may not be so readily available). To prove Theorem 2.14, we use the following result (cf. Lemma 10.4 of Chapter II).

Theorem 2.13. *Let $V \subset \mathbb{C}_{X_1,\ldots,X_n}$ be a nonempty irreducible variety. Suppose V's coordinate ring $\mathbb{C}[x_1,\ldots,x_n]$ has transcendence base $\{x_1,\ldots,x_d\}$. Let* (0) *be a typical point of V, and suppose that $V \cap \mathbb{C}_{X_{d+1},\ldots,X_n}$ consists of only finitely many points. Recall that any product of disks a* **polydisk**. *Then for each polydisk $\Delta^{n-d} \subset \mathbb{C}_{X_{d+1},\ldots,X_n}$ centered at* (0), *there is a polydisk $\Delta^d \subset \mathbb{C}_{X_1,\ldots,X_d}$ centered at* (0) *such that above each point a in Δ^d there is a point of V in $a \times \Delta^{n-d}$.*

PROOF. Note that if V is of codimension 1, the theorem follows immediately from Lemma 10.4 of Chapter II. For arbitrary codimension, the proof can easily be reduced to the codimension-one case, as follows: First, the standard proof of the theorem of the primitive element (as given, for instance, in [van der Waerden, Vol. I, Section 40]) shows that some $\mathbb{C}$-linear combination of $x_{d+1},\ldots,x_n$ is a primitive element for the extension $\mathbb{C}(x_1,\ldots,x_n)$ over $\mathbb{C}(x_1,\ldots,x_d)$. Without loss of generality, assume that coordinates in $\mathbb{C}_{X_{d+1},\ldots,X_n}$ have been chosen so that x_{d+1} is such a primitive element. Then each of $x_{d+2},\ldots,x_n$ is a rational function of $x_1,\ldots,x_{d+1}$. If V' is the variety in $\mathbb{C}_{X_1,\ldots,X_{d+1}}$ with generic point $(x_1,\ldots,x_{d+1})$, then over each point of V' near (0) there $\pi_{X_{d+2},\ldots,X_n}$-lies just one point of V. We are thus led back to the codimension-one case. □

We may now prove Theorem 2.14, which translates dimension into purely algebraic terms. First, if R is any integral domain containing a field k, the **transcendence degree over** k **of** R is the usual transcendence degree over k of R's quotient field, and is denoted by tr deg R/k. In this book, k will be $\mathbb{C}$ unless noted otherwise. We denote the transcendence degree over $\mathbb{C}$ of R by tr deg R.

Theorem 2.14. *Let the irreducible variety $V \subset \mathbb{C}_{X_1,\ldots,X_n}$ have coordinate ring $\mathbb{C}[x_1,\ldots,x_n]$. Then*

$$\dim V = \operatorname{tr\,deg} \mathbb{C}[x_1,\ldots,x_n].$$

PROOF. Let $\{x_1,\ldots,x_d\}$ be a transcendence base of $\mathbb{C}[x_1,\ldots,x_n]$ over $\mathbb{C}$. Let $q_{d+1}(X_{d+1}),\ldots,q_n(X_n)$ be minimal polynomials over $\mathbb{C}(x_1,\ldots,x_d)$ of $x_{d+1},\ldots,x_n$, respectively, with coefficients in $\mathbb{C}[x_1,\ldots,x_d]$. Then since V consists of the set of specializations of $(x_1,\ldots,x_n)$, it is clear that V is contained in $V^\dagger = \mathsf{V}(q_{d+1},\ldots,q_n)$. Since each of $q_{d+1},\ldots,q_n$ is irreducible and nonconstant, each discriminant polynomial

$$\mathscr{D}_{X_i}(q_i) = \mathscr{R}_{X_i}\left(q_i, \frac{\partial q_i}{\partial X_i}\right) \in \mathbb{C}[x_1,\ldots,x_d]$$

is nonzero; hence all the $\mathsf{V}(\mathscr{D}_{X_i}(q_i))$ are proper subvarieties of $\mathbb{C}_{X_1,\ldots,X_d}$, as is their union D. Let (0) be a typical point of V $\pi_{X_{d+1},\ldots,X_n}$-lying above $\mathbb{C}_{X_1,\ldots,X_d}\backslash D$. Then since $(\partial q_i/\partial X_i)(0) \neq 0$ and $(\partial q_j/\partial X_i) \equiv 0$ for $j \neq i$ and $i, j = d+1,\ldots,n$, we see that at each point of $V^\dagger$ near (0), the rank of the

Jacobian matrix $J(\partial q_i/\partial X_j)$ $(d + 1 \leqslant i \leqslant n,\ 1 \leqslant j \leqslant n)$ is $n - d$. Let V^* be any irreducible component of $V^\dagger$ containing V; V^* then contains a point of dimension d. If we could show that for some $\mathbb{C}^n$-open neighborhood U, $V^* \cap U = V \cap U$, then Theorems 2.9 and 2.11 would imply that V has dimension d, thus proving our theorem.

First, we obviously have $V \subset V^* \subset V^\dagger$; hence for any $\mathbb{C}^n$-open neighborhood U of (0),

$$V \cap U \subset V^* \cap U \subset V^\dagger \cap U.$$

We show that these inclusions cannot be strict as follows: First, for some U about (0), $V^\dagger \cap U$ is the graph of a function defined on $\mathbb{C}_{X_1,\ldots,X_d} \cap U$ since the rank of $J(V^\dagger)$ is $n - d$ at all points of $V^\dagger$ near (0); therefore for some U, the natural set-theoretic projection of $V^\dagger \cap U$ on $\mathbb{C}_{X_1,\ldots,X_d}$ is $\mathbb{C}_{X_1,\ldots,X_d} \cap U$. By the single-valuedness of functions, no proper subset of $V^\dagger \cap U$ projects onto all of $\mathbb{C}_{X_1,\ldots,X_d} \cap U$. Next observe that since the Jacobian matrix of $X_1, \ldots, X_d$, $q_{d+1}, \ldots, q_n$ has rank n, (0) has dimension zero in $V^\dagger \cap \mathbb{C}_{X_{d+1},\ldots,X_n}$, and is therefore isolated in $V^\dagger \cap \mathbb{C}_{X_{d+1},\ldots,X_n}$; hence (0) is isolated in $V \cap \mathbb{C}_{X_{d+1},\ldots,X_n}$, too; hence Theorem 2.13 implies that the natural projection of $V \cap U$ on $\mathbb{C}_{X_1,\ldots,X_n}$ is $\mathbb{C}_{X_1,\ldots,X_n} \cap U$. Thus $V \cap U$ cannot be a proper subset of $V^\dagger \cap U$, and therefore not of $V^* \cap U$, either. □

We now turn to the third approach to dimension; it is purely ring theoretic and based on sequences of prime ideals. The main result is Theorem 2.18; our proof of it depends on Theorem 2.15, which we prove next, and which is interesting in its own right.

Theorem 2.15. *If V is an irreducible subvariety of $\mathbb{P}^n(\mathbb{C})$ or of $\mathbb{C}^n$, and if W is any proper subvariety of V, then* $\dim W < \dim V$.

PROOF. We may assume without loss of generality that V is affine, and that $W \neq \emptyset$. Also, since $\dim W$ is the largest of the dimensions of its irreducible components, we may assume that W is irreducible. Let $\mathbb{C}[x_1, \ldots, x_n]$ and $\mathbb{C}[y_1, \ldots, y_n]$ be the coordinate rings of V and W, respectively. We want to show that the transcendence degree of $\mathbb{C}[y_1, \ldots, y_n]$ is strictly less than that of $\mathbb{C}[x_1, \ldots, x_n]$.

Now if $W = \mathsf{V}(\mathfrak{p})$, then $\mathbb{C}[y_1, \ldots, y_n] = \mathbb{C}[x_1, \ldots, x_n]/\mathfrak{p}$, where $\mathfrak{p}$ is a nonzero proper prime ideal of $\mathbb{C}[x_1, \ldots, x_n]$. Let $\{x_1, \ldots, x_d\}$ be a transcendence base of $\mathbb{C}[x_1, \ldots, x_n]$ over $\mathbb{C}$ $(d = \dim V)$; suppose that under the natural homomorphism induced by $\mathfrak{p}$, x_i maps to y_i for $i = 1, \ldots, n$. Now $x_{d+1}, \ldots, x_n$ satisfy algebraic equations over $\mathbb{C}[x_1, \ldots, x_d]$; it is easy to check that each of $y_{d+1}, \ldots, y_n$ must be algebraic over $\mathbb{C}[y_1, \ldots, y_d]$. Hence the transcendence degree of $\mathbb{C}[y_1, \ldots, y_n]$ is no greater than that of $\mathbb{C}[x_1, \ldots, x_n]$. To show it must be strictly less, suppose it is the same; we derive a contradiction. Hence suppose $\{y_1, \ldots, y_d\}$ is a transcendence base of $\mathbb{C}[y_1, \ldots, y_n]$. This implies that $\mathbb{C}[x_1, \ldots, x_d]$ is isomorphic to $\mathbb{C}[y_1, \ldots, y_d]$. Thus the homomorphism $\mathbb{C}[x_1, \ldots, x_n] \to \mathbb{C}[y_1, \ldots, y_n]$ extends to a homomorphism $\mathbb{C}(x_1, \ldots, x_d)[x_{d+1}, \ldots, x_n] \to \mathbb{C}(y_1, \ldots, y_d)[y_{d+1}, \ldots, y_n]$. But both these last two rings are fields, so the kernel of this homomorphism is the zero ideal (0). This kernel obviously contains $\mathfrak{p}$, so $\mathfrak{p} = (0)$, whereas above we saw that $\mathfrak{p} \neq (0)$. We thus have our contradiction. □

There are existence results which add to the information supplied by Theorem 2.15, and which are useful in formulating a purely ring-theoretic definition of dimension. For example, for any variety V, there is always a subvariety of V of exactly one less dimension. This implies that there is a chain of varieties starting at V and descending one dimension at a time, to a zero-dimensional variety. This in effect allows us to get the dimension of a variety by counting the varieties in such a chain. Because the dimension of a variety is equal to the maximum of the dimensions of its components, we may assume all these proper subvarieties are irreducible. Notice that if V and the other varieties in the descending chain are all irreducible, then the chain is *maximal* in the sense that no further (nonempty) irreducible subvarieties of V can be added to the chain, still keeping it strict.

Now the usual distance in $\mathbb{R}$ from the first to the last of any $n + 1$ consecutive integers is n; in a similar spirit, we say that the *length l of a chain $V_0 \supsetneqq V_1 \supsetneqq \ldots \supsetneqq V_l$ of irreducible varieties* is one less than the number of varieties in that chain. Hence our maximal chain above has length $d = \dim V$. One can prove even more: *All* maximal strict chains of nonempty irreducible subvarieties of V have the same length. We prove these facts for affine varieties in the following theorem; this theorem may be extended to include projective varieties, too (cf. Definition 2.7).

Theorem 2.16. *Let $V \subset \mathbb{C}^n$ be any nonempty irreducible variety of dimension d, and let $V_{d_1} \supsetneqq \ldots \supsetneqq V_{d_2}$ be any strict chain of nonempty irreducible subvarieties of V. This chain can be extended* (or refined) *to a maximal chain of irreducible varieties*

$$V = V'_0 \supsetneqq V'_1 \supsetneqq \ldots \supsetneqq V'_d \qquad (V'_d \neq \emptyset),$$

where each variety in the original sequence appears in the extended sequence. Furthermore, any two such maximal chains have the same length.

Remark 2.17. One may recognize an analogy with the Jordan–Hölder refinement theorem for groups or modules.

Theorem 2.16 tells us that we could equally well define the dimension of a nonempty irreducible variety V as the length of any maximal strict chain of nonempty irreducible subvarieties of V. Of course we may use instead a maximal chain of prime ideals; if $R_V = \mathbb{C}[x_1, \ldots, x_n]$ is the coordinate ring of V, then the dimension of V is the length l of any maximal strict chain of prime ideals

$$0 = \mathfrak{p}_0 \subsetneqq \mathfrak{p}_1 \subsetneqq \ldots \subsetneqq \mathfrak{p}_l \qquad (\mathfrak{p}_l \neq R_V).$$

(Note that just as we used only nonempty varieties in Theorem 2.16, we use here only prime ideals of R_V different from R_V.)

Since $R_V = \mathbb{C}[X_1, \ldots, X_n]/\mathsf{J}(V)$, this length is, in turn, the same as the length of any maximal strictly ascending chain of prime ideals in $\mathbb{C}[X_1, \ldots, X_n]$ which starts with the prime ideal $\mathsf{J}(V)$. Now since the dimension of an arbitrary variety $V \subset \mathbb{C}^n$ is the maximum dimension of its irreducible components, if we write $\mathsf{J}(V)$ as the unique irredundant decomposition $\mathsf{J}(V) = \mathfrak{q}_1 \cap \ldots \cap \mathfrak{q}_r$, then V's dimension is the maximum length of all those strictly ascending chains of prime ideals in $\mathbb{C}[X_1, \ldots, X_n]$ which start from any of $\mathfrak{q}_1, \ldots, \mathfrak{q}_r$. Since by irredundancy, any prime ideal smaller than $\mathfrak{q}_i$ properly intersects $\mathsf{J}(V)$, we see that the dimension of V is thus just the length of the longest strictly ascending chain of prime ideals in $\mathbb{C}[X_1, \ldots, X_n]$ containing $\mathsf{J}(V)$. Finally, we know that for any ideal $\mathfrak{a} \subset \mathbb{C}[X_1, \ldots, X_n]$, $\sqrt{\mathfrak{a}}$ is the intersection of those prime ideals which contain $\mathfrak{a}$ (hence also of those minimal prime ideals which contain $\mathfrak{a}$). We thus have this fact:

Theorem 2.18. *Let* $\mathfrak{a} \subset \mathbb{C}[X_1, \ldots, X_n]$. *Then* $\dim \mathsf{V}(\mathfrak{a})$ *is the length of the longest strict chain of* $\mathfrak{a}$*-containing prime ideals in* $\mathbb{C}[X_1, \ldots, X_n]$.

Let us now prove Theorem 2.16.

PROOF OF THEOREM 2.16. It suffices to show that if $W_1 \subset W_2$ are irreducible nonempty subvarieties of V of dimension d_1 and d_2 respectively, then there is a strict chain of irreducible varieties from W_2 to W_1 of length $d_2 - d_1$; or what is the same, that there is a strict chain of prime ideals of length $d_2 - d_1$ in the coordinate ring $R_{W_2} = \mathbb{C}[x_1, \ldots, x_n]$, starting from (0) and ending in $\mathfrak{p}$, where $R_{W_2}/\mathfrak{p} = R_{W_1} = \mathbb{C}[y_1, \ldots, y_n]$. (This will ensure maximality, since for any irreducible variety V, any strict chain of irreducible varieties of length $d = \dim V$ starting with V and ending in a point must be maximal; otherwise, from Theorem 2.15 $\dim V$ would be greater than d.) Now the transcendence degree of R_{W_2} is d_2, and that of R_{W_1} is d_1; we assume without loss of generality that $d_2 > d_1$, and that $\{x_1, \ldots, x_{d_2}\}$ and $\{y_1, \ldots, y_{d_1}\}$ are transcendence bases of R_{W_2} and R_{W_1} respectively. We may also assume that the elements x_i and y_j have been numbered so that the image ring of the homomorphism

$$\mathbb{C}[x_1, \ldots, x_{d_1+1}, x_{d_1+2}, \ldots, x_{d_2}] \to \mathbb{C}[y_1, \ldots, y_{d_1+1}, x_{d_1+2}, \ldots, x_{d_2}] \quad (2)$$

has transcendence degree $d_2 - 1$ over $\mathbb{C}$. (Note that y_{d_1+1} is algebraic over $\mathbb{C}[y_1, \ldots, y_{d_1}]$, therefore also over $\mathbb{C}[y_1, \ldots, y_{d_1}, x_{d_1+2}, \ldots, x_{d_2}]$). Now for $i = d_2 + 1, \ldots, n$, let a minimal polynomial over $\mathbb{C}[x_1, \ldots, x_{d_2}]$ of x_i be $q_i(x_1, \ldots, x_{d_2}, X_i)$; since $\mathbb{C}[x_1, \ldots, x_n] \to \mathbb{C}[y_1, \ldots, y_n]$ is a ring homomorphism, $q_i(y_1, \ldots, y_{d_2}, X_i)$ has positive degree in X_i. Therefore so does $q_i(y_1, \ldots, y_{d_1+1}, x_{d_1+2}, \ldots, x_{d_2}, X_i)$; hence (2) extends to a homomorphism ϕ of $\mathbb{C}[x_1, \ldots, x_n]$; the transcendence degree of this image ring $R_{W_1'}$ of R_{W_1} is of course still $d_2 - 1$. Let $\mathfrak{p}$ be the kernel of this homomorphism;

certainly $\mathfrak{p} \neq (0)$. We have now completed the first step in an induction argument: We similarly construct a homomorphism ϕ' of $R_{W'_1}$ so the image $\phi'(R_{W'_1})$ has transcendence degree $d_2 - 2$ over $\mathbb{C}$; the kernel of $\phi' \circ \phi$ is a prime ideal $\mathfrak{p}' \subset R_{W_1}$, with $\mathfrak{p}' \supsetneqq \mathfrak{p}$. Continuing in this manner, we get the desired chain of prime ideals, hence also of varieties. □

We now make a few observations about dimension which we use in the sequel.

Definition 2.19. A variety in $\mathbb{P}^n(\mathbb{C})$ or $\mathbb{C}^n$ is said to have **pure dimension** if the variety has the same dimension at each of its points.

Definition 2.20. A variety in $\mathbb{P}^n(\mathbb{C})$ or $\mathbb{C}^n$ of pure dimension 1 is called a **curve**.

Remark 2.21. From Theorem 2.18 we see that an irreducible variety V in $\mathbb{C}^n$ is an irreducible curve iff every nonzero proper prime ideal of R_V is maximal. We use this in Section V,5.

Definition 2.22. A variety is a **hypersurface** in $\mathbb{P}^n(\mathbb{C})$ (or in $\mathbb{C}^n$) if it can be defined by a single nonconstant homogeneous polynomial in $\mathbb{C}[X_1, \ldots, X_{n+1}]$ (or by a single nonconstant polynomial in $\mathbb{C}[X_1, \ldots, X_n]$).

Theorem 2.23. *A variety in $\mathbb{P}^n(\mathbb{C})$ or $\mathbb{C}^n$ is a hypersurface $\Leftrightarrow$ it is of pure dimension $n - 1$.*

PROOF. Since any variety in $\mathbb{P}^n(\mathbb{C})$ is represented by a homogeneous variety in $\mathbb{C}^{n+1}$, it suffices to prove the result in the affine case.

$\Rightarrow$: Suppose $V = \mathsf{V}(p) \subset \mathbb{C}_{X_1, \ldots, X_n}$, where p is nonconstant in $\mathbb{C}[X_1, \ldots, X_n]$. Assume first that p is irreducible. Then $\mathsf{V}(p)$ has pure dimension, and for some i, $\partial p/\partial X_i$ is not identically zero; hence $\partial p/dX_i$ cannot vanish on V, for otherwise it would have to be in the prime ideal (p) (that is, a multiple of p), while $\deg \partial p/\partial X_i < \deg p$. Therefore the rank of $J(V) = (\partial p/\partial X_1, \ldots, \partial p/\partial X_n)$ attains the maximum of 1 at a point of V; hence $\dim V = n - 1$. Since any hypersurface is a union of irreducible hypersurfaces, the dimension is pure.

$\Leftarrow$: Suppose $V \subset \mathbb{C}^n$ has pure dimension $n - 1$; we want to show that $V = \mathsf{V}(p)$ for some polynomial p. If this is true for irreducible varieties of dimension $n - 1$, then it is true for arbitrary varieties of pure dimension $n - 1$. Therefore assume V is irreducible, say

$$V = \mathsf{V}(p_1, \ldots, p_r), \quad \text{where all } p_i \text{ are nonconstant.}$$

Now consider p_1. If $p_1 = p_{11} \cdot \ldots \cdot p_{1s}$ is a factorization of p_1 into irreducibles, then $\mathsf{V}(p_1) = \mathsf{V}(p_{11}) \cup \ldots \cup \mathsf{V}(p_{1s})$. Hence $V \subset \mathsf{V}(p_{1i})$ for some i. Since p_{1i} is irreducible, we have $V = \mathsf{V}(p_{1i})$ (Exercise 4.3 of Chapter III). Since p_{1j} is nonconstant, $V = \mathsf{V}(p_{1j})$ is a hypersurface. □

Just as one considers products of sets in set theory and products of spaces in topology, one also has products of varieties. Later on we shall need them, together with a basic dimensionality property of "product varieties." We begin with products of affine varieties.

Theorem 2.24. *Let $V \subset \mathbb{C}_{X_1,\ldots,X_m}$ and $W \subset \mathbb{C}_{Y_1,\ldots,Y_n}$ be two varieties.*

(2.24.1) *The set-theoretic product $V \times W \subset \mathbb{C}_{X_1,\ldots,X_m,Y_1,\ldots,Y_n}$ is a variety.* (*We call it a* **product variety**.)

(2.24.2) *Let V and W be irreducible with generic points $(x) = (x_1, \ldots, x_m)$ and $(y) = (y_1, \ldots, y_n)$, respectively, and suppose that $\mathbb{C}[x] \cap \mathbb{C}[y] = \mathbb{C}$. Then $V \times W$ is irreducible and has (x, y) as a generic point.*

(2.24.3) $\dim V \times W = \dim V + \dim W$.

PROOF. The proof of (2.24.1) may be reduced to the case when V and W are both irreducible, since obviously $(\bigcup_i V_i) \times (\bigcup_j W_j) = \bigcup_{i,j} V_i \times W_j$. This case then follows at once from (2.24.2) which is itself obvious. (2.24.3) is immediate from Theorem 2.14. □

It is natural to also ask about products of projective spaces and varieties. Just as with affine spaces, we can form the set product $\mathbb{P}^m(\mathbb{C}) \times \mathbb{P}^n(\mathbb{C})$, and endow it with the product topology. One might guess that this is in some sense the same as $\mathbb{P}^{m+n}(\mathbb{C})$. But it turns out that except when m or n is zero, $\mathbb{P}^m(\mathbb{C}) \times \mathbb{P}^n(\mathbb{C}) \neq \mathbb{P}^{m+n}(\mathbb{C})$. In fact, at a purely topological level, it turns out that the product of any two spaces homeomorphic to $\mathbb{P}^m(\mathbb{C})$ and $\mathbb{P}^n(\mathbb{C})$, where $m, n > 0$, is never homeomorphic to any $\mathbb{P}^k(\mathbb{C})$. We indicate the gist of a proof for those who know some homology theory. It is known (see, for instance, [Vick, Prop 2.7, p. 49]) that the homology groups (over the integers) of $\mathbb{P}^k(\mathbb{C})$ are:

$$H_i(\mathbb{P}^k(\mathbb{C})) = \begin{cases} \mathbb{Z} & \text{for } i = 0, 2, 4, \ldots, 2k; \\ 0 & \text{otherwise.} \end{cases}$$

The Künneth formula then tells us that

$$H_2(\mathbb{P}^m(\mathbb{C}) \times \mathbb{P}^n(\mathbb{C})) = \sum_{i+j=2} (H_i(\mathbb{P}^m(\mathbb{C})) \otimes H_j(\mathbb{P}^n(\mathbb{C}))) = \mathbb{Z} \oplus \mathbb{Z}$$

(where $\sum$ and $\oplus$ denote direct sum, and $\otimes$ denotes tensor product over $\mathbb{Z}$); but this is not a homology group of any $\mathbb{P}^k(\mathbb{C})$.

Yet products of projective spaces and varieties do naturally arise, as we will see later in this chapter when we use them (or what is the same, "multihomogeneous varieties") in defining at the variety-theoretic level notions like *order* and *multiplicity*, and in proving Bézout's theorem.

We call any product $\mathbb{P}^{n_1}(\mathbb{C}) \times \ldots \times \mathbb{P}^{n_s}(\mathbb{C})$ a **multiprojective space**, an s-**way projective space**, or most precisely, an $(n_1, \ldots, n_s)$-**projective space**, this product being looked at as the set of all $((n_1 + 1) + \ldots + (n_s + 1))$-tuples

$$((a_{11}, \ldots, a_{1,n_1+1}), \ldots, (a_{s1}, \ldots, a_{s,n\ +1})); \tag{3}$$

each such point is identified with

$$((c_1 a_{11}, \ldots, c_1 a_{1,n_1+1}), \ldots, (c_s a_{s1}, \ldots, c_s a_{s,n\ +1})), \tag{4}$$

where $c_1, \ldots, c_s$ are arbitrary elements of $\mathbb{C}\backslash\{0\}$. In analogy with homogeneous sets, we say that a subset S of $\mathbb{C}^{(n_1+1)+\ldots+(n_s+1)}$ is **multihomogeneous** (or s-**way homogeneous**, or $(n_1 + 1, \ldots, n_s + 1)$-**homogeneous**) if whenever a point of the form in (3) is in S, then the corresponding point in (4) is also in S.

In Theorem 2.6 of Chapter II, we proved that a variety $V \subset \mathbb{C}^n$ is homogeneous iff it is definable by a set of homogeneous polynomials. A proof analogous to that of Theorem 2.6, Chapter II, shows that an algebraic variety is $(n_1, \ldots, n_s)$-homogeneous in $\mathbb{C}_{X_{11},\ldots,X_{sn_s}}$ iff it is defined by polynomials $p(X_{11}, \ldots, X_{sn_s})$ which are $(n_1, \ldots, n_s)$-**homogeneous**—that is, for each of $i = 1, \ldots, s$, it is homogeneous in the set of indeterminants $\{X_{i1}, \ldots, X_{in_i}\}$. An $(n_1 + 1, \ldots, n_s + 1)$-homogeneous variety in $\mathbb{C}^{(n_1+1)+\ldots+(n_s+1)}$ then defines a set in $\mathbb{P}^* = \mathbb{P}^{n_1}(\mathbb{C}) \times \ldots \times \mathbb{P}^{n_s}(\mathbb{C})$ which we call a **variety** (if no confusion can arise), a **multiprojective**, s-**way**, or $(n_1, \ldots, n_s)$-**projective**, **variety** in $\mathbb{P}^*$. The reader may check that the basic lattice and decomposition properties of ordinary varieties continue to hold for multiprojective varieties. Note that for varieties $V_i \subset \mathbb{P}^{n_i}(\mathbb{C})$ where $i = 1, \ldots, s$, $V_1 \times \ldots \times V_s$ is s-way projective in $\mathbb{P}^*$.

One may also "multidehomogenize" in the obvious way. If $\mathbb{C}^{n_i}$ denotes a particular dehomogenization of $\mathbb{P}^{n_i}(\mathbb{C})$, then $\mathbb{C}^{n_1} \times \ldots \times \mathbb{C}^{n_s}$ is the corresponding multidehomogenization of $\mathbb{P}^*$; any variety V in $\mathbb{P}^*$ then has a corresponding multi-dehomogenization which we call an **affine representative of** V. (This includes the case when V is a point P.)

Definition 2.25. Let V be multiprojective. The **dimension at** P **of** V, written $\dim_P V$, is the dimension of any affine representative of V at an affine representative of P. The **dimension of** V, written $\dim V$, is $\max_P \dim_P V$. It is clear that the above notion of dimension is well defined.

In terms of the multihomogeneous variety $\mathsf{H}(V)$ in $\mathbb{C}^{(n_1+1)} \times \ldots \times \mathbb{C}^{(n_s+1)}$ corresponding to $V \subset \mathbb{P}^*$, we clearly have $\dim V = \dim \mathsf{H}(V) - s$. Finally, by using multi-dehomogenizations, we get at once the basic dimensionality property:

Theorem 2.26. *If V and W are multiprojective varieties, then*

$$\dim V \times W = \dim V + \dim W.$$

EXERCISES

2.1 Show that a variety $V \neq \emptyset$ in $\mathbb{P}^n(\mathbb{C})$ or $\mathbb{C}^n$ consists of finitely many points iff $\dim V = 0$.

2.2 If for varieties V_1 and V_2 in $\mathbb{P}^n(\mathbb{C})$ or $\mathbb{C}^n$ we have $V_1 \subset V_2$ and $\dim V_1 = \dim V_2$, show that V_1 and V_2 must have an irreducible component in common. What can one conclude if in addition V_1 is of pure dimension?

2.3 Show that for any nonempty variety V in $\mathbb{P}^n(\mathbb{C})$ or $\mathbb{C}^n$, there is a subvariety W with $\dim W < \dim V$, such that V is locally an analytic manifold at each point $P \in V \backslash W$.

2.4 Let $\mathfrak{a}_1, \mathfrak{a}_2$ be any two homogeneous ideals of $\mathbb{C}[X, Y, Z]$ such that neither $\mathfrak{a}_1$ nor $\mathfrak{a}_2$ properly contains a nonzero prime ideal. Show that $\mathfrak{a}_1 + \mathfrak{a}_2 \subsetneqq (X, Y, Z)$.

2.5 Let $V \subset \mathbb{C}^n$ be a variety; let $\bar{\pi}$ be the closed projection of V along $\mathbb{C}_{X_1, \ldots, X_r}$ to $\mathbb{C}_{X_{r+1}, \ldots, X_n}$. Show that $\dim \bar{\pi}(V) \leqslant \dim V$. Reformulate this result in ring-theoretic terms.

2.6 Prove Lemma 10.9 of Chapter III.

2.7 Let V be a variety in $\mathbb{P}^n(\mathbb{C})$, and let $\mathbb{P}^{n-1}(\mathbb{C})$ be a choice of hyperplane at infinity not containing any component of V. Show that V is the topological closure in $\mathbb{P}^n(\mathbb{C})$ of $V \backslash \mathbb{P}^{n-1}(\mathbb{C})$. (Thus the projective completion of an affine variety in $\mathbb{P}^n(\mathbb{C}) \backslash \mathbb{P}^{n-1}(\mathbb{C})$ is its topological closure in $\mathbb{P}^n(\mathbb{C})$.)

2.8 Let V be any nonempty variety in $\mathbb{P}^n(\mathbb{C})$ or $\mathbb{C}^n$, and let P be an arbitrary point of V. Find a ring-theoretic characterization of $\dim_P V$.

3 The dimension theorem

In Theorem 6.1 of Chapter II we proved that any two curves C_1 and C_2 in $\mathbb{P}^2(\mathbb{C})$ have a nonempty intersection. Equivalently, this says

$$\operatorname{cod}(C_1 \cap C_2) \leqslant \operatorname{cod} C_1 + \operatorname{cod} C_2. \tag{5}$$

Our proof of this ultimately depended on showing that there is a nonzero root of a resultant polynomial of p_1 and p_2, where C_i defines the homogeneous ideal $(p_i) \subset \mathbb{C}_{X_1 X_2 X_3}$. Now in extending results, it is often helpful to look for possible mild extensions, and then to generalize bit by bit. As an example, one can easily extend the proof of (5) to arbitrary hypersurfaces of $\mathbb{P}^n(\mathbb{C})$. If these hypersurfaces are defined by nonconstant homogeneous polynomials $p_1, p_2 \in \mathbb{C}[X_1, \ldots, X_{n+1}]$, then in a way analogous to the proof in Chapter II, we may assume coordinates in $\mathbb{C}^{n+1}$ are such that $\mathscr{R}_{X_{n+1}}(p_1, p_2)$ is homogeneous of degree $\deg p_1 \cdot \deg p_2 > 0$, meaning that $\mathsf{V}(\mathscr{R}_{X_{n+1}})$ is a homogeneous hypersurface in $\mathbb{C}_{X_1, \ldots, X_n}$ (in which case $\mathsf{V}(\mathscr{R}_{X_{n+1}})$ will have dimension $n - 1$). The argument may then be completed by noting that above each point of $\mathsf{V}(\mathscr{R}_{X_{n+1}})$ there is a point of $\mathsf{V}(p_1) \cap \mathsf{V}(p_2)$, and that this intersection must then be of dimension at least $n - 1$—that is, it must have codimension at most 2 in $\mathbb{C}^{n+1}$. (The fact that $\mathscr{R}_{X_{n+1}}$ is of degree $\deg p_1 \cdot \deg p_2$ will fit in with a general Bézout theorem in Section 7.)

One may also ask if there is an extension of this projective result to include *affine* hypersurfaces as well. Surely many sets of affine varieties do not satisfy the codimension relation by virtue of the fact that the varieties fail to intersect, while in fact there are points of intersection in their projective completions. But if they *do* intersect in affine space, must this intersection then satisfy the codimension relation? The answer is *yes*. In fact Theorem 3.1, which we prove next, extends this even further to irreducible affine varieties of arbitrary dimension in $\mathbb{C}^n$ (therefore also to arbitrary affine varieties V_1 and V_2 in $\mathbb{C}^n$ provided some highest-dimensional irreducible component of V_1 intersects some highest-dimensional irreducible component of V_2). This fact at once implies that any pair of varieties V_1, V_2 in $\mathbb{P}^n(\mathbb{C})$ satisfies the codimension relation

$$\operatorname{cod}(V_1 \cap V_2) \leqslant \operatorname{cod} V_1 + \operatorname{cod} V_2,$$

since the irreducible components of the corresponding homogeneous varieties always intersect at the origin (hence *they* must satisfy the codimension relation), and since the codimension of a variety in $\mathbb{C}^n$ is the same as that of its homogenization in $\mathbb{C}^{n+1}$. In the projective case, if any of V_1, V_2, or $V_1 \cap V_2$ are empty, we define dim $\varnothing$ to be -1 so that the inequality still holds.

It turns out that working with the affine form gives us a little more flexibility since we need not remain within the domain of homogeneous varieties.

In this section we begin by proving

Theorem 3.1. *Let V_1 and V_2 be any two irreducible varieties in $\mathbb{C}^n$, and suppose $V_1 \cap V_2 \neq \varnothing$. Then*

$$\operatorname{cod}(V_1 \cap V_2) \leqslant \operatorname{cod} V_1 + \operatorname{cod} V_2.$$

In view of the discussion above, we have at once the

Corollary 3.2. *If V_1 and V_2 are any two varieties in $\mathbb{P}^n(\mathbb{C})$, then*

$$\operatorname{cod}(V_1 \cap V_2) \leqslant \operatorname{cod} V_1 + \operatorname{cod} V_2.$$

Remark 3.3. It turns out (Exercise 6.6) that in a certain sense, intersecting pairs of varieties V_1, V_2 usually give equality in Theorem 3.1 and Corollary 3.2. The assumption of equality will be used often in the sequel; we formalize it here:

Definition 3.4. If V_1 and V_2 are any two intersecting irreducible varieties in $\mathbb{P}^n(\mathbb{C})$ or in $\mathbb{C}^n$, then V_1 and V_2 **intersect properly** provided

$$\operatorname{cod}(V_1 \cap V_2) = \operatorname{cod} V_1 + \operatorname{cod} V_2.$$

Arbitrary varieties V_1 and V_2 in $\mathbb{P}^n(\mathbb{C})$ or in $\mathbb{C}^n$ **intersect properly** provided that each irreducible component of V_1 properly intersects each irreducible component of V_2.

PROOF OF THEOREM 3.1. In attempting to prove this theorem, one might naturally begin by trying to generalize to ideal theory our earlier polynomial concept of resultant. Let us see where this leads us. Suppose V_1 and V_2 define the $\mathbb{C}[X_1, \ldots, X_n]$-ideals $\mathfrak{a}_1 = (p_1, \ldots, p_r)$ and $\mathfrak{a}_2 = (q_1, \ldots, q_s)$, respectively. Then a point is in $V_1 \cap V_2$ if and only if it is in each of $\mathsf{V}(p_i, q_j)$ where $i = 1, \ldots, r$ and $j = 1, \ldots, s$. With respect to appropriate coordinates this means that there is a point of $V_1 \cap V_2$ π_{X_n}-lying over a given point $P \in \mathbb{C}_{X_1, \ldots, X_{n-1}}$ only if P is in the variety determined by the ideal generated by $\{\mathscr{R}_{X_n}(p_i, q_j) | i = 1, \ldots, r \text{ and } j = 1, \ldots, s\}$. One thus might define $\mathscr{R}_{X_n}(\mathfrak{a}_1, \mathfrak{a}_2)$ to be the ideal generated by $\{\mathscr{R}_{X_n}(p, q) | p \in \mathfrak{a}_1 \text{ and } q \in \mathfrak{a}_2\}$.

Unless $\mathfrak{a}_1$ and $\mathfrak{a}_2$ are both principal, we meet two serious problems in this approach. First, it may happen that there lie no components of $V_1 \cap V_2$ over any highest-dimensional components of $\mathsf{V}(\mathscr{R}_{X_n}(\mathfrak{a}_1, \mathfrak{a}_2))$, and therefore we cannot directly use $\mathsf{V}(\mathscr{R}_{X_n}(\mathfrak{a}_1, \mathfrak{a}_2))$ to get a lower bound on $\dim V_1 \cap V_2$. But even if this problem didn't arise, we still don't know very much about $\dim \mathsf{V}(\mathscr{R}_{X_n}(\mathfrak{a}_1, \mathfrak{a}_2))$. In the case of intersecting hypersurfaces defined by nonconstant polynomials p_1 and $p_2 \in \mathbb{C}[X_1, \ldots, X_n]$, the all-important fact is that $\mathscr{R}_{X_n}(p_1, p_2)$ is also a polynomial (either nonconstant, or the zero polynomial). To prove the codimension relation in this case, we then capitalized on the fact that such a polynomial defines a subvariety of codimension 0 or 1 in $\mathbb{C}_{X_1, \ldots, X_{n-1}}$.

This strongly suggests trying to arrange things so that our generalized resultant turns out to be a polynomial. For instance, let us begin by assuming that only one of the varieties $V_1 \subset \mathbb{C}^n$ has arbitrary dimension d, and that $V_2 = \mathsf{V}(p) \subset \mathbb{C}^n$ is a hypersurface. Then by an appropriate choice of coordinates, we may assume that V_1 set-theoretically projects onto some d-subspace $\mathbb{C}^d$ of $\mathbb{C}^n$ (using the normalization lemma (Lemma 11.5 of Chapter III)). Our theorem asserts that if V_1 and V_2 intersect, they do so in dimension d or $d - 1$; $V_1 \cap V_2$ projects onto $\mathbb{C}^d$ or a hypersurface of $\mathbb{C}^d$, and it is $\mathbb{C}^d$ or this hypersurface of $\mathbb{C}^d$ which ought to end up being the variety of a generalized resultant polynomial.

Before pursuing this idea let us satisfy ourselves that we will actually be able to push this further to a full proof of Theorem 3.1—that is, that we can prove the theorem once we know this:

(3.5) If $V \subset \mathbb{C}^n$ intersects a hypersurface of $\mathbb{C}^n$, then some component of intersection has dimension $\dim V$ or $\dim V - 1$.

First, note that if each hypersurface cuts down the dimension by at most one, then Theorem 3.1 holds for any varieties of dimension d which are the intersection of $n - d$ hypersurfaces. However, there are d-dimensional varieties which are *not* the intersection of any set of $n - d$ hypersurfaces; sometimes more than $n - d$ hypersurfaces are required to get the given variety. (For instance, $\mathbb{C}_{X_1X_2} \cup \mathbb{C}_{X_3X_4} \subset \mathbb{C}_{X_1X_2X_3X_4}$ is not the intersection of any two hypersurfaces in $\mathbb{C}_{X_1X_2X_3X_4}$. See [Eisenbud and Evans].) But the following trick allows us to get around this difficulty: Let V_1 and V_2 be any two varieties in $\mathbb{C}^n$. Regard V_1 as a variety in $\mathbb{C}_{X_1,\ldots,X_n}$, and V_2 as a variety in $\mathbb{C}_{Y_1,\ldots,Y_n}$. Then $V_1 \times V_2 \subset \mathbb{C}_{X_1,\ldots,X_n,Y_1,\ldots,Y_n}$ has dimension $\dim V_1 + \dim V_2$.

Now the diagonal variety

$$\Delta = \{(a_1, \ldots, a_n, a_1, \ldots, a_n) | (a_1, \ldots, a_n) \in \mathbb{C}^n\} \subset \mathbb{C}^{2n}$$

is an n-dimensional subspace of $\mathbb{C}^{2n}$, and is indeed the intersection of $2n - n$ hypersurfaces, namely

$$\Delta = \mathsf{V}(X_1 - Y_1, \ldots, X_n - Y_n) = \mathsf{V}(X_1 - Y_1) \cap \cdot \ldots \cdot \cap \mathsf{V}(X_n - Y_n).$$

Now looking at $V_1 \cap V_2$ as a subvariety of Δ in the natural way, we see that a point is in $V_1 \cap V_2$ if and only if it is in $(V_1 \times V_2) \cap \Delta$. This is the essential idea, for then $V_1 \cap V_2$ is the intersection of $V_1 \times V_2$ with the n hypersurfaces $\mathsf{V}(X_i - Y_i)$; if each hypersurface cuts down the dimension by at most one, then

$$\dim(V_1 \cap V_2) \geqslant \dim V_1 + \dim V_2 - n,$$

or

$$\begin{aligned}\operatorname{cod}(V_1 \cap V_2) &\leqslant n - (\dim V_1 + \dim V_2 - n)\\ &= (n - \dim V_1) + (n - \dim V_2) = \operatorname{cod} V_1 + \operatorname{cod} V_2.\end{aligned}$$

Hence we will have proved Theorem 3.1 if we can establish (3.5); we do this now.

First, we may without loss of generality assume that V in (3.5) is irreducible. Let $(x) = (x_1, \ldots, x_n)$ be a generic point of V. We may also assume that coordinates have been chosen so that $\{x_1, \ldots, x_d\}$ is a transcendence base of V's coordinate ring $\mathbb{C}[x] = \mathbb{C}[x_1, \ldots, x_n]$, and so that $\mathbb{C}[x]$ is integral over $\mathbb{C}[x_1, \ldots, x_d]$. Let the nonconstant polynomial $p \in \mathbb{C}[X_1, \ldots, X_n]$ define our hypersurface. We now find a "resultant polynomial" $\mathscr{R}_{X_{d+1},\ldots,X_n}(V, p) = \mathscr{R}$, which we also write as $\mathscr{R}(x_1, \ldots, x_d)$), such that $\mathscr{R}(a_1, \ldots, a_d)$ is 0 if and only if there is a point of $V \cap \mathsf{V}(p)$ $\pi_{X_{d+1},\ldots,X_n}$-lying above $(a_1, \ldots, a_d)$.

Since $V \cap \mathsf{V}(p)$ consists of precisely those points of V where p vanishes, let us consider the restriction of p to V—that is, let us consider $p(x) \in \mathbb{C}[x]$, which represents $p(X_1, \ldots, X_n)$ restricted to V. We want those points of $\mathbb{C}_{X_1,\ldots,X_d}$ above which $p(x) = 0$ has at least one root. Of course $p(x) = 0$ has at least one root above $(a) = (a_1, \ldots, a_d) \in \mathbb{C}_{X_1,\ldots,X_d}$ if and only if the product of all $p(x)$'s values at points of V above (a), is equal to zero. But since $p(x) \in \mathbb{C}[x]$ is integral over $\mathbb{C}[x_1, \ldots, x_d]$, the product of these values is just the value at $(a_1, \ldots, a_d)$ of the zero-degree term of $p(x)$'s monic minimal polynomial over $\mathbb{C}(x_1, \ldots, x_d)$. (Note that the zero-degree term of p's monic minimal polynomial over $\mathbb{C}(X_1, \ldots, X_d)$ is actually in $\mathbb{C}[X_1, \ldots, X_d]$. For if it were in $\mathbb{C}(X_1, \ldots, X_d) \setminus \mathbb{C}[X_1, \ldots, X_d]$, there would be points in $\mathbb{C}_{X_1,\ldots,X_d}$ above which some zeros of p would "escape to infinity." This can never happen for p integral over $\mathbb{C}[X_1, \ldots, X_d]$, as the reader can easily verify.) This zero-degree term is then our desired resultant polynomial $\mathscr{R} = \mathscr{R}(x_1, \ldots, x_d)$. (See Remark 3.7.) The assumption that V and $\mathsf{V}(p)$ intersect implies that $\mathscr{R}$ is not a nonzero constant; hence $\mathsf{V}(\mathscr{R}) \subset \mathbb{C}_{X_1,\ldots,X_d}$ is either all of $\mathbb{C}_{X_1,\ldots,X_d}$, or a hypersurface of $\mathbb{C}_{X_1,\ldots,X_d}$, and the projection of $V \cap \mathsf{V}(p)$ on $\mathbb{C}_{X_1,\ldots,X_d}$ is precisely $\mathsf{V}(\mathscr{R})$. Since projection does not increase dimension (Exercise 2.5), $V \cap \mathsf{V}(p)$ has dimension at least $d - 1$—that is, $\mathsf{V}(p)$ cuts down V's dimension by at most one. We have therefore proved Theorem 3.1. □

Remark 3.6. It is reasonable to ask how the resultant of this section compares with the resultant of two polynomials in Chapter II. First, note the following lemma, which is important in its own right. Its proof is straightforward and is indicated in Exercise 3.6.

Lemma 3.7. *For any two nonconstant polynomials in* $\mathbb{C}[X]$

$$p_1(X) = (X - b_1) \cdot \ldots \cdot (X - b_r) \quad \text{and} \tag{6}$$

$$p_2(X) = (X - c_1) \cdot \ldots \cdot (X - c_s), \tag{7}$$

the resultant of p_1 *and* p_2 *is the "difference product"*

$$\mathscr{R}_X(p_1, p_2) = \prod_{\substack{i=1,\ldots,r \\ j=1,\ldots,s}} (b_i - c_j). \tag{8}$$

(*Notice that* $\mathscr{R}_X(p_1, p_2) = -\mathscr{R}_X(p_2, p_1)$.)

Now let p_1 and p_2 be two nonconstant polynomials in $\mathbb{C}[X_1, \ldots, X_n]$ defining hypersurfaces $V_1 = \mathsf{V}(p_1)$ and $V_2 = \mathsf{V}(p_2)$; we assume that p_1 is irreducible, and that coordinates have been chosen so that $X_n^{\deg p_1}$ and $X_n^{\deg p_2}$ are terms of p_1 and p_2, respectively. Let $(a) = (a_1, \ldots, a_{n-1})$ be any point of $\mathbb{C}_{X_1,\ldots,X_{n-1}}$, and let the $r = \deg p_1$ points of V_1 lying over (a) be $(a, b_1), \ldots, (a, b_r)$. From the definition of $\mathscr{R}$ in this section, we see that at (a), $\mathscr{R}_{X_n}(\mathsf{V}(p_1), p_2)$ is the product $p_2(a, b_1) \cdot \ldots \cdot p_2(a, b_r)$. Now for a fixed b_i, $p_2(a, b_i)$ is just the polynomial $p_2(a, X_n)$ evaluated at $X_n = b_i$. But $p_2(a, X_n)$ factors in the form

$$p_2(a, X_n) = (X_n - c_1) \cdot \ldots \cdot (X_n - c_s),$$

(where $s \geqslant 1$ and $c_i \in \mathbb{C}$), so we have $p_2(a, b_i) = k(b_i - c_1) \cdot \ldots \cdot (b_i - c_s)$. Therefore

$$p_2(a, b_1) \cdot \ldots \cdot p_2(a, b_r) = \prod_{i,j} (b_i - c_j).$$

From this, we see that this last difference product is just the difference product in (8). This of course means that the resultant $\mathscr{R}_{X_n}(p_1, p_2)$ of Chapter II vanishes precisely when $\mathscr{R}_{X_n}(\mathsf{V}(p_1), p_2)$ of this section does.

A natural question arises in connection with Theorem 3.1: Though we have shown that $\operatorname{cod}(V_1 \cap V_2) \leqslant \operatorname{cod} V_1 + \operatorname{cod} V_2$, is it necessarily true that *every component* of $V_1 \cap V_2$ must have codimension $\leqslant \operatorname{cod} V_1 + \operatorname{cod} V_2$? In the case of curves this is trivially true since the component varieties of $V_1 \cap V_2$ are no smaller than points. But in varieties V_1 and V_2 of higher dimensions, it is conceivable that some components of $V_1 \cap V_2$ might have codimension *larger* than $\operatorname{cod} V_1 + \operatorname{cod} V_2$. It turns out, however, that this can never happen. This is the strongest result we prove in this section.

Theorem 3.8. *If V_1 and V_2 are irreducible varieties in $\mathbb{C}^n$, then each component of $V_1 \cap V_2$ has codimension at most* $\operatorname{cod} V_1 + \operatorname{cod} V_2$.

Our proof will essentially consist in looking at one component of $V_1 \cap V_2$ at a time; we do this by "removing" all but the one under consideration. "Removing" a subvariety V' from an affine variety V will, for us, mean mapping V to another affine variety in such a way that V' is mapped into the hyperplane at infinity, thus "escaping" from the affine part, but so that the remaining points of V do *not* escape to infinity. An example will help to clarify this idea.

EXAMPLE 3.9. Let $X = c$ be any point of $\mathbb{C}$; then there is a natural map from the variety $\mathbb{C}$ having generic point (x), to the complex hyperbola in $\mathbb{C}^2$ having generic point $(x, 1/(x - c))$; this map is defined by $(a) \to (a, 1/(a - c))$. It is 1:1 and onto between $\mathbb{C}\backslash\{c\}$ and the hyperbola; c itself has no image in $\mathbb{C}^2$. More generally, if $c_1, \ldots, c_m$ are m distinct points of $\mathbb{C}$, then we may send any number of these to infinity, while keeping all the rest finite. For instance, for $k \leqslant m$, $(x) \to (x, 1/(x - c_1) \cdot \ldots \cdot (x - c_k))$ maps the first k of these points to infinity, the remaining points mapping to points of $\mathbb{C}^2$. Similarly, one can remove the parabola $\mathsf{V}(Y - X^2)$ from $\mathbb{C}_{XY}$ using the map

$$(x, y) \to \left(x, y, \frac{1}{y - x^2}\right);$$

note that in the subvariety of $\mathbb{C}_{XYZ}$ having generic point $(x, y, 1/(y - x^2))$, none of its points π_Z-lie above the parabola in $\mathbb{C}_{XY}$.

With this as background, let us now turn to the

PROOF OF THEOREM 3.8. First, in view of our argument about writing $V_1 \cap V_2$ as $(V_1 \times V_2) \cap \Delta$, we see that it suffices to let only one variety $V \subset \mathbb{C}^n$ be of arbitrary dimension d, and the other a hypersurface $\mathsf{V}(p)$, where $V \cap \mathsf{V}(p) \neq \varnothing$. Without loss of generality, assume that V is irreducible; let $(x) = (x_1, \ldots, x_n)$ be a generic point of V. Suppose the components of $V \cap \mathsf{V}(p)$ are $W_1, \ldots, W_s$; we show that an arbitrary W_i, say W_1, has dimension $d - 1$. Let a generic point of W_1 be $(z) = (z_1, \ldots, z_n)$, and let $q_2, \ldots, q_s$ be polynomials identically zero on $W_2, \ldots, W_s$ respectively, but not identically zero on W_1 (that is, $q_i \in \mathsf{J}(W_i)\backslash\mathsf{J}(W_1)$). Then $(x, 1/q_2(x) \cdot \ldots \cdot q_s(x))$ is a generic point of a variety V^* in $\mathbb{C}^{n+1} = \mathbb{C}_{X_1, \ldots, X_{n+1}}$. Note that since

$1/q_2(x)\cdot\ldots\cdot q_s(x) \in \mathbb{C}(x)$, we have tr deg $\mathbb{C}[x] =$ tr deg $\mathbb{C}[x, 1/q_2(x)\cdot\ldots\cdot q_s(x)]$ so dim $V =$ dim V^*.

Now since $\mathbb{C}[z]$ is a homomorphic image of $\mathbb{C}[x]$, it follows that $\mathbb{C}[z, 1/q_2(z)\cdot\ldots\cdot q_s(z)]$ is a homomorphic image of $\mathbb{C}[x, 1/q_2(x)\cdot\ldots\cdot q_s(x)]$; this is so since for $i = 2, \ldots, s$, $q_i(z)$ does not vanish on all of W_1. Therefore $q_i(z)$ is a nonzero element of $\mathbb{C}[z]$—that is, $q_2(z)\cdot\ldots\cdot q_s(z) \neq 0$. Thus $(z, 1/q_2(z)\cdot\ldots\cdot q_s(z))$ is a generic point of a variety $W^* \subset \mathbb{C}_{X_1,\ldots,X_{n+1}}$ such that dim $W^* =$ dim W_1. Now $p \in \mathbb{C}[X_1, \ldots, X_n]$ can be looked at as an element of $\mathbb{C}[X_1, \ldots, X_{n+1}]$; p then defines the hypersurface $\mathsf{V}(p)^* = \mathsf{V}(p) \times \mathbb{C}_{X_{n+1}}$ in $\mathbb{C}_{X_1,\ldots,X_{n+1}}$. Since each of $W_2, \ldots, W_s$ is "mapped to infinity" under $V \to V^*$, the intersection of V^* with $\mathsf{V}(p)^*$ consists of only the single component W^*. Thus dim $W^* = \dim(V^* \cap \mathsf{V}(p)^*) \geqslant d - 1$. But dim $W^* =$ dim W_1, so dim $W_1 \geqslant d - 1$, which is what we wanted to prove. □

Exercises

3.1 Is the assumption that V is irreducible in Theorems 3.1 and 3.8 necessary?

3.2 Let $V \subset \mathbb{C}^n$ be irreducible. Show that there is a complex subspace L of $\mathbb{C}^n$ with dim $L =$ cod V so that every parallel translate of L intersects V properly. Can we replace "dim $L =$ cod V" by "dim $L \geqslant$ cod V"?

3.3 In the fifth sentence of Theorem 3.1's proof, suppose that "only if" is replaced by "provided that." Is this converse statement true?

3.4 Generalize Exercise 2.4 to homogeneous ideals in $\mathbb{C}[X_1, \ldots, X_n]$.

3.5 Rephrase Theorems 3.1 and 3.8 and Corollary 3.2 as statements about ideals.

3.6 Prove Lemma 3.7. [*Hint*: Replace the constants b_i and c_j in (6) and (7) by indeterminates Y_i and Z_j, obtaining

$$(X - Y_1)\cdot\ldots\cdot(X - Y_r) \in \mathbb{C}[X, Y_1, \ldots, Y_r] \tag{9}$$

$$(X - Z_1)\cdot\ldots\cdot(X - Z_s) \in \mathbb{C}[X, Z_1, \ldots, Z_s]. \tag{10}$$

Show that the resultant with respect to X of these new polynomials is homogeneous of degree rs. Now substitute Z_j in (10) for an arbitrary Y_i in (9), and conclude that an appropriate resultant is divisible by $Y_i - Z_j$, and therefore also by $\prod_{1 \leqslant i \leqslant r,\, 1 \leqslant j \leqslant s} (Y_i - Z_j)$. Obtain equality by comparing suitable terms on each side of (8).]

3.7 Suppose that in Theorems 3.1 and 3.8 "$\mathbb{C}^n$" is replaced by "an irreducible variety V," and suppose "cod W" means "dim $V -$ dim W." Are the new statements still true? Is the similar analogue of Corollary 3.2 true?

4 A Jacobian criterion for nonsingularity

In this section we prove Theorems 4.1 and 4.3, which give a "Jacobian" characterization of smoothness at a point P of an affine or projective variety. We prove it for irreducible varieties; the extension to arbitrary varieties is straightforward (Exercise 4.6). These results generalize the one for curves

(Theorem 7.4 of Chapter II). Recall the definition of smoothness, Definition 7.3 of Chapter II. With notation as in Notation 2.2, we have for the affine case:

Theorem 4.1. *Let $V \subset \mathbb{C}^n$ be an irreducible variety.*

$$V \text{ is smooth at } P \in V \Leftrightarrow \operatorname{rank}(J(V)_P) = \operatorname{cod} V.$$

Before proving this theorem, we note that it easily implies a projective analogue. For this, consider $P \in V \subset \mathbb{P}^n(\mathbb{C})$; if some affine representative $W \subset \mathbb{C}^n$ of V is smooth at P (that is, smooth at the corresponding affine representative Q of P), then any affine representative of V containing P is smooth at P. We see this as follows: Let $\mathsf{H}(V)$ be the homogeneous variety in $\mathbb{C}^{n+1}$ corresponding to V. Without loss of generality, assume coordinates $X_1, \ldots, X_{n+1}$ in $\mathbb{C}^{n+1}$ are such that $W = \mathsf{H}(V) \cap \mathsf{V}(X_{n+1} - 1)$ and $Q = \mathsf{H}(P) \cap \mathsf{V}(X_{n+1} - 1)$. Then, as in Lemma 2.6, $\operatorname{rank}(J(\mathsf{H}(V))_Q) = \operatorname{rank}(J(W)_Q)$. Also, $\operatorname{cod} \mathsf{H}(V)$ (in $\mathbb{C}^{n+1}$) is equal to $\operatorname{cod} W$ (in $\mathsf{V}(X_{n+1} - 1)$), for one may assume that the intersection with $\mathsf{V}(X_{n+1} - 1)$ is proper. (If $\mathsf{H}(V) \cap \mathsf{V}(X_{n+1} - 1) = \mathsf{V}(X_{n+1} - 1)$, then $V = \mathbb{P}^n(\mathbb{C})$; the result is trivial in this case.)

Assuming Theorem 4.1, we thus see that W is smooth at Q iff $\mathsf{H}(V)$ is smooth at Q. But then clearly $\mathsf{H}(V)$ is smooth at any nonzero point on the 1-subspace of $\mathbb{C}^{n+1}$ through Q. Similarly, for any affine representative W' of V at the corresponding affine representative Q' of P, $\operatorname{rank}(J(\mathsf{H}(V))_{Q'}) = \operatorname{rank}(J(W')_{Q'})$, and $\operatorname{cod} \mathsf{H}(V)$ (in $\mathbb{C}^{n+1}$) $= \operatorname{cod} W'$ (in a copy of $\mathbb{C}^n$); hence Theorem 4.1 implies that W' is smooth at Q'. Thus Theorem 4.1 implies that the following notion of smoothness is well defined:

Definition 4.2. An irreducible variety $V \subset \mathbb{P}^n(\mathbb{C})$ is **smooth at** $P \in V$ if some affine representative of V is smooth at the corresponding affine representative of P.

Then Theorem 4.1 implies the following projective analogue:

Theorem 4.3. *Let P be any point of an irreducible variety $V \subset \mathbb{P}^n(\mathbb{C})$, let $\mathsf{H}(V)$ be the corresponding homogeneous variety in $\mathbb{C}^{n+1}$, and let Q be any nonzero point on the* 1*-subspace $\mathsf{H}(P)$ of $\mathbb{C}^{n+1}$. Then V is smooth at P iff*

$$\operatorname{rank}(J(\mathsf{H}(V))_Q) = \operatorname{cod} \mathsf{H}(V).$$

In view of Theorems 4.1 and 4.3 we make this definition, which extends Definition 7.5 of Chapter II:

Definition 4.4. With notation as in Theorem 4.1, any irreducible variety $V \subset \mathbb{C}^n$ is **nonsingular at** P (or P is **nonsingular in** V) if $\operatorname{rank}(J(V)_P) = \operatorname{cod} V$; if $V \subset \mathbb{P}^n(\mathbb{C})$ is irreducible, then V is **nonsingular at** P if it is nonsingular in some affine representative of V containing P. Such a V in $\mathbb{C}^n$

or $\mathbb{P}^n(\mathbb{C})$ is **singular at** P (or P is **singular in** V) if it is not nonsingular there; V is **nonsingular** if it is nonsingular at each of its points.

PROOF OF THEOREM 4.1. We assume without loss of generality that P is the origin $(0) \in \mathbb{C}^n$.

$\Leftarrow$: If $\operatorname{rank}(J(V)_{(0)}) = \operatorname{cod} V$, then there are cod V rows in $J(V)$ which are linearly independent at (0). Since cod V equals the largest rank of $J(V)$ at points of V, its rank is never larger than cod V; and at all points of V near (0), its rank is never smaller than cod V since the entries of $J(V)$ are continuous functions. Hence the hypotheses of the general implicit mapping theorem (Theorem 3.5 of Chapter II) hold. Since we have already established Theorem 3.5 of Chapter II, we have also established $\Leftarrow$.

$\Rightarrow$: Our general strategy is this: We reduce the problem to the case when V is a hypersurface, for then one can proceed with exactly the same kind of argument used in Theorem 7.4 of Chapter II. If V is not already a hypersurface, we shall see that we may take as our hypersurface the (closed) projection of V on an appropriate subspace of $\mathbb{C}^n$ having dimension $\dim V + 1$.

Denote cod V by r, and $\operatorname{rank}(J(V)_{(0)})$ by s. We use a contrapositive argument. Therefore assume $s < r$ (we never have $s > r$), and assume V is smooth at (0)—that is, relative to coordinates $(X, Y) = (X_1, \ldots, X_{n-r}, Y_1, \ldots, Y_r)$ and neighborhoods $U_X \subset \mathbb{C}_X$ and $U_{Y_i} \subset \mathbb{C}_{Y_i}$ about (0), there are smooth, complex-valued functions $f_i : U_X \to U_{Y_i}$ $(i = 1, \ldots, r)$ such that the part of V near (0) is the zero-set of $F_1 = Y_1 - f_1, \ldots, F_r = Y_r - f_r$. We will obtain a contradiction.

First, it can be easily verified, just as in the proof of Theorem 7.4 of Chapter II, that the tangent space T to V at (0) is a complex subspace of $\mathbb{C}^n$ (rather than only a real subspace of $\mathbb{R}^{2n}$), for T is the limit of tangent planes T_{Q_i} at nonsingular points $Q_i \in V$ as $Q_i \to (0)$. Also, as in the proof of Theorem 7.4 of Chapter II, we may assume our coordinates (X, Y) in $\mathbb{C}^n$ have been chosen so that $\mathbb{C}_X$ is T. In fact, we may write the $2r \times 2n$ Jacobian matrix at (0) of the real and imaginary parts of the functions F_i with respect to the $2n$ real and imaginary axes of the X and Y coordinates, so that the last $2r$ columns are the "Y" columns, which furthermore form a $2r \times 2r$ identity matrix. These columns are thus linearly independent. Note that the derivative at (0) of $(F_1, \ldots, F_r)$ along any real 1-subspace of $\mathbb{C}_Y$ is a nonzero vector, while the derivative along any real 1-subspace of T is the zero vector. (This last statement is true because any real 1-subspace of T is the limit of real secant lines L_j through (0) and points $P_j \in V$ $(P_j \to (0))$, where the $F_i(P_j)$ all have the constant value zero.) Therefore if g is any complex-valued differentiable function on $\mathbb{C}^n$ such that g's zero-set includes V, then the derivative of g at (0) along any complex 1-subspace of T must be zero. This is, of course, true for each polynomial $p \in \mathsf{J}(V)$; hence all the vectors

$$\left(\frac{\partial p}{\partial X_1}, \ldots, \frac{\partial p}{\partial X_{n-r}}, \frac{\partial p}{\partial Y_1}, \ldots, \frac{\partial p}{\partial Y_r}\right)_{(0)}$$

lie in $\mathbb{C}_Y$. But since $s < r$, even more is true—by an appropriate choice of coordinates $Y_1, \ldots, Y_r$ in $\mathbb{C}_Y$, all these vectors may be taken to lie in $\mathbb{C}_{Y_2, \ldots, Y_r}$. Hence for any polynomial p vanishing on V, we have

$$\frac{\partial p}{\partial X_1} = \cdots = \frac{\partial p}{\partial X_{n-r}} = \frac{\partial p}{\partial Y_1} = 0.$$

We now assume our coordinates satisfy the above conditions; we may assume in addition that

$$\mathbb{P}^{r-1}(\mathbb{C}) \cap V = (0), \tag{11}$$

where $\mathbb{P}^{r-1}(\mathbb{C})$ denotes the projective completion of $\mathbb{C}_{Y_2, \ldots, Y_r}$ in $\mathbb{P}^n(\mathbb{C})$. With respect to these coordinates, let $\bar{\pi}(V)$ denote the closed projection of V on $\mathbb{C}_X \times \mathbb{C}_{Y_1}$. (Therefore $\bar{\pi}(V)$ is a hypersurface in $\mathbb{C}_X \times \mathbb{C}_{Y_1}$.) Since V is represented near (0) as the graph of $Y_1 - f_1, \ldots, Y_r - f_r$, for a sufficiently small neighborhood U of (0) in $\mathbb{C}_X \times \mathbb{C}_{Y_1}$, $Y_1 - f_1$ describes $\bar{\pi}(V) \cap U$. The condition (11) on our coordinates ensures that the graph of $Y_1 - f_1$ really does describe all of $\bar{\pi}(V) \cap U$; that is, there is no part of V in $U \times \mathbb{C}_{Y_2, \ldots, Y_r}$ other than that given by the graph of $Y_1 - f_1, \ldots, Y_r - f_r$, so V's projection into U consists of exactly the graph of $Y_1 - f_1$.

Since $\bar{\pi}(V)$ is a hypersurface in $\mathbb{C}_X \times \mathbb{C}_{Y_1}$, it is of the form $\bar{\pi}(V) = \mathsf{V}(q)$ for some polynomial $q \in \mathbb{C}[X, Y_1]$. We may assume that q has no nonconstant repeated factors. Now we are at a point analogous to the third paragraph from the end in the proof of Theorem 7.4 of Chapter II. On the one hand the hypersurface $\mathsf{V}(q)$ is locally the graph of a function; on the other hand,

$$\frac{\partial q}{\partial X_1} = \cdots = \frac{\partial q}{\partial X_{n-r}} = \frac{\partial q}{\partial Y_1} = 0.$$

This last means that the order of q at (0) must be $\geqslant 2$; one may now easily extend the argument for curves in Theorem 7.4 of Chapter II to show that in these coordinates V could not be the graph of a function, a contradiction. □

Corollary of Theorems 4.1 and 4.3. *The set of all singular points in an irreducible variety V in $\mathbb{C}^n$ or in $\mathbb{P}^n(\mathbb{C})$ is a proper subvariety of V.*

PROOF. The rank of any matrix is the largest of the ranks of its square submatrices, and the determinant of any square matrix is a polynomial in the entries of that matrix. □

EXERCISES

4.1 Let $p_1, \ldots, p_r \in \mathbb{C}[X_1, \ldots, X_n]$ define an irreducible variety V in $\mathbb{C}_{X_1, \ldots, X_n}$, and suppose that the $r \times n$ Jacobian $(\partial p_i/\partial X_j)$ has at $P \in V$ rank strictly less than cod V. Why does this not imply that V must be singular at P?

4.2 Let V_1 and V_2 be irreducible varieties in $\mathbb{C}^n$ or in $\mathbb{P}^n(\mathbb{C})$, and suppose neither variety is contained in the other. If $P \in V_1 \cap V_2$, show that $V_1 \cup V_2$ is not smooth at P.

4.3 Show that a pure-dimensional reducible variety $V \subset \mathbb{P}^n(\mathbb{C})$ cannot be smooth if $\dim V \geqslant n/2$.

4.4 If $V \subset \mathbb{C}^n$ and $W \subset \mathbb{C}^m$ are nonsingular, is $V \times W \subset \mathbb{C}^{n+m}$ nonsingular? Can the product of complex affine varieties with singular points be nonsingular?

4.5 Let P be a point in an irreducible variety $V \subset \mathbb{C}^n$. We say that a complex line in $\mathbb{C}^n$ through P is **tangent to** V **at** P if it is, in the obvious sense, the limit of some sequence of complex lines through P and $Q_i \in V$, as $Q_i \to P$ ($Q_i \neq P$). If V is nonsingular at P, show that the set of all complex lines tangent to V at P forms a linear variety in $\mathbb{C}^n$ of dimension equal to $\dim_P V$. We call this linear variety the **tangent space to** V **at** P.

4.6 Generalize the definitions and results in this section to include reducible varieties. [*Hint*: Use an appropriate definition of *local codimension.*] Test your results on different types of concrete examples, such as $\mathbb{C}_{XY} \cup \mathbb{C}_Z \subset \mathbb{C}_{XYZ}$, $\mathsf{V}(Y - X^2) \cup \mathsf{V}(Y) \subset \mathbb{C}_{XY}$, etc.

5 Connectedness and orientability

In Section 1 we stated that arbitrary irreducible complex varieties are connected and orientable. We prove these two facts in this section. In this section and throughout the remainder of this book, we will use phrases like "Property A **holds almost everywhere** (or **at almost each point**) **on an irreducible** (or, more generally **a pure-dimensional**) **variety** V" if Property A holds at all points of V off some subvariety W of V, where $\dim W < \dim V$.

Theorem 5.1. *Let V be any irreducible variety in $\mathbb{C}^n$ or in $\mathbb{P}^n(\mathbb{C})$. Then V is connected.*

PROOF. We prove the theorem by showing that for any two points $P_1, P_2 \in V$, there is a connected subset of V containing P_1 and P_2 (Lemma 8.8 of Chapter II.) If V is projective, we may dehomogenize $\mathbb{P}^n(\mathbb{C})$ at a hyperplane containing neither P_1 nor P_2; therefore without loss of generality, *we assume V is affine.* We also assume $\dim V \geqslant 1$, since an irreducible variety of dimension zero consists of only one point, and $\varnothing$ is trivially connected.

We begin by showing this:

> **(5.2)** Any irreducible curve C in $\mathbb{C}^n = \mathbb{C}_{X_1, \ldots, X_n}$ is connected.

To prove (5.2), let the coordinate ring of C be $R_C = \mathbb{C}[x_1, \ldots, x_n]$; R_C's transcendence degree over $\mathbb{C}$ is 1. We may assume that $x_1 = X_1$ is transcendental over $\mathbb{C}$, and that x_2 is a primitive element in R_C of $\mathbb{C}(x_1, \ldots, x_n)$ over $\mathbb{C}(x_1)$; thus each of $x_3, \ldots, x_n$ is a rational function of x_1, x_2. Therefore if C' is the irreducible curve in $\mathbb{C}_{X_1 X_2}$ with generic point (x_1, x_2), then over almost each point of C', there $\pi_{X_3, \ldots, X_n}$-lies just one point of C.

Now if q_i is a minimal polynomial of x_i over $\mathbb{C}[x_1]$ for $i = 2, \ldots, n$, then the union of the discriminant varieties, $\bigcup_{i=2}^{n} \mathsf{V}(\mathscr{D}_{X_i}(q_i))$, consists of only finitely many points of $\mathbb{C}_{X_1}$. Also note that over *no* point of $\mathbb{C}_{X_1}$ are there infinitely many points of C or C', for then all of C or C' (they are both irreducible) would lie above that point, and x_1 would not be transcendental over $\mathbb{C}$. Finally, note that for each q_i, the conditions of Corollary 3.9 of Chapter II are satisfied at any point of $\mathsf{V}(q_i) \subset \mathbb{C}_{X_1X_i}$ not lying over a point of $\bigcup_{i=2}^{n} \mathsf{V}(\mathscr{D}_{X_i}(q_i))$. These facts imply that at almost every point P of C, the part of C near P is the graph of an analytic function

$$X_i = \psi_i(X_1) \qquad (i = 2, \ldots, n).$$

We shall write $\psi = (\psi_2, \ldots, \psi_n)$. Likewise, at almost every point Q of C', the part of C' near Q is given by the graph of

$$X_2 = \psi_2(X_1).$$

There is thus a homeomorphism h, given by $(a_1, a_2) \to \psi(a_1)$ from $C' \backslash \{\text{finitely many points}\}$ to $C \backslash \{\text{finitely many points}\}$. Since C' is an irreducible plane curve, $C' \backslash \{\text{finitely many points}\}$ is connected. (See Exercise 8.2 of Chapter II.) Since a connected set has connected closure (Lemma 8.3 of Chapter II), and since C is the closure in $\mathbb{C}^n$ of the image under h of $C' \backslash \{\text{finitely many points}\}$, we see C is connected. Thus (5.2) is proved.

We may now easily complete the proof of Theorem 5.1. Let P_1 and P_2 be any two points of an irreducible variety $V \subset \mathbb{C}^n = \mathbb{C}_{X_1, \ldots, X_n}$. We shall prove our theorem by showing that there is an irreducible curve $C \subset V$ containing P_1 and P_2; we do this by finding a generic point of a curve in V such that P_1 and P_2 are specialization points of that generic point. First, choose $\mathbb{C}_{X_1}$ so it passes through P_1 and P_2, with $\mathbb{C}_{X_1}$-coordinates 0 and 1, respectively. Let $\mathbb{C}[x_1, \ldots, x_n]$ be the coordinate ring of V; then x_1 is transcendental over $\mathbb{C}$. (If x_1 were algebraic over $\mathbb{C}$, then $x_1 \in \mathbb{C}$, and x_1 could not attain both the values 0 and 1.) Now assume the axes $\mathbb{C}_{X_2}, \ldots, \mathbb{C}_{X_n}$ have been chosen so that $\{x_1, \ldots, x_d\}$ forms a transcendence base of $\mathbb{C}[x_1, \ldots, x_n]$ over $\mathbb{C}$; by tipping the axes $\mathbb{C}_{X_{d+1}}, \ldots, \mathbb{C}_{X_n}$ a bit as in the proof of the normalization lemma (Lemma 11.5, of Chapter III), we may further assume that $\mathbb{C}[x_1, \ldots, x_n]$ is *integral* over $\mathbb{C}[x_1, \ldots, x_d]$. Now the map $x_1 \to x_1, x_2 \to 0, \ldots, x_d \to 0$ defines a homomorphism of $\mathbb{C}[x_1, \ldots, x_d]$. The argument used in proving Theorem 11.1.2 of Chapter III shows that this map can be successively extended to a homomorphism of $\mathbb{C}[x_1, \ldots, x_n]$. Let the image of this extended homomorphism be $R = \mathbb{C}[x_1, 0, \ldots, 0, y_{d+1}, \ldots, y_n]$. This ring has transcendence degree 1 over $\mathbb{C}$ (each y_i is algebraic over $\mathbb{C}(x_1)$; since $P_1 = 0 \in \mathbb{C}_{X_1}$ and $P_2 = 1 \in \mathbb{C}_{X_1}$ are in V, we know that $(0, \ldots, 0)$ and $(1, 0, \ldots, 0)$ define specializations of R. Thus the generic point $(x_1, 0, \ldots, 0, y_{d+1}, \ldots, y_n)$ defines an irreducible curve C in V which passes through P_1 and P_2, as promised. □

We now turn to the question of orientability. We prove that irreducible nonsingular varieties are orientable. As noted in Remark 9.4 of Chapter II, the definition of a smooth orientable manifold of dimension n is exactly that for dimension 2 (Definition 9.3 of Chapter II) with 2 replaced by n.

We see that any irreducible nonsingular variety of dimension d is a smooth, real $2d$-manifold.

Theorem 5.3. *Let V be an irreducible d-dimensional nonsingular variety in $\mathbb{P}^n(\mathbb{C})$ or $\mathbb{C}^n$. Then V is orientable as a real $2d$-manifold.*

PROOF. The proof is a generalization of that for curves in Chapter II. First, we know from Theorem 4.1 that the part of V about an arbitrary point $P \in V$ is locally the graph of an analytic function. Then, just as in the proof for curves, there are V-neighborhoods $U(Q)$ and $U(Q')$ containing P, and associated analytic maps Φ_Q and $\Phi_{Q'}$, defining homeomorphisms from neighborhoods of $\mathbb{R}^{2d}$ to $U(Q)$ and to $U(Q')$, respectively; we want to show that for any such neighborhoods and maps, $\Phi = \Phi_{Q'}{}^{-1} \circ \Phi_Q$ is orientation preserving. That is, writing $Z_1 = X_1 + iX_2, \ldots, Z_d = X_{2d-1} + iX_{2d}$, and $\Phi = \Phi(X_1, \ldots, X_{2d}) = (\Phi_1 + i\Phi_2), \ldots, (\Phi_{2d-1} + i\Phi_{2d})$, we want to show that $\det(\partial\Phi_i/\partial X_j) > 0$ at each point of $\Phi_Q{}^{-1}(U(Q) \cap U(Q'))$. As in the case of curves, this determinant is nonzero at each such point, since Φ is invertible. To prove it is positive, we put $(\partial\Phi_i/\partial X_j)$ into a different form, without changing its determinant. First, d pairwise interchanges of $(\partial\Phi_i/\partial X_j)$'s columns, and another d such interchanges of the rows, totalling an even number of pairwise switches, leave the value of $\det(\partial\Phi_i/\partial X_j)$ unchanged. We may therefore assume that the Φ_i and the X_j appear in this order: $(\Phi_1, \Phi_3, \ldots, \Phi_{2d-1}, \Phi_2, \Phi_4, \ldots, \Phi_{2d})$, and $(X_1, X_3, \ldots, X_{2d-1}, X_2, X_4, \ldots, X_{2d})$. Hence the real parts of the Φ_i and X_j are in the first d rows and d columns, respectively. The Cauchy–Riemann equations directly show that the matrix is now of the form

$$\begin{pmatrix} A & B \\ -B & A \end{pmatrix},$$

where A and B are $d \times d$ matrices. If I is the identity matrix of order d, then

$$\begin{pmatrix} I & -iI \\ 0 & I \end{pmatrix} \quad \text{and} \quad \begin{pmatrix} I & iI \\ 0 & I \end{pmatrix}$$

both have determinant 1. Multiplying by these elementary matrices defines row and column operations on $2d \times 2d$ matrices. In particular, we have

$$\begin{pmatrix} I & -iI \\ 0 & I \end{pmatrix}\begin{pmatrix} A & B \\ -B & A \end{pmatrix}\begin{pmatrix} I & iI \\ 0 & I \end{pmatrix} = \begin{pmatrix} A + iB & 0 \\ -B & A - iB \end{pmatrix};$$

The determinant of this is

$$\det(A + iB) \cdot \det(A - iB) = \det(A + iB) \cdot \overline{\det(A + iB)} > 0. \qquad \square$$

6 Multiplicity

In this section we consider the notion of *degree of a variety* and the related concept, *multiplicity of intersection of properly intersecting varieties*. In the next section we prove the basic *Bézout Theorem* which relates these notions.

The word *degree* perhaps brings to mind the degree of a polynomial. For a nonzero polynomial of one variable $p \in \mathbb{C}[X]$, the fundamental theorem of algebra tells us that p has deg p zeros, counted with multiplicity; we may say, geometrically, that the graph of $Y = p(X)$ intersects $\mathbb{C}_X$ in deg p points "counted with multiplicity." We may think of the multiple points as becoming separated by translating the line $Y = 0$ a bit. Thus, rather than intersecting the graph of $Y - p(X)$ with the line $Y = 0$, if we intersect it with a translate $Y = c$, we are then looking at the zeros of $p(X) = c$. Now $p(X) - c$ has a multiple zero at a point $a \in \mathbb{C}_X$ iff both $p(a) - c = 0$ and $p'(a) = 0$. But there are only finitely many points a such that $p'(a) = 0$, and, of course, for each such a, there is only one c such that $p(a) = c$. Hence for all but finitely many points c, $p(X) - c$ has distinct zeros—that is, almost all lines $Y = c$ intersect the variety $\mathsf{V}(Y - p(X))$ in deg p distinct points.

There are a number of generalizations of this. For instance, one can suitably parametrize all complex lines in $\mathbb{C}_{XY}$, and then prove that almost all these lines intersect $\mathsf{V}(Y - p(X))$ in exactly deg p points; one can also prove various higher-dimensional generalizations of this, as well as projective analogues. In this way we will be led to a geometric definition of degree for any variety. An analogous route will lead us to a way of counting multiple components of intersection of properly-intersecting varieties.

In the example above we translated (or in a sense "perturbed") one of the intersecting varieties to separate multiple points. A basic idea that we use again and again is to appropriately perturb varieties having a zero-dimensional intersection so that any multiple points of intersection are separated and can be counted, thus allowing us to make notions such as *degree* and *multiplicity* precise. We shall modify varieties using linear changes in the variables of the polynomials defining them. Such linear changes, when nonsingular, are so mild that they don't change the degree of any polynomial; the "singular perturbations" are important too, for they can simplify varieties by changing them into unions of linear varieties, where counting intersection points is an easy matter. (We use this last idea in Section 7.)

In this section, we first briefly describe these linear changes; we then give a sequence of definitions of "order" and "multiplicity," each based on a corresponding theorem. For an arbitrary irreducible complex variety, we will have both "local" and "global" definitions which respectively generalize the local notion of order at a point $X = x_0$ of $p(X) \in \mathbb{C}[X]$, and the global notion of total degree (or total order) of $p(X)$. We then define *multiplicity of intersection* for components of properly-intersecting varieties, which leads to a fundamental homomorphism property of degree (Bézout's theorem, Theorem 7.1). Our definitions are essentially geometric and, as one might expect, they can all be translated into purely algebraic terms.

We now turn to the *linear perturbations*. Since many of our considerations will take place in projective space, we work in a projective or homogeneous variety setting; it will be seen that most of these results hold in a general affine setting, too.

First, consider $\mathbb{P}^n(\mathbb{C})$ as the set of 1-subspaces in $\mathbb{C}_{X_1,\ldots,X_{n+1}} = \mathbb{C}_X$. Any $(n + 1) \times (n + 1)$ matrix A, with coefficients in $\mathbb{C}$, defines a linear transformation $X \to XA$ of the variables $X = (X_1, \ldots, X_{n+1})$, which in turn transforms any homogeneous polynomial $p(X)$ into another homogeneous polynominal $p(XA) = p^*(X)$. This transforms any homogeneous variety $V = \mathsf{V}(I)$ into another homogeneous variety $V^* = \mathsf{V}(I^*)$, where

$$I^* = \{p^*(X) = p(XA) | p(X) \in I\}. \tag{12}$$

In this way A induces a transformation on any projective variety in $\mathbb{P}^n(\mathbb{C})$. If A is nonsingular, it is easily checked that for any homogeneous variety $V \subset \mathbb{C}^{n+1}$,

$$V^* = VA^{-1} = \{x \in \mathbb{C}^{n+1} | xA \in V\}. \tag{13}$$

Note that the two $(n + 1) \times (n + 1)$ matrices (a_{ij}) and (ca_{ij}) $(c \in \mathbb{C}\backslash\{0\})$ induce the same transformation on projective varieties in $\mathbb{P}^n(\mathbb{C})$.

Example 6.1

(6.1.1) Any projective subspace $L^r \subset \mathbb{P}^n(\mathbb{C})$ (given by an $(r + 1)$-subspace of $\mathbb{C}^{n+1}$) is transformed by any $(n + 1) \times (n + 1)$ matrix A into a projective subspace of dimension $\geqslant r$. If A is nonsingular, the transformed space has dimension exactly r.

(6.3.2) In $\mathbb{P}^2(\mathbb{C})$, the circle defined by $X_1^2 + X_2^2 - X_3^2$ is transformed by

$$A = \begin{pmatrix} 1 & 0 & 0 \\ 0 & 1 & 1 \\ 0 & 1 & -1 \end{pmatrix},$$

into another nondegenerate quadratic curve (a parabola relative to $\mathbb{C}_{X_1X_2}$).

Note that the circle can be degenerated into two lines by a matrix such as

$$\begin{pmatrix} 0 & 0 & 0 \\ 0 & 1 & 0 \\ 0 & 0 & 1 \end{pmatrix},$$

which changes $X_1^2 + X_2^2 - X_3^2$ into $X_2^2 - X_3^2 = (X_2 + X_3)(X_2 - X_3)$, whose variety is the union of two projective 1-subspaces of $\mathbb{P}^2(\mathbb{C})$. Of course this method of degenerating a circle can easily be extended to any quadratic hypersurface in $\mathbb{P}^n(\mathbb{C})$; for example, any such variety can be degenerated to a union of two hyperplanes in $\mathbb{P}^n(\mathbb{C})$.

Often we want to consider all possible "linear changes" of a variety, or at least all "small" changes (those whose matrices are entrywise near the identity matrix). For this purpose, it is natural to use matrices $U = (U_{ij})$ with $(n + 1)^2$ algebraically independent indeterminant entries U_{ij}. Any homogeneous variety $V \subset \mathbb{C}^{n+1}$ together with all its "linear transforms" of the type we are considering, then forms a 2-way, or **bihomogeneous** variety in $\mathbb{C}^{(n+1)^2} \times \mathbb{C}^{n+1}$

(homogeneous in $(X_1, \ldots, X_{n+1})$, and homogeneous in the entries U_{ij}); over each point of $\mathbb{C}^{(n+1)^2}$ lies a transform of V. When considering intersections of several varieties, we often want to perturb each V_i independently in order to separate any multiple components of intersection. (By counting the separated components, we then arrive at the *multiplicity of intersection*.) For instance, to consider all independent transforms of each of two intersecting varieties V_1 and $V_2 \subset \mathbb{P}^n(\mathbb{C})$, we may use $2(n+1)^2$ indeterminants ($(n+1)^2$ in each of two matrices), thus getting a 3-way homogeneous subvariety of $\mathbb{C}^{(n+1)^2} \times \mathbb{C}^{(n+1)^2} \times \mathbb{C}^{n+1}$; over each point of $\mathbb{C}^{(n+1)^2} \times \mathbb{C}^{(n+1)^2}$ lies the intersection of two independent transforms V_1' and V_2'', respectively.

Suppose varieties $V_1, \ldots, V_s \subset \mathbb{P}^n(\mathbb{C})$ are independently transformed, these independent transformations being specified by points of $\mathbb{C}^{s(n+1)^2}$. It often happens that a property holds for *almost all* these transforms, in the sense that it holds for all s-tuples of varieties corresponding to points off a proper subvariety of $\mathbb{C}^{s(n+1)^2}$. When we use a phrase such as **almost all linear transforms**, or **perturbations, of a set of varieties** $V_1, \ldots, V_s$, we shall mean it in the above sense. In the special case of all transforms of a projective subspace $L^r \subset \mathbb{P}^n(\mathbb{C})$, we may use without ambiguity the phrase **almost all transforms of any r-dimensional projective subspace of** $\mathbb{P}^n(\mathbb{C})$, since any two projective r-subspaces of $\mathbb{P}^n(\mathbb{C})$ are related by some nonsingular $(n+1) \times (n+1)$ matrix.

We now turn to our first theorem which allows us to define the degree of an arbitrary variety (Definition 6.3).

Theorem 6.2

(6.2.1) *For any variety V in $\mathbb{P}^n(\mathbb{C})$ or in $\mathbb{C}^n$ of pure dimension r, almost all linear transforms of any projective subspace $L^{n-r} \subset \mathbb{P}^n(\mathbb{C})$, or of any affine subspace $L^{n-r} \subset \mathbb{C}^n$, intersect V in a common, fixed number of distinct points.*

(6.2.2) *If V in (6.2.1) is a hypersurface defined by a product p of distinct irreducible polynomials in $\mathbb{C}[X_1, \ldots, X_{n+1}]$, or in $\mathbb{C}[X_1, \ldots, X_n]$, then this common number is* $\deg p$.

Definition 6.3. For any affine or projective variety of pure dimension, the number given in Theorem 6.2 is called the **degree of** V; we denote this degree by $\deg V$.

Proof of Theorem 6.2

(6.2.1). Let $\{U_{ij}\}$ $(i, j = 1, \ldots, n+1)$ be $(n+1)^2$ algebraically independent indeterminates. Then the bihomogeneous variety consisting of all transforms of some projective subspace L^{n-r} (say $L^{n-r} = \mathsf{V}(l_\alpha(X))$, where the l_α are linear in $X = (X_1, \ldots, X_{n+1})$) is

$$V^\dagger = \mathsf{V}(l_\alpha(X(U_{ij}))). \tag{14}$$

If V is projective, let $\mathsf{V}(X_{n+1} - 1)$ in $\mathbb{C}^{n+1}$ define a hyperplane in $\mathbb{P}^n(\mathbb{C})$ intersecting V properly, and let V' be the dehomogenization

$\mathsf{H}(V) \cap \mathsf{V}(X_{n+1} - 1)$ in $\mathbb{C}^{n+1}$ where $\mathsf{H}(V)$ is the homogeneous variety in $\mathbb{C}^{n+1}$ corresponding to V; if $V \subset \mathbb{C}^n$ is affine, let V' be a copy of this variety in $\mathsf{V}(X_{n+1} - 1)$. Then the subvariety

$$V^\dagger \cap (\mathbb{C}^{(n+1)^2} \times V') \tag{15}$$

of $\mathbb{C}^{(n+1)^2} \times \mathbb{C}^{n+1}$ has only finitely many points over almost each point of $\mathbb{C}^{(n+1)^2}$ (Exercise 6.1). At almost every point of $\mathbb{C}^{(n+1)^2}$, these finitely many points correspond to all points of V's intersection with the corresponding transform of L^{n-r}. Now, applying Theorem 11.1.1 of Chapter III to each component of $V^\dagger \cap (\mathbb{C}^{(n+1)^2} \times V')$ projecting onto $\mathbb{C}^{(n+1)^2}$, we see that over almost every point of $\mathbb{C}^{(n+1)^2}$ there is a fixed number of points, which means that almost every transform of L^{n-r} intersects V in a fixed number of points.

(6.2.2). It suffices to show that over each point in some open neighborhood of $\mathbb{C}^{(n+1)^2}$ there lie exactly $\deg p$ points of $V^\dagger \cap (\mathbb{C}^{(n+1)^2} \times V')$. In this case, L^{n-r} is a complex line L. Let

$$X_1 = c_1T - a_1, \quad \ldots \quad, X_n = c_nT - a_n$$

be a parametrization of any line L in $\mathbb{C}^n = \mathsf{V}(X_{n+1} - 1) \subset \mathbb{C}^{n+1}$. The zeros of $p^\times(T) = p(c_1T - a_1, \ldots, c_nT - a_n)$ (or of $p(c_1T - a_1, \ldots, c_nT - a_n, 1)$ in the homogeneous case) give those points in which L intersects $\mathsf{V}(p)$. We may choose $(c_1, \ldots, c_n, a_1, \ldots, a_n)$ so that these conditions are satisfied:

(a) $\deg p^\times(T) = \deg p$. (Obvious.)
(b) $p^\times(T)$ has $\deg p$ distinct zeros.
(The coefficient of $T^{\deg p}$ in $p^\times(T)$ is a nonzero polynomial in $c_1, \ldots, c_n$, $a_1, \ldots, a_n$; the discriminant $\mathscr{D}_T(p^\times(T))$ is a polynomial in $c_1, \ldots, c_n$, $a_1, \ldots, a_n$ too, and is nonzero since p is a product of distinct irreducibles.)

Clearly, if a fixed $2n$-tuple $(c_1, \ldots, c_n, a_1, \ldots, a_n)_0$ satisfies (a) and (b), so do all nearby $2n$-tuples. If P_0 is any point in $\mathbb{C}^{(n+1)^2}$ over which lies the line L_0 determined by $(c_1, \ldots, c_n, a_1, \ldots, a_n)_0$, then all points in $\mathbb{C}^{(n+1)^2}$ near P_0 correspond to lines "near" L_0, which therefore also intersect $\mathsf{V}(p)$ in $\deg p$ distinct points. Hence above each point in an open neighborhood of $\mathbb{C}^{(n+1)^2}$ there lie exactly $\deg p$ points of $V^\dagger \cap (\mathbb{C}^{(n+1)^2} \times V)$. □

Theorem 6.2 and Definition 6.3 are global in the sense that they refer to a property of the entire variety. Now, for example, note that although the graph in $\mathbb{C}_{XY}$ of $Y - [(X - c_1)^{m_1} \cdot \ldots \cdot (X - c_s)^{m_s}]$ ($c_1, \ldots, c_s$ distinct) intersects the line $Y = 0$ in one point at each point $(c_i, 0)$, small translates $Y = c$ intersect the graph in m_i distinct points near $(c_i, 0)$. We thus see a geometric meaning to "the order of $p(X)$ at c_i." It is therefore natural to ask if, for arbitrary varieties of pure dimension, there is a local analogue of the global notion of degree defined above.

For such a notion, we would want this: Given a variety V^r of pure dimension in $\mathbb{P}^n(\mathbb{C})$ or $\mathbb{C}^n$, given a point $P \in V^r$, and given any projective or affine subspace $L_0{}^{n-r}$ properly intersecting V^r at P, for almost all subspaces L^{n-r} near $L_0{}^{n-r}$, there is a common, constant number of distinct points of intersection near P. This would be the local degree or *order* of V at P in the direction $L_0{}^{n-r}$. One could then further consider *all* subspaces L^{n-r} of $\mathbb{P}^n(\mathbb{C})$ or $\mathbb{C}^n$ through P and ask if in fact we get the same order in the direction L^{n-r} for almost all these subspaces L^{n-r}. This would then represent *the* order at P of V (rather than the order in a particular direction. (Example: At (0, 0), the variety $\mathsf{V}(Y - X^2) \subset \mathbb{C}_{XY}$ has order 2 in the $\mathbb{C}_X$-direction, but order 1 in all other directions, so the order at (0, 0) of $\mathsf{V}(Y - X^2)$ is 1.)

The answer to each of the above queries is "yes": There is a local order at each point of any variety relative to a direction as well as one in an absolute sense. Our treatment will be essentially parallel to that of the (global) degree defined above: Theorems 6.6 and 6.8 allow us to extend the local definitions of relative and absolute order of a polynomial to local relative and absolute order of a variety (Definitions 6.7 and 6.9).

We begin with algebraic definitions of *relative* and *absolute* order of a polynomial at a point; we then will have local results (Theorems 6.6.2 and 6.8.2) which translate these definitions into geometric terms.

Definition 6.4. If $p \in \mathbb{C}[X_1, \ldots, X_m]$ is expanded about $(a_1, \ldots, a_m)$ so that a typical term of p is a nonzero constant times $(X_1 - a_1)^{d_1} \cdot \ldots \cdot (X_m - a_m)^{d_m}$, then the lowest total degree of all such terms of p is called the **total order**, or **order in** $X_1, \ldots, X_m$, or simply the **order of** p **at** $(a_1, \ldots, a_m)$. More generally, p has at each point $(a_1, \ldots, a_m)$ an order with respect to any affine subspace A of $\mathbb{C}^m$ through $(a_1, \ldots, a_m)$, in the following way: If A has dimension r and is parametrized by

$$X_i = a_i + \sum_{j=1}^{r} c_{ij} T_j \qquad (i = 1, \ldots, m),$$

then the **order with respect to** A (or **in the direction of** A) **of** $p = p(X_1, \ldots, X_m)$ **at** $(a_1, \ldots, a_m)$ is the order in $T_1, \ldots, T_r$ of

$$p\left(a_1 + \sum_{j=1}^{r} c_{1j} T_j, \ldots, a_m + \sum_{j=1}^{r} c_{mj} T_j\right)$$

at $(0, \ldots, 0)$. (Given A and $(a_1, \ldots, a_m)$, this order is easily seen to be independent of the choice of the coefficients c_{ij} parametrizing A.) In the special case when A lies in the direction of a coordinate subspace, for instance if A is given by $X_{r+1} = a_{r+1}, \ldots, X_m = a_m$ (or parametrically, by $X_i = a_i + T_i$ for $i = 1, \ldots, r$ and $X_i = a_i$ for $i = r + 1, \ldots, m$), then the order with respect to A of p at $(a_1, \ldots, a_m)$ is called the **order in** $X_1, \ldots, X_r$ **of** p **at** $(a_1, \ldots, a_m)$. It is evidently equal to the order in $X_1, \ldots, X_r$ of $p(X_1, \ldots, X_r, a_{r+1}, \ldots, a_m)$ at $(a_1, \ldots, a_r)$. (Cf. Definition 10.3 of Chapter

II.) Finally, if $p \in \mathbb{C}[X_1, \ldots, X_{m+1}]$ is homogeneous and if A is any linear subspace of $\mathbb{C}_{X_1,\ldots,X_{m+1}}$ through $(a_1, \ldots, a_{m+1})$, it is easily checked that the order with respect to A of p at $(a_1, \ldots, a_{m+1})$ is the same as the order at $(ca_1, \ldots, ca_{m+1})$ $(c \in \mathbb{C}\backslash\{0\})$. Since $(a_1, \ldots, a_{m+1})$ and $(ca_1, \ldots, ca_{m+1})$ both define the same point in $\mathbb{P}^m(\mathbb{C})$, there is, in $\mathbb{P}^m(\mathbb{C})$ with homogeneous coordinates $(X_1, \ldots, X_{m+1})$, a well-defined notion of **order with respect to a projective subspace** $L \subset \mathbb{P}^m(\mathbb{C})$ **of a homogeneous polynomial** $p \in \mathbb{C}[X_1, \ldots, X_{m+1}]$ **at** $P \in \mathbb{P}^m(\mathbb{C})$. It is a straightforward exercise to verify that the order in the projective setting is the same as the order in any fixed dehomogenization of P, $\mathbb{P}^m(\mathbb{C})$ and p (such that P is not a point at infinity).

EXAMPLE 6.5

(6.5.1) The polynomial $X^2 - X \in \mathbb{C}[X]$ has degree 2, and has order 1 in X at (0); since $X^2 - X = (X - 1)^2 + (X - 1)$, it also has order 1 in X at (1). Expansion about any other point (a) yields order 0 at (a).

(6.5.2) The polynomial $p(X, Y) = Y^2 - X^3 - X^2$ (defining an alpha curve) has degree 3, and has (total) order 2 at (0, 0); it has order 1 at any other point of the curve, and order 0 at any point off the curve. It has order 3 in X at (0, 0) in the directions $Y = \pm X$ and order 2 at (0, 0) with respect to any other direction.

We next state the local form of Theorem 6.2. First, note that if V is any projective or affine variety in $\mathbb{P}^n(\mathbb{C})$ or $\mathbb{C}^n$, then in the variety of all linear transforms of V, V lies above the identity matrix. We shall say that V^T is *near* V if V^T is the transform of V by a matrix T, all of those entries are close to the corresponding entries of the identity matrix. Since all matrices near the identity matrix are nonsingular, any V^T close to V has the same dimension as V.

Theorem 6.6

(6.6.1) *Let P be a point of a pure-dimensional variety V^r in $\mathbb{P}^n(\mathbb{C})$ or $\mathbb{C}^n$, and let L be a projective or affine $(n - r)$-subspace of $\mathbb{P}^n(\mathbb{C})$ or $\mathbb{C}^n$ properly intersecting V at P. Then for almost every linear transform L' of L^{n-r} sufficiently near L, there is a common fixed number of distinct points of $V \cap L'$ arbitrarily near P.*

(6.6.2) *If V is a projective or affine hypersurface in $\mathbb{P}^n(\mathbb{C})$ or $\mathbb{C}^n$ defined by a product p of distinct, irreducible polynomials (homogeneous in $\mathbb{C}[X_1, \ldots, X_{n+1}]$ in the projective case, or ordinary polynomials in $\mathbb{C}[X_1, \ldots, X_n]$ in the affine case), and if L is any projective or affine line properly intersecting V at P, the number given in* (6.6.1) *is the order with respect to L of p at P.*

Definition 6.7. The number in Theorem 6.6 is called the **order with respect to** L **of** V **at** P, or the **multiplicity of intersection of** V **and** L **at** P; we denote it by $i(V, L; P)$.

Note that for any fixed point $P \in \mathbb{P}^n(\mathbb{C})$, the set S of all matrices transforming a given projective subspace $L^r \subset \mathbb{P}^n(\mathbb{C})$ to a subspace containing P (or *P-containing* subspace) forms a subspace of $\mathbb{C}^{(n+1)^2}$. For if A and B are matrices in S, if $a = (a_1, \ldots, a_{n+1})$ are homogeneous coordinates for P, and if $\ell(X_1, \ldots, X_{n+1})$ is any linear function in the homogeneous ideal defined by L^r, we have $\ell(a(A + B)) = \ell(aA) + \ell(aB) = 0 + 0 = 0$. Hence S is closed under addition. Similarly, it is closed under scalar multiplication by elements in $\mathbb{C}$. Thus S is a subspace of $\mathbb{C}^{(n+1)^2}$. It is obvious that S is proper if $L^r \subsetneq \mathbb{P}^n(\mathbb{C})$. (Similar statements hold in the affine case.) One thus sees that there is a well-defined concept of "almost all P-containing transforms" of a projective or affine subspace.

Corresponding to the relative Theorem 6.6, we have this absolute result:

Theorem 6.8

(6.8.1) *Let P be any point of a pure r-dimensional variety $V = V^r$ in $\mathbb{P}^n(\mathbb{C})$ or $\mathbb{C}^n$. For almost all P-containing transforms L' of an $(n - r)$-dimensional projective or affine subspace L of $\mathbb{P}^n(\mathbb{C})$ or $\mathbb{C}^n$, $i(V, L'; P)$ is defined and has a common, fixed value.*

(6.8.2) *If V in $\mathbb{P}^n(\mathbb{C})$ or $\mathbb{C}^n$ is defined by a product p of distinct, irreducible polynomials (in $\mathbb{C}[X_1, \ldots, X_{n+1}]$ or $\mathbb{C}[X_1, \ldots, X_n]$, respectively), this common number is the order of p at P.*

We then have

Definition 6.9. The number given by Theorem 6.8 is called the **order of V at P**, or the **multiplicity of V at P**; we denote it by $m(V; P)$.

Our proofs of Theorems 6.6 and 6.8 will run along the same lines as that of Theorem 6.2, and for this reason we need corresponding local forms of Theorem 11.1.1 of Chapter III.

The analogue of Theorem 11.1.1 of Chapter III we use in proving Theorem 6.6 is contained in the following

Theorem 6.10. *One may, in Theorem 2.13, replace the concluding phrase "above each point a in Δ^d there is a point of V in $a \times \Delta^{n-d}$" by the phrase "above almost each point a in Δ^d there is a common, fixed (positive) number of points of V in $a \times \Delta^{n-d}$."*

The proof is the same as the proof of Theorem 2.13.

PROOF OF THEOREM 6.6

(6.6.1) The proof is essentially the same as the proof of Theorem 6.2; one need only replace the reference to Theorem 11.1.1 of Chapter III by a reference to Theorem 6.10, applied at a point of $\mathbb{C}^{(n+1)^2} \times \mathbb{C}^{n+1}$ corresponding to the point P of $V \cap L$.

(6.6.2) We may assume that V is affine. Therefore let $P = (a_1, \ldots, a_n) \in \mathbb{C}^n$; then with respect to any of L's parametrizations $X_i - a_i = c_i T$ $(i = 1, \ldots, n)$, $p(c_1 T, \ldots, c_n T)$ has a zero of multiplicity m at $T = 0$, where m is the order with respect to L of p at $(a_1, \ldots, a_n)$. But since $p(X_1, \ldots, X_n)$ is a product of distinct irreducibles, its zeros are all distinct on almost every line in $\mathbb{C}^n$ (Theorem 6.2), hence also on almost every line near L. Applying Lemma 10.4 of Chapter II together with Theorem 6.10 to the variety corresponding to (15) shows that for every L' sufficiently near L, there are exactly m distinct points of $V \cap L'$ arbitrarily near P. □

Let us now see what is involved in proving Theorem 6.8. In Theorem 6.6 we showed that the number of points near P of $V^r \cap L'$ is the same for all L' near a fixed $L = L^{n-r}$ properly intersecting V^r (that is, the same for all points near that point of $\mathbb{C}^{(n+1)^2}$ corresponding to L). We want to generalize from *one* space L through P, to *all* transforms of L^{n-r} passing through P (that is, from one point of $\mathbb{C}^{(n+1)^2}$ to a whole subspace of $\mathbb{C}^{(n+1)^2}$). For this, we shall appropriately generalize Theorem 6.10 so "$(0) \in V$" can be replaced by "irreducible subvariety of V." First, from Theorem 6.10, we know that if $V^s \subset \mathbb{C}_{X_1, \ldots, X_n}$ is an irreducible variety of dimension s variety-theoretically projecting onto $\mathbb{C}_{X_1, \ldots, X_s}$, then at every point $P \in V$, it is true that for each sufficiently small polydisk $\Delta^{n-s}(P) \subset \mathbb{C}_{X_{s+1}, \ldots, X_n}$ centered at P, there is a polydisk $\Delta^s(P) \subset \mathbb{C}_{X_1, \ldots, X_s}$ centered at P so that over almost each point $Q \in \Delta^s(P)$, there is a common, fixed number $n(P)$ of points of $V \cap (\Delta^s(P) \times \Delta^{n-s}(P))$. We shall use the following result:

Theorem 6.11. *Let V^s and $n(P)$ be as immediately above, and let W be any irreducible subvariety of V^s. The numbers $n(P)$ assume the same value at almost all points $P \in W$.*

Theorem 6.11 is an immediate consequence of

Theorem 6.12. *Let $V^s \subset \mathbb{C}_{X_1, \ldots, X_n}$ be an irreducible variety of dimension s variety-theoretically projecting onto $\mathbb{C}_{X_1, \ldots, X_s}$, and let W be an irreducible subvariety of V^s. Then for each integer $k \geqslant 0$, the set of points Q of W such that $n(P) \geqslant k$ forms a subvariety W_k of W.*

PROOF. Suppose, first, that V^s is a hypersurface. Without loss of generality, let $p \in \mathbb{C}[X_1, \ldots, X_n]$ be irreducible; then from Theorem 6.6.2, the order with respect to $\mathbb{C}_{X_n}$ of $V^s = \mathsf{V}(p)$ at $(a_1, \ldots, a_n) \in \mathsf{V}(p)$ is just the order in X_n of p at $(a_1, \ldots, a_n)$. Now it is easily seen that the order in X_n of $p \in \mathbb{C}[X_1 \ldots, X_n]$ at $(a_1, \ldots, a_n)$, is $\geqslant k$ iff its first $k - 1$ partial derivatives with respect to X_n vanish there. This condition obviously defines a subvariety of $\mathsf{V}(p)$, and therefore also of W.

The proof for arbitrary irreducible V^s is very similar to the proof of Theorem 6.10 for arbitrary V; we therefore leave it as an easy exercise (Exercise 6.2). □

PROOF OF THEOREM 6.8

(6.8.1) The proof is basically the same as that of Theorem 6.2.1, except that we apply Theorem 6.11 (instead of Theorem 11.1.1 of Chapter III). For the variety V^s in Theorem 6.11, we take the variety $V^\dagger \cap (\mathbb{C}^{(n+1)^2} \times V')$ appearing in (15). $V^s \subset \mathbb{C}^{(n+1)^2} \times \mathbb{C}^n \times \{1\} \subset \mathbb{C}^{(n+1)^2} \times \mathbb{C}^{n+1})$ is "algebraic" over $\mathbb{C}^{(n+1)^2}$ (in the sense that the coordinate ring of V^s is algebraic over that of $\mathbb{C}^{(n+1)^2}$). Now consider the subspace S of $\mathbb{C}^{(n+1)^2}$ parametrizing those P-containing transforms L' of L; the subvariety in $\mathbb{C}^{(n+1)^2} \times \mathbb{C}^{n+1}$ consisting of those points of the variety in (15) which lie above S is easily seen to have as an irreducible component the translate $S \times \{P\} \times \{1\}$ of S. (All our transforms of the given L contain the *fixed* point $P \in V^r$.) Let this translate be W in Theorem 6.11. For almost every transformation $T \in S$, $\dim(L)^T = n - r$, and for each such T, $i(V^r, (L)^T; P)$ equals $n(Q)$ (as defined immediately before Theorem (6.11), where $Q \in W$ corresponds to $(L)^T$. This completes the proof of (6.8.1).

(6.8.2) We assume without loss of generality that V is affine and that $P = (0) \in \mathbb{C}_{X_1,\ldots,X_n}$. We may write $p = p_m + p_{m+1} + \ldots$, where p_i is 0 or homogeneous in $X_1, \ldots, X_n$ of degree i, and where $p_m \neq 0$. The order of p at (0) is then m, and under the substitution $X_i = c_i T$ parametrizing a typical line L through (0),

$$p_m(c_1 T, \ldots, c_n T) = T^m p_m(c_1, \ldots, c_n),$$

which is thus either zero or still homogeneous of degree m. It is zero only at points $(c_1, \ldots, c_n) \in \mathsf{V}(p_m)$, and $\mathsf{V}(p_m)$ is proper in $\mathbb{C}^n$; when it is of degree m, $i(V, L; (0)) = m$. Hence for almost all (0)-containing transforms $(L)^T$ of some L, $i(V, L)^T; (0))$ is the order of p at (0). Thus (6.8.2) is proved, and therefore also Theorem 6.8. □

We can generalize the notion of order or multiplicity of a variety at a point, to order or multiplicity of a variety at, or along, an irreducible subvariety.

Theorem 6.13. *Let X be an irreducible subvariety of a pure-dimensional variety V in $\mathbb{P}^n(\mathbb{C})$ or $\mathbb{C}^n$. For almost every point P on X, $m(V; P)$ has a common, fixed value.*

PROOF. The proof is essentially the same as the proof of Theorem 6.8.1; assume without loss of generality that X is affine, and in place of the translate $S \times \{P\} \times \{1\})$, use $S \times X \times \{1\}$. This gives an "almost all" statement on points of X instead on only P. □

Definition 6.14. The number in Theorem 6.13 is called the **multiplicity of V along W**, denoted by $m(V; W)$. More generally, if any V has multiplicity k at almost every point of a pure-dimensional subvariety W, then k, denoted by $m(V; W)$, is the **multiplicity of V along W**.

We now turn to the definitions of degree and of intersection multiplicity of properly-intersecting varieties. As before, we use linear transformations. Thus, let V_1 and V_2 in $\mathbb{P}^n(\mathbb{C})$ or $\mathbb{C}^n$ be properly-intersecting varieties of pure dimensions r and s, respectively. We assume $r + s \geqslant n$. (Therefore $\dim(V_1 \cap V_2) \geqslant 0$.) Then $\mathbb{C}^{(n+1)^2}$ parametrizes the set of linear transforms of V_1, another copy of $\mathbb{C}^{(n+1)^2}$ independently parametrizes those of V_2 and a third parametrizes transforms of some $L^{(n-r)+(n-s)}$. By the dimension theorem, any subspace of dimension $(n - r) + (n - s)$ intersecting $V_1 \cap V_2$ properly does so in dimension zero. (Note that we are independently changing all three varieties, whereas earlier, with two varieties V and L, we transformed only L. We could in fact just as well have independently transformed both V and L earlier, and we could here independently transform only two of the three varieties, leaving the third one fixed. For example if $T = (a_{ij})$, $T' = (a'_{ij})$, and $T'' = (a''_{ij})$ are nonsingular transformations on V_1, V_2, and L respectively, then for the transformations $S' = (a'_{ij})(a_{ij})^{-1}$ and $S'' = (a''_{ij})(a_{ij})^{-1}$, the number of points in $V_1{}^T \cap V_2{}^{T'} \cap L^{T''}$ is the same as the number of points in $V_1 \cap V_2{}^{S'} \cap L^{S''}$.)

Let $\mathsf{V}(\{p_{\alpha k}(X)\})$, $(k = 1, 2, 3)$ be the respective homogeneous varieties in $\mathbb{C}^{n+1}$ corresponding to V_1 and V_2, and a projective $(n - r) + (n - s)$-subspace of $\mathbb{P}^n(\mathbb{C})$. Then in self-explanatory notation, the variety

$$V^{\dagger\dagger} = \mathsf{V}(\{p_{\alpha k}(XU_{ijk})\}) \qquad (i, j = 1, \ldots, n + 1; k = 1, 2, 3) \tag{16}$$

is 4-way homogeneous in $\mathbb{C}^{3(n+1)^2} \times \mathbb{C}^{n+1}$. The variety $V^{\dagger\dagger}$ thus generalizes to three independent transformations the variety $V^{\dagger}$ in (14).

Now $\mathsf{V}(X_{n+1})$ in $\mathbb{C}_{X_1, \ldots, X_{n+1}}$ defines a hyperplane in $\mathbb{P}^n(\mathbb{C})$. Then, just as in the proof of Theorem 6.2.1,

$$V^{\dagger\dagger} \cap (\mathbb{C}^{3(n+1)^2} \times \mathsf{V}(X_{n+1} - 1))$$

variety-theoretically projects onto $\mathbb{C}^{3(n+1)^2}$, and every component variety-theoretically projecting onto $\mathbb{C}^{3(n+1)^2}$ is algebraic over $\mathbb{C}^{3(n+1)^2}$. Hence above almost each point of $\mathbb{C}^{3(n+1)^2}$, there is a common, fixed number of points. Translating this back to V_1, V_2, and L gives us

Theorem 6.15. *Let V_1 and V_2 in $\mathbb{P}^n(\mathbb{C})$ or $\mathbb{C}^n$ be of pure dimension r and s, respectively, and let L be any $(2n - r - s)$-dimensional subspace of $\mathbb{P}^n(\mathbb{C})$. Then for almost every transform $V_1{}^T$ of V_1, $V_2{}^{T'}$ of V_2, and $L^{T''}$ of L, $V_1{}^T \cup V_2{}^{T'} \cap L^{T''}$ consists of a common, fixed number of points.*

Definition 6.16. Let pure-dimensional varieties V_1 and V_2 in $\mathbb{P}^n(\mathbb{C})$ or $\mathbb{C}^n$ intersect properly; the fixed number in Theorem 6.15 is the **degree of intersection of V_1 and V_2**, written as $\deg(V_1 \cdot V_2)$.

Remark 6.17. Note that $\deg(V_1 \cdot V_2)$ is not in general the same as $\deg(V_1 \cap V_2)$, in Definition 6.3. See Example 6.26. The notation $\deg(V_1 \cdot V_2)$ will be further illuminated in Remark 6.25.

Theorem 6.18. *Let V_1 and V_2 in $\mathbb{P}^n(\mathbb{C})$ or $\mathbb{C}^n$ be of pure dimensions r and s, respectively, and let $L^{(n-r)+(n-s)} = L$ be linear of dimension $2n - r - s$. If V_1, V_2, and L intersect properly at a point P, then for almost every transform $V_1{}^T$ near V_1, $V_2{}^{T'}$ near V_2, and $L^{T''}$ near L, there is a common, fixed number of distinct points of $V_1{}^T \cap V_2{}^{T'} \cap L^{T''}$ near P.*

Proof. The proof is entirely analogous to that of Theorem 6.15, except we use Theorem 6.10 instead of Theorem 11.1.1 of Chapter III. □

Definition 6.19. The fixed number of Theorem 6.18 is the **intersection multiplicity**, or **multiplicity of intersection, of V_1, V_2, and L at P**; it is denoted by $i(V_1, V_2, L; P)$.

Theorem 6.20. *Let V_1 and V_2 in $\mathbb{P}^n(\mathbb{C})$ or $\mathbb{C}^n$ be of pure dimensions r and s, respectively. If they intersect properly at a point P, then for almost every P-containing transform L' of a linear variety $L^{(n-r)+(n-s)}$, $i(V_1, V_2, L'; P)$ is defined and has a common, fixed value.*

Proof. The proof is similar to that of Theorem 6.8.1. The "V^s" used in that proof is now

$$V^{\dagger\dagger} \cap (\mathbb{C}^{3(n+1)^2} \times \mathsf{V}(X_{n+1} - 1)),$$

which is algebraic over a copy of $\mathbb{C}^{3(n+1)^2}$. As noted just before the statement of Theorem 6.8, there is a proper subspace S of $\mathbb{C}^{(n+1)^2}$ parametrizing the P-containing transforms of $L^{(n-r)+(n-s)}$; we again take W to be the translate $S \times \{P\} \times \{1\}$. The proof may now be completed, making the obvious changes in the proof of Theorem 6.8.1. □

Definition 6.21. The fixed number in Theorem 6.20 is the **intersection multiplicity**, or **multiplicity of intersection, of V_1 and V_2 at P**; it is denoted by $i(V_1, V_2; P)$.

Theorem 6.22. *Let V_1 and V_2 in $\mathbb{P}^n(\mathbb{C})$ or $\mathbb{C}^n$ be of pure dimension, and suppose they intersect properly. If C is an irreducible component of $V_1 \cap V_2$, then at almost every point $P \in C$, $i(V_1, V_2; P)$ has a common, fixed value.*

Proof. The proof is like that of Theorem 6.20; assume C is affine, and in place of $S \times \{P\} \times \{1\}$ for W, use $S \times C \times \{1\}$. This gives the "almost all" statement over C instead of just at P. □

Definition 6.23. The fixed number in Theorem 6.22 is the **multiplicity of intersection of V_1 and V_2 along C**, and is denoted by $i(V_1, V_2; C)$.

Definition 6.24. Let V_1 and V_2 be two properly-intersecting pure-dimensional varieties in $\mathbb{P}^n(\mathbb{C})$ or $\mathbb{C}^n$. The formal sum $\sum_{j=1}^n i(V_1, V_2; C_j)C_j$ of the distinct irreducible components $C_1, \ldots, C_n$ of $V_1 \cap V_2$ is called the **intersection product** of V_1 and V_2, and is denoted by $V_1 \cdot V_2$.

Remark 6.25. It is natural to define the degree of $V_1 \cdot V_2$ to be $\sum_{j=1}^{n} i(V_1, V_2; C_j) \cdot \deg C_j$. If we do this, then we see that the symbol "$\deg(V_1 \cdot V_2)$" in Definition 6.16, is in fact what its notation indicates—it is the degree of $V_1 \cdot V_2$.

EXAMPLE 6.26. Rotating the circle $\mathsf{V}((X - r)^2 + Z^2 - s^2) \subset \mathbb{R}_{XZ}$ ($r, s \in \mathbb{R}$, $r > s > 0$) about $\mathbb{R}_Z$ in $\mathbb{R}_{XYZ}$ describes a real torus defined by the fourth-degree polynomial

$$p(X, Y, Z) = (X^2 + Y^2 + Z^2 + r^2 - s^2)^2 - 4r^2(X^2 + Y^2);$$

it "rests on a tabletop" in the sense that it is tangent to $\mathsf{V}(Z - s) \subset \mathbb{R}_{XYZ}$. The corresponding complex varieties $\mathsf{V}(p)$ and $\mathsf{V}(Z - s)$ in $\mathbb{C}_{XYZ}$ are surfaces of degree 4 and 1, respectively. The variety $C = \mathsf{V}(p) \cap V(Z - s)$ has degree 2 (since it is a circle), and $i(\mathsf{V}(p), \mathsf{V}(Z - s); P) = 2$ at each $P \in C$, so $i(\mathsf{V}(p), \mathsf{V}(Z - s); C) = 2$. Thus

$$\mathsf{V}(p) \cdot \mathsf{V}(Z - s) = 2C,$$

and

$$\deg(\mathsf{V}(p) \cdot \mathsf{V}(Z - s)) = 2 \deg(\mathsf{V}(p) \cap \mathsf{V}(Z - s)) = 4.$$

Note that C and $\mathsf{V}(p)$ are nonsingular. Thus $m(C; P) = 1$ for each $P \in C$, $m(\mathsf{V}(p); Q) = 1$ for each $Q \in \mathsf{V}(p)$, and $m(\mathsf{V}(p); C) = 1$.

EXERCISES

6.1 With notation as in the proof of Theorem 6.2, show that for almost every point $P \in \mathbb{C}^{(n+1)^2}$, there lie above P only finitely many points of the variety in (15).

6.2 Prove Theorem 6.12 for any irreducible variety $V^s \subset \mathbb{C}^n$.

6.3 Let V and W be properly-intersecting varieties in $\mathbb{C}^n$, and let P be any point of $V \cap W$. If T is a nonsingular linear transformation of $\mathbb{C}^n$, show that $i(V, W; P) = i(T(V), T(W); T(P))$.

6.4 For any variety V in $\mathbb{C}^n$ or $\mathbb{P}^n(\mathbb{C})$, show that V is nonsingular at $P \in V$ iff $m(V; P) = 1$. Generalize to the case where P is replaced by an irreducible subvariety of V.

6.5 Let $V \subset \mathbb{C}^n$ be irreducible of dimension $r \geqslant n/2$, let $L \subset \mathbb{C}^n$ be a linear variety of dimension r properly intersecting V, and let $P \in V \cap L$ be a nonsingular point of V. Show that L is the tangent space to V at P iff $i(V, L; P) > 1$ (cf. Exercise 4.5).

6.6 (a) Let V_1 and V_2 be varieties in $\mathbb{P}^n(\mathbb{C})$ such that $\dim V_1 + \dim V_2 \geqslant n$. Show that for almost every linear transform V_1^T of V_1, V_1^T and V_2 intersect properly.
(b) State and prove an analogous result in the affine setting.

6.7 The class of perturbations considered in this section is not the only one that can be used to arrive at multiplicity and multiplicity of intersection. For instance, if $V = \mathsf{V}(p) \subset \mathbb{C}_{X_1, \ldots, X_n} = \mathbb{C}^n$ is a hypersurface, it turns out that one can use the one-dimensional family of level surfaces $\{\mathsf{V}(p(X) - c) | c \in \mathbb{C}_Z\}$ to replace the set of linear transforms of V. Assume and use this fact in (a) and (b) below.
(a) Show that the multiplicity of intersection at (0, 0) of $\mathsf{V}(XY) \subset \mathbb{C}_{XY}$ with any 1-subspace other than $\mathbb{C}_X$ or $\mathbb{C}_Y$, is two.

(b) Let $(X) = (X_1, \ldots, X_n)$, let $q(X) \in \mathbb{C}[X]\backslash\mathbb{C}$, and let a linear variety $L \subset \mathbb{C}_X$ properly intersect $\mathsf{V}(q)$. Show that at an arbitrary point $(0) \in \mathsf{V}(q) \cap L$, intersecting the level curves of the hypersurface $\mathsf{V}(Y - q(X)) \subset \mathbb{C}_{XY}$ with L yields the order with respect to L of q at (0).

6.8 Let C and $\mathsf{V}(p)$ ($p \in \mathbb{C}[X, Y]\backslash\mathbb{C}$) be two properly-intersecting curves in $\mathbb{C}_{XY} = \mathbb{C}^2$. For almost every $c \in \mathbb{C}$ $\mathsf{V}(p - c) \cap C$ is a finite set A_c; the number of points in A_c clustered near an arbitrary point $(0) \in \mathsf{V}(p) \cap C$, for c arbitrarily small, turns out to be $i(\mathsf{V}(p), C; (0))$. Let S_j be any one of the subsets of C near $(x_0, y_0) = (0)$ in Theorem 4.13 of Chapter II, and suppose that the points of S_j are parametrized near (0) by, say, $X = X(T)$ and $Y = Y(T)$. Those points near (0) of A_c on that S_j are then given by the set of all T such that $p(X(T), Y(T)) = c$; the number of such points as $c \to 0$ is of course just the order with respect to T of $p(X(T), Y(T))$. Since each factor in (20) of Chapter II gives a parametrization of (a representative of) a branch of C through (0) (namely, a parametrization of the form $X = T^m$ and $Y = f(T)$), we have another way of finding the multiplicity of intersection of two properly-intersecting plane curves. Assume and use this method in the following.

(a) In $\mathbb{C}_{XY} = \mathbb{C}^2$, let $C_1 = \mathsf{V}(Y^2 - X^3)$, $C_2 = \mathsf{V}(X^2 - Y^3)$, and $C_3 = \mathsf{V}(Y^2 - 2X^3)$. Find $i(C_1, C_2; (0))$ and $i(C_1, C_3; (0))$.

(b) Let C_1, C_2, and C_3 be as in (a). By homogenizing and dehomogenizing to take care of points at infinity, directly verify Bézout's theorem for the completions in $\mathbb{P}^2(\mathbb{C})$ of C_1 and C_2, and also for the completions of C_1 and C_3.

(c) Find the multiplicity of intersection at the origin of the two curves in Exercise 4a, b, Section II, 4.

6.9 (a) Let $\sum_{i,j=1}^3 c_{ij} X_i X_j$ define a conic in $\mathbb{P}^2(\mathbb{C})$, where $c_{ij} = c_{ji} \in \mathbb{C}$. Show that the conic consists of either two distinct lines or one "double" line (in the obvious sense) iff the determinant $|c_{ij}| = 0$. (Such a conic in $\mathbb{P}^2(\mathbb{C})$ is called **reducible**.)

(b) Let $F = F(X_1, X_2, X_3)$ be homogeneous, let $(a) = (a_1, a_2, a_3)$ be a point of $\mathbb{P}^2(\mathbb{C})$, and let $X_i = a_i S + b_i T$ ($b_i \in \mathbb{C}$, $i = 1, 2, 3$) be parametric equations for a line L in $\mathbb{P}^2(\mathbb{C})$ through (a). Show that the order with respect to L of F at (a) is at least three iff

$$F(a) = \sum_{i=1}^{3} \left[\frac{\partial F}{\partial X_i}(a)\right] b_i = \sum_{i,j=1}^{3} \left[\frac{\partial^2 F}{\partial X_i \partial X_j}(a)\right] b_i b_j = 0. \tag{17}$$

(See Definition 6.4.)

For any c in $\mathbb{C}\backslash\{0\}$, (17) obviously holds iff it holds with cb_i in place of b_i ($i = 1, 2, 3$). Thus if (17) holds, then for any value assigned to (X_1, X_2, X_3) such that

$$\sum_{i=1}^{3} \left[\left(\frac{\partial F}{\partial X_i}\right)(a)\right] X_i = 0,$$

we also have at that same value, $\sum_{i,j=1}^3 [(\partial^2 F/\partial X_i \partial X_j)(a)] X_i X_j = 0$. Geometrically, this says that the line

$$\sum_{i=1}^{3} \left[\left(\frac{\partial F}{\partial X_i}\right)(a)\right] X_i = 0$$

is a component of the conic

$$\sum_{i,j=1}^{3} \left[\left(\frac{\partial^2 F}{\partial X_i \partial X_j}\right)(a)\right] X_i X_j = 0;$$

that is, this conic is reducible. Thus by (a), $\det[(\partial^2 F/\partial X_i \partial X_j)(a)] = 0$. We therefore conclude that any point P of $V(F) \subset \mathbb{P}^2(\mathbb{C})$ having order 3 along some line through P, is a point of $\mathsf{V}(F, \det(\partial^2 F/\partial X_i \partial X_j)) \subset \mathbb{P}^2(\mathbb{C})$. Prove that this can happen for only finitely many points P and only finitely many lines L in $\mathbb{P}^2(\mathbb{C})$.

7 Bézout's theorem

In this section we prove Bézout's theorem for varieties in $\mathbb{P}^n(\mathbb{C})$.

Theorem 7.1 (Bézout's theorem). *Suppose two pure-dimensional varieties V and W in $\mathbb{P}^n(\mathbb{C})$ intersect properly. Then*

$$\deg(V \cdot W) = \deg V \cdot \deg W.$$

We shall prove this theorem by showing that it can be reduced to the case when one variety, say V, is of an especially simple form, namely when it is a union of $\deg V$ distinct projective subspaces of dimension $\dim V$. From Theorem 6.2 and Definition 6.3 we know that W, almost every projective subspace of dimension $\dim V$, and almost every projective subspace of dimension $n - \dim V - \dim W$ intersect in $\deg W$ points, counted with multiplicity; replacing the one subspace of dimension $\dim V$ by a union of $\deg V$ such subspaces then yields $\deg V \cdot \deg W$ points, counted with multiplicity.

As an example, the variety $V \subset \mathbb{C}_{X_1 X_2}$ consisting of the two lines $X_1 = \pm 1$ has order 2; since the parabola $W = \mathsf{V}(X_2{}^2 - X_1)$ intersects each line $X_1 = 1$ and $X_1 = -1$ in two distinct points, it intersects V in $\deg V \cdot \deg W = 4$ distinct points. (Note that the completions of V and W don't intersect at the line at infinity, so there are still exactly four points of intersection in $\mathbb{P}^2(\mathbb{C})$.)

If V is not of such a simple form, we will show that it can be changed, by means of projective transformations, so that it is; of course we need to show that in so modifying V, its degree of intersection with W doesn't change, so that computing the order of intersection using our simpler variety really does yield the right number, $\deg(V \cdot W)$.

Let us begin by looking at the question of simplifying varieties via projective transformations.

Example 7.2. Consider the circle in $\mathbb{P}^2(\mathbb{C})$ defined by $\mathsf{V}(X_1{}^2 + X_2{}^2 - X_3{}^2)$. Replacing (X_1, X_2, X_3) in $X_1{}^2 + X_2{}^2 - X_3{}^2$ by $(X_1, X_2, X_3)(U_{ij})$, where (U_{ij}) is a 3×3 matrix of indeterminants, yields a variety $V^\dagger$ in $\mathbb{C}^9 \times \mathbb{C}^3$ (Cf. (14)); an arbitrary projective transformation of V is obtained by evaluating (U_{ij}) at an arbitrary point $(a_{ij}) \in \mathbb{C}^9$. If $U_{ij} = 0$ whenever $i \neq j$, then (X_1, X_2, X_3) maps to $(U_{11}X_1, U_{22}X_2, U_{33}X_3)$; any choice a_{ii} of the U_{ii} simply amounts to a change in the coefficients of X_1, X_2, and X_3 in $X_1{}^2 + X_2{}^2 - X_3{}^2$. If exactly one of these coefficients a_{ii} is zero, the polynomial breaks up into distinct factors. For instance when $a_{11} = 0$ and

$a_{22} = a_{33} = 1$, the polynomial becomes $X_2{}^2 - X_3{}^2 = (X_2 + X_3)(X_2 - X_3)$, so the circle is transformed into two lines in $\mathbb{P}^2(\mathbb{C})$ under

$$\begin{pmatrix} 0 & 0 & 0 \\ 0 & 1 & 0 \\ 0 & 0 & 1 \end{pmatrix}.$$

Often it is necessary to make more than one coefficient zero to fully reduce, or "degenerate" a variety to a union of linear varieties. For instance $\mathsf{V}(X_1{}^2 + X_2{}^2 + X_3{}^2 - X_4{}^2)$, which defines a complex sphere in $\mathbb{P}^3(\mathbb{C})$, becomes the cylinder $\mathsf{V}(X_2{}^2 + X_3{}^2 - X_4{}^2)$ with $a_{11} = 0$ and $a_{22} = a_{33} = a_{44} = 1$, but it reduces to two planes $\mathsf{V}(X_3 - X_4) \cup \mathsf{V}(X_3 + X_4)$ upon also setting $a_{22} = 0$.

The general result we shall use is this:

Lemma 7.3. *With respect to appropriate coordinates, any pure-dimensional variety $V \subset \mathbb{P}^n(\mathbb{C})$ can be reduced to a union of* $\deg V$ *distinct subspaces of dimension $r = \dim V$ by means of a projective transformation defined by an $(n + 1) \times (n + 1)$ matrix*

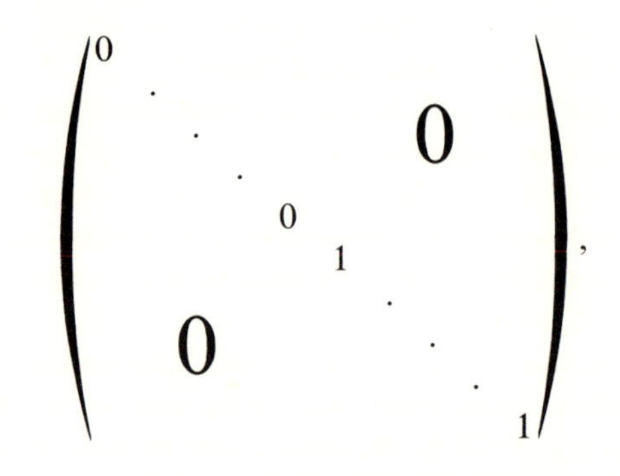

where r of the entries on the main diagonal are 0 *and $n - r + 1$ entries are* 1.

PROOF. Let $\mathsf{H}(V) \subset \mathbb{C}_{X_1,\ldots,X_{n+1}}$ be the homogeneous variety representing V. Without loss of generality we may suppose coordinates have been chosen so that $\mathbb{C}_{X_{r+1},\ldots,X_{n+1}}$ intersects $\mathsf{H}(V)$ in $\deg V$ distinct 1-subspaces $L_1, \ldots, L_{\deg V}$. For each $i = 1, \ldots, \deg V$, let S_i be the $(r + 1)$-subspace of $\mathbb{C}^{(n+1)}$ spanned by L_i and $\mathbb{C}_{X_1,\ldots,X_r}$. These $\deg V$ subspaces are all distinct; their union V' will turn out to be our simplified variety.

Note that $\mathbb{C}_{X_{r+1},\ldots,X_{n+1}} = \mathsf{V}(X_1, \ldots, X_r)$, so if $\mathsf{H}(V) = \mathsf{V}(\mathfrak{a})$, then the union of the subspaces L_i is the common zero-set of $\mathfrak{a}$ and $X_1, \ldots, X_r$. Thus

$$\bigcup_i L_i = \mathbb{C}_{X_{r+1},\ldots,X_{n+1}} \cap \mathsf{V}(\{p(0, \ldots, 0, X_{r+1}, \ldots, X_{n+1}) \mid p \in \mathfrak{a}\}),$$

and V' is $\mathsf{V}(\{p(0, \ldots, 0, X_{r+1}, \ldots, X_{n+1}) \mid p \in \mathfrak{a}\})$.

We now see that the union V' of the S_i is the transform of V under the $(n + 1) \times (n + 1)$ matrix

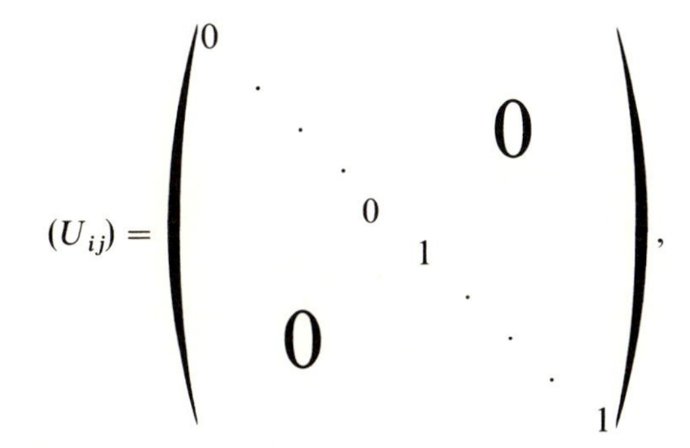

where r of the entries on the main diagonal are 0 and $n - r + 1$ entries are 1. □

We show next that the reduced variety V' actually does yield $\deg(V \cdot W)$ points when intersected with W and an appropriate linear variety.

Notation 7.4. Let $V\,(\subset \mathbb{P}^n(\mathbb{C}))$ and coordinates $X_1, \ldots, X_{n+1}$ of $\mathbb{C}^{n+1}$, be as in Lemma 7.3. Let $P \in \mathbb{C}^{(n+1)^2}$ denote the matrix of Lemma 7.3, and let V' denote the transform of V under P. Let $W \subset \mathbb{P}^n(\mathbb{C})$ be of pure dimension, let L be a subspace of $\mathbb{P}^n(\mathbb{C})$ for which $V' \cap W \cap L$ is proper and consists of $\deg V \cdot \deg W$ distinct points, and let I be the identity matrix of order $n + 1$. (Then $P \times I \times I \in \mathbb{C}^{3(n+1)^2}$ corresponds to the set of varieties $\{V', W, L\}$.) As in (16), let the set $\{U_{ijk}\}$ ($i, j = 1, \ldots, n + 1$ and $k = 1, 2, 3$) be $3(n + 1)^2$ (algebraically independent) indeterminants parametrizing the set of independent projective transformations of V, W, and L; denote $(X_1, \ldots, X_{n+1})$ by X. If V, W, and L are defined by ideals $\mathfrak{a}_k = \{p_{\alpha k}\}$ ($k = 1, 2, 3$) respectively, let $V^{\dagger\dagger} = \mathsf{V}(\{p_{\alpha k}(X(U_{ijk}))\})$.

According to Definition 6.16, we see that Bézout's theorem says this:

(7.5) For almost every point $Q \in \mathbb{C}^{3(n+1)^2}$, there are $\deg V \cdot \deg W$ distinct points of $V^{\dagger\dagger}$ above Q.

The matrix P is a point of $\mathbb{C}^{(n+1)^2}$, so $\{P\} \times \mathbb{C}^{2(n+1)^2}$ is an irreducible subvariety of $\mathbb{C}^{3(n+1)^2}$. Now above almost each point of $\{P\} \times \mathbb{C}^{2(n+1)^2}$ there are exactly $\deg V \cdot \deg W$ points of $V^{\dagger\dagger}$. However, it is conceivable that $\{P\} \times \mathbb{C}^{2(n+1)^2}$ is itself an exceptional subvariety of $\mathbb{C}^{3(n+1)^2}$; for instance, *a priori*, there might be more than $\deg V \cdot \deg W$ points of $V^{\dagger\dagger}$ above points of $\mathbb{C}^{3(n+1)^2}$ near $\{P\} \times \mathbb{C}^{2(n+1)^2}$. We show this cannot happen by proving

Lemma 7.6. *Let V, W, P, and $V^{\dagger\dagger}$ be as in Notation 7.4. There is a point Q_0 in $\{P\} \times \mathbb{C}^{2(n+1)^2}$ so that there are exactly* deg $V \cdot$ deg W *distinct points of $V^{\dagger\dagger}$ π_X-lying above Q_0, and so that if Q_0^* is an arbitrary one of these* deg $V \cdot$ deg W *points, then for each $Q \in \mathbb{C}^{3(n+1)^2}$ sufficiently near Q_0, there is exactly one point of $V^{\dagger\dagger}$ arbitrarily near Q_0^* which π_X-lies above Q.*

EXAMPLE 7.7. Let the completions in $\mathbb{P}^2(\mathbb{C})$ of a complex circle and parabola in $\mathbb{C}_{X_1X_2}$ be $V = \mathrm{V}(X_1{}^2 + X_2{}^2 - X_3{}^2)$ and $W = \mathrm{V}(X_2X_3 - X_1{}^2)$. At

$$P = \begin{pmatrix} 0 & 0 & 0 \\ 0 & 1 & 0 \\ 0 & 0 & 1 \end{pmatrix} \in \mathbb{C}^{(2+1)^2},$$

the transform of V is $\mathrm{V}(X_2{}^2 - X_3{}^2)$, which intersects W in deg $V \cdot$ deg $W = 4$ points. We may therefore let $P \times I \times I$ be the point Q_0 in Lemma 7.6. Then $P \times I \times I$ describes the reduced circle V' (two lines) and W. A point Q near Q_0 corresponds to "slightly perturbed curves"; the elongated ellipse and tilted parabola in Figure 1 indicate a possibility in the real affine part

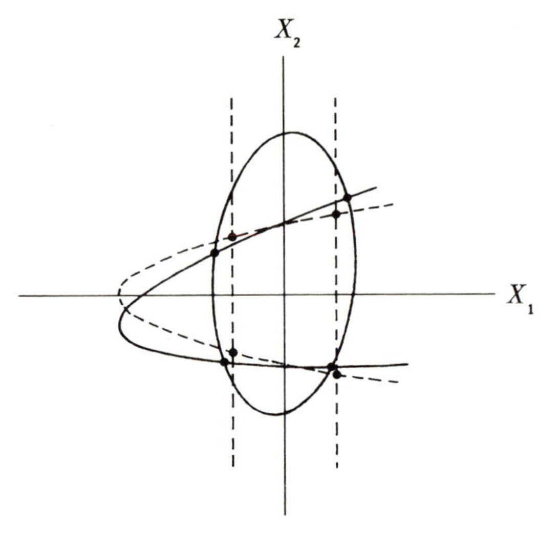

Figure 1

$\mathbb{R}_{X_1X_2}$. We want to show that arbitrarily near any point of $V' \cap W$, there is exactly one intersection point of W with any projective transform of V sufficiently near V', as suggested in Figure 1. If this is true, then each intersection corresponding to points in an open neighborhood of $\mathbb{C}^{3(n+1)^2}$ (and therefore over almost all points of $\mathbb{C}^{3(n+1)^2}$) consists of the same number of distinct points as does the intersection of the reduced circle with the original parabola, so the degree of $V \cdot W$ is indeed deg $V \cdot$ deg W, as desired.

PROOF OF LEMMA 7.6. Let $\Delta = \Delta^{3(n+1)^2} \times \Delta^{n+1}$ be a neighborhood in $\mathbb{C}^{3(n+1)^2} \times \mathbb{C}^{n+1}$ about Q_0^*. We prove the lemma by showing that for some such Δ, $V^{\dagger\dagger} \cap \Delta$ is the graph of an analytic map from $\Delta^{3(n+1)^2}$ to Δ^{n+1}. For this, we apply the Implicit mapping theorem (Theorem 3.5 of Chapter II).

To show the hypotheses of Theorem 4.1 are satisfied, we proceed as follows (cf. Notation 7.4). First, it is easy to see that we may choose coordinates in $\mathbb{C}^{n+1}$ so that above the point $P \times I \times I \in \mathbb{C}^{3(n+1)^2}$, W is nonsingular at each of the $\deg V \cdot \deg W$ distinct points in which V' intersects $W \cap L$. Now let $J(V')$ be an "$\infty \times (n+1)$ matrix" $(\partial p/\partial X_i)$ (where p runs through all elements of $\mathrm{J}(V')$ for $i = 1, \ldots, n+1$) with analogous meanings for $J(W)$ and $J(L)$. Let J^* be a matrix whose set of rows consists of the sets of rows of $J(V')$, $J(W)$, and $J(L)$.

By Theorem 4.1, at each point Q_0^* of $V' \cap W \cap L$ we have $\operatorname{rank}(J(W)_{Q_0^*}) = \operatorname{cod} W$ and $\operatorname{rank}(J(L)_{Q_0^*}) = \operatorname{cod} L$. One may further choose L and V' so that at each $Q_0^* \in V' \cap W \cap L$, V', W and L intersect "transversally" in the sense that $\operatorname{rank}(J^*_{Q_0^*})$ is $\operatorname{cod} V' + \operatorname{cod} W + \operatorname{cod} L = n+1$. Write $Y_{ik} = \sum_j U_{ijk} X_j$, and $Y_k = (Y_{1k}, \ldots, Y_{n+1,k})$. In analogy with $J(V')$, $J(W)$, and $J(L)$, let $J(V^{\dagger\dagger})$ be an "$\infty \times (3(n+1)^2 + n + 1)$ matrix," where one takes partials of all the $p_{\alpha k}(Y_k)$ with respect to all variables X and U. At each Q_0^*, the $X_1, \ldots, X_{n+1}$-columns of $J(V^{\dagger\dagger})$ coincide with the columns of J^* (for an appropriate ordering of the rows of J^*); this follows from the chain rule. Writing $p = p_{\alpha k}$, we have:

$$\left.\frac{\partial p(Y_k)}{\partial X_i}\right|_{Q_0^*} = \left[\sum_j \left.\frac{\partial p(X)}{\partial X_j}\right|_{X=Y_k} \cdot U_{jik}\right]_{Q_0^*} = \left.\frac{\partial p(X)}{\partial X_i}\right|_{Q_0^*}$$

$(Q_0^* \in P \times I \times I \times \mathbb{C}_{X_1, \ldots, X_{n+1}})$.

We have seen that $\operatorname{rank}(J^*_{Q_0^*}) = n+1$, so $\operatorname{rank}(J(V^{\dagger\dagger})_{Q_0^*}) \geqslant n+1$. We actually have equality, for the remaining $3(n+1)^2$ columns of $J(V^{\dagger\dagger})$ are linearly dependent on the $X_1, \ldots, X_{n+1}$-columns; this again follows from the chain rule:

$$\left.\frac{\partial p(Y_k)}{\partial U_{ijk}}\right|_{Q_0^*} = \left[\left.\frac{\partial p(X)}{\partial X_i}\right|_{X=Y_k} \cdot X_j\right]_{Q_0^*},$$

which is a constant times $\partial p(X)/\partial X_i|_{Q_0^*}$. By continuity of the entries of $J(V^{\dagger\dagger})$, we see that $\operatorname{rank}(J(V^{\dagger\dagger})) = n+1$ throughout a sufficiently small Δ^{n+1}.

The hypotheses of Theorem II,3.5 are now satisfied; hence for $\Delta = \Delta^{3(n+1)^2} \times \Delta^{n+1}$ sufficiently small, $V^{\dagger\dagger} \cap \Delta$ is the graph of an analytic map from $\Delta^{3(n+1)^2}$ to Δ^{n+1}, as desired. □

Corollary 7.8. *If two curves in $\mathbb{P}^2(\mathbb{C})$ of degrees n and m intersect in more than nm points (counted with multiplicity), then the two curves have at least one irreducible component in common.*

EXERCISES

7.1 Let two "concentric spheres" S_1 and S_2 in $\mathbb{P}^n(\mathbb{C})$ be defined by

$$\mathsf{V}(X_1^2 + \ldots + X_n^2 - 1) \quad \text{and} \quad \mathsf{V}(X_1^2 + \ldots + X_n^2 - 4).$$

In what variety and with what multiplicity do S_1 and S_2 intersect?

7.2 (a) Let $F_1, F_2 \in \mathbb{C}[X, Y, Z]\backslash\mathbb{C}$ be homogeneous of degree n. Show that for every $c_1, c_2 \in \mathbb{C}$, $\mathsf{V}(c_1F_1 + c_2F_2)$ contains $\mathsf{V}(F_1) \cap V(F_2)$.

(b) Suppose $\mathsf{V}(F_1) \cap \mathsf{V}(F_2) \subset \mathbb{P}^2(\mathbb{C})$ consists of n^2 distinct points; suppose that $H \in \mathbb{C}[X, Y, Z]\backslash\mathbb{C}$ is homogeneous and irreducible of degree $m \leqslant n$, and that $\mathsf{V}(H) \subset \mathbb{P}^2(\mathbb{C})$ contains mn points of $\mathsf{V}(F_1) \cap \mathsf{V}(F_2)$. Show that for some choice of c_1 and c_2, $\mathsf{V}(c_1F_1 + c_2F_2)$ contains $\mathsf{V}(H)$ as a component.

7.3 Using Exercise 7.2 above, prove *Pascal's theorem*: Let C be an irreducible conic in $\mathbb{P}^2(\mathbb{R})$, and let $L_1, \ldots, L_6$ be the real projective lines extending successive sides of a (not necessarily convex) hexagon inscribed in C; then the three points $L_1 \cap L_4$, $L_2 \cap L_5$, and $L_3 \cap L_6$ are collinear.

7.4 Using Exercise 6.9, complete the proof of the genus formula (Theorem 10.1 of Chapter II) by proving the existence of coordinates in $\mathbb{C}^2$ described in the last paragraph of Section II, 10.

7.5 In this exercise we use Definition 5.14 of Chapter V, which gives mild generalizations of the notions of curve and intersection multiplicity. Let V and W ($\subset \mathbb{C}^2$) consist, respectively, of n and m lines (counted with multiplicity) through an arbitrary, fixed point $P \in \mathbb{C}^2$. If each line in V is different from each line in W, then obviously $i(V, W; P) = nm$. Show that this can be generalized as follows:

(a) Let $q \in \mathbb{C}[X, Y]\backslash\mathbb{C}$ have order m at P—that is, suppose that the homogeneous polynomial q^* consisting of all the lowest-degree terms of q when expanded about P, has (total) degree m. ($\mathsf{C}(q^*)$ then consists of m lines with multiplicity through P; $V(q^*)$ is often called the **tangent cone to** $V(q)$ **at** P.) Show that if $\mathsf{V}(q^*)$ and the above V intersect properly, then $i(V, C(q); P) = mn$.

(b) Let notation be as in (a). We show that V can be replaced by n "smooth analytic arcs" in the following sense. Let $p \in \mathbb{C}[X, Y]\backslash\mathbb{C}$; suppose that $\operatorname{ord}_P p = n$, and that each of p's n fractional-power series expansions about P (as in Corollary 4.17 of Chapter II) is actually an ordinary power series in X. Under these conditions, show that if $\mathsf{V}(p^*)$ and $\mathsf{V}(q^*)$ intersect properly, then $i(\mathsf{C}(p), \mathsf{C}(q); P) = nm$. (If $\mathsf{V}(p^*)$ and $\mathsf{V}(q^*)$ intersect properly, we say that $\mathsf{V}(p)$ **and** $\mathsf{V}(q)$ **intersect transversally at** P.)

(c) Let notation be as in (a) and (b), and let p satisfy the conditions in (b). Suppose that P is a singular point of $\mathsf{V}(p)$, and that all the linear factors of p^* are distinct. Then $\mathsf{V}(p)$ is said to have an **ordinary singularity at** P. Show that if $\mathsf{V}(p)$ has an ordinary singularity at P, and if $\mathsf{V}(p)$ and $\mathbb{C}_Y$ intersect transversally at P, then

$$i\left(\mathsf{C}(p), \mathsf{C}\left(\frac{\partial p}{\partial Y}\right); P\right) = n(n-1).$$

7.6 It is natural to ask if the genus formula for nonsingular plane curves can be extended to more general curves. It can; in this exercise we prove one such generalization. First, we have the following basic definition:

Definition. Let S be any topological space obtained from a compact connected orientable topological 2-manifold M by identifying finitely many points to finitely many points. The **genus of** S is defined to be the genus of M.

It is easily checked that this definition assigns a well-defined integer to S.

Now, using Exercise 7.5, prove this generalization of Theorem II, 10.1:

Let $C \subset \mathbb{P}^2(\mathbb{C})$ be an irreducible curve of degree n. Suppose that C has only ordinary singularities, say $P_1, \ldots, P_m$, and suppose that the multiplicity of C at each such P_i is r_i. Then the genus g of C is equal to

$$\frac{(n-1)(n-2)}{2} - \sum_{i=1}^{m} \frac{r_i(r_i-1)}{2}.$$

7.7 Let g be an arbitrary nonnegative integer. Find a curve $C \subset \mathbb{P}^2(\mathbb{C})$ having genus g.

7.8 Let $C \subset \mathbb{P}^n(\mathbb{C})$ and $C' \subset \mathbb{P}^m(\mathbb{C})$ be irreducible curves which are birationally isomorphic. Show that C and C' have the same genus. [*Hint*: Establish a homeomorphism between $C \backslash \{\text{finitely many points}\}$ and $C' \backslash \{\text{finitely many points}\}$.] (This result says that the genus of an irreducible curve is a "birational invariant"; see Exercise 6.12 of Chapter V.)

CHAPTER V

Some elementary mathematics on curves

1 Introduction

So far our efforts have principally been directed toward getting information about algebraic varieties. But one can regard varieties themselves as "spaces on which one does mathematics." The reader has met this idea before. For instance one can transfer analysis on $\mathbb{R}$ to analysis on the more general real variety $\mathbb{R}^n$, getting, for example, multivariable calculus. And in complex analysis one regards the variety $\mathbb{C}$ (or more generally $\mathbb{C}^n$) as a carrier of analytic functions; one then studies differentiation and integration of these functions, and so on. One can also do analysis on real differentiable manifolds or complex analytic manifolds. (See, for instance, [Narasimhan], [Spivak], or [Wells].) Such a study in turn often sheds new light on the underlying space; algebraic geometry is no exception in this respect. In such questions, algebraic varieties occupy a special position; they have so much structure that one can transfer to them many ideas not only from analysis, but also from number theory, and these generalizations interconnect with and enrich each other. Transferring differentials and integration to nonsingular complex varieties is particularly natural, and in fact in its early days, algebraic geometry was regarded as a part of complex analysis (by Abel (1802–1829), Riemann (1826–1866), Weierstrass (1815–1897), etc.). More recently, attempts to transfer notions from topology and analysis to appropriate generalizations of algebraic varieties have met with varying degrees of success, and have for example already shed new light on some old problems in number theory. In applying algebraic geometry to other fields in this way, one often needs to translate classical geometric results into ring-theoretic terms.

We begin, in Section 2, with the notion of the field K_V of rational functions on an irreducible variety V, together with evaluation of these functions. Evaluation turns out to be an essentially new idea, since there may be points on V at which the value of a rational function is "indeterminate." We resolve this question geometrically (using *modes of approach* to a point of V); we also translate into ring-theoretic terms obtaining *valuation subrings* of K_V. Closely related to valuation rings are *local rings* which essentially extend consideration from *modes of approach* to a point, to *arbitrarily small neighborhoods* of a point; the properties which we need are outlined in Section 3.

In Section 4 we give a ring-theoretic characterization of nonsingularity which, besides being important in its own right, leads us to Section 5 where we apply the idea of abstract algebraic variety to arrive at *abstract nonsingular curves.* Here we present a fundamental decomposition theorem which generalizes both the fundamental theorem of algebra and the fundamental theorem of arithmetic. This decomposition theorem affords a very nice example of a geometry-versus-ring theory dictionary—it yields an isomorphism between all point chains on the curve, and all nonzero ideals of the corresponding coordinate ring.

In Section 6 we extend to irreducible curves in $\mathbb{P}^2(\mathbb{C})$ some of the familiar global results of analytic function theory on the Riemann sphere ($\mathbb{C} \cup \{\infty\}$); we also introduce differentials, and establish some of the differential analogues of these function-theoretic results. This lays the groundwork for Section 7, where we prove the famous Riemann–Roch theorem for nonsingular curves in $\mathbb{P}^2(\mathbb{C})$. This theorem provides a measure of how many rational functions and differentials there are having at most a prescribed set of poles on the curve. This result spotlights the close connection between function theory on the curve and the structure of the curve itself for, conversely, knowing how many such rational functions and differentials there are determines the genus (that is, the topology) of the curve.

2 Valuation rings

Let $V \subset \mathbb{C}_{X_1,\ldots,X_n}$ be an irreducible variety with coordinate ring $\mathbb{C}[x_1,\ldots,x_n] = \mathbb{C}[x]$. The value of a polynomial $p(x) \in \mathbb{C}[x]$ at a point $x_0 \in V$ is $p(x_0)$ which, ring-theoretically, is just the image of $p(x)$ in $\mathbb{C}[x]/\mathfrak{m}$, where $\mathfrak{m}$ is the maximal ideal of $\mathbb{C}[x]$ corresponding to x_0. We also know that a variety $\mathsf{V}(\mathfrak{p}) \subset \mathbb{C}^n$ has $\mathbb{C}[X_1, \ldots, X_n]/\mathfrak{p}$ as coordinate ring and that two affine varieties are (polynomially) isomorphic iff their coordinate rings are isomorphic. Just as we regard $\mathbb{C}[x]$ as the ring of polynomials on V, we may regard $\mathbb{C}(x)$ as a field of rational functions on V. The field $\mathbb{C}(x)$ is called the **function field** of V, and is also denoted by K_C. It is natural to seek $\mathbb{C}(x)$-analogues of the properties of $\mathbb{C}[x]$ just mentioned. These properties are not merely trivial extensions of those of $\mathbb{C}[x]$, as our next examples show.

Example 2.1. Let $V = \mathbb{C}_{XY}$. The function field of V is $\mathbb{C}(X, Y)$. What is the value of Y/X at the origin? We cannot directly assign a definite value, even infinity, to 0/0. The origin is then a point of indeterminancy for Y/X. We can, however, approach (0, 0) along different directions in $\mathbb{C}_{XY}$; for various directions we get different values. For instance, approaching (0, 0) along $\mathbb{C}_X$ (parametrized by $X = T$ and $Y = 0$) gives the value $\lim_{T \to 0} (0/T) = 0$; approaching along $\mathbb{C}_Y$ yields ∞, and approaching (0, 0) along the line $X = aT$, $Y = bT$ ($a \neq 0$) yields the value b/a.

Of course one can approach (0, 0) in a much more arbitrary way than along lines; however one might conjecture that approaching (0, 0) along smooth curves all having the same tangent at (0, 0) would yield the same value at a point of indeterminancy. Consider this example:

EXAMPLE 2.2. The origin $(0, 0) \in \mathbb{C}_{XY}$ is a point of indeterminancy for the rational function $Y/X^2 \in \mathbb{C}(X, Y)$. Approaching (0, 0) along $\mathbb{C}_X$ again yields the value 0. Every parabola $\mathsf{V}(Y - cX^2)(c \in \mathbb{C}\backslash\{0\})$ is tangent to $\mathbb{C}_X$. Along $Y = cX^2$, Y/X^2 becomes $cX^2/X^2 = c$, and *we get a different value for each of these tangent parabolas.*

Example 2.2 shows the importance of considering modes of approach other than lines. Yet we cannot admit just *any* mode of approach, for we can again end up with indeterminancy.

EXAMPLE 2.3. Consider $Y/X \in \mathbb{R}(X, Y)$; suppose that we approach $(0, 0) \in \mathbb{R}_{XY}$ along the curve $\mathbb{C}$ defined by

$$C: \begin{cases} Y = X \sin \dfrac{1}{X} & X \neq 0 \\ Y = 0 & X = 0. \end{cases}$$

Values of Y/X oscillate between $+1$ and -1 as we approach $(0, 0) \in C$, and Y/X is again assigned no unique value at (0, 0).

From these examples, it is evident that we must be judicious in how we approach a point of indeterminancy to arrive at a well-defined value. Since we are dealing with algebraic varieties, it might seem reasonable to restrict our approach along points in an irreducible algebraic curve. (Approaching on a subvariety of dimension $\geqslant 2$ can obviously again lead to indeterminancy, as Example 2.1 shows.) Note that C in Example 2.3 is not algebraic. (No algebraic curve intersects a line in infinitely many discrete points by Bézout's theorem.) Restricting our approach along irreducible algebraic curves is still not quite good enough, as we see next:

EXAMPLE 2.4. The alpha curve $\mathsf{V}(Y^2 - X^2(X + 1)) \subset \mathbb{C}_{XY}$ is irreducible, but near the origin it "splits up" into two parts which turn out to have $\mathsf{V}(Y - X)$ and $\mathsf{V}(Y + X)$ as tangents at (0, 0). Approaching (0, 0) along these parts separately (even within $\mathbb{R}^2$) assigns the two values $+1$ and -1 to $y/x \in \mathbb{C}(x, y)$, where $\mathbb{C}[x, y] = \mathbb{C}[X, Y]/(Y^2 - X^2(X + 1))$.

We shall see that one always gets single-valuedness by restricting attention to parts of algebraic curves which in a certain sense "cannot be split up." In the example above, each of the two "parts" or "branches" through (0, 0) of the alpha curve defines a *mode of approach to* (0, 0); selecting one of these branches will determine a unique limit for every element of $\mathbb{C}(x, y)$.

In the plane, such branches turn out to be just the parts of algebraic curves given by the fractional-power series representations of Theorem 4.13 of Chapter II, where we split up a neighborhood of a point on a curve into the one-point union of topological disks. In the plane, we will see that these fractional-power series determine, on a function field, the most general map satisfying the basic properties one would expect "evaluation at a point" to have. These branches may have singularities (for instance $\mathsf{V}(Y^2 - X^3)$ at (0, 0)).

We now turn to a more precise description of *mode of approach*. Let $\mathbb{C}(x_1, \ldots, x_n)$ be the rational function field of an irreducible variety $V \subset \mathbb{C}_{X_1, \ldots, X_n}$, and let $(0) \in V$ be an aribtrary point of V. In the familiar case of $V = \mathbb{C}_X$ (X a single indeterminate), when the function field is $\mathbb{C}(X)$, the evaluation at (0) is well defined and presents no problems—each $f \in \mathbb{C}(X)$, when written in reduced form p/q, assumes the unique value $p(0)/q(0) \in \mathbb{C} \cup \{\infty\}$. (We never have $p(0) = q(0) = 0$ if f is in reduced form.) If $f, g \in \mathbb{C}(X)$ and $f(0), g(0) \in \mathbb{C}$, then $(f + g)(0)$ is of course $f(0) + g(0)$; $(f \cdot g)(0) = f(0) \cdot g(0)$; and $f(0) = \infty$ iff $1/f(0) = 0$. For $V \subset \mathbb{C}^n$ with coordinate ring $\mathbb{C}[x_1, \ldots, x_n]$, in any way of extending the natural evaluation of elements of $\mathbb{C}[x_1, \ldots, x_n]$ at $(0) \in V$ to elements of $\mathbb{C}(x_1, \ldots, x_n)$, we would expect at the very least this same kind of behavior—that is, each $f \in \mathbb{C}(x_1, \ldots, x_n)$ would be assigned a unique value in $\mathbb{C} \cup \{\infty\}$ satisfying these two conditions:

(2.5.1) If f and g are assigned finite values $a, b \in \mathbb{C}$, then $f + g$ and $f \cdot g$ are assigned values $a + b$ and $a \cdot b$, respectively;
(2.5.2) f is assigned the value ∞ iff $1/f$ is assigned the value 0.

It turns out that even these rudimentary assumptions on an "evaluation" of elements in $\mathbb{C}(x_1, \ldots, x_n)$ yield important information. First, it is immediate that the set of all elements of $\mathbb{C}(x_1, \ldots, x_n)$ which are given finite values forms a subring R. We know that any coordinate ring $\mathbb{C}[x_1, \ldots, x_n] = \mathbb{C}[x]$ determines an affine variety V (unique up to isomorphism) whose coordinate ring R_V is, in turn, isomorphic to $\mathbb{C}[x]$. Analogously, one may ask:

(2.6) Does R determine in a natural way some geometric object having a "coordinate" ring isomorphic to R?

The answer to this is *yes* in many important cases, and the object turns out to isolate the essential idea of *mode of approach*. To get a more precise notion of this, consider an example: Let f be a real-valued function on $\mathbb{R}\backslash\{0\}$ and suppose $\lim_{X\to 0} f(X)$ exists. Now f surely need not have all of $\mathbb{R}\backslash\{0\}$ as domain to evaluate $\lim_{X\to 0} f(X)$. Even for a fixed neighborhood N of 0, $N\backslash\{0\}$ is more than is needed. Yet since f is not defined at 0, more than the single point 0 is required. We want an object expressing "arbitrary nearness to a point." This leads us to the *germ* of a set at a point (Definition 2.7). In the above example, one object containing the basic notion of arbitrary nearness to (0) is the set of all neighborhoods about (0). This is the *germ of* $\mathbb{R}$ *at* (0). Another example: In $\mathbb{R}^2$, let L be a line through (0); the germ of L at (0) is the set of all subsets of $\mathbb{R}^2$ coinciding with L throughout some $\mathbb{R}^2$-neighborhood of (0). Note that if S is any one of these subsets, then the germ of L at (0) is the same as the germ of S at (0)—that is, we identify S and L iff they agree at all points sufficiently close to (0). This, of course, is all that matters in asking for the limit of a function along S or L. It is in this way that *germ* incorporates the idea of "nearness," and will be used in making precise our idea of mode of approach.

Definition 2.7. Let $\mathscr{T}$ be a topological space, let S be a subset of $\mathscr{T}$, and let P be a point of $\mathscr{T}$. The **germ of S at P** is the set of all subsets T of $\mathscr{T}$ which coincide with S on some $\mathscr{T}$-neighborhood of P; that is, T is a member of the germ iff there is some open neighborhood U about P such that $S \cap U = T \cap U$. This germ is denoted by $S_P{}^{\sim}$. P is called the **center of $S_P{}^{\sim}$**, and any member of $S_P{}^{\sim}$ is called a **representative of $S_P{}^{\sim}$**. If there is some open neighborhood U about P for which $S \cap U = \varnothing$, then the germ of S at P is called the **empty germ at P**.

It is clear that any point $P \in \mathscr{T}$ partitions the set of all subsets of $\mathscr{T}$ into equivalence classes, any two subsets of $\mathscr{T}$ being equivalent iff they have the same germ at P. Note that for any $P \in \mathscr{T}$ and any two subsets $S, S' \subseteq \mathscr{T}$, $S_P{}^{\sim} = S_P'{}^{\sim}$ iff $S \cap U = S' \cap U$ for some neighborhood U of P.

Just as on sets we may define functions, on germs of sets we may define *germs of functions*:

Definition 2.8. Let $\mathscr{T}$ be a topological space, let $P \in \mathscr{T}$, and let f be any function on some $\mathscr{T}$-neighborhood of P. The **germ of f at P** is the set of all functions g defined on neighborhoods of P which coincide with f on some neighborhood of P. We denote the germ of f at P by $f_P{}^{\sim}$; we call $f_P{}^{\sim}$ a **function germ**.

As with germs of sets, function germs may be looked at as equivalence classes of functions, two functions being equivalent iff they have the same germ at P. Note that for functions f and g, $f_P{}^{\sim} = g_P{}^{\sim}$ iff f and g agree in a neighborhood of P. Operations on sets and on functions induce analogous operations on germs of sets and on function germs. With notation as in Definition 2.7, we have

Definition 2.9. The germs $S_P^{\sim} \cup S_P'^{\sim}$ and $S_P^{\sim} \cup S_P'^{\sim}$ are defined to be $(S \cup S')_P^{\sim}$ and $(S \cap S')_P^{\sim}$, respectively. (It is easy to see that these notions are well defined.)

Similarly, we have

Definition 2.10. Let f and g be any two functions defined on a neighborhood of $P \in \mathscr{T}$ with values in a field; we define $f_P^{\sim} + g_P^{\sim}$ by $(f + g)_P^{\sim}$, $f_P^{\sim} \cdot g_P^{\sim}$ by $(f \cdot g)_P^{\sim}$, and $-(f_P^{\sim})$ by $(-f)_P^{\sim}$. If there is a neighborhood of P on which f is never zero, then we define $1/f_P^{\sim}$ by $(1/f)_P^{\sim}$. (These are clearly well defined.)

Definitions 2.8–2.10 are very general; one can consider important classes of germs at different "levels," for instance the set of subsets S of $\mathscr{T}$ closed at P (that is, S closed within a sufficiently small neighborhood of P); if furthermore $\mathscr{T}$ is $\mathbb{C}$ supplied with the usual topology, we can replace *closed at P* by *analytic at P* (that is, S coincides throughout some neighborhood of P with the zero-set of a function analytic at $P \in \mathbb{C}$). One then speaks of *closed germs*, *analytic germs*, etc. And for functions, one may, for instance, consider functions continuous, differentiable, or analytic at a point. Definition 2.10 shows that the set of all function germs in any such fixed level in general forms a ring. We may, more generally, consider any ring R of functions, each function being defined on some neighborhood of a fixed point P of a topological space $\mathscr{T}$. The set $R_P^{\sim} = \{f_P^{\sim} \mid f \in R\}$ forms the induced set of function germs at P. In view of Definition 2.10 we see that $f \to f_P^{\sim}$ is a ring homomorphism, and $R_P^{\sim}$ is in general a proper homomorphic image of R.

EXAMPLE 2.11. Let R be the ring of all real-valued functions on $\mathbb{R}$ which are constant in a neighborhood of (0). The elements of $R_P^{\sim}$ are collections of functions, and each collection can be represented by a constant function. $R_P^{\sim}$ is in this case isomorphic to $\mathbb{R}$. As another example, consider polynomials or functions analytic at $(0) \in \mathbb{C}$. There is essentially only *one* function in each germ (from the familiar "identity theorem" for power series).

The main applications of this section in the rest of the chapter, are to curves. In Theorem 2.28 we prove the following: Suppose (i) P is any point of an irreducible curve $C \subset \mathbb{C}_{X_1, \ldots, X_n}$ with coordinate ring $\mathbb{C}[x_1, \ldots, x_n]$ and function field $\mathbb{C}(x_1, \ldots, x_n) = K_C$, (ii) there is given an evaluation at P of each function $f \in K_C$ coinciding with the natural evaluation of $\mathbb{C}[x_1, \ldots, x_n]$ at P and satisfying properties (2.5.1) and (2.5.2), and (iii) R is the subring of K_C assigned finite values. Then we can conclude that there is canonically associated with R a germ $B_P^{\sim}$ (B for "branch") such that R can be regarded in a natural way as a ring of function germs on $B_P^{\sim}$, and such that for any $f \in R$, its initially-given evaluation at P will be the value of $f_P^{\sim}$ at P. Before proving Theorem 2.28 we establish a number of basic results.

We first look at some properties of the above ring R. We begin with an example. Consider $\mathbb{C} = \mathbb{C}_X$; at (0) the corresponding ring $R \subset \mathbb{C}(X)$ of functions assigned finite values at (0) consists of the functions p/q, where p and q are polynomials and $q(0) \neq 0$. The ring R can be naturally regarded as a ring of functions on $\mathbb{C}_{(0)}{}^{\sim}$. There is but one point of $\mathbb{C}$ common to all sets in $\mathbb{C}_{(0)}{}^{\sim}$. If, as suggested earlier, R is to act like a "coordinate ring" on $\mathbb{C}_{(0)}{}^{\sim}$, then we might conjecture that R has but one maximal ideal, corresponding to the point (0). This is indeed so; in fact we show, more generally (Theorem 2.12 and its corollary), that the ring R of all elements in any field K assigned values satisfying properties (2.5.1) and (2.5.2), has a unique maximal ideal.

We start with the following simple characterization, which leads to Definition 2.13.

Theorem 2.12. *Let K and k be fields; if each element in K is assigned a value in $k \cup \{\infty\}$, and if this assignment satisfies properties* (2.5.1) *and* (2.5.2), *then the set of elements assigned finite values forms a subring R of K, and for each $a \in K$, $a \notin R$ implies $1/a \in R$. Conversely, let K be a field; if R is any subring of K such that for each $a \in K$, $a \notin R$ implies $1/a \in R$, then there is a field k such that each element of K is assigned a value in $k \cup \{\infty\}$, this assignment satisfying* (2.5.1) *and* (2.5.2).

Proof. The first half is obvious. For the converse, assume without loss of generality that $R \neq K$, and let $\mathfrak{m}$ be the set of elements a of R such that $1/a \notin R$. We show that $\mathfrak{m}$ is a maximal ideal in R. Then properties (2.5.1) and (2.5.2) follow at once for the field $R/\mathfrak{m}$.

We first show that $\mathfrak{m}$ is an ideal. $\mathfrak{m}$ is closed under addition. For let $a, b \in \mathfrak{m}$. If a or b is 0, then $a + b \in \mathfrak{m}$. Therefore assume $a \neq 0$, $b \neq 0$. We show $a + b \in \mathfrak{m}$. By hypothesis on R, either $a/b \in R$ or $(a/b)^{-1} = b/a \in R$. Suppose $a/b \in R$. Now $1 \in R$ (if not, then $1/1 = 1 \in R$); hence because R is assumed to be a ring, $1 + (a/b) = (b + a)/b \in R$. To show $a + b \in \mathfrak{m}$, suppose $a + b \notin \mathfrak{m}$. Then $1/(a + b) \in R$, and since R is a ring, $[(b + a)/b][1/(a + b)] = 1/b \in R$. But this is impossible, since we assumed from the outset that $b \in \mathfrak{m}$; by $\mathfrak{m}$'s definition, $b \in \mathfrak{m}$ implies $1/b \notin R$. Next, $\mathfrak{m}$ has the absorbing property, for suppose $a \in R$, $b \in \mathfrak{m}$ but that $ab \notin \mathfrak{m}$. Then $1/ab \in R$, hence $a/ab = 1/b \in R$, again giving us a contradiction.

It is now easy to see that $\mathfrak{m}$ is maximal, since any ideal $\mathfrak{a} \subset R$ containing an element $c \in R \backslash \mathfrak{m}$ must also contain $c \cdot (1/c) = 1$ ($c \in R \backslash \mathfrak{m}$ implies $1/c \in R$), which implies that $\mathfrak{a} = R$. Thus there can be no proper ideal of R larger than $\mathfrak{m}$. □

Note that if a field K is given an evaluation satisfying (2.5.1) and (2.5.2), the subring R of elements assigned finite values in turn determines the same evaluation (up to an isomorphism of $k = R/\mathfrak{m}$) via $a \to a + \mathfrak{m}$ (if $a \in R$), and $a \to \infty$ (if $a \notin R$).

In view of Theorem 2.12 we make the following

Definition 2.13. A subring R of a field K is called a **valuation ring** if, for each $a \in K$, either $a \in R$ or $1/a \in R$. If R contains a subfield k of K, then R is called a **valuation ring over** k.

Remark 2.14. Henceforth, the term *valuation ring* in this book will always mean valuation ring over $\mathbb{C}$ unless stated otherwise.

Corollary to Theorem 2.12. *Every valuation ring R has a unique maximal ideal.*

PROOF. Any maximal ideal other than $\mathfrak{m} = \{a \in R \mid 1/a \notin R\}$ would have to contain an element of $R \backslash \mathfrak{m}$, and we saw this is impossible. $\square$

There is an important side to evaluation which we have not touched upon yet. We begin with an example. Consider the field $K = \mathbb{C}(X)$. In addition to assigning at a point P a value $f(P) \in \mathbb{C} \cup \{\infty\}$ to each $f \in \mathbb{C}(X)$, we may also assign an *order*, $\text{ord}_P(f)$; it is a straightforward generalization of the definition for polynomials: We define, for $f = p/q$ ($p, q \in \mathbb{C}[X]$),

$$\text{ord}_P(f) = \text{ord}_P\left(\frac{p}{q}\right) = \text{ord}_P(p) - \text{ord}_P(q), \quad \text{for any } P \in \mathbb{C}. \tag{1}$$

It is obvious that $\text{ord}_P(f)$ is well defined. Observe that if p/q (written in lowest terms) is expanded about P (e.g., expand p and q about P and use "long division"), the exponent of the lowest-degree term is just $\text{ord}_P(p/q)$; this fits in with the term *order*. (For $c \in \mathbb{C}\backslash\{0\}$, $\text{ord}_P(c) = 0$; $\text{ord}_P(0) = \infty$, by definition. We assume ∞ is greater than any element of $\mathbb{Z}$.) As with polynomials, ord_P for elements of $\mathbb{C}(X)$ satisfies

(a) $\text{ord}_P(f + g) \geqslant \min(\text{ord}_P(f), \text{ord}_P(g))$,
(b) $\text{ord}_P(f \cdot g) = \text{ord}_P(f) + \text{ord}_P(g)$ for arbitrary $f, g \in \mathbb{C}(X)\backslash\{0\}$.

These two properties may be taken as basic for a more general definition.

Definition 2.15. Let K be any field and let ord be a function from $K\backslash\{0\}$ *onto* the set of all integers $\mathbb{Z}$ such that for any $a, b \in K\backslash\{0\}$, we have

(2.15.1) $\text{ord}(a + b) \geqslant \min\{\text{ord}(a), \text{ord}(b)\}$
(2.15.2) $\text{ord}(a \cdot b) = \text{ord}(a) + \text{ord}(b)$.

Then ord is called a **discrete rank one valuation of** K; if k is a subfield K and $\text{ord}(a) = 0$ for all $a \in k\backslash\{0\}$, we say ord is a **discrete rank one valuation of** K **over** k. For a given ord on K, we say $a \in K$ **has order** n if $\text{ord}(a) = n$.

As the reader might guess from the above terminology, there are more general valuations, not necessarily discrete or rank one; although at the end of this section we show how discrete valuations of rank >1 arise in evaluations on more general fields, in the sequel we will not use valuations more general than discrete rank one.

Note that for any ord on a field K,

$$\mathrm{ord}(1) = 0 \tag{2}$$

since $\mathrm{ord}(1) = \mathrm{ord}(1 \cdot 1) = \mathrm{ord}(1) + \mathrm{ord}(1)$. Also, for any nonzero $a \in K$, $\mathrm{ord}(a(1/a)) = \mathrm{ord}(a) + \mathrm{ord}(1/a) = 0$; hence

$$\mathrm{ord}\left(\frac{1}{a}\right) = -\mathrm{ord}(a) \qquad (0 \neq a \in K). \tag{3}$$

This in turn implies that

$$\mathrm{ord}\left(\frac{a}{b}\right) = \mathrm{ord}(a) - \mathrm{ord}(b) \tag{4}$$

which generalizes (1).

For a given ord on K, $R = \{0\} \cup \{a \in K \mid \mathrm{ord}(a) \geqslant 0\}$ is a ring (immediate from (2.15.1) and (2.15.2)); if $a \notin R$, then $1/a \in R$ (since $\mathrm{ord}(1/a) = -\mathrm{ord}(a)$). Hence ord defines a valuation subring of K.

Definition 2.16. Let K be any field and let ord be any discrete rank one valuation of K. The ring $R = \{0\} \cup \{a \in K \mid \mathrm{ord}(a) \geqslant 0\}$ is the **valuation ring of** ord; it is called a **discrete rank one valuation ring**.

Remark 2.17. It is easy to see that all functions ord determining a given discrete rank one valuation ring R of K, may be identified in a canonical way with the function defined by $K \backslash \{0\} \to (K \backslash \{0\})/\mathscr{U}$, where $\mathscr{U} = R \backslash \mathfrak{m}$ is the multiplicative subgroup of elements of K having order 0. In this way any discrete rank one valuation ring itself determines an essentially unique discrete rank one valuation on K.

Discrete rank one valuation rings have many nice properties. In the case of the function field $\mathbb{C}(x_1, \ldots, x_n)$ of a curve C, there is a fundamental connection between discrete rank one valuation rings of $\mathbb{C}(x_1, \ldots, x_n)$, and modes of approach to points of C. Basic to this is the following

Lemma 2.18. *Any proper valuation ring R ($\mathbb{C} \subset R$) of a field $K = \mathbb{C}(x_1, \ldots, x_n)$ having transcendence degree one over $\mathbb{C}$ is discrete rank one.*

The idea of the proof is quite simple, and is given in Exercise 2.2.

Definition 2.19. *Let R be a discrete rank one valuation ring of a field, and let $\mathfrak{a}$ be any ideal of R. Define* $\mathrm{ord}(\mathfrak{a})$ *by*

$$\mathrm{ord}(\mathfrak{a}) = \min\{\mathrm{ord}(a) \mid a \in \mathfrak{a}\}.$$

Lemma 2.20. *Let R be a discrete rank one valuation ring of a field. For any ideal $\mathfrak{a} \subset R$,*

$$\mathfrak{a} = \{a \in R \mid \operatorname{ord}(a) \geqslant \operatorname{ord}(\mathfrak{a})\}.$$

Proof. First, for each nonnegative integer n, $\mathfrak{a}$ contains at least one element of order $\operatorname{ord}(\mathfrak{a}) + n$. For let a be any element of $\mathfrak{a}$ such that $\operatorname{ord}(a) = \operatorname{ord}(\mathfrak{a})$, and let b be any element of R of order 1. (The number 1 is the smallest positive ord-value of elements in K, since ord is onto $\mathbb{Z}$.) Then $\operatorname{ord}(ab^n) = \operatorname{ord}(\mathfrak{a}) + n$.

Second, if $\mathfrak{a}$ contains at least one element c of order m, it contains all the elements of R of order m. For if d is any element of R of order m, then $\operatorname{ord}(d/c) = 0$, hence $d/c \in R$ by the definition of R. Therefore $c(d/c) = d \in \mathfrak{a}$ since $\mathfrak{a}$ is an ideal. □

Corollary 2.21. *Every discrete rank one valuation ring is a principal ideal ring.*

Proof. Let $\mathfrak{a}$ be any nonzero ideal of R, and let a be an element of least order in $\mathfrak{a}$. From the proof of Lemma 2.20 we see that (a) consists of all elements of R of order $\geqslant \operatorname{ord}(\mathfrak{a})$; but so does $\mathfrak{a}$, so $\mathfrak{a} = (a)$. □

Corollary 2.22. *Every discrete rank one valuation ring is Noetherian.*

Proof. Every principal ideal ring R is Noetherian. (Let $\mathfrak{a}_1 \subseteq \mathfrak{a}_2 \subseteq \cdots$ be an ascending sequence of ideals of R. Then $\bigcup_i \mathfrak{a}_i = (a)$ for some $a \in R$. But $a \in \mathfrak{a}_i$, for some i. Hence $\mathfrak{a}_i = \mathfrak{a}_{i+1} = \cdots$.) □

Corollary 2.23. *Let $\mathfrak{a}$ be any proper ideal in a discrete rank one valuation ring R. Then $\bigcap_{m=1}^{\infty} \mathfrak{a}^m = (0)$.*

Proof. If $\bigcap_{m=1}^{\infty} \mathfrak{a}^m \neq (0)$, then $\bigcap_{m=1}^{\infty} \mathfrak{a}^m = (b)$ for some $b \neq 0$. If $\operatorname{ord}(\mathfrak{a}) = r$ and $\operatorname{ord}(b) = s$, then $mr > s$ for sufficiently large m; hence $b \notin \mathfrak{a}^m$, a contradiction. □

Remark 2.24. More generally, for any Noetherian domain R, if $\mathfrak{a}$ is a proper ideal of R, then $\bigcap_{m=1}^{\infty} \mathfrak{a}^m = (0)$. This is Krull's theorem. For a proof see, for example, [Zariski and Samuel, Vol. I, p. 216].

Corollary 2.25. *The unique maximal ideal of a discrete rank one valuation ring R is generated by any element of order* 1, *and consists of all the elements of R having order* $\geqslant 1$.

Our next result, Theorem 2.26, will lead us to a geometric interpretation of valuation subrings of fields of curves. We know that any function $f(X) = p(X)/q(X) \in \mathbb{C}(X)$ where X is a single indeterminant, is analytic at any point P where $q(P) \neq 0$. Theorem 2.26 generalizes this, in that for any discrete rank one valuation subring R of the function field of any curve, Theorem 2.26

allows us to regard each element of R as an appropriate analytic function germ.

To identify elements of R with infinite sums of elements of R, we need a topology on R. We use a natural topology induced by R's maximal ideal $\mathfrak{m}$. This topology is rather weak—the "formal" power series which we construct will automatically converge relative to this weaker topology. We will show separately that the power series obtained this way do in fact converge in the usual, stronger sense of complex analysis.

We define the natural topology on a discrete, rank one valuation ring R by means of a basis for open subsets of R, as follows: Let R's maximal ideal be $\mathfrak{m}$. For each $a \in R$ and positive integer n, $a + \mathfrak{m}^n$ is a subset of R which we take to be a typical open set of our basis.

Since $\bigcap_{n=1}^{\infty} \mathfrak{m}^n = (0)$, we may translate equality "$a = b$" in R into the useful topological form, "$a - b \in \mathfrak{m}^n$ for all sufficiently large n." Also, note that for arbitrary $a, b \in R$,

$$(a + \mathfrak{m}^n) + (b + \mathfrak{m}^m) = (a + b) + \mathfrak{m}^n \qquad (n \leqslant m);$$
$$(a + \mathfrak{m}^n) \cdot (b + \mathfrak{m}^m) = ab + \mathfrak{m}^n \qquad (n \leqslant m).$$

From this it is easy to check that addition and multiplication are continuous with respect to this topology. (R thus forms a "topological ring.") This topology may be extended to a topology on R's quotient field K by replacing the above $a, b \in R$ by $a, b \in K$.

We say that an element $x \in R$ has a **power series representation** $\sum_{i=0}^{\infty} c_i r^i$ if, given any $N > 0$, we have $x - \sum_{i=0}^{n} c_i r^i \in \mathfrak{m}^N$ for all sufficiently large n. In this sense we may consider that x equals $\sum_{i=0}^{\infty} c_i r^i$—that is, it is identified with an element of R. It is easy to verify that if $x = \sum_i c_i r^i$ and $y = \sum_j d_j r^j$, then $x + y$, $x - y$, xy and x/y ($y \neq 0$) are the sum, difference, product, and quotient of the corresponding power series. Operations on power series are performed in a way similar to those on polynomials (cf. Section II, 4).

We may now state our theorem.

Theorem 2.26. *Let R be a valuation subring of a field $\mathbb{C}(x_1, \ldots, x_n)$ having transcendence degree 1 over $\mathbb{C}$ ($\mathbb{C} \subset R$). Then R is discrete, rank one, and for any element $T \in R$ of order one, each $x \in R$ has a power series representation $x = \sum_{i=0}^{\infty} c_i T^i$. If the symbol T is considered as a complex variable, then $\sum_{i=0}^{\infty} c_i T^i$ is analytic at $(0) \in \mathbb{C}_T$.*

PROOF. Let $\mathfrak{m}$ be R's maximal ideal. Then $R/\mathfrak{m}$ is a field containing $\mathbb{C}$. Now T is a generator of $\mathfrak{m}$; since $T \notin \mathbb{C}$ and $\mathbb{C}$ is algebraically closed, T is transcendental over $\mathbb{C}$. Therefore $R/\mathfrak{m}$ is algebraic over $\mathbb{C}$ (hence is $\mathbb{C}$ itself) as the following argument shows: If $u \in R$ is any element whose image is a given $c \in R/\mathfrak{m}$, then since $\mathbb{C}(x_1, \ldots, x_n)$'s transcendence degree over $\mathbb{C}$ is 1, u is algebraic over $\mathbb{C}(T)$, and therefore satisfies an irreducible polynomial equation

$$p_0(T)u^n + \ldots + p_n(T) = 0 \qquad (p_i(T) \in \mathbb{C}[T]).$$

Under $R \to R/\mathfrak{m}$, T maps to 0 and u maps to c, hence $p_0(0)c^n + \ldots + p_n(0) = 0$; since each $p_i(0) \in \mathbb{C}$, and since not every $p_i(0) = 0$, c is algebraic over $\mathbb{C}$, hence in $\mathbb{C}$. Thus $R/\mathfrak{m} = \mathbb{C}$.

This shows that any element y of R may be written as $y = c + rT$, for some $c \in \mathbb{C}$ and $r \in R$; in fact it shows that if $x \in \mathfrak{m}^n$, then $x = cT^n + rT^{n+1}$ ($c \in \mathbb{C}$ and $r \in R$), for surely $x = y'T^n$ where $y' \in R$, and y' in turn is $c + rT$ for some $c \in \mathbb{C}$ and $r \in R$. It is now easy to prove the existence of the power series representation. Let x be any element of R. Then there exist $c_i \in \mathbb{C}$ and $r_i \in R$ such that

$$\begin{aligned} x &= c_0 + r_0 T & &(\text{hence } x - c_0 \in \mathfrak{m}) \\ x - c_0 &= c_1 T + r_1 T^2 & &(\text{hence } x - c_0 - c_1 T \in \mathfrak{m}^2) \\ x - c_0 - c_1 T &= c_2 T^2 + r_2 T^3 & &(\text{hence } x - c_0 - c_1 T - c_2 T^2 \in \mathfrak{m}^3) \\ &\ \vdots \end{aligned}$$

The limit of the sequence $x - c_0 - c_1 T - c_2 T^2 - \ldots$ is therefore in $\bigcap_{n=1}^{\infty} \mathfrak{m}^n = (0)$, whence we have $x = \sum_{i=0}^{\infty} c_i T^i$. Now let us show that this formal power series is actually analytic at (0). Surely T is transcendental over $\mathbb{C}$. Let $p(T, X)$ be the minimal polynomial for x over $\mathbb{C}(T)$. Now $p(T, \sum_{i=0}^{\infty} c_i T^i) = 0$; we see this as follows: $\sum_{i=0}^{n} c_i T^i = x - \sum_{i>n} c_i T^i$, hence $p(T, \sum_{i=0}^{n} c_i T^i) = p(T, x - \sum_{i>n} c_i T^i)$; the right-hand side is easily checked to be $p(T, x) + \alpha_n\,(=\alpha_n)$, where for n sufficiently large, α_n is in any preassigned power of $\mathfrak{m}$. By the uniqueness part of Corollary 4.17 of Chapter II it follows that $X - \sum_{i=0}^{\infty} c_i T^i$ corresponds to one of the factors in (20) appearing in that corollary. By Corollary 4.18 of Chapter II the series $\sum_{i=0}^{\infty} c_i T^i$ is analytic at $T = (0)$. □

With notation as in Theorem 2.26, we have

Corollary 2.27. *Each element x of $\mathbb{C}(x_1, \ldots, x_n)$ has a Laurent* series *development $x = \sum_{i=n_0}^{\infty} c_i T^i$ ($n_0 \in \mathbb{Z}$) convergent, except possibly at zero, in some neighborhood of* $(0) \in \mathbb{C}_T$.

PROOF. Write x as y/z, where y and z are in R. For some $n \geqslant 0$, $z = c_n T^n + c_{n+1} T^{n+1} + \ldots = T^n(c_n + c_{n+1} T + \ldots)$ where $c_n \neq 0$; since the reciprocal of $c_n + c_{n+1} T + \ldots$ is analytic at (0), y/z is the product of T^{-n} and a power series analytic at (0). □

Let $\mathbb{C}(x_1, \ldots, x_n)$ have transcendence degree 1 over $\mathbb{C}$, let R be a valuation ring of $\mathbb{C}(x_1, \ldots, x_n)$ over $\mathbb{C}$, and suppose the maximal ideal $\mathfrak{m}$ of R intersects $\mathbb{C}[x_1, \ldots, x_n]$ in a maximal ideal. Let C be an affine curve with coordinate ring $\mathbb{C}[x_1, \ldots, x_n]$, and let $P \in C$ correspond to $\mathfrak{m} \cap \mathbb{C}[x_1, \ldots, x_n]$. Then Theorem 2.26 implies that R yields a subset $B_P \subset C$ (a *representative of a branch of C at P*) containing P so that for each element $x \in \mathbb{C}(x_1, \ldots, x_n)$, this holds: At all points of B_P sufficiently near P, R assigns to x a unique value in $\mathbb{C} \cup \{\infty\}$, this assignment satisfying (2.5.1) and (2.5.2). The set B_P

will serve as a *mode of approach* to P. This is done as follows: Let $C \subset \mathbb{C}^n$ have $(x_1, \ldots, x_n)$ as generic point. Since $x_i \in R$, for any $T \in R$ of order 1, each x_i has a power series representation $x_i = \sum_{j=0}^{\infty} c_{ij} T^j$ $(i = 1, \ldots, n)$, convergent in some common neighborhood N of $(0) \in \mathbb{C}_T$. Any $t_0 \in N$ thus determines a specialization of $(x_1, \ldots, x_n)$, namely

$$(x_1, \ldots, x_n) \to \left(\sum_j c_{1j} t_0{}^j, \ldots, \sum_j c_{nj} t_0{}^j\right),$$

and therefore defines a subset of C, consisting of all specialization points of C as t_0 ranges throughout N. Let B_P be this subset. We say that B_P is "analytic at P," in the sense that it is, within a sufficiently small $\mathbb{C}^n$-neighborhood U of P, the zero set of a set of functions analytic throughout U. The point $(c_{10}, \ldots, c_{n0})$ corresponding to $t_0 = 0$ is P. Note that the germ $B_P{}^{\sim}$ of B_P is independent of the choice of the order-1 element of R. (If T' is any other order-1 element of R, we may write $T = \sum_{k=1}^{\infty} a_k (T')^k$ where $a_1 \neq 0$; this establishes a homeomorphism between neighborhoods of (0) in $\mathbb{C}_T$ and $\mathbb{C}_{T'}$, so $\sum_j c_{ij} T^j$ and $\sum_j c_{ij} (\sum_k a_k (T')^k)^j$ describe the same set in C near P as T and T' vary near (0) in $\mathbb{C}_T$ and $\mathbb{C}_{T'}$, respectively.) Now let x be any element of $\mathbb{C}(x_1, \ldots, x_n)$. If $x \in R$, then it has a power series representation $\sum_j c_j T^j$, and the value at P, corresponding to $T = 0$, is just c_0.

The set of elements of R having constant term 0 in the power expansion evidently forms $\mathfrak{m} \subset R$. The value of any element $x \in R$ at $(c_{10}, \ldots, c_{n0})$ is thus $x + \mathfrak{m} \in R/\mathfrak{m} = \mathbb{C}$. An element $x \in \mathbb{C}(x_1, \ldots, x_n)$ is therefore assigned the value 0 iff $x \in \mathfrak{m}$, and is assigned the value ∞ iff $x \notin R$. Thus the properties (2.5.1) and (2.5.2) are satisfied.

The neighborhood of P in B_P throughout which we guarantee the above assignment depends on x. The natural, well-defined object determined by R is thus the *germ* $B_P{}^{\sim}$; the elements of R are then analytic function germs on $B_P{}^{\sim}$. The center P of $B_P{}^{\sim}$ is the image $(c_{10}, \ldots, c_{n0})$ of the center of the power series expansion. In view of these observations, we now have:

Theorem 2.28. *Let P be a point of an irreducible curve $C \subset \mathbb{C}^n$ with coordinate ring $\mathbb{C}[x_1, \ldots, x_n]$ and function field $K_C = \mathbb{C}(x_1, \ldots, x_n)$; let the elements of K_C be given an evaluation at P extending the natural evaluation at P of elements in $\mathbb{C}[x_1, \ldots, x_n]$, and satisfying* (2.5.1) *and* (2.5.2). *Then the associated valuation ring R defines an analytic germ $B_P{}^{\sim}$ at P, and the elements of R may be regarded as analytic function germs on $B_P{}^{\sim}$, the evaluation of any $f \in R$ at P coinciding with the value of $f_P{}^{\sim}$ at P.*

The following definitions will be used in the sequel:

Definition 2.29. The maximal ideal $\mathfrak{m}$ of a valuation ring R is called the **center** of R. If R is a valuation ring in a field $\mathbb{C}(x_1, \ldots, x_n)$ of transcendence degree one over $\mathbb{C}$, and if $\mathfrak{m}$ intersects the coordinate ring $\mathbb{C}[x_1, \ldots, x_n]$ of an irreducible curve $C \subset \mathbb{C}_{X_1, \ldots, X_n}$ in a maximal ideal, then the associated point $P \in C$ is the center of the germ $B_P\tilde{}$ determined by R, and is called the **center of** R **on** C. Any such $B_P\tilde{}$ is called a **branch of** C **centered at** P (or **through** P, or **at** P).

Definition 2.30. If $C \subset \mathbb{C}_{X_1, \ldots, X_n}$ is an irreducible curve, then an element U in C's function field is called a **uniformizing parameter** (or **uniformizing variable**) **at** $P \in C$ if the part of C about P can be represented by power series in U:

$$\left(\sum_j c_{1j} U^j, \ldots, \sum_j c_{nj} U^j\right),$$

where $P = (c_{10}, \ldots, c_{n0})$.

This next result will be needed later; it gives a simple geometric way of getting uniformizing parameters of a *plane* curve at any nonsingular point. It is essentially a corollary of Theorem 3.6 of Chapter II.

Theorem 2.31. *Let* $P = (0, 0)$ *be an arbitrary nonsingular point of an irreducible plane curve* $C = \mathsf{V}(p) \subset \mathbb{C}_{XY}$, *and suppose* $\mathsf{V}(X)\,(= \mathbb{C}_Y \subset \mathbb{C}_{XY})$ *is not tangent to* C *at* P. *Then* X *(or more precisely, the image of* X *in the coordinate ring* $\mathbb{C}[X, Y]/(p(X, Y))$ *of* C*) is a uniformizing parameter for* C *at* P.

Proof. Assume without loss of generality that coordinates have been chosen so the tangent line to C at $(0, 0)$ is $\mathbb{C}_X$ $(= \mathsf{V}(Y))$. Then $p_X(0, 0) = 0$, so by nonsingularity, $p_Y(0, 0) \neq 0$. By Theorem 3.6 of Chapter II, C is locally described by $Y = g(X)$ where g is analytic at 0. Hence X is a uniformizing parameter for C at $(0, 0)$, since the part of C near $(0, 0)$ is represented by the power series $(X, g(X))$. □

So far we have not considered any concrete examples of valuation rings, except very simple ones—those in $\mathbb{C}(X)$ which contain $\mathbb{C}$. A natural question is this: Are there other valuation rings? To answer this, note that we have shown so far that at a point P of an irreducible affine curve C having coordinate ring $\mathbb{C}[x_1, \ldots, x_n]$ and function field $K_C = \mathbb{C}(x_1, \ldots, x_n)$, *if* one has an evaluation on K_C satisfying properties (2.5.1) and (2.5.2) and coinciding with the natural one on $\mathbb{C}[x_1, \ldots, x_n]$ at some $P \in C$, *then* there is an associated branch $B_P\tilde{}$ serving as a mode of approach to P along which each element of K_C has a well-defined limiting value in $\mathbb{C} \cup \{\infty\}$. However, it is conceivable that there could be points on C at which there is no such evaluation of K_C (points at which one can of course evaluate elements in the coordinate ring, but cannot extend the evaluation to all of K_C). *This can never happen. At each point of* C, *any evaluation on* $\mathbb{C}[x_1, \ldots, x_n]$ *always extends in at least one way to an evaluation on* K_C. We prove this fundamental fact for plane curves in Theorem 2.32. This confirms the geometrically intuitive guess that valuation rings abound in algebraic geometry.

Theorem 2.32. *Let P be an arbitrary point of an irreducible plane curve $C \subset \mathbb{C}_{XY}$ having coordinate ring $\mathbb{C}[x, y]$ and function field $K_C = \mathbb{C}(x, y)$. Then there is an evaluation of the elements of K_C satisfying* (2.5.1) *and* (2.5.2) *and coinciding with the natural evaluation of $\mathbb{C}[x, y]$ at P.*

PROOF. If at P we could express both x and y as convergent power series in an element $T \in K_C$ of order one, then every element $f \in K_C$ would be a Laurent series in T, $f(x, y) = c_m T^m + c_{m+1} T^{m+1} + \ldots$. If (x, y) evaluated at $T = 0$ is $P \in C$, then assigning to f the value

$$\begin{cases} c_m & \text{if } m = 0 \\ 0 & \text{if } m > 0 \\ \infty & \text{if } m < 0 \end{cases}$$

defines an evaluation of the required kind.

We can get power series representations for x and y as follows: Let $p(X, Y)$ be an irreducible polynomial (monic in Y) defining C; suppose $\deg_Y p = n$, and take $P = (0, 0)$. Let the distinct zeros of $p(0, Y)$ be $y_1, \ldots, y_s$, and let Y_{jik} denote the "Y_{ji}" extending (16) in Section II,4 corresponding to the zero y_k. Then the product

$$\prod_{j, i, k} (Y - Y_{jik}) \tag{5}$$

is a polynomial in Y of degree n with fractional-power series coefficients; from the discussion following (16) in Section II,4 we see that for each value $x \in \mathbb{C}$ sufficiently near (0), the product agrees with $p(x, Y)$. Since the coefficients of $p(x, Y)$ are polynomials in the zeros of $p(x, Y)$, the fractional-power series coefficients of this product agree in a neighborhood of $(0) \in \mathbb{C}_X$ with the polynomial coefficients of the Y^i in $p(X, Y)$. Thus (5) represents, near $(0) \in \mathbb{C}_X$, a factorization of $p(X, Y)$ into fractional-power series. If the fractional-power series in any factor $Y - Y_{jik}$ is a power series in $X^{1/m}$ then this factor yields, upon setting $T = X^{1/m}$, the parametrization

$$\begin{aligned} x &= X = T^m \\ y &= g(T), \quad g \text{ analytic at } (0) \in \mathbb{C}_T. \end{aligned} \tag{6}$$

For at least one of the n such parametrizations, $T = 0$ corresponds to (0, 0). Each such parametrization defines an extension to K of the natural evaluation of $\mathbb{C}[x, y]$ at P. □

Theorem 2.32 can be generalized to a purely ring-theoretic setting—namely, if R is any subring of any field K, and $h: R \to k$ is a ring homomorphism of R into any algebraically closed field k (thus h assigns values in k to elements in R), then h may be extended to an evaluation of the elements of K satisfying (2.5.1) and (2.5.2). Such an evaluation is also called a *place*;

since this more general theorem extends the evaluation h to a place, the theorem is often referred to as the *place extension theorem*. For proofs, see, for example, [Lang, Chapter I, Theorem 1] or [Zariski and Samuel, vol II, Chapter VI, Theorem 5′].

The power series representation in (6) is not so special as it appears; by a change of uniformizing parameter, every power series representation for (x, y) (cf. Theorem 2.26) can be put into this form. This is important, for it allows us to connect up *arbitrary* power series representations with the factorization of the irreducible polynomial defining C.

Theorem 2.33. *Let $C \subset \mathbb{C}_{XY}$ be an irreducible curve with coordinate ring $\mathbb{C}[x, y]$ and function field $\mathbb{C}(x, y)$. Let P be a given point of C, let R be a valuation ring with center P on C, and let $B_P^{\sim}$ be the associated branch at P of C. Then a representative of $B_P^{\sim}$ (and therefore, in an obvious sense, $B_P^{\sim}$ itself) can be represented by power series of the form*

$$\begin{aligned} x &= T^m \qquad (m \text{ a positive integer}) \\ y &= f(T) \qquad (f \text{ analytic at } (0) \in \mathbb{C}_T) \end{aligned} \tag{7}$$

Proof. Assume without loss of generality that $P = (0, 0) \in \mathbb{C}_{XY}$ and that $x \notin \mathbb{C}$. Then the order in R of x is $m \geqslant 1$. From Theorem 2.26 we have, for some U of order 1 in R,

$$\begin{aligned} x &= U^m(c_0 + c_1 U + \ldots) \qquad (c_0 \neq 0), \\ y &= d_n U^n + \ldots \qquad (n > 0). \end{aligned}$$

For some power series $a_0 + a_1 U + \ldots$ analytic at $U = 0$, we have

$$c_0 + c_1 U + \ldots = (a_0 + a_1 U + \ldots)^m.$$

(Equating coefficients of like powers of U one gets $a_0{}^m = c_0, ma_0{}^{m-1}a_1 = c_1, ma_0{}^{m-1}a_2 + (m(m-1)/2)a_0{}^{m-2}a_1{}^2 = c_2, \ldots$. It is easily checked that for each $i > 0$, a_i can be expressed rationally in terms of c_i and the preceding $a_0, \ldots, a_{i-1}$. The series $a_0 + a_1 U + \ldots$ converges in a neighborhood of $(0) \in \mathbb{C}_U$, for otherwise its mth power $c_0 + c_1 U + \ldots$ would not.) If we write $T = a_0 U + a_1 U^2 + \ldots$, we then have $x = U^m(c_0 + c_1 U + \ldots) = (a_0 U + a_1 U^2 + \ldots)^m = T^m$. Because $a_0 \neq 0$, T is order 1 in U, so by Theorem 3.6 of Chapter II, we can write $U = g(T)$ where g is analytic at $(0) \in \mathbb{C}_T$; hence y is analytic in T. Thus x and y can be written in the simple form

$$\begin{aligned} x &= T^m \\ y &= f(T) \qquad (f \text{ analytic at } (0) \in \mathbb{C}_T). \end{aligned}$$ □

We now summarize our results connecting evaluations at P of elements of $\mathbb{C}(x, y)$, with modes of approach to P.

Let $C \subset \mathbb{C}_{XY}$ be an irreducible curve defined by an irreducible polynomial $p(X, Y)$; let (x, y) be a generic point of C, and let P be any point of C. Then:

> **(2.34.1)** There is always at least one evaluation of $\mathbb{C}(x, y)$ extending the natural evaluation of $\mathbb{C}[x, y]$ at P, the evaluation being defined by power series of the simple form given in (7).
> **(2.34.2)** Every evaluation of $\mathbb{C}(x, y)$ extending the natural one on $\mathbb{C}[x, y]$ at P gives rise to a discrete rank one valuation ring, hence to a power series representation of (x, y) (Theorem 2.26) and therefore also to a power series representation of $B_P^{\sim}$; this representation can be furthermore assumed to be of the simple form given in (7).
> **(2.34.3)** By uniqueness of the fractional-power series factorization in (5), we see that the extensions to $\mathbb{C}(x, y)$ of the natural evaluation on $\mathbb{C}[x, y]$ at P are given by precisely the fractional-power series factors $Y - Y_{jik}$ of $p(X, Y)$ (as in (5)).

So far in this section, all our considerations have been in the affine setting. Since there is often much important geometry at infinity (including evaluating at infinity the elements of a variety's function field), it is important to extend the definitions of *function field*, *center of valuation ring*, and the like, to the projective case.

To begin, we define the **function field** $K_{\mathbb{P}^n(\mathbb{C})}$ *of* $\mathbb{P}^n(\mathbb{C})$ to be the zero element together with the 0-forms of $\mathbb{C}(X_1, \ldots, X_{n+1})$—that is, elements $p(X_1, \ldots, X_{n+1})/q(X_1, \ldots, X_{n+1})$, where p and q are forms of equal degree. The value of each such quotient is constant along "subspaces-minus-the-origin" of $\mathbb{C}_{X_1, \ldots, X_{n+1}}$, and therefore yields a well-defined function on the points of $\mathbb{P}^n(\mathbb{C})$. The set consisting of the zero element together with all such 0-forms constitutes a subfield of $\mathbb{C}(X_1, \ldots, X_{n+1})$ isomorphic to $\mathbb{C}(X_1, \ldots, X_n)$. (Note that

$$\frac{p(X_1, \ldots, X_{n+1})}{q(X_1, \ldots, X_{n+1})} = \frac{p(X_1/X_{n+1}, \ldots, 1)}{q(X_1/X_{n+1}, \ldots, 1)} = \frac{p(Y_1, \ldots, Y_n, 1)}{q(Y_1, \ldots, Y_n, 1)}$$

for indeterminates $Y_i = X_i/X_{n+1}$.) Since this subfield of $\mathbb{C}(X_1, \ldots, X_{n+1})$ is unchanged by nonsingular linear transformations of $X_1, \ldots, X_{n+1}$, dehomogenizing $\mathbb{P}^n(\mathbb{C})$ at any hyperplane still yields a field isomorphic to $\mathbb{C}(X_1, \ldots, X_n)$.

One can next define the *function field* K_V *of an irreducible variety* $V \subset \mathbb{P}^n(\mathbb{C})$ to be the field of restrictions to V of functions on $K_{\mathbb{P}^n(\mathbb{C})}$. If $\mathbb{C}[x_1, \ldots, x_{n+1}] = \mathbb{C}[X_1, \ldots, X_{n+1}]/\mathsf{J}(V)$ where $\mathsf{J}(V) = V$'s homogeneous ideal, then one can easily check that K_V is isomorphic to the field consisting of 0 together with quotients $p(x_1, \ldots, x_{n+1})/q(x_1, \ldots, x_{n+1})$, where $p(X_1, \ldots, X_{n+1})$ and $q(X_1, \ldots, X_{n+1})$ are forms of equal degree. And, as with $K_{\mathbb{P}^n(\mathbb{C})}$, we see that K_V is equal to the quotient field of any affine representative of V.

We can generalize to the projective setting the notion in Definition 2.29 of center of a valuation ring R of an affine curve $C \subset \mathbb{C}_{X_1,\ldots,X_n}$. Definition 2.29 is somewhat restrictive in that it applies only when R's maximal ideal $\mathfrak{m}$ intersects C's coordinate ring in a maximal ideal. Although it is obvious that $\mathfrak{m}$ intersects C's coordinate ring in a prime ideal, it is not always true that it intersects it in a maximal ideal.

EXAMPLE 2.35. Consider $\mathbb{C}$ with coordinate ring $\mathbb{C}[X]$; 0 and the set of all $p(X)/q(X)$ such that $\deg p \leqslant \deg(q)$, form a valuation subring $R \subset \mathbb{C}(X)$. The maximal ideal $\mathfrak{m}$ of R consists of 0 and those $p(X)/q(X) \in R$ for which $\deg p < \deg q$. Clearly $\mathfrak{m} \cap \mathbb{C}[X] = (0)$, which is not maximal in $\mathbb{C}[X]$. Thus Definition 2.29 fails to assign to R a center on C. Its "center" turns out to be at infinity; we make this precise by next extending the notion of center of a valuation ring to include the projective case.

For our purposes, let $C \subset \mathbb{P}^n(\mathbb{C})$ be any irreducible curve, let its homogeneous variety have coordinate ring $\mathbb{C}[x_1, \ldots, x_{n+1}]$, and assume R is a valuation ring of K_C. Then ord is well defined by R. Assume without loss of generality that x_{n+1} is such that for each $i = 1, \ldots, n+1$, $\operatorname{ord}(x_i/x_{n+1}) \geqslant 0$—that is, each $x_i/x_{n+1} \in R$. Denote the image of x_i/x_{n+1} under $R \to R/\mathfrak{m}$ by a_i. Then $(x_1/x_{n+1}, \ldots, 1)$ specializes to a point $(a) = (a_1, \ldots, a_n, 1) \in \mathbb{C}^{n+1}\backslash\{(0)\}$; (a) defines a point $P_R \in V$, for if p is any homogeneous polynomial of $\mathsf{J}(V)$, then $p(x_1, \ldots, x_{n+1}) = 0$, hence $p(x_1/x_{n+1}, \ldots, 1) = 0$, therefore $p(a_1, \ldots, 1) = 0$. Any other x_j such that $\operatorname{ord}(x_i/x_j) \geqslant 0$ for $i = 1, \ldots, n+1$ will determine the same point P_R in V—that is, it will determine (a) up to a nonzero multiple, because

$$\left(\frac{x_1}{x_j}, \ldots, \frac{x_{n+1}}{x_j}\right) = \left(\frac{x_1}{x_{n+1}} \cdot \frac{x_{n+1}}{x_j}, \ldots, 1 \cdot \frac{x_{n+1}}{x_j}\right);$$

since $\operatorname{ord}(x_{n+1}/x_j) = 0$, its image in $R/\mathfrak{m}$ is a nonzero constant. It is easily seen that the point $P_R \in V$ thus depends only on R; we call it the **center of** R **on** V. If $\operatorname{ord}(x_i/x_j) \geqslant 0$ $(i = 1, \ldots, n+1)$, then P_R may obviously be regarded as the center of R on $\mathsf{D}(V)$, where $\mathsf{D}(V)$ is the dehomogenization of V relative to X_j. The maximal ideal of R intersects each such coordinate ring $\mathbb{C}[x_1/x_j, \ldots, x_{n+1}/x_j]$ in a maximal ideal. For any X_k such that $\operatorname{ord}(x_k/x_{n+1}) > 0$, the center of R on V lies on the hyperplane at infinity, $\mathsf{V}(X_k)$.

Now let us return to Example 2.35. The function ord is "$-\deg$" ($\deg p/q = \deg p - \deg q \in \mathbb{Z}$), so R is a discrete rank one valuation ring of $\mathbb{C}(X)$. Let the homogeneous variety corresponding to $\mathbb{P}^1(\mathbb{C})$ be $\mathbb{C}_{X_1X_2}$, and let $\mathbb{C}_{X_1}$ represent the point at infinity of $\mathbb{P}^1(\mathbb{C})$. Since $\mathbb{C}_{X_1X_2}$'s coordinate ring is $\mathbb{C}[X_1, X_2]$, the affine part of $\mathbb{P}^1(\mathbb{C})$ has coordinate ring $\mathbb{C}[X_1/X_2, 1]$, or $\mathbb{C}[X]$ (denoting X_1/X_2 by X). With the order of Example 2.35 on $\mathbb{C}(X) = \mathbb{C}(X_1/X_2)$, we see that $X_1/X_2 = X \notin R$. This means that $\operatorname{ord}(X_2/X_1) > 0$;

the center of R on $\mathbb{P}^1(\mathbb{C})$ lies on the point at infinity corresponding to $\mathsf{V}(X_2) = \mathbb{C}_{X_1}$.

So far, we have dealt with extending from coordinate rings to function fields certain notions related to *evaluation*. We can do the same for *isomorphism*. We know that two coordinate rings R_{V_1} and R_{V_2} of irreducible affine varieties V_1 and V_2 are isomorphic iff V_1 and V_2 are polynomially isomorphic (Theorem 8.7 of Chapter III). If *function fields* K_{V_1} and K_{V_2} of irreducible varieties V_1 and V_2 are isomorphic one can ask, analogously, whether V_1 and V_2 are then "isomorphic." Surely they cannot in general be polynomially isomorphic—for instance, $\mathbb{C}_X$ and the curve $V(Y^2 - X^3) \subset \mathbb{C}_{XY}$ are not polynomially isomorphic since $\mathbb{C}_X$ is nonsingular and the cusp curve is not. However, the corresponding function fields $\mathbb{C}(X)$ and $\mathbb{C}(X, X^{1/2}) = \mathbb{C}(X^{1/2})$ *are* isomorphic. But one can ask for a more relaxed notion of isomorphism of varieties in which irreducible affine or projective varieties are isomorphic iff their function fields are isomorphic over $\mathbb{C}$. This broader equivalence is called **birational equivalence**, since if $\mathbb{C}(x_1, \ldots, x_n)$ is $\mathbb{C}$-isomorphic to $\mathbb{C}(y_1, \ldots, y_m)$, then each y_i corresponds to a rational function of the quantities $\{x_1, \ldots, x_n\}$, and each x_j corresponds to a rational function of the quantities $(y_1, \ldots, y_m\}$. As in the case of isomorphic coordinate rings, this leaves the obvious question: What is the translation of birational equivalence into geometric terms? Our connection between valuation rings and germs will help to answer this.

First of all, an isomorphism between K_{V_1} and K_{V_2} induces in a natural way a 1:1-onto correspondence between the set of all valuation rings of K_{V_1} and all those of K_{V_2}. Now suppose V_1 and V_2 are irreducible curves C_1 and C_2 respectively. An isomorphism $K_{C_1} \simeq K_{C_2}$ then induces a 1 : 1-onto correspondence between the branches of C_1 and those of C_2. Since several branches may be centered at the same point P, this correspondence does not imply a 1 : 1 correspondence between the points of C_1 and the points of C_2. However, if P is a given point in C_1, there are only finitely many valuation rings having P as center on C_1 (Exercise 2.4). Each of these valuation rings has a well-defined center on C_2. Hence the 1 : 1 correspondence between valuation rings of K_{C_1} and K_{C_2} induces a correspondence between finite sets of points in C_1 and finite sets of points in C_2 (that is, between 0-dimensional subvarieties of C_1 and of C_2). An important special case is when the isomorphism is the identity map (that is, when $K_{C_1} = K_{C_2} = K$). In this case, let affine coordinate rings of C_1 and C_2 be R_1 and R_2, respectively, let R be a given valuation ring in K, and let $\mathfrak{m}$ be R's maximal ideal. Then ($R_1 \cap \mathfrak{m}$ and $R_2 \cap \mathfrak{m}$ maximal in R) $\Rightarrow$ ($R_1 \cap \mathfrak{m}$ and $R_2 \cap \mathfrak{m}$ are in corresponding 0-dimensional varieties).

In evaluating elements of function fields, we have thus far worked mainly with curves. We now briefly look at evaluations of function field elements on varieties having arbitrary dimension. Recall that endowing a field with a discrete rank one valuation yields a discrete rank one valuation ring R of K; furthermore, if K has transcendence degree 1 over $\mathbb{C}$, we saw that $R/\mathfrak{m} \simeq \mathbb{C}$, so evaluation may be thought of as taking place in $\mathbb{C}$. However, in the case of varieties of arbitrary dimension, we cannot conclude that a discrete rank one valuation ring yields values in $\mathbb{C}$. In fact, here is an example to the contrary:

EXAMPLE 2.36. Let $\operatorname{ord}(p(X, Y))$ denote the total order at $(0, 0)$ of a polynomial $p \in \mathbb{C}[X, Y]$; as in (1), if $f(X, Y) = p(X, Y)/q(X, Y)$ is any element of $\mathbb{C}(X, Y)$, we define $\operatorname{ord}(f)$ to be $\operatorname{ord}(p) - \operatorname{ord}(q)$, this obviously being well defined. This definition satisfies (2.15.1) and (2.15.2), thus defining on $\mathbb{C}(X, Y)$ a discrete rank one valuation. Now not only does every nonzero element $c \in \mathbb{C}$ have order 0, but so also does X/Y, for example. The elements X/Y and c cannot possibly represent the same coset in $R/\mathfrak{m}$, for their difference $(X/Y) - c$ would then be in $\mathfrak{m}$; this is not so since $\operatorname{ord}((X/Y) - c) = \operatorname{ord}((X - cY)/Y) = 1 - 1 = 0$. One easily checks that $R/\mathfrak{m}$ is $\mathbb{C}(X/Y)$. Hence the discrete rank one valuation ring takes elements in $\mathbb{C}(X, Y)$ and assigns values in a field of transcendence degree one over $\mathbb{C}$.

Geometrically, discrete rank one valuation rings can be regarded as giving an evaluation *at* or *along* an entire irreducible subvariety of codimension 1 in an irreducible variety (thus generalizing evaluation at points in the case of an irreducible curve). In the above example, $\mathbb{C}(X/Y)$ is the function field of that codimension-1 subvariety.

This suggests repeating the process. For instance, a discrete rank one valuation subring of $\mathbb{C}(X/Y)$ will yield values in $\mathbb{C}$. We illustrate the situation:

EXAMPLE 2.37. In $\mathbb{C}(X, Y)$, let $R = \{p(X, Y)/q(X, Y) \mid q(X, 0) \neq 0\}$ (p, q relatively prime in $\mathbb{C}[X, Y]$). The ring R consists of the set of all elements in $\mathbb{C}(X, Y)$ having nonnegative order in Y at $Y = 0$. Thus, looking at elements of $\mathbb{C}(X, Y)$ as elements of $K(Y)$ ($K = \mathbb{C}(X)$), we see that every element $f(X, Y) \in \mathbb{C}(X, Y)$ has at $Y = 0$ a well-defined value $f(X, 0)$ in $\mathbb{C}(X) \cup \{\infty\}$, and we may regard $f(X, 0)$ as $\lim_{Y \to 0} f(X, Y)$. We may suggestively look at this as letting the line $Y = a$ approach the line $Y = 0$ in $\mathbb{C}_{XY}$; the line $Y = a$ gives the value $f(X, a)$ to $f(X, Y)$, and $f(X, 0)$ is the limit of these values. To arrive at a value in $\mathbb{C}$ at $(0, 0)$, the natural thing to do now is to write $f(X, 0)$ in reduced form (as a quotient of relatively prime polynomials), then let X approach 0. This will assign to $f(X, Y)$ at $(0, 0)$ a well-defined value in $\mathbb{C} \cup \{\infty\}$.

In the above example, we approached *first* in the Y-direction, *then* in X-direction to get a value in $\mathbb{C} \cup \{\infty\}$. It is reasonable to next ask whether we get the same value by approaching first in the X-direction, then in the Y-direction. In general, we do not.

EXAMPLE 2.38. For $Y/X \in \mathbb{C}(X, Y)$, we have $\lim_{X \to 0}(Y/X) = \infty$ and $\lim_{Y \to 0}(\infty) = \infty$. But $\lim_{Y \to 0}(Y/X) = 0$ and $\lim_{X \to 0}(0) = 0$.

We want a kind of "order" that reflects the asymmetry of the above situation. If for instance we approach in the X-direction first, we are looking at $f(X, Y) \in \mathbb{C}(X, Y)$ as an element of $K(X)$, where $K = \mathbb{C}(Y)$. Hence any nonzero $f(X, Y)$ has an order in X, and any nonzero element of $\mathbb{C}(Y)$ has order 0 in X. Thus in this case the order of X is greater than that of Y, hence Y/X has negative order, and its assigned value at $(0, 0)$ is indeed ∞. More generally, we see that if the order at $(0, 0)$ in X of $p(X, Y)$ is greater than that of $q(X, Y)$, then p/q is assigned the value 0 when $(0, 0)$ is approached first along the X-direction. But what about two elements having the *same* order in X at $(0, 0)$ say $X + Y$ and $X + Y^2$? If the order in X were all that mattered, then $(X + Y^2)/(X + Y)$ would have order 0 at $(0, 0)$, and we would expect the value assigned at $(0, 0)$ to be a nonzero element of $\mathbb{C}$. But in fact $\lim_{X\to 0}(X + Y^2)/(X + Y) = Y$ and $\lim_{Y\to 0} Y = 0$; the same method of evaluation gives the value ∞ to $(X + Y)/(X + Y^2)$. Hence, relative to this method of evaluation, we should consider that the order of $X + Y^2$ is *greater* than that of $X + Y$. Relative to this particular method of evaluation, we may more generally infer that if $X\text{-ord}(p) > X\text{-ord}(q)$ (where "X-ord" denotes order in X at $(0, 0)$), then the order of p is strictly greater than the order of q. But if $X\text{-ord}(p) = X\text{-ord}(q)$, then the order of p is equal to or greater than the order of q iff $Y\text{-ord}(p) \geqslant Y\text{-ord}(q)$. Hence when comparing two elements of $\mathbb{C}(X, Y)$, it is only when their X-orders agree that the Y-order becomes important. One thus gets an order with values not in $\mathbb{Z}$, but in countably many copies of $\mathbb{Z}$, these copies being strung out, one after the other, to form a big totally ordered set. It is natural to assign coordinates (X, Y) to points in this big set, as follows:

(i) The X-coordinate answers, "What copy does it belong to?"
(ii) The Y-coordinate answers "Where is it in that copy?"

The points in our set may just as well be represented as the product set $\mathbb{Z}_X \times \mathbb{Z}_Y$. It forms a group with componentwise subtraction, and is ordered as follows:

For any $(n_1, n_2), (m_1, m_2) \in \mathbb{Z}_X \times \mathbb{Z}_Y$ we have $(n_1, n_2) \geqslant (m_1, m_2)$ provided either

$$n_1 > m_1, \text{ or}$$
$$n_1 = m_1 \text{ and } n_2 \geqslant m_2.$$

This type of total order is called *lexicographic order*. (A lexicographer who alphabetically arranges words in a dictionary is essentially assigning coordinates to each word; the first coordinate of a word—that is, its first letter—is the most important, and so on.)

If we approach along the Y-direction first, then the lexicographic order would be $\mathbb{Z}_Y \times \mathbb{Z}_X$, where the Y-component takes precedence over the X-component.

To get evaluation in $\mathbb{C} \cup \{\infty\}$ at points in $\mathbb{C}_{X_1, \ldots, X_n}$, the corresponding order function would take values in a product of n copies of $\mathbb{Z}$, say $\mathbb{Z}_{i_1} \times \ldots \times \mathbb{Z}_{i_n}$ (supplied with the lexicographic order), where we approach first in the X_{i_1}-direction, then in the X_{i_2}-direction, etc.

EXERCISES

2.1 Prove that for ord in Definition 2.15, $\operatorname{ord}(a) < \operatorname{ord}(b)$ implies $\operatorname{ord}(a + b) = \operatorname{ord}(a)$.

2.2 Prove Lemma 2.18 as follows:

(a) First consider $\mathbb{C}(X)$, where X is an element of K transcendental over $\mathbb{C}$. Show that $R \cap \mathbb{C}(X)$ is a valuation ring in $\mathbb{C}(X)$.

(b) Show that $((\mathbb{C}(X)\backslash\{0\})/\mathscr{U}, \cdot)$ is group-isomorphic to $(\mathbb{Z}, +)$, where $\mathscr{U}$ is the multiplicative subgroup $(R\backslash\mathfrak{m}) \cap \mathbb{C}(X)$ of $\mathbb{C}(X)\backslash\{0\}$, and hence conclude that $R \cap \mathbb{C}(X)$ is a discrete rank one valuation ring of $\mathbb{C}(X)$.

(c) Show that $\mathbb{Z}$ in Part (b) is in a natural way a subgroup of $G = ((K\backslash\{0\})/(R - \mathfrak{m}), \cdot)$, and that there are only a finite number N of cosets of $\mathbb{Z}$ in G. [*Hint*: Let $y_1, \ldots, y_m$ be elements in K representing different cosets of $\mathbb{Z}$ in G. Since K is finite algebraic over $\mathbb{C}(X)$, it suffices to show that the y_i are linearly independent over $\mathbb{C}(X)$, for then there could be only finitely many elements y_i representing these different cosets. In $\sum_i c_i y_i = 0$ $(c_i \in K)$, if some $c_i \neq 0$, then at least two of $y_1, \ldots, y_m$ have the same order. (Use the above Exercise 2.1.) Conclude from this that these two elements represent the same coset, which gives a contradiction.]

(d) Show that $(G, \cdot)$ is isomorphic to $(\mathbb{Z}, +)$. [*Hint*: Consider the map $g \to g^N$, $g \in G$.]

2.3 Is the condition in Theorem 2.31 that $\mathsf{V}(X)$ not be tangent to C at P necessary? Why?

2.4 If P is a point of an irreducible curve C, show that there are only finitely many valuation rings of K_C having P as center on C, and thus show that there are only finitely many branches of C through P.

2.5 Find a valuation subring R of $\mathbb{C}(X, Y)$ which is not Noetherian. Exhibit explicitly an infinite strictly increasing sequence of ideals in R.

3 Local rings

In the last section, the search for a way to evaluate rational functions at a point P on an affine curve C led in a natural way to valuation rings. A valuation ring corresponds to an *analytic mode of approach to P*. The elements of C's rational function field may then be regarded in a natural way as function germs, these function germs taking on values in $\mathbb{C} \cup \{\infty\}$; the ones assuming only finite values constitute the valuation ring.

We saw that such a mode of approach on a curve C did not in general correspond to all the points of C near P, but only to an "analytic arc" in C through P. But often it is important to consider *all* points of C (or more generally, of any variety V) about a point—for instance, in asking whether the variety is singular or nonsingular there, or what its order is there, or for the multiplicity of intersection with another variety at that point, and so on.

The analytic arcs of the last section may not tell the whole story. For example, any representative of each of the two branches through (0, 0) of the curve $\mathsf{V}(Y^2 - X^2(X + 1)) \subset \mathbb{C}_{XY}$ turns out to be nonsingular at (0, 0) in the sense that it is *smooth* there (Definition 7.3 of Chapter II, yet their union, giving all points of the curve near (0, 0), is singular there.

We thus ask: Given a germ $V_P{}^{\sim}$ of V, is there in V's function field a subring whose elements naturally form "the coordinate ring of function germs on $V_P{}^{\sim}$"? There indeed is; such a ring will express properties of a variety in a neighborhood of a point while, in effect, throwing away the excess baggage corresponding to the local behavior at other points of the variety which do not concern us at the moment. Such *local rings* allow us to get useful ring-theoretic characterizations of local data such as nonsingularity, order, and the like.

In the case we are considering in this section, that is, germs of the set of all points in V near a point P, the germ notion can be put into a slightly simpler form. We first note this fact:

Theorem 3.1. *Let V_1 and V_2 be varieties in $\mathbb{P}^n(\mathbb{C})$ or in $\mathbb{C}^n$, each of whose irreducible components contains a given point P; if there is an open neighborhood U of $\mathbb{P}^n(\mathbb{C})$ or $\mathbb{C}^n$ about P such that $V_1 \cap U = V_2 \cap U$, then $V_1 = V_2$.*

Proof. If V_1 and V_2 are both irreducible, then the theorem follows at once from Theorem 2.11 of Chapter IV. In the general case, we note that each irreducible component of V_1 contains a point $P \in U$ which is in no other irreducible component of V_1, and in precisely one of V_2's irreducible components; these two components agree near P, so again they are identical. Hence every irreducible component of V_1 coincides with one of V_2's, and conversely (by symmetry). □

Theorem 3.1 implies that arbitrary varieties V_1 and V_2 have the same germ at P iff the set of irreducible components of V_1 through P is the same as the set of irreducible components of V_2 through P. There is thus a smallest variety in $\mathbb{P}^n(\mathbb{C})$ (or in $\mathbb{C}^n$) having a given germ at P; the set of unions of irreducible varieties through P may thus be identified in a natural way with the set of germs at P of algebraic varieties. Lattice and decomposition structures are immediately seen to be the same. Given a variety V in $\mathbb{P}^n(\mathbb{C})$ (or in $\mathbb{C}^n$) and a point $P \in V$, we may therefore think of $V_P{}^{\sim}$ as an ordered pair $(V_{(P)}, P)$, where $V_{(P)}$ is the subvariety of V consisting of the union of those irreducible components of V which contain P. We shall denote this ordered pair by V_P, and we may say that $V_P = V'_{P'}$ iff $P = P'$ and $V_{(P)} = V'_{(P')}$, and that $V_P \subset V'_{P'}$ iff $P = P'$ and $V_{(P)} \subset V'_{(P')}$; we define $V_P \cap V'_P$ to be $(V \cap V')_P$ and $V_P \cup V'_P$ to be $(V \cup V')_P$. (We formalize these notions in Definition 3.3.)

Our ring associated with V_P (that is, with $V_P{}^{\sim}$) will be analogous to the coordinate ring of V. Notice that for any irreducible $V \subset \mathbb{C}_{X_1, \ldots, X_n}$, V's coordinate ring $R = \mathbb{C}[x_1, \ldots, x_n]$ consists precisely of those elements of

$\mathbb{C}(x_1, \ldots, x_n)$ which are defined and finite at all points of V—that is, it consists of all those rational functions which can be written in the form p/q $(p, q \in R)$ such that q is never zero on V. (PROOF : For any element of $\mathbb{C}(x_1, \ldots, x_n)$ there is either a way of writing it as a quotient p/q $(p, q \in R)$ where $1/q \in R$, or for every representation as a quotient p/q $(p, q \in R)$, we have $1/q \notin R$. In the first case $p/q \in R$ and p/q is everywhere well defined and finite on V. In the second case, q is not a unit in R, so for the ideal (q), we have $(q) \subsetneqq R$; hence it has a zero in V, by the *Nullstellensatz* (Theorem 5.1 of Chapter III). Thus q itself has a zero in V, so p/q is not both well defined and finite at all points of V.)

Now let P be a point of $V \subset \mathbb{C}^n$. Let $R = \mathbb{C}[x_1, \ldots, x_n]$ be the coordinate ring of V and $\mathfrak{m}$, the maximal ideal corresponding to P. The set of all those elements of $\mathbb{C}(x_1, \ldots, x_n)$ which are defined and finite at P is

$$\left\{\frac{p}{q}\,\middle|\, p \in R, q \in R\backslash\mathfrak{m}\right\}.$$

We call this the *localization of* R *at* P, or *at* $\mathfrak{m}$, and denote it by $R_{\mathfrak{m}}$.

Just as valuation rings contain exactly one maximal ideal (corresponding to the center of the valuation ring), so also $R_{\mathfrak{m}}$ has just one maximal ideal $\mathfrak{M}$, and it corresponds to P. This ideal is the set of all nonunits of $R_{\mathfrak{m}}$ (that is, all elements of $R_{\mathfrak{m}}$ which do not have multiplicative inverses in $R_{\mathfrak{m}}$), namely

$$\mathfrak{M} = \left\{\frac{p}{q}\,\middle|\, p \in \mathfrak{m}, q \in R\backslash\mathfrak{m}\right\}.$$

It is clear that since any other element of $R_{\mathfrak{m}}$ is of the form r/s where r and $s \in R\backslash\mathfrak{m}$, any ideal containing such an r/s must contain $(r/s)(s/r) = 1$, that is, it must be $R_{\mathfrak{m}}$ itself. Hence since $\mathfrak{M}$ is obviously an ideal, it is maximal. Any other maximal ideal $\mathfrak{N}$ would, of course, have to contain an element not in $\mathfrak{M}$, hence $\mathfrak{N}$ would have to be $R_{\mathfrak{m}}$ itself, which is not maximal.

A particularly important generalization of this idea is when $\mathfrak{m}$ is replaced by any prime ideal $\mathfrak{p}$ of R. The following definition is basic.

Definition 3.2 Let R be any domain. For any prime ideal $\mathfrak{p}$ of R, let $R_{\mathfrak{p}}$ be

$$R_{\mathfrak{p}} = \left\{\frac{p}{q}\,\middle|\, p \in R, q \in R\backslash\mathfrak{p}\right\}.$$

$R_{\mathfrak{p}}$ is called **localization of R at** $\mathfrak{p}$.

As before, R has a unique maximal ideal.

In just the same way that we considered an irreducible subvariety as generalizing the notion of point (hence we speak of order of a variety "at" or "along" an irreducible subvariety), so here, too, we will see that localizing a coordinate ring R to $R_{\mathfrak{p}}$ will geometrically correspond to restricting our attention to those irreducible components of V which contain $\mathsf{V}(\mathfrak{p})$. The definitions given earlier centering around V_P can be correspondingly generalized. We now make this formal.

Definition 3.3. Let V and V' be two affine or two projective varieties; let subvarieties $W \subset V$ and $W' \subset V'$ be irreducible. Then V_W denotes the ordered pair $(V_{(W)}, W)$, where $V_{(W)}$ is the union of those irreducible components of V containing W. ($V_{(W)} = \emptyset$ if there are no irreducible components of V containing W.) We define $=$, $\subset$, $\cap$, and $\cup$ as follows: $V_W = V'_{W'}$ iff $W = W'$ and $V_{(W)} = V'_{(W')}$; $V_W \subset V'_{W'}$ iff $W = W'$ and $V_{(W)} \subset V'_{(W')}$; $V_W \cap V'_W = (V \cap V')_W$; and $V_W \cup V'_W = (V \cup V')_W$. (It is clear that $V_W \cap V'_W$ and $V_W \cup V'_W$ are well defined.) Also, V_W is **irreducible** iff $V_W = V'_W \cup V''_W$ implies $V_W = V'_W$ or $V_W = V''_W$.

An example of the way in which $R_{\mathfrak{p}}$ expresses a property of V along $\mathsf{V}(\mathfrak{p})$ will be given in the next section, where we give a local ring characterization of nonsingularity along an irreducible subvariety. Rings having a unique maximal ideal are useful in studying local properties of geometric objects at many different levels (topological, differentiable, analytic, algebraic), and have been given special names. For instance, any ring (commutative, with identity) having a unique maximal ideal is called a *quasi-local ring*. Many (though not all) quasi-local rings occuring in algebraic geometry are Noetherian (Cf. Exercise 2.5). We make the following

Definition 3.4. A Noetherian ring (commutative with identity) having a unique maximal ideal is called a **local ring**.

We shall show presently (Lemma 3.9) that for any Noetherian ring R and prime ideal $\mathfrak{p}$ of R, $R_{\mathfrak{p}}$ is a local ring. (Also Cf. Exercise 3.1.) For purposes of exposition

we shall for the remainder of this section assume that R is a coordinate ring (hence Noetherian), and that $\mathfrak{p}$ is a fixed prime ideal of R. (Hence R is embedded in $R_{\mathfrak{p}}$.)

We include for future use the following important definitions:

Definition 3.5. If R is the coordinate ring of an irreducible variety $V \subset \mathbb{C}^n$, and if $\mathfrak{p} = \mathsf{J}(W)$ is the prime ideal of an irreducible subvariety W of V, then the local ring $R_{\mathfrak{p}}$ is called the **localization of V at W** (or **along W**), or the **local ring of V at W**; in this case $R_{\mathfrak{p}}$ is also denoted by $\mathfrak{o}(W; V)$.

Definition 3.6. Let $V \subset \mathbb{P}^n(\mathbb{C})$ be irreducible, and let K_V be V's function field—that is, the set of quotients of equal-degree forms in $x_1, \ldots, x_{n+1}$, where $\mathbb{C}[x_1, \ldots, x_{n+1}] = \mathbb{C}[X_1, \ldots, X_{n+1}]/\mathsf{J}(V)$. If W is an irreducible subvariety of V, then the set of all elements of K_V which can be written as p/q, where p and q are forms in $x_1, \ldots, x_{n+1}$ of the same degree, and where q is not identically zero on W, forms a subring of K_V; it is called the **local ring of V at W**, and is denoted by $\mathfrak{o}(W; V)$.

Remark 3.7. If $W \subset V$ are irreducible varieties in $\mathbb{P}^n(\mathbb{C})$, and if R is the coordinate ring of any dehomogenization $\mathsf{D}(V)$ of V (where W is not contained in the hyperplane at infinity), then $\mathfrak{o}(W; V)$ is the localization $R_\mathfrak{p} = \mathfrak{o}(\mathsf{D}(W); \mathsf{D}(V))$ of R at $\mathsf{D}(W) = \mathsf{V}(\mathfrak{p})$; this follows from the fact that if we without loss of generality dehomogenize at X_{n+1}, then

$$\frac{p(x_1, \ldots, x_{n+1})}{q(x_1, \ldots, x_{n+1})} = \frac{p(x_1/x_{n+1}, \ldots, 1)}{q(x_1/x_{n+1}, \ldots, 1)}.$$

The left-hand side is an element of $\mathfrak{o}(W; V)$, while the right-hand side belongs to $R_\mathfrak{p}$.

Many of the basic algebraic and geometric relations between R and $R_\mathfrak{p}$ may be compactly expressed using a double sequence, as in Diagrams 2 and 3 of Chapter III. We explore this next. Again, for expository purposes we select a fixed variety $V \subset \mathbb{C}_{X_1, \ldots, X_n}$ having $R = \mathbb{C}[x_1, \ldots, x_n]$ as coordinate ring, and we let $W = \mathsf{V}(\mathfrak{p})$ be an arbitrary, fixed irreducible subvariety of V.

Our sequence is given in Diagram 1.

$$\begin{array}{ccccc}
\mathscr{I}(R) & \underset{\mathsf{i}}{\overset{\sqrt{}}{\rightleftarrows}} & \mathscr{J}(R) & \underset{\mathsf{J}}{\overset{\mathsf{V}}{\rightleftarrows}} & \mathscr{V}(R) \\
(\ )^e \downarrow \uparrow (\ )^c & & (\ )^e \downarrow \uparrow (\ )^c & & (\ )_W \downarrow \uparrow i \\
\mathscr{I}(R_\mathfrak{p}) & \underset{\mathsf{i}^*}{\overset{\sqrt{}}{\rightleftarrows}} & \mathscr{J}(R_\mathfrak{p}) & \underset{\mathsf{J}^*}{\overset{\mathsf{G}^*}{\rightleftarrows}} & \mathscr{G}(R_\mathfrak{p})
\end{array}$$

Diagram 1.

In this diagram, $\mathscr{I}(R_\mathfrak{p})$ denotes the lattice $(\mathscr{I}(R_\mathfrak{p}), \subset, \cap, +)$ of ideals of $R_\mathfrak{p}$ and $\mathscr{J}(R_\mathfrak{p})$ denotes the lattice $(\mathscr{J}(R_\mathfrak{p}), \subset, \cap, +)$ of closed ideals of $R_\mathfrak{p}$. Closure in $\mathscr{I}(R_\mathfrak{p})$ is with respect to the radical of Definition 1.1 of Chapter III; by Lemma 5.7 of Chapter III the radical of an ideal $\mathfrak{a}$ in $R_\mathfrak{p}$ will be seen to be the intersection of all prime ideals of $R_\mathfrak{p}$ which contain $\mathfrak{a}$, since $R_\mathfrak{p}$ is Noetherian (Lemma 3.9). This radical is not in general the intersection of the $\mathfrak{a}$-containing maximal ideals of $R_\mathfrak{p}$, since $R_\mathfrak{p}$ has but one maximal ideal. Continuing the explanation of symbols in Diagram 1, $\mathscr{G}(R_\mathfrak{p})$ denotes the lattice $(\mathscr{G}(R_\mathfrak{p}), \subset, \cap, \cup)$ of all V_W where $V \in \mathscr{I}$ and W is fixed, with $\subset$, $\cap$, and $\cup$ as in Definition 3.3. The letter $\mathscr{G}$ reminds us that these ordered pairs V_W are identified with germs (We remark that there exists an analogous sequence at the analytic level, where one uses germs instead of representatives, since there is not in general a canonical representative of each "analytic germ," as is the case with algebraic varieties, where there is a unique smallest algebraic variety representing a given "algebraic germ." One can even push certain aspects to the differential level.) It is easily seen that $\mathscr{G}(R_\mathfrak{p})$ actually is a lattice, using Definition 3.3 together with the fact that $\varnothing$ and the subvarieties of V containing $\mathsf{V}(\mathfrak{p})$ form a lattice.

As for the various maps, $(\ )^c$ and $(\ )^e$ are just contraction and extension of ideals. Since $R \to R_\mathfrak{p}$ is an embedding, $(\ )^c$ reduces to intersection with R. In contrast to extension in Section III,10, we shall see that $(\ )^e$ maps closed ideals in $\mathscr{I}(R)$ to closed ideals in $\mathscr{I}(R_\mathfrak{p})$. The map $(\ )_W$ sends V into V_W, and $\imath$ assigns to each V_W the variety $\imath(V_W) = V_{(W)}$. (Thus $\imath$ simply removes from V_W reference to the "center" W.) Finally, the bottom horizontal maps i* and $\sqrt{\ }$ are the embedding and radical maps; G^* and J^* will be defined in terms of the other maps, and will turn out to be mutually inverse lattice-reversing isomorphisms.

In establishing properties of these maps, extension and contraction between $\mathscr{I}(R)$ and $\mathscr{I}(R_\mathfrak{p})$ play a basic part; we look at them first.

$(\ )^e: \mathscr{I}(R) \to \mathscr{I}(R_\mathfrak{p})$

This map is onto $\mathscr{I}(R_\mathfrak{p})$; in particular, each ideal $\mathfrak{a}^* \subset R_\mathfrak{p}$ comes from the ideal $\mathfrak{a}^{*c} \subset R$—that is,

For each $\mathfrak{a}^* \in R_\mathfrak{p}$,

$$\mathfrak{a}^* = \mathfrak{a}^{*ce} \tag{8}$$

Proof. That $\mathfrak{a}^{*ce} \subset \mathfrak{a}^*$ is obvious, since $a^* \in \mathfrak{a}^{*ce}$ implies that $a^* = a/m$ for some $a \in \mathfrak{a}^{*c}$ and some $m \in R\backslash\mathfrak{p}$. To show $\mathfrak{a}^* \subset \mathfrak{a}^{*ce}$, let $a^* \in \mathfrak{a}^*$. Then $a^* \in R_\mathfrak{p}$, which implies $a^* = a/m$ for some $a \in R$ and $m \in R\backslash\mathfrak{p}$; also $a = ma^*$, so $a \in \mathfrak{a}^*$, which means $a \in \mathfrak{a}^* \cap R = \mathfrak{a}^{*c}$. Hence $\mathfrak{a}^* = a/m \in \mathfrak{a}^{*ce}$, □

Next note that $(\ )^e$ is not necessarily 1 : 1, since

$$\mathfrak{a}^e = R_\mathfrak{p} \quad \text{for every ideal } \mathfrak{a} \not\subset \mathfrak{p}. \tag{9}$$

($\mathfrak{a} \subset \mathfrak{p}$ implies that there is an $m \in \mathfrak{a} \cap (R\backslash\mathfrak{p})$, hence $m/m = 1 \in \mathfrak{a}^e$.)

However,

(3.8) $(\ )^e$ is 1 : 1 on the set of contracted ideals of $\mathscr{I}(R)$.

For if $\mathfrak{a} = \mathfrak{a}^{*c}$ and $\mathfrak{b} = \mathfrak{b}^{*c}$, and if $\mathfrak{a}^e = \mathfrak{a}^{*ce} = \mathfrak{b}^e = \mathfrak{b}^{*ce}$, then $\mathfrak{a}^* = \mathfrak{b}^*$, so $\mathfrak{a} = \mathfrak{a}^{*c} = \mathfrak{b} = \mathfrak{b}^{*c}$.

$(\ )^c: \mathscr{I}(R_\mathfrak{p}) \to \mathscr{I}(R)$

This map is *not* necessarily onto, because $\mathfrak{a}^{*c}$ is either R or is contained in $\mathfrak{p}$. (If $\mathfrak{a}^{*c}$ is not contained in $\mathfrak{p}$, then $\mathfrak{a}^* = \mathfrak{a}^{ce} = R_\mathfrak{p}$, whence $\mathfrak{a}^{*c} = R$.)

Next note that $(\ )^c$ is 1 : 1, for if $\mathfrak{a}^{*c} = \mathfrak{b}^{*c}$, then $\mathfrak{a}^{*ce} = \mathfrak{b}^{*ce} = \mathfrak{a}^* = \mathfrak{b}^*$.

In general $\mathfrak{a} \neq \mathfrak{a}^{ec}$, but we always have

$$\mathfrak{a} \subset \mathfrak{a}^{ec}. \tag{10}$$

(Theorem 3.14 will supply geometric meaning to (10), and also to Theorem 3.10 below.)

The following characterization of $\mathfrak{a}^{ec}$ is useful:

$$\mathfrak{a}^{ec} = \{a \in R \mid am \in \mathfrak{a}, \text{ for some } m \in R \backslash \mathfrak{p}\}. \tag{11}$$

PROOF

$\subset$: Each element of $\mathfrak{a}^e$ is a sum of quotients of elements in $\mathfrak{a}$ by elements in $R \backslash \mathfrak{p}$; obviously such a sum is itself such a quotient. Hence an element a is in $\mathfrak{a}^{ec}$ iff it is in R and is of the form $a = a'/m$ where $a' \in \mathfrak{a}$. Hence $am = a' \in \mathfrak{a}$, proving the inclusion.

$\supset$: Any a on the right-hand side of (11) can be written as $a = am/m = a'/m$ where $a' \in \mathfrak{a}$, hence $a \in \mathfrak{a}^e$; but also $a \in R$, so $a \in \mathfrak{a}^{ec}$. □

An immediate corollary of the injectivity of $(\ )^c$ is this basic fact, referred to earlier:

Lemma 3.9. *The ring $R_{\mathfrak{p}}$ is Noetherian.*

PROOF. $(\ )^c$ is 1 : 1 onto the set of contracted ideals of R; since $(\ )^c$ preserves inclusion, any infinite strictly ascending sequence of ideals in $R_{\mathfrak{p}}$ would map, under $(\ )^c$, to an infinite strictly ascending sequence in R, which is not possible. □

In establishing lattice properties of $(\ )^e$ and $(\ )^c$ we shall use the next result, which gives a case where $\mathfrak{a} = \mathfrak{a}^{ec}$, instead of only $\mathfrak{a} \subset \mathfrak{a}^{ec}$.

Theorem 3.10. *Let R and $\mathfrak{p}$ be as above. Then $\mathfrak{q} = \mathfrak{q}^{ec}$ for any irreducible ideal $\mathfrak{q} \subset \mathfrak{p}$.*

PROOF. Let x be any element of R, and let (x) be the principal ideal of R generated by x; define the quotient ideal $\mathfrak{q}:(x)$ to be $\mathfrak{q}:(x) = \{r \in R \mid xr \in \mathfrak{q}\}$. (This is a special case of the quotient ideal in Exercise 4.5 of Chapter III.) It then follows at once from (11) together with the definition of $\mathfrak{q}:(x)$ that we may express the conclusion $\mathfrak{q} = \mathfrak{q}^{ec}$ in the form "For each $x \in R \backslash \mathfrak{p}, \mathfrak{q} = \mathfrak{q}:(x)$." We therefore prove that $\mathfrak{q} = \mathfrak{q}:(x)$ for each $x \in R$.

$\mathfrak{q} \subset \mathfrak{q}:(x)$: This is obvious from the definition of quotient.

$\mathfrak{q}:(x) \subset \mathfrak{q}$: First, we see that from the definition of $\mathfrak{q}:(x)$, we have

$$(\mathfrak{q}:(x))(x) \subset \mathfrak{q}. \tag{12}$$

Suppose $\mathfrak{q}:(x) \not\subset \mathfrak{q}$; let $y \in (\mathfrak{q}:(x)) \backslash \mathfrak{q}$, and let $z \in (x) \backslash \sqrt{\mathfrak{q}}$. ($\sqrt{\mathfrak{q}}$ is prime from Exercise 5.3, of Chapter III, and $\sqrt{\mathfrak{q}} \subset \mathfrak{p}$.) Then from (12) we have $yz \in \mathfrak{q}$. Now $\mathfrak{q}$ is primary (Exercise 4.5 of Chapter III), so since $y \notin \mathfrak{q}$, we have $z^m \in \mathfrak{q}$ for some m. But $z \notin \sqrt{\mathfrak{q}}$; because $\sqrt{\mathfrak{q}}$ is prime, this means $z^m \notin \mathfrak{q}$, a contradiction. □

We now look at how much lattice structure is preserved by $(\ )^e$ and $(\ )^c$. We first consider $(\ )^e : \mathscr{I}(R) \to \mathscr{I}(R_{\mathfrak{p}})$.

It is immediate from definitions that extension from any ring into any other ring preserves sums.

Though extension need not preserve intersections for arbitrary rings, in our case it does:

$$(\mathfrak{a} \cap \mathfrak{b})^e = \mathfrak{a}^e \cap \mathfrak{b}^e. \tag{13}$$

PROOF. It suffices to prove that for any finite intersection of irreducibles $\mathfrak{q}_i$ in R,

$$(\mathfrak{q}_1 \cap \ldots \cap \mathfrak{q}_r)^e = \mathfrak{q}_1{}^e \cap \ldots \cap \mathfrak{q}_r{}^e. \tag{14}$$

For if this holds, then if

$$\mathfrak{a} = \bigcap_{i=1}^{r} \mathfrak{q}_i, \qquad \mathfrak{b} = \bigcap_{j=1}^{s} \mathfrak{q}'_j \quad \text{and} \quad \mathfrak{a} \cap \mathfrak{b} = \left(\bigcap_{i=1}^{r} \mathfrak{q}_i\right) \cap \left(\bigcap_{j=1}^{s} \mathfrak{q}'_j\right)$$

are decompositions into irreducibles, we have

$$(\mathfrak{a} \cap \mathfrak{b})^e = \left(\left(\bigcap_{i=1}^{r} \mathfrak{q}_i\right) \cap \left(\bigcap_{j=1}^{s} \mathfrak{q}'_j\right)\right)^e = \left(\bigcap_{i=1}^{r} \mathfrak{q}_i{}^e\right) \cap \left(\bigcap_{j=1}^{s} \mathfrak{q}_j{}'^e\right) = \mathfrak{a}^e \cap \mathfrak{b}^e.$$

We now prove (14). The inclusion "$\subset$" follows at once from the definition of extension. For "$\supset$," let a^* be an arbitrary element of $\mathfrak{q}_1{}^e \cap \ldots \cap q_r{}^e$; assume that $q_i \subset \mathfrak{p}$ for $i = 1, \ldots, k$, and that $\mathfrak{q}_i \not\subset \mathfrak{p}$ for $i = k + 1, \ldots, r$. Since by (9), $\mathfrak{q}_i{}^e = R_\mathfrak{p}$ for $i = k + 1, \ldots, r$, we see that

$$\mathfrak{q}_1{}^e \cap \ldots \cap \mathfrak{q}_r{}^e = \mathfrak{q}_1{}^e \cap \ldots \cap \mathfrak{q}_k{}^e.$$

("$\supset$" in (14) holds trivially if every $\mathfrak{q}_i \not\subset \mathfrak{p}$.) Since $\mathfrak{q}_1{}^e \cap \ldots \cap \mathfrak{q}_k{}^e \subset R_\mathfrak{p}$, $a^* \in \mathfrak{q}_1{}^e \cap \ldots \cap \mathfrak{q}_k{}^e$ is of the form $a^* = a/m$, for some $a \in R$ and $m \in R\backslash\mathfrak{p}$. Using the fact that each ideal in $\mathscr{I}(R_\mathfrak{p})$ is the extension of its contraction (from (8)), a may be further assumed to be in $(\mathfrak{q}_1{}^e \cap \ldots \cap \mathfrak{q}_k{}^e)^c$; this last ideal is $\mathfrak{q}_1{}^{ec} \cap \ldots \cap \mathfrak{q}_k{}^{ec}$ since $(\ \ )^c$, being intersection with R, preserves intersections. Now apply Theorem 3.10: For each $i = 1, \ldots, k$, we have $\mathfrak{q}_i \subset \mathfrak{p}$, so $\mathfrak{q}_i{}^{ec}$ $(i = 1, \ldots, k)$. It follows that $a^* = a/m$ for some $a \in \mathfrak{q}_1 \cap \ldots \cap \mathfrak{q}_k$ and $m \in R\backslash\mathfrak{p}$. Now $(R\backslash\mathfrak{p}) \cap \mathfrak{q}_{k+1} \cap \ldots \cap \mathfrak{q}_r \neq \varnothing$; for any m' in this intersection, $am' \in \mathfrak{q}_1 \cap \ldots \cap \mathfrak{q}_r$. Thus $a^* = am'/mm'$, which means $a^* \in (\mathfrak{q}_1 \cap \ldots \cap \mathfrak{q}_r)^e$; thus "$\supset$" is proved, and therefore also (13). □

We next consider $(\ \ )^c : \mathscr{I}(R_\mathfrak{p}) \to \mathscr{I}(R)$.

This map obviously preserves intersections, being just intersection with R.

It does not in general preserve sums, though from the definition of $(\ \ )^c$ we see at once that

$$\mathfrak{a}^{*c} + \mathfrak{b}^{*c} \subset (\mathfrak{a}^* + \mathfrak{b}^*)^c.$$

Thus at the geometric level (notation as in Definition 3.3), if $V_P = (V_{(P)}, P)$ and $V'_P = (V'_{(P)}, P)$ then, although by definition $V_P \cap V'_P = (V \cap V')_P$, it may happen that $V_{(P)} \cap V'_{(P)} \supsetneqq (V \cap V')_{(P)}$. This can occur since $V_{(P)} \cap V'_{(P)}$ may have components not containing P, whereas every component of $(V \cap V')_{(P)}$ contains P. Guided by geometry, one can now easily construct many examples in which $(\ )^c$ does not preserve sums. We give one here.

EXAMPLE 3.11. In $\mathbb{C}_{XY}$ if $P = (0, 0)$, if $V = \mathsf{V}(Y - X)$, and if $V' = \mathsf{V}(Y - X^2)$, then $V_P = (\mathsf{V}(Y - X), (0, 0))$ and $V'_P = (\mathsf{V}(Y - X^2), (0, 0))$. Thus $(V \cap V')_P = (\{(0, 0)\}, (0, 0)) \subsetneqq (\mathsf{V}(Y - X) \cap \mathsf{V}(Y - X^2), (0, 0)) = (\{0, 0\} \cup \{1, 1\}, (0, 0))$. We can then translate this geometric fact into ideal language (Cf. Theorem 3.14): Let $\mathfrak{a}^*$ and $\mathfrak{b}^*$ be the principal ideals $(Y - X)$ and $(Y - X^2)$ in the localization $\mathbb{C}[X, Y]_{(X, Y)}$ of $\mathbb{C}[X, Y]$ at $(0, 0)$. Then

$$\mathfrak{a}^{*c} = (Y - X) \subset \mathbb{C}[X, Y] \quad \text{and} \quad \mathfrak{b}^{*c} = (Y - X^2) \subset \mathbb{C}[X, Y];$$

then

$$(\mathfrak{a}^* + \mathfrak{b}^*)^c = (X, Y) \subset \mathbb{C}[X, Y],$$

since $(\mathfrak{a}^* + \mathfrak{b}^*)^c$ contains

$$X = \frac{(Y - X^2) - (Y - X)}{1 - X} \quad \text{and} \quad Y = \frac{(Y - X^2) - X(Y - X)}{1 - X}.$$

Thus $\mathfrak{a}^{*c} + \mathfrak{b}^{*c} = (Y - X) + (Y - X^2)$ is strictly smaller than $(\mathfrak{a}^* + \mathfrak{b}^*)^c$, since

$$\mathsf{V}(X, Y) \subsetneqq \mathsf{V}((Y - X) + (Y - X^2)).$$

We conclude this discussion of our two maps between $\mathscr{I}(R)$ and $\mathscr{I}(R_{\mathfrak{p}})$ by observing that although $(\ )^e$ preserves sums and intersections, and although it defines a 1 : 1-onto map from the set of contracted ideals of $\mathscr{I}(R)$ to $\mathscr{I}(R_{\mathfrak{p}})$ (from (3.8)) it is not in general a lattice isomorphism, since the set of contracted ideals in $\mathscr{I}_R$ is not itself always a lattice—it is not in general closed under addition. For instance, the ideal $(Y - X) + (Y - X^2)$, in the example above, is not contracted, for its variety contains a component other than $\{(0, 0)\}$.

We now turn to extension and contraction at the $\mathscr{J}$-level. Our first task is to check that $(\ )^e$ and $(\ )^c$ actually do map into $\mathscr{J}(R_{\mathfrak{p}})$ and into $\mathscr{J}(R)$, respectively. For $(\ )^e$, this of course says that the extension of a closed ideal is still closed. To see this, note that for $\mathfrak{a} \subset R$, $\mathfrak{a}^e = \{a/m \mid a \in \mathfrak{a} \text{ and } m \in R\backslash\mathfrak{p}\}$. If $\mathfrak{q}$ is any prime ideal of R, then either $\mathfrak{q}$ intersects $R\backslash\mathfrak{p}$ (in which case $\mathfrak{q}^e = R_{\mathfrak{p}}$), or $\mathfrak{q} \cap (R\backslash\mathfrak{p}) = \varnothing$. In this last case one easily shows that any quotient r/m, where $r \in R$ and $m \in R\backslash\mathfrak{p}$, is in $\mathfrak{q}^e$ iff $r \in \mathfrak{q}$. (Note that $r \notin \mathfrak{q}$ and $r/m = q/m'$ imply that $rm' = qm$; then $rm' \notin \mathfrak{q}$, but $qm \in \mathfrak{q}$, a contradiction.) This implies at once that $\mathfrak{q}^e$ is prime in $R_{\mathfrak{p}}$. Since extension preserves intersection, any finite intersection of primes in R extends to either $R_{\mathfrak{p}}$ or an intersection of primes. In either case the extended ideal is closed.

It is immediate that $(\ )^c$ maps from $\mathscr{J}(R_{\mathfrak{p}})$ to $\mathscr{J}(R)$, since contraction is just intersection with R, and therefore preserves intersection and primality; hence the contraction of any intersection of prime ideals is an intersection of prime ideals.

We next consider 1 : 1 and onto properties.

$(\ )^e$ maps $\mathscr{J}(R)$ onto $\mathscr{J}(R_\mathfrak{p})$ since (8) tells us that any $\mathfrak{a}^* \in \mathscr{J}(R_\mathfrak{p})$ is an extension of a closed ideal—that is, $\mathfrak{a}^* = (\mathfrak{a}^{*c})^e$, and $\mathfrak{a}^{*c} \in \mathscr{J}(R)$.

$(\ )^e$ is not necessarily 1 : 1 from $\mathscr{J}(R)$, just as it wasn't from $\mathscr{I}(R)$, and for the same reasons. (It is still 1 : 1 on the *contracted* ideals.)

$(\ )^c$ is not necessarily onto $\mathscr{J}(R)$ for the same reasons as in the $\mathscr{I}$-case.

$(\ )^c$ is 1 : 1 on $\mathscr{J}(R_\mathfrak{p})$ since it is 1 : 1 on $\mathscr{I}(R_\mathfrak{p})$.

Now we come to the lattice properties.

(3.12) $(\ )^e : \mathscr{J}(R) \to \mathscr{J}(R_\mathfrak{p})$ *preserves both sums and intersections.*

PROOF. $(\ )^e$ preserves intersections since it does so on $\mathscr{I}(R)$ (from (13)).

$(\ )^e$ preserves sums—that is, $(\mathfrak{a} + \mathfrak{b})^e = \mathfrak{a}^e + \mathfrak{b}^e$, or

$$(\sqrt{\mathfrak{a} + \mathfrak{b}})^e = \sqrt{\mathfrak{a}^e + \mathfrak{b}^e}. \tag{15}$$

We show this as follows:

$\subset$: Since $\mathfrak{a}^e + \mathfrak{b}^e = (\mathfrak{a} + \mathfrak{b})^e$, the inclusion $(\sqrt{\mathfrak{a} + \mathfrak{b}})^e \subset \sqrt{\mathfrak{a}^e + \mathfrak{b}^e}$ becomes $(\sqrt{\mathfrak{a} + \mathfrak{b}})^e \subset \sqrt{(\mathfrak{a} + \mathfrak{b})^e}$. But this last inclusion is easily established, since for any ideal $\mathfrak{c}$ in a ring we have $(\sqrt{\mathfrak{c}})^e \subset \sqrt{\mathfrak{c}^e}$. ($c \in \sqrt{\mathfrak{c}}$ implies that $c^n \in \mathfrak{c}$ for some n. Then $m \notin \mathfrak{p}$ implies that $m^n \notin \mathfrak{p}$, so $c^n/m^n \in \mathfrak{c}^e$, i.e., $c/m \in \sqrt{\mathfrak{c}^e}$.)

$\supset$: Certainly $\sqrt{\mathfrak{a} + \mathfrak{b}} \supset \mathfrak{a} + \mathfrak{b}$, hence $(\sqrt{\mathfrak{a} + \mathfrak{b}})^e \supset (\mathfrak{a} + \mathfrak{b})^e = \mathfrak{a}^e + \mathfrak{b}^e$, therefore

$$\sqrt{(\sqrt{\mathfrak{a} + \mathfrak{b}})^e} \supset \sqrt{\mathfrak{a}^e + \mathfrak{b}^e}.$$

Since $(\ )^e$ maps closed ideals into closed ideals,

$$\sqrt{(\sqrt{\mathfrak{a} + \mathfrak{b}})^e} = (\sqrt{\mathfrak{a} + \mathfrak{b}})^e,$$

so "$\supset$" is established, and therefore also (3.12). □

(3.13) $(\ )^c : \mathscr{J}(R_\mathfrak{p}) \to \mathscr{J}(R)$ preserves intersections but not necessarily sums.

This is obvious from our comments in the $\mathscr{I}$-case.

Finally, as in the $\mathscr{I}$-case, $(\ )^e$ defines an onto homomorphism from $\mathscr{J}(R)$ to $\mathscr{J}(R_\mathfrak{p})$ which is 1 : 1 on the set of contracted ideals of $\mathscr{J}(R)$; but it is not generally a lattice isomorphism since the contracted ideals of $\mathscr{J}(R)$ need not form a lattice.

We now consider the remaining two vertical maps, $(\ )_W$ and $\imath$. The following properties are all immediate:

$(\ )_W : \mathscr{V}(R) \to \mathscr{G}(R_\mathfrak{p})$ is onto, but not in general 1 : 1. Its restriction to the image $\imath(\mathscr{G}(R_\mathfrak{p})) \subset \mathscr{V}(R)$ is the inverse of $\imath$. $(\ )_W$ preserves $\cap$ and $\cup$.

$\imath : \mathscr{G}(R_\mathfrak{p}) \to \mathscr{V}(R)$ is 1 : 1, but not in general onto. It preserves $\cup$, but need not preserve $\cap$, as Example 3.11 shows.

As for the horizontal maps, it is easily seen that the embedding $\mathrm{i}^*: \mathscr{J}(R_\mathfrak{p}) \to \mathscr{I}(R_\mathfrak{p})$ is 1 : 1, that $\sqrt{\ \ } : \mathscr{I}(R_\mathfrak{p}) \to \mathscr{J}(R_\mathfrak{p})$ is onto, and that both these maps preserve intersections. Sums are preserved by $\sqrt{\ \ }$, but not in general by i^* (see Exercises 3.3 and 3.4.).

The two maps left to consider are G^* and J^*. They turn out to be mutually inverse lattice-reversing isomorphisms, and are the local analogues of V and J. In dealing with these maps we shall use this basic fact:

Theorem 3.14. *For each ideal $\mathfrak{a} \in \mathscr{J}(R)$, $\mathfrak{a}^{ec}$ is the intersection of those prime ideals which contain $\mathfrak{a}$ and which are contained in $\mathfrak{p}$. (The intersection of an empty set of prime ideals is defined to be R.) Thus if $\mathfrak{a} \subset R = R_V$ defines the subvariety X of V (and $\mathfrak{p}$ defines W), then $\mathfrak{a}^{ec}$ defines $X_{(W)}$—that is, the mapping $\mathfrak{a} \to \mathfrak{a}^{ec}$ geometrically corresponds to taking the germ at W of X. In particular, the image of $\mathscr{J}(R_\mathfrak{p})$ under contraction consists precisely of R together with those ideals in $\mathscr{J}(R)$ which are intersections of prime ideals contained in $\mathfrak{p}$.*

Proof. Write $\mathfrak{a} = \mathfrak{p}_1 \cap \ldots \cap \mathfrak{p}_r$ where each $\mathfrak{p}_i$ is prime in R. It follows at once from Theorem 3.10 that $\mathfrak{p}_i^{ec} = \mathfrak{p}_i$ iff $\mathfrak{p}_i \subset \mathfrak{p}$. If $\mathfrak{p}_i \not\subset \mathfrak{p}$, then clearly $\mathfrak{p}_i^e = R_\mathfrak{p}$, so $\mathfrak{p}_i^{ec} = R$. Since $(\ \ )^e$ and $(\ \ )^c$ preserve intersections, we have

$$\mathfrak{a}^{ec} = \bigcap_{i=1}^{r} \mathfrak{p}_i^{ec} = \bigcap_{\mathfrak{p}_i \subset \mathfrak{p}} \mathfrak{p}_i. \qquad \square$$

We now define G^* and J^*; we shall do this using already-established "paths" in Diagram 1.

Definition 3.15. $\mathsf{G}^*: \mathscr{J}(R_\mathfrak{p}) \to \mathscr{G}(R_\mathfrak{p})$ is the composition of the maps

$$\mathscr{J}(R_\mathfrak{p}) \xrightarrow{(\ \ )^c} \mathscr{J}(R) \xrightarrow{\mathsf{V}} \mathscr{V}(R) \xrightarrow{(\ \ )_W} \mathscr{G}(R_\mathfrak{p});$$

$\mathsf{J}^*: \mathscr{G}(R_\mathfrak{p}) \to \mathscr{J}(R_\mathfrak{p})$ is the composition of the maps

$$\mathscr{G}(R_\mathfrak{p}) \xrightarrow{\ i\ } \mathscr{V}(R) \xrightarrow{\mathsf{J}} \mathscr{J}(R) \xrightarrow{(\ \ )^e} \mathscr{J}(R_\mathfrak{p}).$$

That is, $\mathsf{G}^*(\mathfrak{a}^*) = (\mathsf{V}(\mathfrak{a}^{*c}))_W$ and $\mathsf{J}^*(X_W) = (\mathsf{J}(i(X_W)))^e$ for $\mathfrak{a}^* \in \mathscr{J}(R_\mathfrak{p})$ and $X_W \in \mathscr{G}(R_\mathfrak{p})$.

The maps G^* and J^* are both 1 : 1-onto and mutual inverses; this is easily verified by using the characterization in Theorem 3.14 and facts already established concerning 1 : 1 and onto properties of the maps used to define G^* and J^*.

We will have established the lattice-reversing isomorphism between $\mathscr{G}(R_\mathfrak{p})$ and $\mathscr{J}(R_\mathfrak{p})$ once we prove

Theorem 3.16. *Let* $\mathfrak{a}^*$, $\mathfrak{b}^*$ *be any two ideals in* $\mathcal{J}(R_\mathfrak{p})$, *and* X_W, Y_W *any two elements of* $\mathcal{G}(R_\mathfrak{p})$. *Then*

(3.16.1) $\mathsf{G}^*(\mathfrak{a}^* \cap \mathfrak{b}^*) = \mathsf{G}^*(\mathfrak{a}^*) \cup \mathsf{G}^*(\mathfrak{b}^*)$
(3.16.2) $\mathsf{G}^*(\mathfrak{a}^* + \mathfrak{b}^*) = \mathsf{G}^*(\mathfrak{a}^*) \cap \mathsf{G}^*(\mathfrak{b}^*)$
(3.16.3) $\mathsf{J}^*(X_W \cup Y_W) = \mathsf{J}^*(X_W) \cap \mathsf{J}^*(Y_W)$
(3.16.4) $\mathsf{J}^*(X_W \cap U_W) = \mathsf{J}^*(X_W) + \mathsf{J}^*(Y_W)$

PROOF (3.16.1): This is easy, since $(\ \ )^c$ preserves $\cap$, V is lattice-reversing, and $(\ \ )_W$ preserves $\cup$.

(3.16.3): This holds, since $\imath$ preserves $\cup$, J is lattice-reversing, and $(\ \ )^e$ preserves $\cap$.

(3.16.4) For this, we want to show that $\imath(X_W \cap Y_W)$ and $\imath(X_W) \cap \imath(Y_W)$, which will in general be different, nonetheless have the same image in $\mathcal{J}(R_\mathfrak{p})$ under J followed by $(\ \ )^e$. In fact, from the definition of $X_W \cap Y_W$ we see that

$$\imath(X_W) \cap \imath(Y_W) = \imath(X_W \cap Y_W) \cup Z,$$

where $Z \subset V$ is some variety which does not contain W. Thus

$$\mathsf{J}(\imath(X_W) \cap \imath(Y_W)) = \mathsf{J}(\imath(X_W \cap Y_W)) \cap \mathfrak{c} \tag{16}$$

where $\mathfrak{c} \in \mathcal{J}(R)$ is an ideal not contained in $\mathfrak{p}$. We know $(\ \ )^e$ preserves intersections on these ideals (from 3.12), and that $\mathfrak{c}^e = R_\mathfrak{p}$; hence applying $(\ \ )^e$ to both sides of (16) gives

$$(\mathsf{J}(\imath(X_W) \cap \imath(Y_W)))^e = \mathsf{J}^*(X_W \cap Y_W).$$

But $\mathsf{J}(\imath(X_W) \cap \imath(Y_W)) = \mathsf{J}(\imath(X_W)) + \mathsf{J}(\imath(Y_W))$; since $(\ \ )^e$ preserves $+$ (from (3.12)), we then have (3.16.4).

(3.16.2): We want to show that $(\mathfrak{a}^* + \mathfrak{b}^*)^c$ and $\mathfrak{a}^{*c} + \mathfrak{b}^{*c}$ (which may be different) have the same image in $\mathcal{G}(R_\mathfrak{p})$ under V followed by $(\ \ )_W$. It clearly is enough to show that in the irredundant decompositions into prime ideals of $(\mathfrak{a}^* + \mathfrak{b}^*)^c$ and of $\mathfrak{a}^{*c} + \mathfrak{b}^{*c}$, those prime ideals contained in $\mathfrak{p}$ are the same for both these ideals. In view of Theorem 3.14 it suffices to show that

$$((\mathfrak{a}^* + \mathfrak{b}^*)^c)^{ec} = (\mathfrak{a}^{*c} + \mathfrak{b}^{*c})^{ec}.$$

We do this by showing

$$(\mathfrak{a}^* + \mathfrak{b}^*)^{ce} = (\mathfrak{a}^{*c} + \mathfrak{b}^{*c})^e. \tag{17}$$

From (8), we see that the left-hand side of (17) is

$$(\mathfrak{a}^* + \mathfrak{b}^*)^{ce} = \mathfrak{a}^* + \mathfrak{b}^* = \sqrt{\mathfrak{a}^{*ce} + \mathfrak{b}^{*ce}} = \sqrt{(\mathfrak{a}^{*c} + \mathfrak{b}^{*c})^e}.$$

This will equal the right-hand side of (17) if we show

$$\sqrt{\mathfrak{c}^e} = (\sqrt{\mathfrak{c}})^e \quad \text{for any } \mathfrak{c} \in \mathcal{J}(R). \tag{18}$$

For this, let $\mathfrak{c} = \mathfrak{q}_1 \cap \ldots \cap \mathfrak{q}_s$ be an irredundant decomposition into irreducibles. Then

$$\sqrt{\mathfrak{c}^e} = \sqrt{(\mathfrak{q}_1 \cap \ldots \cap \mathfrak{q}_s)^e} = \sqrt{\mathfrak{q}_1{}^e \cap \ldots \cap \mathfrak{q}_s{}^e} = \sqrt{\mathfrak{q}_1{}^e} \cap \ldots \cap \sqrt{\mathfrak{q}_s{}^e}.$$

If we can show $\sqrt{\mathfrak{q}_i^e} = (\sqrt{\mathfrak{q}_i})^e$, (18) will follow at once. If $\mathfrak{q}_i$ is not contracted, then Theorem 3.10 implies that it intersects $R\backslash\mathfrak{p}$, because if $\sqrt{\mathfrak{q}_i}$ intersects $R\backslash\mathfrak{p}$, say $x \in \sqrt{\mathfrak{q}_i} \cap (R\backslash\mathfrak{p})$, then so does every x^n, hence also $\mathfrak{q}_i$ intersects $R\backslash\mathfrak{p}$. Hence $\sqrt{\mathfrak{q}_i^e} = (\sqrt{\mathfrak{q}_i})^e = R_\mathfrak{p}$. If $\mathfrak{q}_i$ is contracted, write $\mathfrak{q}_i = (\mathfrak{q}_i^*)^c$. Then

$$\sqrt{\mathfrak{q}_i^e} = \sqrt{\mathfrak{q}_i^{*ce}} = \sqrt{\mathfrak{q}_i^*} = (\sqrt{\mathfrak{q}_i^*})^{ce} = (\sqrt{\mathfrak{q}_i^{*c}})^e = (\sqrt{\mathfrak{q}_i})^e.$$

This establishes (18), and therefore (3.16.2). We have now completed the proof of Theorem 3.16. □

Exercises

3.1 Let R be a Noetherian domain, and let M be a nonempty subset of $R\backslash\{0\}$ which is closed under multiplication. (M is then called a **multiplicative system in** R.) Is the ring $R_M = \{r/m \mid r \in R, m \in M\}$ Noetherian?

3.2 Use Theorem 3.14 to give a geometric interpretation to (10) and to Theorem 3.10.

3.3 Show that for any ideals $\mathfrak{a}$ and $\mathfrak{b}$ in any ring R, we have $\sqrt{\mathfrak{a} + \mathfrak{b}} = \sqrt{\sqrt{\mathfrak{a}} + \sqrt{\mathfrak{b}}}$. [*Hint*: Observe that $\mathfrak{a} + \mathfrak{b} \subset \sqrt{\mathfrak{a}} + \sqrt{\mathfrak{b}}$, that $\sqrt{\mathfrak{a}} + \sqrt{\mathfrak{b}} \subset \sqrt{\mathfrak{a} + \mathfrak{b}}$, and that $\sqrt{}$ is a closure map.]

3.4 Show that the embedding $\mathrm{i}^* : \mathscr{I}(R_\mathfrak{p}) \to \mathscr{I}(R_\mathfrak{p})$ does not in general preserve sums.

3.5 Let $W \subset V$ be two irreducible varieties in $\mathbb{P}^n(\mathbb{C})$ or $\mathbb{C}^n$. Let $l(\mathfrak{o}(W; V))$ be the length of the longest chain of prime ideals in $\mathfrak{o}(W; V)$. Show that the local ring of V at W "regards W as a point" (i.e., regards W as having dimension zero) in the sense that $l(\mathfrak{o}(W; V)) = \dim V - \dim W$.

3.6 Show that the local ring at any point of $\mathbb{C}^1$ is a valuation subring of $\mathbb{C}(X)$. Show that this is not true for arbitrary plane curves; give a geometric justification of this fact.

3.7 Let $K_2 = \mathbb{C}(x_1, \ldots, x_n)$ be an algebraic extension of $K_1 = \mathbb{C}(x_1, \ldots, x_m)$; let $R = \mathbb{C}[x_1, \ldots, x_m]_\mathfrak{p}$ be a local ring in K_1, and let $R^*(\subset K_2)$ be a finitely-generated integral extension of R. Show that although R^* may not be a local ring, it has only finitely many maximal ideals, and these all lie over the maximal ideal of R. Interpret this result geometrically. What can happen if R^* is only a finitely-generated *algebraic* extension of R?

3.8 Let $Y^2 - X$ define an integral extension K of $\mathbb{C}(X)$ (where X is a single indeterminate); find a ring R^* in K containing a local subring R of $\mathbb{C}(X)$, such that R^* has exactly two maximal ideals lying over a maximal ideal of $\mathbb{C}[X]$. Find generators for these two maximal ideals.

3.9 Is a transcendental extension of a local ring $R(\subset \mathbb{C}(x_1, \ldots, x_n))$ still local? If K is a subfield of $\mathbb{C}(x_1, \ldots, x_n)$ containing $\mathbb{C}$, is $R \cap K$ local? Interpret your answers geometrically.

3.10 Although a local ring R has only one maximal ideal, it may have infinitely many prime ideals. Find a geometric interpretation of this fact.

4 A ring-theoretic characterization of nonsingularity

In this section we give a ring-theoretic characterization of nonsingularity. Singularity or nonsingularity of a point P in a variety V is a *local property*, that is, whether V is singular or nonsingular at a point $P \in V$ can be determined by looking at the part of V within an arbitrarily small $\mathbb{P}^n(\mathbb{C})$- or $\mathbb{C}^n$-open neighborhood about P; correspondingly, our rings will be local, too. Such a purely ring-theoretic characterization is useful for a number of reasons. For instance, it allows us to generalize the ideas of singularity and nonsingularity in important ways. We shall see an application of this in the next section, where we give a generalization of the fundamental theorem of arithmetic. Also, for any nonsingular curve C, such a characterization gives us a way of connecting arbitrary nonzero ideals of C's coordinate ring with "point chains" on C. (Cf. Exercise 2.4 of Chapter III.) In another direction, we have seen how an irreducible subvariety can serve as a kind of "generalized point," and we speak of an irreducible subvariety as being singular or nonsingular; again, local rings come into play. Finally, such a characterization gives a very easy way of showing that polynomial isomorphism of varieties preserves nonsingularity (Exercise 4.2).

Our first goal is to get a satisfactory definition of nonsingular irreducible subvariety W of an irreducible variety V. Although we state our definitions and results for irreducible varieties V and W, they can be extended to include arbitrary varieties. Let us begin by recalling that the set of all points which are singular in a variety V forms a proper subvariety $\mathsf{S}(V)$ of V (this is a corollary of Theorems 4.1 and 4.3 of Chapter IV). Any irreducible subvariety W of V not contained in $\mathsf{S}(V)$ intersects $\mathsf{S}(V)$ in a proper subvariety of W, so that almost every point of W is nonsingular in V. In general we must expect W to contain a proper subvariety of points singular in V. If we do not wish to rule out too many subvarieties, we should consider a subvariety W of V to be nonsingular in V if almost every point of W is nonsingular in V. The following definition reflects this idea. (See Theorem 4.2.)

Definition 4.1. Let $V \subset \mathbb{C}^n$ be an irreducible variety of dimension r, and let $W \subseteq V$ be an irreducible subvariety with generic point $(y) = (y_1, \ldots, y_n)$. Then **W is nonsingular in V** provided

$$\operatorname{rank}(J(V)_{(y)}) = n - r.$$

$J(V)_{(y)}$ is the "$\infty \times n$ matrix" $(\partial p/\partial X_i)_{(y)}$, where p runs through $\mathsf{J}(V)$ for $i = 1, \ldots, n$. (Cf. Notation 2.2 of Chapter III.) If $V \subset \mathbb{P}^n(\mathbb{C})$ is irreducible, and W is an irreducible subvariety of V, then **W is nonsingular in V** if some nonempty affine part $\mathsf{D}(W)$ of it is nonsingular in $\mathsf{D}(V)$ (see Definition 4.2 of Chapter IV).

This definition reduces to Definition 4.4 of Chapter IV when dim $W = 0$, that is, when W is a point. To tie Definition 4.1 in with the discussion before it, we have

Theorem 4.2. *Let W be an irreducible subvariety of an r-dimensional irreducible variety V in $\mathbb{C}^n$ or $\mathbb{P}^n(\mathbb{C})$. Then W is nonsingular in V iff almost every point $P \in W$ is nonsingular in V.*

PROOF. It suffices to assume that V is affine. Let (y) be a generic point of W.

$\Rightarrow$: Some $(n - r) \times (n - r)$ submatrix of $J(V)_{(y)}$ has nonzero determinant. This determinant is an element of $\mathbb{C}[y]$—that is, it is a polynomial $q(X) \in \mathbb{C}[X]$ evaluated at (y). Now $\mathsf{J}(W)$ consists of all those polynomials $p(X) \in \mathbb{C}[X_1, \ldots, X_n]$ such that $p(y) = 0$; hence $q \notin \mathsf{J}(W)$, so $\mathsf{V}(q)$ intersects W properly. Consequently q is nonzero at almost every point $P \in W$, so rank $J(V)_{(P)} = n - r$ for almost every $P \in W$. Thus V is nonsingular at almost every $P \in W$.

$\Leftarrow$: If the determinant $q(y)$ of some $(n - r) \times (n - r)$ submatrix of $J(V)_{(y)}$ is nonzero at almost every point of W (in fact, even at one point of W), then $q(y) \neq 0$, so the rank of $J(V)_{(y)}$ is at least $n - r$. But $n - r$ is the maximum rank of $J(V)_{(P)}$ over points $P \in W$; if $J(V)_{(y)}$ had rank $> n - r$, there would be points of W where $J(V)$ has rank $> n - r$. Thus rank $J(V)_{(y)} = n - r$, and W is nonsingular on V. □

We next turn to the question of a ring-theoretic characterization of nonsingularity. The following idea leads to an understanding of this characterization. First, given any s-dimensional subvariety W of an r-dimensional $V \subset \mathbb{C}^n$, one can find $r - s$ hypersurfaces $\mathsf{V}(p_1), \ldots, \mathsf{V}(p_{r-s})$ whose intersection with V is an s-dimensional variety containing W (cf. the proof of Lemma 4.6). When V and W are irreducible, then for the local ring $\mathfrak{o}(W; V) = R_{\mathsf{J}(W)}$ of V at W, we have the following:

Let (x) be a generic point of V. If W' is any s-dimensional variety containing W, then $W'_W = W_W = (W, W)$. Thus the ideal $\mathfrak{n} = (p_1(x), \ldots, p_{r-s}(x)) \subset R_{\mathsf{J}(W)}$ defines $W_{(W)}$, and $\sqrt{\mathfrak{n}}$ is just the maximal ideal $\mathfrak{m}$ of $R_{\mathsf{J}(W)}$. Now any irreducible subvariety W of V is nonsingular in V iff it has multiplicity 1 in V (Exercise 6.4 of Chapter IV). Based on a hope that our correspondence between ideals and geometric objects (chains) is faithful enough, we might conjecture that W is nonsingular in V iff for some choice of $p_1, \ldots, p_{r-s}$, we have $\mathfrak{n} = \mathfrak{m}$. This conjecture turns out to be true, and its local-ring formulation is our characterization of nonsingularity in Theorem 4.8.

Before turning to the formal statements, we consider some examples to clarify the above idea.

EXAMPLE 4.3. Consider the parabola $V = \mathsf{V}(Y - X^2) \subset \mathbb{C}_{XY}$, and the nonsingular point $W = (0, 0) \in V$. Here $r = 1$, $s = 0$. The 1-hypersurface $\mathsf{V}(X)$ intersects V in $W = (0, 0)$ $(1 = r - s)$. From the standpoint of the local ring of V at (0, 0), we have the following: The coordinate ring of V is $\mathbb{C}[X, X^2] = \mathbb{C}[X]$, its local ring at (0, 0) is $\mathbb{C}[X]_{(X)}$; the unique maximal ideal is clearly principal since it is generated by the single element X, so $\mathfrak{n} = \mathfrak{m}$.

EXAMPLE 4.4. In contrast to the parabola above, consider the curve $V = \mathsf{V}(Y^2 - X^3) \subset \mathbb{C}_{XY}$, and the singular point (0, 0). Again, the hypersurface $\mathbb{C}_Y$ intersects V in $W = (0, 0)$. The curve's coordinate ring is $\mathbb{C}[X, X^{3/2}]$, the maximal ideal in $\mathbb{C}[X, X^{3/2}]$ corresponding to (0, 0) is $\mathfrak{p} = (X, X^{3/2})$, and the local ring at (0, 0) is $\mathbb{C}[X, X^{3/2}]_{\mathfrak{p}}$. In this local ring, $\mathfrak{m} = (X, X^{3/2})$ and $\mathfrak{n} = (X)$; clearly $\mathfrak{n} \subsetneqq \mathfrak{m}$. Our characterization in Theorem 4.8 will show that for any variety $\mathsf{V}(p)$ having (0, 0) as an isolated point of intersection with $\mathsf{V}(Y^2 - X^3)$ we always have $\mathfrak{n} \subsetneqq \mathfrak{m}$.

EXAMPLE 4.5. For a higher-dimensional example, we may consider the cusp curve's "cylindrization" $V^* = \mathsf{V}(Y^2 - X^3) \subset \mathbb{C}_{XYZ}$. The set of singular points is $\mathbb{C}_Z$. Since we are now dealing with a surface, we need *two* hypersurfaces $\mathsf{V}(p_1)$ and $\mathsf{V}(p_2)$ to intersect V^* in a set of dimension 0. For instance, $V^* \cap \mathsf{V}(X) \cap \mathsf{V}(Z) = (0, 0, 0)$; V^*'s local ring at (0, 0, 0) is $\mathbb{C}[X, X^{3/2}, Z]_{(X, X^{3/2}, Z)}$. In this local ring we have $(X, Z) \subsetneqq \sqrt{(X, Z)} = (X, x^{3/2}, Z)$. One has an analogous result when the 0-dimensional singular subvariety $W = (0, 0, 0)$ is replaced by a higher-dimensional subvariety, for example $\mathbb{C}_Z$. Then one can intersect V^* down to $\mathbb{C}_Z$ using only one hypersurface, for instance $\mathbb{C}_{YZ} = \mathsf{V}(X)$. As in Example 4.4, $\mathfrak{n} = (X) \subsetneqq \mathfrak{m} = (X, X^{3/2})$.

We now make precise the general idea expressed just before Example 4.3. We begin with

Lemma 4.6. *Let $W \subset V$ be irreducible varieties of $\mathbb{C}^n$ of dimensions s and r respectively, and let $\mathfrak{m}$ be the maximal ideal of the local ring of W in V. Then there exist $r - s$ elements $a_1, \ldots, a_{r-s}$ of $\mathfrak{m}$ such that*

$$\sqrt{(a_1, \ldots, a_{r-s})} = \mathfrak{m}.$$

PROOF. It suffices to find $r - s$ polynomials $p_1, \ldots, p_{r-s} \in \mathbb{C}[X_1, \ldots, X_n]$ such that $V \cap \mathsf{V}(p_1) \cap \ldots \cap \mathsf{V}(p_{r-s})$ is an s-dimensional variety containing W. For this, write $\mathfrak{p} = \mathsf{J}(V)$ and $\mathfrak{q} = \mathsf{J}(W) \subset \mathbb{C}[X_1, \ldots, X_n]$. If $W \subsetneqq V$, then $\mathfrak{p} \subsetneqq \mathfrak{q}$; in this case choose for p_1 any polynomial in $\mathfrak{q} \backslash \mathfrak{p}$. Then each component $V_1, \ldots, V_t$ of $V \cap \mathsf{V}(p_1)$ is of dimension $r - 1$; let their associated prime ideals be $\mathfrak{p}_1, \ldots, \mathfrak{p}_t$. We want to choose for p_2 any polynomial of $\mathfrak{q}$ which is not in any of $\mathfrak{p}_1, \ldots, \mathfrak{p}_t$, for $\mathsf{V}(p_2)$ would then intersect each of $V_1, \ldots, V_t$ in dimension $r - 2$, and the lemma's proof could easily be completed using induction. But such a choice is easy: Since there are no proper containment relations among $\mathfrak{p}_1, \ldots, \mathfrak{p}_t$, we may, for each pair $\mathfrak{p}_i, \mathfrak{p}_j$ of distinct prime ideals, find an element p_{ij} of $\mathfrak{q}$ not in $\mathfrak{p}_i$ but in $\mathfrak{p}_j$. Then

$$p_i^* = p_{i1} \cdot \ldots \cdot p_{i,i-1} \cdot p_{i,i+1} \cdot \ldots \cdot p_{it}$$

is not in $\mathfrak{p}_i$, but is in every one of the other $t - 1$ ideals. It is easily seen that $p_2 = \sum_{i=1}^{t} p_i^*$ is in $\mathfrak{q}$, but in none of $\mathfrak{p}_1, \ldots, \mathfrak{p}_t$. This proves the lemma. □

Remark 4.7. In the language of germs, Lemma 4.6 implies that any pure r-dimensional "germ" V_W (where dim V_W means dim $V_{(W)}$) is the proper intersection of $n - r$ $(n - 1)$-dimensional germs at W. It is a kind of germ-theoretic converse to the already-established projective result that any proper intersection of r hypersurfaces of $\mathbb{P}^n(\mathbb{C})$ has pure dimension $n - r$. As noted in Section IV,3, it is not true that this projective result has a converse—that is, not every projective (or even affine) variety of pure dimension r is the intersection of $n - r$ appropriately-chosen hypersurfaces. It may happen that there are always extra components in the intersection, additional hypersurfaces being needed to remove them. At the level of germs, we are in effect ignoring these extra components.

We now come to the promised local-ring characterization of nonsingularity.

Theorem 4.8. *Let $W \subset V$ be irreducible varieties in $\mathbb{C}^n$ of dimensions s and r, respectively, and let $\mathfrak{m}$ be the maximal ideal of the local ring $R = \mathfrak{o}(W; V)$ of V at W. Then W is nonsingular in V iff $\mathfrak{m}$ is R-generated by some set of $r - s$ elements. (The local ring $\mathfrak{o}(W; V)$ is then said to be* **regular***.)*

Remark 4.9. Theorem 4.8 easily implies a projective analogue. See Exercise 4.1.

PROOF OF THEOREM 4.8. We first establish the theorem for $V = \mathbb{C}^n$; we then use this result to prove the full theorem.

First, note from Definition 4.1 that any irreducible variety $W \subset \mathbb{C}^n$ is always nonsingular in $\mathbb{C}^n$; in fact, each point of W is nonsingular in $\mathbb{C}^n$, because $\mathrm{J}(\mathbb{C}^n)$ is the zero ideal, so rank $J(\mathbb{C}^n)_P = n - n = 0$ for each $P \in \mathbb{C}^n$. (Of course there may be points of W singular in W.) In this case "$\Leftarrow$" of Theorem 4.8 is trivial. We now prove "$\Rightarrow$"; i.e., that $\mathfrak{m}$ is generated by a set of $n - s$ elements of R.

Assume $(x) = (x_1, \ldots, x_s)$ is a transcendence base of W's function field; we may clearly choose the first s coordinates of a generic point for $\mathbb{C}^n$ to be (x), too. We shall write (x, y) and (x, z) for generic points of $\mathbb{C}^n$ and W, respectively, where $(y) = (y_1, \ldots, y_{n-s})$ and $(z) = (z_1, \ldots, z_{n-s})$. Our local ring is

$$R = \mathbb{C}[x, y]_{\mathrm{J}(W)} = \left\{ \frac{p(x, y)}{q(x, y)} \,\middle|\, q(x, z) \neq 0 \right\}.$$

Any nonzero $q(x, y)$ involving only (x) of course satisfies $q(x, z) \neq 0$; hence if we denote $\mathbb{C}(x)$ by k, we can rewrite R as

$$R = \left\{ \frac{p^*(y)}{q^*(y)} \,\middle|\, p^*, q^* \in k[y] \text{ and } q^*(z) \neq 0 \right\}.$$

We shall choose for our $n - s$ elements, polynomials $p_1^*, \ldots, p_{n-s}^* \in \mathfrak{m} \cap k[y]$; clearly it suffices to show that these polynomials generate $\mathfrak{m} \cap k[y]$ over $k[y]$.

We take p_{n-s}^* to be any polynomial in $k[y_1, \ldots, y_{n-s}]$ monic in y_{n-s}, and such that $p_{n-s}^*(z_1, \ldots, z_{n-s-1}, y_{n-s})$ is the irreducible polynomial of z_{n-s} over $k[z_1, \ldots, z_{n-s-1}]$. (This last ring is a field, since each of $z_1, \ldots, z_{n-s-1}$ is algebraic over k.) Then $p_{n-s}^*(z_1, \ldots, z_{n-s}) = 0$, so $p_{n-s}^*(y) \in \mathfrak{m} \cap k[y]$, since $\mathfrak{m} \cap k[y] = \{p^* \in k[y] \mid p^*(z) = 0\}$.

Now let $q^*(y)$ be any element of $\mathfrak{m} \cap k[y]$. We may write

$$q^*(y) = p_{n-s}^*(y)t^*(y) + r^*(y) \tag{4.10}$$

for some $t^*, r^* \in k(y_1, \ldots, y_{n-s-1})[y_{n-s}]$, where either $r^* = 0 \in k(y_1, \ldots, y_{n-s-1})[y_{n-s}]$ or $\deg r^* < \deg p_{n-s}^*$ (degree in y_{n-s}).

Now $r^*(z_1, \ldots, z_{n-s-1}, y_{n-s})$ is the zero polynomial in y_{n-s}. This is so since we do not have $\deg r^* < \deg p_{n-s}^*$ which in turn is true because (a) $r^*(z) = 0$ (from (4.10) and the fact that $q^*(z) = p_{n-s}^*(z) = 0$), and (b) $p_{n-s}^*(z_1, \ldots, z_{n-s-1}, y_{n-s})$ is already a polynomial in $k[z_1, \ldots, z_{n-s-1}, y_{n-s}]$ of least degree in y_{n-s} such that $p_{n-s}^*(z) = 0$. Thus each coefficient

$$p_{n-sj}^* \in k[y_1, \ldots, y_{n-s-1}]$$

of the term $y_{n-s}{}^j$ in $r^*(y)$ satisfies

$$p_{n-sj}^*(z_1, \ldots, z_{n-s-1}) = 0,$$

so each $p_{n-sj}^*(y_1, \ldots, y_{n-s-1})$ is in $\mathfrak{m} \cap k[y_1, \ldots, y_{n-s-1}]$. In view of this, the problem of showing that $p_1^*, \ldots, p_{n-s}^*$ generate $\mathfrak{m} \cap k[y_1, \ldots, y_{n-s}]$ over $k[y_1, \ldots, y_{n-s}]$ has been reduced to showing that $p_1^*, \ldots, p_{n-s-1}^*$ generate $\mathfrak{m} \cap k[y_1, \ldots, y_{n-s-1}]$ over $k[y_1, \ldots, y_{n-s-1}]$. In this way we complete the proof of the case $V = \mathbb{C}^n$ using induction.

Now let us indicate the basic strategy for the proof of the full theorem. We shall prove that if W is nonsingular in V, then $\mathfrak{m}$ is R-generated by some set of $r - s$ elements. The proof of the converse is essentially just the reverse of the half we prove; we leave it for the exercises (Exercise 4.4).

Our proof basically consists in changing the problem from one concerning the number of generators of $\mathfrak{m}$, to one concerning the dimension of a vector space in which finding generators is easier. We do this in several steps. The overall goal is to show there are elements $x_1, \ldots, x_{r-s} \in \mathfrak{m}$ such that $\mathfrak{m} = x_1R + \ldots + x_{r-s}R$. The following result describes the first transformation of the problem. Notation is as in Theorem 4.8.

Lemma 4.11. *For any elements* $x_1, \ldots, x_{r-s} \in \mathfrak{m}$,

$$(\mathfrak{m} = x_1R + \ldots + x_{r-s}R) \Leftrightarrow (\mathfrak{m} = x_1R + \ldots + x_{r-s}R + \mathfrak{m}^2).$$

We give the proof of this lemma after indicating the basic idea of Theorem 4.8's proof.

This changes the problem from requiring that $x_1, \ldots, x_{r-s}$ generate $\mathfrak{m}$, to requiring only that $x_1, \ldots, x_{r-s}$ *together with* $\mathfrak{m}^2$, generate $\mathfrak{m}$.

Next, note that $\mathfrak{m}/\mathfrak{m}^2$ forms in a natural way a vector space over $R/\mathfrak{m}$. In view of this, the reader may easily verify the next lemma; it describes the next transformation of our problem.

Lemma 4.12. *For any elements* $x_1, \ldots, x_{r-s} \in \mathfrak{m}$, $\mathfrak{m} = x_1 R + \ldots + x_{r-s} R + \mathfrak{m}^2$ *iff* $\mathfrak{m}/\mathfrak{m}^2$ *has dimension* $r - s$ *over* $R/\mathfrak{m}$, *with basis elements* $x_1 + \mathfrak{m}^2, \ldots, x_{r-s} + \mathfrak{m}^2$.

Assuming Lemma 4.12, we see that our task is to show that $\mathfrak{m}/\mathfrak{m}^2$ has dimension over $R/\mathfrak{m}$ no greater than $r - s$. We further transform our problem by expressing $\mathfrak{m}/\mathfrak{m}^2$ in terms of ideals in the local ring R^* of W in $\mathbb{C}^n$ (rather than in V), which will in effect bring us back to the case of our theorem already proved. For this, let the prime ideals of V and W be $\mathfrak{p}$ and $\mathfrak{q}$ $(\subset \mathbb{C}[X_1, \ldots, X_n])$, respectively. Assume without loss of generality that the first s components of the generic points of V and of W are the same—say the generic point of V is $(x_1, \ldots, x_s, y_1, \ldots, y_{n-s})$, and that of W is $(x_1, \ldots, x_s, z_1, \ldots, z_{n-s})$. The local ring of W in $\mathbb{C}^n$ is then $R^* = \mathbb{C}[X_1, \ldots, X_n]_{\mathfrak{q}}$. The extended ideal $\mathfrak{p}^* \subset R^*$ generated by $\mathfrak{p}$ is prime; the map $(X_1, \ldots, X_n) \to (x_1, \ldots, x_s, y_1, \ldots, y_{n-s})$ then defines an isomorphism from $R^*/\mathfrak{p}^*$ to R. If $\mathfrak{m}^*$ is the maximal ideal of R^*, then

$$\mathfrak{m} = \mathfrak{m}^*/\mathfrak{p}^* \tag{19}$$

As for $\mathfrak{m}^2$, (19) shows that a typical element of $\mathfrak{m}^2$ is $(m + \mathfrak{p}^*) \cdot (m' + \mathfrak{p}^*)$ (where m and $m' \in \mathfrak{m}^*$), which is $mm' + \mathfrak{p}^*$; hence $\mathfrak{m}^2 = ((\mathfrak{m}^*)^2 + \mathfrak{p}^*)/\mathfrak{p}^*$. Our final transformations are then given by:

$$\mathfrak{m}/\mathfrak{m}^2 = \frac{\mathfrak{m}^*/\mathfrak{p}^*}{(\mathfrak{m}^* + \mathfrak{p}^*)/\mathfrak{p}^*} \simeq \mathfrak{m}^*/(\mathfrak{m}^* + \mathfrak{p}^*) \simeq \frac{\mathfrak{m}^*/(\mathfrak{m}^*)^2}{((\mathfrak{m}^*)^2 + \mathfrak{p}^*)/(\mathfrak{m}^*)^2} \tag{20}$$

Note that all of these are vector spaces over

$$R/\mathfrak{m} \simeq \frac{R^*/\mathfrak{p}^*}{\mathfrak{m}^*/\mathfrak{p}^*} \simeq R^*/\mathfrak{m}^*.$$

The two isomorphisms in (20) follow from the familiar *second law of isomorphism* for groups (with operators). Thus to prove Theorem 4.8, we want to show that the last expression in (20) has dimension at most $r - s$ over $R^*/\mathfrak{m}^*$.

As for the numerator $\mathfrak{m}^*/(\mathfrak{m}^*)^2$, applying the case of Theorem 4.8 already proved to W (which is nonsingular in $\mathbb{C}^n$) shows that $\dim(\mathfrak{m}^*/(\mathfrak{m}^*)^2) = n - s$. Hence our final task is to prove:

(4.13) If W is nonsingular in V, then

$$\dim[((\mathfrak{m}^*)^2 + \mathfrak{p}^*)/(\mathfrak{m}^*)^2] \geqslant n - r.$$

For if we know (4.13), then the final quotient in (20) has dimension no greater than $(n - s) - (n - r) = r - s$.

Now that we've given the overall strategy, let us fill in the missing pieces (the proofs of Lemma 4.11 and (4.13)).

PROOF OF LEMMA 4.11

$\Rightarrow$: Obvious.

$\Leftarrow$: We prove this by showing that $\mathfrak{m}/(Rx_1 + \dots + Rx_{r-s})$ is the zero module over R. For brevity, denote $Rx_1 + \dots + Rx_{r-s}$ by $\mathfrak{n}$. Assume $\mathfrak{m} = \mathfrak{m}^2 + \mathfrak{n}$. This implies that $\mathfrak{m}/\mathfrak{n} \simeq (\mathfrak{m}^2 + \mathfrak{n})/\mathfrak{n} \simeq (\mathfrak{m} + \mathfrak{n})(\mathfrak{m} + \mathfrak{n})/\mathfrak{n} = ((\mathfrak{m} + \mathfrak{n})/\mathfrak{n})^2 = (\mathfrak{m}/\mathfrak{n})^2$. The ring $R/\mathfrak{n}$ is Noetherian, since R is; therefore the ideal $\mathfrak{m}/\mathfrak{n}$ is R-generated by some finite set $\{y_1, \dots, y_t\}$ of elements in $\mathfrak{m}/\mathfrak{n}$. Since $\mathfrak{m}/\mathfrak{n} = (\mathfrak{m}/\mathfrak{n})^2$, we can write

$$y_t = \sum_{i=1}^{t} y_i z_i \qquad (y_i, z_i \in \mathfrak{m}/\mathfrak{n}).$$

Therefore, with obvious notation, we have

$$y_t(1 - z_t) = \sum_{i=1}^{t-1} y_i z_i.$$

We know $1/(1 - z_t) \in R/\mathfrak{n}$, so we may write

$$y_t = \sum_{i=1}^{t-1} \frac{z_i}{1 - z_t} \cdot y_i.$$

Hence $\{y_1, \dots, y_{t-1}\}$ is a generating set for $\mathfrak{m}/\mathfrak{n}$, too. Repeating this argument shows us that $\{y_1\}$ is a generating set, and for this we have $y_1(1 - z_1) = 0$. Thus $y_1 = 0$, hence $\mathfrak{m}/\mathfrak{n} = (0)$; therefore $\mathfrak{m} = \mathfrak{n}$, as desired. □

PROOF OF 4.13

Let W have generic point (y); since W is nonsingular in V, Definition 4.1 tells us that after appropriately reindexing $X_1, \dots, X_n$ if necessary, there are polynomials $p_1, \dots, p_{n-r}$ in $\mathfrak{p}$ so that

$$\det\left(\frac{\partial p_i}{\partial X_j}\right)_{(y)} \neq 0 \qquad (i, j = 1, \dots, n - r). \tag{21}$$

We claim that $\{p_1 + (\mathfrak{m}^*)^2, \dots, p_{n-r} + (\mathfrak{m}^*)^2\}$ are $R^*/\mathfrak{m}^*$-linearly independent elements of $(\mathfrak{p}^* + (\mathfrak{m}^*)^2)/(\mathfrak{m}^*)^2$. First, since each p_i is obviously in the maximal ideal of R^*, its constant term is 0; Thus we may write

$$p_i = a_{i1}X_1 + \dots + a_{in}X_n + \{\text{terms of higher order}\}. \tag{22}$$

Hence $p_i + (\mathfrak{m}^*)^2 = a_{i1}X_1 + \dots + a_{in}X_n + (\mathfrak{m}^*)^2$, where $a_{ij} \in \mathbb{C}$. Now $\{a_{i1}X_1 + \dots + a_{in}X_n + (\mathfrak{m}^*)^2\}$ $(i = 1, \dots, n - r)$ are linearly independent over $R^*/\mathfrak{m}^*$ provided

$$\sum_{i=1}^{n-r} c_i(a_{i1}X_1 + \dots + a_{in}X_n) \in (\mathfrak{m}^*)^2$$

$$\text{implies} \quad c_1 = \dots = c_{n-r} = 0 \qquad (c_i \in R^*/\mathfrak{m}^*) \tag{23}$$

The only linear combination of the X_i which is in $(\mathfrak{m}^*)^2$ is the zero linear combination, so (23) implies that $\{a_{i1}X_1 + \ldots + a_{in}X_n\}$ $(i = 1, \ldots, n - r)$ are linearly independent over $R^*/\mathfrak{m}^*$; this happens, of course, iff the $n - r$ vectors $(a_{i1}, \ldots, a_{in})$ are linearly independent over $R^*/\mathfrak{m}^*$. But from (22) we see that $\partial p_i/\partial X_j$ evaluated at (y) is just a_{ij}, so (21) says the vectors $(a_{i1}, \ldots, a_{in})$ are indeed linearly independent over $R^*/\mathfrak{m}^*$. □

Remark 4.14. At several places in the sequel, we will use the local ring at a point P of a nonsingular curve C (in $\mathbb{C}^2$ or $\mathbb{P}^2(\mathbb{C})$). In this case, the local ring $\mathfrak{o}(W; V) = \mathfrak{o}(P; C)$ of Definition 3.5 or 3.6 is regular, and hence a principal ideal ring. If the maximal ideal $\mathfrak{m}$ of $\mathfrak{o}(P; C)$ is $\mathfrak{m} = (m)$, then assigning the integer n to elements of the form $m^n u$ (u a unit) defines in a natural way an order on K_C, which we denote by ord_P. This order clearly satisfies (2.15.1) and (2.15.2). (Thus $\mathfrak{o}(P; C)$ is a discrete rank one valuation ring.) This order obviously generalizes the ord_P in (1) defined on $K_C = \mathbb{C}(X)$. If $C \subset \mathbb{C}^2$, then (cf. Theorem 2.31) for all but finitely many choices of coordinate axes $\mathbb{C}_X$ and $\mathbb{C}_Y$, the function $x \in \mathbb{C}[x, y] = \mathbb{C}[X, Y]/\mathrm{J}(C)$ is a uniformizing parameter for C at P—say $y = g(x)$. It is then easily seen that for any $q(x, y) \in \mathbb{C}[x, y]$, $\text{ord}_P\, q$ is the smallest exponent of x in the power series $q(x, g(x))$.

Exercises

4.1 Show that Theorem 4.8 implies the corresponding projective theorem.

4.2 Let $\phi: V_1 \to V_2$ be a polynomial isomorphism between two affine varieties V_1 and V_2. Use Theorem 4.8 to show that for any point $P \in V_1$, P is nonsingular in V_1 iff $\phi(P)$ is nonsingular in V_2. Does this generalize to any irreducible subvariety $W \subset V_1$ in place of P?

4.3 Reversing the steps in the proof of "$\Rightarrow$," prove the "$\Leftarrow$" half of Theorem 4.8, assuming "$\Leftarrow$" for $V = \mathbb{C}^n$. [*Hint*: Prove it first when W is a point; for the general case, use W's function field as ground field.]

5 Ideal theory on a nonsingular curve

In building up our algebra-geometry dictionary, we have succeeded in getting an isomorphism only between *closed* ideals and varieties. As hinted in Exercise 2.4 of Chapter III, there is further geometric information hiding in arbitrary ideals. In this section we consider a certain type of coordinate ring for which one can get a faithful geometric interpretation of *all* nonzero ideals.

We briefly looked at zero dimensional "varieties-with-multiplicity" (that is, *point chains*) in Exercise 2.4 of Chapter III; there we saw that for the very simple variety $V = \mathbb{C}$, there is a lattice isomorphism between the lattice of all nonzero ideals and the lattice of all positive point chains. A natural question arises: To what extent can we generalize this result to more general varieties? Another question is suggested by the results of Exercises 2.4 and

2.5(a) of Chapter III, which may be regarded in a natural way as a geometric translation of the fundamental theorem of algebra; the analogy between the fundamental theorems of algebra and arithmetic leads one to speculate that there might possibly be a unified framework into which all these results fit. It is at this point that extending results in the classical $\mathbb{C}$-setting to more general commutative rings begins to pay off.

First note that $\mathbb{C}$, as a variety, is nonsingular of complex dimension one. Now the coordinate ring of $\mathbb{C}$ is $\mathbb{C}[X]$, and $\mathbb{C}$ may be identified in a natural way with the set of all maximal ideals of $\mathbb{C}[X]$. Then from the more general vantage point of commutative rings we have:

> **(5.1.1)** Every nonzero proper prime ideal of $\mathbb{C}[X]$ is maximal—that is, every point of $\mathbb{C}$ has dimension one at each of its points (cf. Theorem 2.18 of Chapter IV);
> **(5.1.2)** Since $\mathbb{C}$ is nonsingular, the local ring in $\mathbb{C}(X)$ of each point of $\mathbb{C}$ is *regular* (cf. Theorem 4.8) which in our case means that the local ring's maximal ideal is principal.

One arithmetic analogue of $\mathbb{C}[X]$ is $\mathbb{Z}$; one may apply the definition of abstract variety (Definition 8.9 of Chapter III) to any commutative Noetherian ring R with identity, getting a very general kind of abstract variety V_R (which is usually supplied with a topology). In this sense $V_{\mathbb{Z}}$ is the set of maximal ideals of $\mathbb{Z}$, which can be identified with a countable set of points corresponding to the set of positive prime integers $\{2, 3, 5, 7, \ldots\}$. (This is an example of the usefulness of the abstract-variety idea. Actually, the set of all prime ideals of R, denoted by Spec R, is another kind of "variety" having even greater flexibility than the set of maximal ideals. In such a broadened concept of variety, notions such as order, multiplicity of intersection, or nonsingularity *along* or *at* an irreducible subvariety (prime ideal) have equal standing with those notions at a point (maximal ideal).)

In $\mathbb{Z}$, just as in $\mathbb{C}[X]$, every nonzero proper prime ideal is maximal; correspondingly, we take the dimension of the variety $V_{\mathbb{Z}}$ to be one. Now the local ring at any maximal ideal (p), where p is a prime, is in analogy with $\mathbb{C}[X]$, the set

$$\mathbb{Z}_{(p)} = \left\{ \frac{n}{m} \,\middle|\, n, m \in \mathbb{Z} \quad \text{and} \quad p \nmid m \right\} \subset \mathbb{Q}.$$

It is clear that the maximal ideal of $\mathbb{Z}_{(p)}$ is generated by p, so $\mathbb{Z}_{(p)}$ is regular; hence, looking at Theorem 4.8 as a criterion for nonsingularity, we may consider $V_{\mathbb{Z}}$ to be *nonsingular*.

Thus the fundamental theorem of arithmetic can be translated into geometric terms in just the same way as in the $\mathbb{C}[X]$ case: $n = p_1^{m_1} \cdot \ldots \cdot p_s^{m_s}$ corresponds to a point chain in $V_{\mathbb{Z}}$, and we again have a lattice isomorphism between the nonzero ideals of $\mathbb{Z}$ and the *positive* point chains of $V_{\mathbb{Z}}$ (that is, point chains all of whose coefficients are nonnegative).

In this section we prove, in Theorem 5.12, that for any "irreducible non-singular abstract curve V_R" defined by a Noetherian domain R satisfying (5.1.1) and (5.1.2) there is a corresponding "fundamental theorem" establishing an isomorphism between the nonzero ideals of R and the positive point chains of V_R.

Such an isomorphism will yield a unique decomposition into irreducibles. Note that we are of course dealing with the whole lattice $\mathscr{I}(R)$ of ideals in R. From the examples of $\mathbb{C}[X]$ and $\mathbb{Z}$, one might wonder if one could simply dispense with ideals and generalize to arbitrary Noetherian domains by seeking a unique decomposition of *elements* into products of irreducibles. For certain special domains, this was, historically, the approach taken (*Euclidean domains* being one example). But without extra assumptions, one cannot in general get uniqueness, once one has existence. For instance, in the domain $\mathbb{Z}[\sqrt{-5}]$, $3 \cdot 7 = (1 + 2\sqrt{-5})(1 - 2\sqrt{-5})$ are two different decompositions of 21 into irreducibles. (See, e.g., [Borevich and Shafarevich, pp. 167, 168].)

Though in the case of $\mathbb{Z}$ or $\mathbb{C}[X]$ we can translate unique decomposition from elements to the associated principal ideals, the ideal-theoretic approach offers no particular advantage since each ideal in $\mathbb{Z}$ and in $\mathbb{C}[X]$ is principal, so that ideals don't introduce anything essentially new. But for instance in $Z[\sqrt{-5}]$, not every ideal is principal, since every principal ideal domain is a unique factorization domain. Since every element of $Z[\sqrt{-5}]$ defines a principal ideal, but not every ideal is principal, the ideals thus represent objects more general than elements or "numbers" of $Z[\sqrt{-5}]$.

Larger ideals, which in a sense correspond to "smaller ideal numbers," will represent the pieces into which we will factor elements of such domains. Thus for $6 = 2 \cdot 3 \in \mathbb{Z}$, (2) and (3) both contain (6), and each of (2) and (3) represents an "irreducible piece" into which 6 factors. This same idea applies equally well to coordinate rings of ordinary nonsingular, irreducible curves in $\mathbb{C}^2$. An example may give the reader a geometric idea of how the introduction of nonprincipal ideals reinstates uniqueness.

Example 5.2. Let us consider the complex circle $C = \mathsf{V}(X^2 + Y^2 - 2) \subset \mathbb{C}_{XY}$, its coordinate ring being $\mathbb{C}[x, y] = \mathbb{C}[X, Y]/(X^2 + Y^2 - 2)$. Any nonzero ideal of $\mathbb{C}[x, y]$ defines $\varnothing$ or a finite set of points in C. The sets of points in C defined by the principal ideals $\mathfrak{a}_1 = (x - 1)$ and $\mathfrak{a}_2 = (x + 1)$ of $\mathbb{C}[x, y]$ are $\{(1, 1), (1, -1)\}$ and $\{(-1, 1), (-1, -1)\}$, respectively. (See Figure 1.)

The totality of these four points is $\mathsf{V}(\mathfrak{a}_1) \cup \mathsf{V}(\mathfrak{a}_2) = \mathsf{V}(\mathfrak{a}_1 \cap \mathfrak{a}_2) = \mathsf{V}(\mathfrak{a}_1 \cdot \mathfrak{a}_2)$, and may also be looked at as the union $\mathsf{V}(\mathfrak{b}_1) \cup \mathsf{V}(\mathfrak{b}_2)$, where $\mathfrak{b}_1 = (1 - y)$ and $\mathfrak{b}_2 = (1 + y)$. It turns out that the elements $x - 1$, $x + 1$, $1 - y$, and $1 + y$ are all irreducible in $\mathbb{C}[x, y]$; since $x^2 + y^2 = 2$, we have $(x - 1)(x + 1) = (1 + y)(1 - y)$, representing two distinct factorizations of an element of $\mathbb{C}[x, y]$. However, though these four elements of $\mathbb{C}[x, y]$

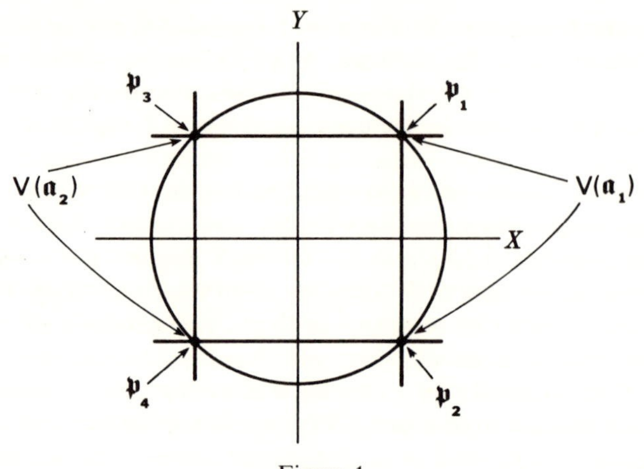

Figure 1

are irreducible in $\mathbb{C}[x, y]$, the associated principal ideals are not irreducible in $\mathbb{C}[x, y]$, in the sense that $(x - 1)$ breaks up into the product $(x - 1, 1 - y) \cdot (x - 1, 1 + y)$ of two ideals corresponding to the points $(1, 1)$ and $(1, -1)$ of $\mathsf{V}(x - 1)$. We thus get four different prime ideals which represent the true building blocks for unique factorization, namely

$$\mathfrak{p}_1 = (x - 1, 1 - y), \qquad \mathfrak{p}_2 = (x - 1, 1 + y)$$

$$\mathfrak{p}_3 = (x + 1, 1 - y), \qquad \mathfrak{p}_4 = (x + 1, 1 + y).$$

The ideal $(x - 1)$ is the same as the ideal

$$\begin{aligned}\mathfrak{c} = \mathfrak{p}_1 \cdot \mathfrak{p}_2 &= (x - 1, 1 - y) \cdot (x - 1, 1 + y) \\ &= ((x - 1)^2, (x - 1)(1 + y), (x - 1)(1 - y), (1 - y)(1 + y)),\end{aligned}$$

because $x - 1 = (1/2)[(x - 1)(1 + y) + (x - 1)(1 - y)] \in \mathfrak{c}$, so $(x - 1) \subseteq \mathfrak{c}$. And $\mathfrak{c} \subseteq (x - 1)$, since $(x - 1)^2$, $(x - 1)(1 + y)$, $(x - 1)(1 - y)$ and $(1 - y) \cdot (1 + y)$ $(=(x - 1)(x + 1))$ are all in the ideal $(x - 1)$. Similarly,

$$(x + 1) = \mathfrak{p}_3 \cdot \mathfrak{p}_4, \qquad (1 - y) = \mathfrak{p}_1 \cdot \mathfrak{p}_3, \quad \text{and} \quad (1 + y) = \mathfrak{p}_2 \cdot \mathfrak{p}_4.$$

Hence, since

$$(\mathfrak{p}_1 \cdot \mathfrak{p}_2) \cdot (\mathfrak{p}_3 \cdot \mathfrak{p}_4) = (\mathfrak{p}_1 \cdot \mathfrak{p}_3) \cdot (\mathfrak{p}_2 \cdot \mathfrak{p}_4),$$

breaking up principal ideals generated by irreducible elements into products of prime ideals, does give us unique factorization in this case.

In our generalizations of the fundamental theorems of algebra and arithmetic to an arbitrary nonsingular irreducible algebraic curve, the nonzero ideals will correspond to point chains on the curve; a nonzero proper prime ideal $\mathfrak{p}$ will correspond to a single point on the curve, and $\mathfrak{a} = \mathfrak{p}_1^{m_1} \cdot \ldots \cdot \mathfrak{p}_s^{m_s}$ will correspond to a chain $P_1^{m_1} + \ldots + P_s^{m_s}$ on the curve (where $P_i = \mathsf{V}(\mathfrak{p}_i)$).

Actually, the fundamental theorems of algebra and arithmetic generalize at once to the quotient fields $\mathbb{Q}$ and $\mathbb{C}(X)$. (Elements of $\mathbb{C}(X)$ are "meromorphic functions" on $\mathbb{C}$; *meros* is a Greek combining form meaning *fraction*.) In each case the elements are quotients of products of irreducibles from $\mathbb{Z}$ or $\mathbb{C}[X]$, each representation being unique when reduced to lowest terms. There is a corresponding generalization of *ideal* to *fractional ideal*; we begin our formal considerations in this section with this concept.

We begin with an example in $\mathbb{Q}$: Given a fraction $a/b \in \mathbb{Q}$, the corresponding "fractional ideal of $\mathbb{Z}$" consists of all multiples of a/b by elements in $\mathbb{Z}$; analogously for $\mathbb{C}(X)$. Note that in $\mathbb{Q}$, each element of a given fractional ideal can be expressed as n/d_0 with $n \in \mathbb{Z}$ and d_0 being a common, fixed element of $\mathbb{Z}$; analogously for $\mathbb{C}(X)$. Note also that each of these fractional ideals forms a module over $\mathbb{Z}$ or $\mathbb{C}[X]$; for instance in the case of $\mathbb{Z}$, a fractional ideal forms a subring of $\mathbb{Q}$ closed under multiplication by arbitrary elements of $\mathbb{Z}$. We now make the following general

Definition 5.3. Let R be any integral domain, and let K be its quotient field. Then a **fractional ideal of** R is any sub-R-module R' of K satisfying these properties:

(5.3.1) There exists a $d_0 \in R$ such that each element of R' can be written in the form r/d_0 where $r \in R$.
(5.3.2) $a \in R'$ and $r \in R$ implies $ar \in R'$.

Any d_0 satisfying Property 5.3.1 is called a **universal denominator** of the fractional ideal. If d_0 can be chosen to be 1 for a fractional ideal, we call the ideal an **integral ideal**. A **principal fractional ideal of** R consists of all R-multiples of a fixed element a of K. Instead of using notation like R', we will continue to denote fractional ideals by German letters $\mathfrak{a}$, $\mathfrak{b}$, $\mathfrak{A}$, etc.

Remark 5.4. It is easily checked that any fractional ideal of R is of the form

$$\left\{\frac{r}{d_0}\,\middle|\, r \in \mathfrak{a},\ d_0 \text{ fixed in } R\right\},$$

where $\mathfrak{a}$ is an ordinary ideal of R. Hence the ordinary ideals of R are just the integral ideals of R, and ordinary principal ideals of R are the integral principal ideals of R.

It follows at once from Definition 5.3 that the intersection of two fractional ideals of R is again a fractional ideal of R; one can define sum and product of fractional ideals $\mathfrak{a}$, $\mathfrak{b} \in R$ by $\mathfrak{a} + \mathfrak{b} = \{a + b \mid a \in \mathfrak{a} \text{ and } b \in \mathfrak{b}\}$ and $\mathfrak{a} \cdot \mathfrak{b} = \{\text{all finite sums of products } ab, \text{ where } a \in \mathfrak{a} \text{ and } b \in \mathfrak{b}\}$. From these definitions it is easy to see that sums and products of fractional ideals are still fractional ideals.

Any element of $\mathbb{Q}$ is a quotient of two integral elements; it is fair to ask if, analogously, any fractional ideal of R is a "quotient" of two integral ideals.

This is indeed so. First, the quotient of any $a \in \mathbb{Q}$ by any $b \in \mathbb{Q}\backslash\{0\}$ is that element x of $\mathbb{Q}$ satisfying $bx = a$; the following definition generalizes this idea:

Definition 5.5. Let $\mathfrak{a}$ and $\mathfrak{b}$ be fractional ideals of R. Then the set of all elements $x \in K$ satisfying $x\mathfrak{b} \subset \mathfrak{a}$ forms a fractional ideal called the **quotient of $\mathfrak{a}$ by $\mathfrak{b}$**, denoted $\mathfrak{a} : \mathfrak{b}$. A fractional ideal $\mathfrak{a}$ is called **invertible** if there exists a fractional ideal $\mathfrak{b}$ such that $\mathfrak{a} \cdot \mathfrak{b} = R$; then $\mathfrak{b}$ is the **inverse** of $\mathfrak{a}$, and is denoted by $\mathfrak{a}^{-1}$ or $1/\mathfrak{a}$.

Remark 5.6. When $\mathfrak{a}$ and $\mathfrak{b}$ are both ordinary ideals of R, Definition 5.5 yields a larger set than the "$\mathfrak{a} : \mathfrak{b}$" of Exercise 4.5 of Chapter III since in that exercise we restrict x to be in R. One thus must make clear in which sense one is taking the quotient. For the remainder of the book, we shall always mean it in the sense of Definition 5.5.

We can now easily see that any fractional ideal is the quotient of two integral ones: Let $\mathfrak{A}$ be any fractional ideal of R, and let $d_0 \in R$ be a universal denominator of $\mathfrak{A}$. Then $\mathfrak{a} = \mathfrak{A} \cap R$ and (d_0) are integral ideals of R, and $\mathfrak{A} = \mathfrak{a} : (d_0)$. (Each of "$\subset$" and "$\supset$" of the last equality follows directly from the definitions of fractional ideal and of $\mathfrak{a} : (d_0)$.)

Using fractional ideals, we will be able to see the essentially identical nature of point chains on an irreducible nonsingular curve and nonzero ideals of its coordinate ring R (Theorem 5.12). Theorem 5.12 says that every integral ideal $\mathfrak{a}$ of R is uniquely the product $\mathfrak{a} = \mathfrak{p}_1^{m_1} \cdot \ldots \cdot \mathfrak{p}_s^{m_s}$ of finitely many (ordinary) prime ideals of R, and every fractional ideal of R can be uniquely written as $\mathfrak{p}_1^{m_1} \cdot \ldots \cdot \mathfrak{p}_s^{m_s}/\mathfrak{q}_1^{n_1} \cdot \ldots \cdot \mathfrak{q}_t^{n_t}$ (where the ideals $\mathfrak{p}_i$ and $\mathfrak{q}_j$ are prime, and no $\mathfrak{p}_i$ equals any $\mathfrak{q}_j$).

Before stating Theorem 5.12 formally, we shall convert the local ring translation of nonsingularity, namely, (5.1.2), to a form which will allow us to state Theorem 5.12 in a somewhat more standard form, and will simplify the proof of the result. Specifically, we convert (5.1.2) to (5.12.2). (We need not convert (5.1.1)—condition (5.12.1) is simply (5.1.1) stated in the more general setting of Theorem 5.12.) The key concept here is that of *normal domain.* We begin with the following result.

Lemma 5.7. *Suppose R is a Noetherian domain with quotient field K, and let $\mathfrak{m}$ be any maximal ideal of R; if the maximal ideal $\mathfrak{M}$ of $R_{\mathfrak{m}}$ is principal, then $R_{\mathfrak{m}}$ is a valuation ring.*

PROOF. Suppose $\mathfrak{M} = (m)$. We first show this:

> **(5.8)** Each element a of $R_{\mathfrak{m}}$ can be written as $a = um^n$ for some unit $u \in R_{\mathfrak{m}}\backslash\mathfrak{M}$ and some nonnegative integer n.

If m does not divide a (that is, if for no $x \in R_{\mathfrak{m}}$ do we have $a = xm$) then a cannot be in $\mathfrak{M}$, hence a is a unit of $R_{\mathfrak{m}}$ ($a = u \cdot m^0$). If m does divide a, write

$a = x_1 m$. If m divides x_1, then $a = x_2 m^2$, etc. Since $R_{\mathfrak{m}}$ is Noetherian, this process must terminate after a finite number n of steps, $a = x_n m^n$, otherwise we would have a strictly ascending sequence of ideals $(a) \subsetneqq (x_1) \subsetneqq (x_2) \subsetneqq \ldots$. (This sequence would be strict, for if there were an i such that $(x_i) = (x_{i+1})$—that is, if $x_{i+1} = rx_i$ for some $r \in R_{\mathfrak{m}}$—then $a = x_i m^i = x_{i+1} m^{i+1} = rx_i m^{i+1}$. Cancelling $x_i m^i$ gives $1 = rm$, so $(m) = R_{\mathfrak{m}}$, a contradiction.) Since a doesn't divide x_n, x_n is a unit of $R_{\mathfrak{m}}$, and (5.8) follows.

It is now easily seen that $R_{\mathfrak{m}}$ is a valuation ring of K, for if b is any nonzero element of K, we may write $b = c/d$, where $c, d \in R$; if $c = um^n$, $d = u'm^{n'}$, then $b = u''m^{n-n'}$ (u, u', and u'' units of $R_{\mathfrak{m}}$). If $n - n' \geqslant 0$, $c \in R_{\mathfrak{m}}$; if $n - n' \leqslant 0$, $1/c \in R_{\mathfrak{m}}$. Hence Lemma 5.7 is proved. □

Now recall the notion of an element being integral over a domain (after the proof of (6.1.1) in Chapter III).

Definition 5.9. Let R be an integral domain with quotient field K. If each element of K integral over R is already in R, then we say R is **integrally closed in** K, or that R is **normal**.

EXAMPLE 5.10. The ring $\mathbb{Z}$ is a canonical example of a normal domain. Every element a of $\mathbb{Q}$ satisfies an equation $bx - c = 0$, with $b, c \in \mathbb{Z}$. Clearly $a = c/b$ is integral over $\mathbb{Z}$ iff b may be taken to be 1. In a similar way we see $\mathbb{C}[X]$ is normal. The coordinate ring $\mathbb{C}[X, Y]/(Y^2 - X^3) = \mathbb{C}[X, X^{3/2}]$ of the cusp curve $\mathbf{V}(Y^2 - X^3) \subset \mathbb{C}_{XY}$ is not normal, for $X^{1/2}$ is not in $\mathbb{C}[X, X^{3/2}]$, yet it is integral over $\mathbb{C}[X, X^{3/2}]$ since $Z = X^{1/2}$ satisfies the integral equation $Z^2 - X = 0$.

Note that the abstract varieties determined by the rings $\mathbb{Z}$ and $\mathbb{C}[X]$ above are nonsingular; the cusp curve, having coordinate ring $\mathbb{C}[X, X^{3/2}]$, has a singularity at (0, 0). We shall see in Exercise 5.3 that a coordinate ring of an irreducible variety V is normal iff V is "nonsingular in codimension 1"—that is, iff each irreducible subvariety of codimension 1 in V is nonsingular in V. (Cf. Definition 4.1.) (Thus the subvariety $\mathbf{S}(V)$ of points singular in V has codimension $\geqslant 2$ in V.)

The next result provides an example at the local level, and will be used in what follows.

Lemma 5.11. *Let $R_{\mathfrak{m}}$ be the valuation ring of Lemma 5.7. Then $R_{\mathfrak{m}}$ is integrally closed in its quotient field K.*

PROOF. If $R_{\mathfrak{m}}$'s maximal ideal is (m), let $a = um^{-n}$, where $n > 0$, be a typical element of $K \backslash R_{\mathfrak{m}}$ (u, a unit in $R_{\mathfrak{m}}$). If a were integral over $R_{\mathfrak{m}}$, there would be an equation of the form $m^{-nr} + c_1 m^{-n(r-1)} + \ldots + c_r = 0$ ($c_i \in R_{\mathfrak{m}}$), which implies that $1/m = d_1 + \ldots + d_{nr-1} m^{nr-1}$ ($d_i \in R_{\mathfrak{m}}$). Since m is not a unit, $1/m \notin R_{\mathfrak{m}}$. Yet $d_1 + \ldots + d_{nr-1} m^{nr-1} \in R_{\mathfrak{m}}$, a contradiction. □

Note that since any element outside all maximal ideals of a domain R is a unit of R, we have that $R = \cap R_{\mathfrak{m}}$, where intersection extends over all maximal ideals of R. Also, it follows immediately from Definition 5.9 that for any collection of subrings R_γ of a field K, if each R_γ is integrally closed in K, then $\bigcap_\gamma R_\gamma$ is too. From this it follows that $R = \cap R_{\mathfrak{m}}$ is integrally closed in K.

If D is a domain in which every nonzero proper prime ideal $\mathfrak{p}$ is maximal, and if $D_{\mathfrak{p}}$ is regular, then $D_{\mathfrak{p}}$ is a principal ideal ring. This fact and Lemmas 5.7 and 5.11 convert (5.1.2) to (5.12.2) below. In Theorem 5.12, (5.12.1) and (5.12.2) together imply that any concrete model of D is a nonsingular curve. Our basic decomposition result is:

Theorem 5.12. *Let D be a Noetherian integral domain satisfying these properties:*

(5.12.1) *Every nonzero proper prime ideal of D is maximal;*
(5.12.2) *D is integrally closed in its quotient field.*

Then: If $\mathfrak{a}$ is a nonzero integral ideal of D, it is a product of finitely many (not necessarily distinct) prime ideals $\mathfrak{p}_i$:

$$\mathfrak{a} = \mathfrak{p}_1 \cdot \ldots \cdot \mathfrak{p}_r. \tag{24}$$

This factorization is unique up to the order of the prime ideals $\mathfrak{p}_i$.

If $\mathfrak{A}$ is a nonzero fractional ideal of D, it is a quotient of products of prime integral ideals—that is, we may write

$$\mathfrak{A} = \frac{\mathfrak{p}_1 \cdot \ldots \cdot \mathfrak{p}_s}{\mathfrak{q}_1 \cdot \ldots \cdot \mathfrak{q}_t}. \tag{25}$$

We may assume that no $\mathfrak{p}_i$ equals any $\mathfrak{q}_j$; then this representation is unique up to the order of the factors. (We agree that the empty product of prime ideals is the unit ideal D.)

Conversely, for a domain D, the above conclusions imply that D is Noetherian and that it satisfies (5.12.1) *and* (5.12.2).

The reader with some familiarity with the classical theory of ideals will recognize Theorem 5.12 as a characterization of *Dedekind domains* (Definition 5.13). Since elementary proofs of this characterization are easily found in the literature, we do not reproduce one here. (See, for instance, [van der Waerden, vol. II, Sections 102–104].) For later use, we make the formal

Definition 5.13. Any Noetherian domain satisfying (5.12.1) and (5.12.2) is a **Dedekind domain**.

Exercises 2.4 and 2.5(a) of Chapter III may be looked at as a translation of the fundamental theorem of algebra into geometric terms using point chains; Theorem 5.12 may similarly be translated into geometric terms.

Consider the abstract curve $\underset{\sim}{V}_D$ corresponding to D. (One could just as well deal with any fixed concrete model V of $\underset{\sim}{V}_D$ here.) A **point chain** on $\underset{\sim}{V}_D$ is a formal finite sum $\sum n_i P_i$ of "points," or proper prime ideals, of $\underset{\sim}{V}_D$, each n_i being an integer. The point chain is **positive** if each $n_i \geqslant 0$. If $D = \sum_{i=1}^N m_i P_i$ and $E = \sum_{i=1}^N n_i P_i$, we say $D \leqslant E$ iff $m_i \leqslant n_i$ for each i such that $1 \leqslant i \leqslant N$.

Define $\sum m_i P_i + \sum n_i P_i$ to be $\sum (m_i + n_i)P_i$, and define the zero chain to be $0 = \sum 0P_i = \varnothing$. Then the set of positive point chains on $\underset{\sim}{V}_D$ becomes in a natural way a commutative semigroup with identity—that is, a set with a binary operation satisfying all axioms for a commutative group except possibly the one guaranteeing inverses, and the set of *all* point chains on $\underset{\sim}{V}_D$ becomes a commutative group. Theorem 5.12 then says that *the set of all integral ideals of D, under multiplication of ideals, forms a commutative semigroup with identity canonically isomorphic with the semigroup of positive point chains on* $\underset{\sim}{V}_D$, *and the set of all fractional ideals of D forms a commutative group isomorphic with the group of all point chains on* $\underset{\sim}{V}_D$. Such point chains are also called **divisors**, and one often refers to this group as the **group of divisors** or **divisor group of** D (or of $\underset{\sim}{V}_D$, or of any concrete model of $\underset{\sim}{V}_D$).

In the remaining two sections, we shall use the notion of **degree of a point chain** (or of an ideal, or fractional ideal). A point chain $C = \sum n_i P_i$ has **degree** $\sum_i n_i$; this degree is denoted deg C. Equivalently, the **degree of the ideal** $\mathfrak{A}$ in Theorem 5.12 is the number $s - t$ of proper prime ideal factors in the numerator of the right-hand side of (25) minus the number of them in the denominator. Note how this generalizes the degree of a polynomial or rational function in $\mathbb{C}(X)$.

We next turn our attention to (5.15) and Theorem 5.16, which will put into perspective the example at the beginning of this section; this result will also be used in Section 7 of this chapter. First, we make the following basic definition, which extends some earlier notions; we use this idea here and in the next section.

Definition 5.14. A **curve with multiplicity** is any finite sum of curves $\sum m_i C_i$ (m_i a nonnegative integer), in either the projective or affine setting. If $q(X_1, X_2) \in \mathbb{C}[X_1, X_2]$ has factorization $q_1{}^{m_1} \cdot \ldots \cdot q_r{}^{m_r}$ where each q_i is irreducible, then on $\mathbb{C}_{X_1 X_2}$, the affine curve with multiplicity associated with q is

$$\mathsf{C}(q) = \sum_{i=1}^{r} m_i \, \mathsf{V}(q_i).$$

(Analogously, a homogeneous polynomial $F \in \mathbb{C}[X_1, X_2, X_3]$ defines a curve with multiplicity in $\mathbb{P}^2(\mathbb{C})$.) For two affine or projective curves with multiplicity $\sum_i m_i C_i$ and $\sum_j m'_j C'_j$ having no common components C_i, C'_j, their **multiplicity of intersection at** P is defined to be

$$i\left(\sum_i m_i C_i, \sum_j m_j C'_j; P\right) = \sum_{i,j} m_i m'_j i(C_i, C'_j; P).$$

The **intersection cycle** of these curves with multiplicity is the formal sum

$$\left(\sum_i m_i C_i\right)\cdot\left(\sum_j m'_j C'_j\right) = \sum_{i,j,P} m_i m'_j i(C_i, C'_j; P)P,$$

where for each ordered pair (i, j), P runs (once) over each of the points of the finite set

$$C_i \cap C'_j.$$

We now assume our Dedekind domain is the coordinate ring of a nonsingular irreducible curve $C \subset \mathbb{C}_{X_1X_2}$. Consider C to be a curve with multiplicity one, and write $C = \mathsf{V}(q) = \mathsf{C}(q)$ for an irreducible $q \in \mathbb{C}[X_1, X_2]$; let $\mathbb{C}[X_1, X_2]/(q) = \mathbb{C}[x_1, x_2]$. In this case, for any $p(X_1, X_2) \in \mathbb{C}[X_1, X_2]$ $(0 \neq p, q \nmid p)$ one can see the factorization of the principal ideal $\mathfrak{a} = (p(x_1, x_2))$ into $\mathfrak{p}_1^{m_1} \cdot \ldots \cdot \mathfrak{p}_r^{m_r}$ in a very geometric way. For this, let P be any point of $\mathsf{V}(p) \cap C$. It is clear that P corresponds to one of $\mathfrak{a}$'s factors $\mathfrak{p}_i$. But even more is true: The exponent m_i of $\mathfrak{p}_i^{m_i}$ is $i(\mathsf{C}(p), C; P)$! That is,

(5.15) The divisor $\mathfrak{p}_1^{m_1} \cdot \ldots \cdot \mathfrak{p}_r^{m_r}$ corresponds to presicely $\mathsf{C}(p) \cdot C$.

This is not only interesting in its own right, but shows us how to extend Example 5.2 (see, for example, Exercise 5.1); it is, furthermore, used in an essential way in Section 7 of this chapter.

The following theorem at once implies (5.15):

Theorem 5.16. *Let* (0) *be an arbitrary point of a nonsingular curve* $C \subset \mathbb{C}_{X_1,X_2}$ *defined by an irreducible polynomial* $q(X_1, X_2) = q \in \mathbb{C}[X_1, X_2]$, *and let* $p(X_1, X_2) = p \in \mathbb{C}[X_1, X_2]$ *be any irreducible polynomial distinct from* q. *Let* $\mathbb{C}[x_1, x_2]$ *be* C*'s coordinate ring. Then the following integers are equal*:

(5.16.1) $\operatorname{ord}_{(0)}(p(x_1, x_2))$ (*Note that* $p(x_1, x_2) \in \mathfrak{o}((0);C)$. *Also, this integer is obviously independent of the choice of linear coordinates about* (0).)

(5.16.2) *The order at* (0) *in* X_1 *of the resultant* $R_{X_2}(p, q)$, *relative to all but finitely many choices of linear coordinates* (X_1, X_2) *about* (0).

(5.16.3) $i(\mathsf{C}(p), C; (0))$.

PROOF. The following conditions are satisfied for all but finitely many choices of linear coordinates about $(0) \in \mathbb{C}_{X_1X_2}$: (a) The leading terms of p and q are $X_2^{\deg p}$ and $X_2^{\deg q}$, respectively; (b) $(0) = C \cap \mathsf{V}(p) \cap \mathbb{C}_{X_2}$; (c) $X_1 = x_1$ is a uniformizing parameter of C at each point P_i of $C \cap \mathbb{C}_{X_2}$ (Thus at each P_i, C is locally represented by $X_2 = \sum_k a_{ik} X_1^{k}$.); (d) For some integer m, $\mathsf{V}(p)$ is locally represented at each point Q_j of $\mathsf{V}(p) \cap \mathbb{C}_{X_2}$ by a fractional-power series $X_2 = \sum_k b_{ik} X_1^{k/m}$.

From Remark 4.19, we may write

$$p(X_1, X_2) = \prod_i \left(X_2 - \sum_k b_{ik} X_1^{k/m} \right),$$

and

$$q(X_1, X_2) = \prod_j \left(X_2 - \sum_k a_{jk} X_1^{k} \right).$$

Thus at $P_0 = (0)$, $\mathrm{ord}_{(0)}(p(x_1, x_2))$ is the leading exponent of

$$p\left(X_1, \sum_k a_{0k} X_1^{k} \right) = \prod_i \left(\sum_k a_{0k} X_1^{k} - \sum b_{ik} X_1^{k/m} \right). \tag{26}$$

This last expression is almost the resultant $R_{X_2}(p, q)$; $R_{X_2}(p, q)$ itself is $\prod_{i,j} (\sum_k a_{jk} X_1^k - \sum_k b_{ik} X_1^{k/m})$. For all but finitely many choices of linear coordinates about (0), the order of $p(x_1, x_2)$ at any $P_i \neq (0)$ is zero, so the leading exponent of each of the factors of $\mathscr{R}_{X_2}(p, q)$ not in (26) is zero; hence $\mathrm{ord}_{(0)}(p(x_1, x_2)) = \mathrm{ord}_{(0)}(\mathscr{R}_{X_2}(p, q))$ = multiplicity of X_1 in $\mathscr{R}_{X_2}(p, q)$.

Now to evaluate $i(\mathbf{C}(p), C; (0))$, we want to find the number of distinct points near (0) in which $\mathbf{V}(p)$ intersects "almost every small linear perturbation of C." If q^* is q's homogenization (with respect to X_3), then the perturbations of C are given by the restriction to $\mathbb{C}_{X_1X_2}$ of $q^*((U_{ij})X)$ (where (U_{ij}) = 3×3 matrix of indeterminants, $X = (X_1, X_2, X_3)$), and is therefore the zero-set in $\mathbb{C}_{X_1X_2}$ of the polynomial $q^\dagger(X, U_{ij})$ obtained from $q(X_1, X_2)$ by replacing each monomial $X_1{}^i X_2{}^j$ in q by

$$(U_{11}X_1 + U_{12}X_2 + U_{13})^i \cdot (U_{21}X_1 + U_{22}X_2 + U_{23})^j \cdot (U_{31}X_1 + U_{32}X_2 + U_{33})^{\deg q - i - j}.$$

The resultant $\mathscr{R}_{X_2}(p, q^\dagger)$ is a polynomial in X_1 and in the U_{ij}, and coincides with $\mathscr{R}_{X_2}(p, q)$ at

$$(U_{ij}) = \begin{pmatrix} 1 & 0 & 0 \\ 0 & 1 & 0 \\ 0 & 0 & 0 \end{pmatrix}.$$

Its order in X_1 at $X = (0)$ and

$$(U_{ij}) = \begin{pmatrix} 1 & 0 & 0 \\ 0 & 1 & 0 \\ 0 & 0 & 0 \end{pmatrix},$$

is, of course, still the multiplicity of X_1 in $\mathscr{R}_{X_2}(p, q)$. Since almost all perturbations of $\mathbf{V}(q)$ intersect $\mathbf{V}(p)$ in $\deg p \cdot \deg q$ distinct points, the discriminant

$\mathscr{D}_{X_1}(\mathscr{R}_{X_2}(p, q^\dagger))$ is a nonzero polynomial in the elements U_{ij}, hence defines a proper subvariety in $\mathbb{C}^9 = \mathbb{C}_{(U_{ij})}$. Thus at almost each point near

$$\begin{pmatrix} 1 & 0 & 0 \\ 0 & 1 & 0 \\ 0 & 0 & 0 \end{pmatrix} \in \mathbb{C}^9,$$

$\mathscr{R}_{X_2}(p, q^\dagger)$ is a polynomial in X_1 and has $\operatorname{ord}_{(0)}(p(x_1, x_2))$ distinct zeros near $X_1 = 0$. It is easily seen that at $X = 0$,

$$(U_{ij}) = \begin{pmatrix} 1 & 0 & 0 \\ 0 & 1 & 0 \\ 0 & 0 & 0 \end{pmatrix},$$

$\mathsf{V}(q^\dagger)$ is locally the graph of an analytic function; hence $\mathsf{V}(q)$ intersects almost every perturbation of C in $\operatorname{ord}_{(0)}(p(x_1, x_2))$ distinct points. Thus the integers in Theorem 5.16 are equal. □

Exercises

5.1 Let $\mathbb{C}[x, y]$ be the coordinate ring of $\mathsf{V}(X^2 + Y^2 - 1) \subseteq \mathbb{C}_{XY}$. Find the unique factorization into prime ideals of the ideal $(y^2 - (x-1)^2(x+1)) \subset \mathbb{C}[x, y]$.

5.2 Let R be the coordinate ring of $\mathsf{V}(Y^2 - X^3) \subset \mathbb{C}_{XY}$. In Example 5.10 we saw that R is not integrally closed in its quotient field. Find a nonzero ideal of R which is not uniquely the product of maximal ideals in R. (Theorem 5.12 ensures that there exists such an ideal.)

5.3 An irreducible variety $V \subseteq \mathbb{C}^n$ is called **normal** if V's coordinate ring R is normal. An irreducible variety $V \subseteq \mathbb{P}^n(\mathbb{C})$ is called **normal** if every affine part of V is normal.

Prove as follows that a normal variety V is "nonsingular in codimension 1," in the sense that every irreducible subvariety W of codimension 1 in V, is nonsingular in V:

(a) Show that if R is integrally closed in its quotient field, and if M is any multiplicative system in R (as in Exercise 3.1), then $R_M = \{r/m \mid r \in R, m \in M\}$ is integrally closed in its quotient field.

(b) Let R be V's coordinate ring, and let W be any irreducible variety of codimension 1 in V. Use (a) to show that $R_\mathfrak{p}$ is a Dedekind domain, where $\mathfrak{p} = \mathsf{J}(W)\ (\subset R)$.

(c) Show that $R_\mathfrak{p}$'s maximal ideal $\mathfrak{m}$ is *principal*. [*Hint*: $\mathfrak{m} \cdot \mathfrak{m}^{-1} = R_\mathfrak{p}$, so $1 = \sum_i m_i m_i'$ ($m_i \in \mathfrak{m}$ and $m_i' \in \mathfrak{m}^{-1}$). Deduce that $1 = m \cdot m'$ for some $m \in \mathfrak{m}$, $m' \in \mathfrak{m}^{-1}$, and hence that $\mathfrak{m} = mR$.]

6 Some elementary function theory on a nonsingular curve

In this and the next section we look at some "function theory" on a nonsingular curve. This can be motivated by a general question: How much complex function theory on the *Riemann sphere*, (that is, $\mathbb{P}^1(\mathbb{C})$) can be carried over to an arbitrary projective nonsingular curve C? It turns out one can go quite far. As an example, in the theory on $\mathbb{P}^1(\mathbb{C})$, there is the familiar cluster of results relating meromorphic functions on $\mathbb{P}^1(\mathbb{C})$ (that is, elements of $\mathbb{P}^1(\mathbb{C})$'s function field) with their zeros and poles. Some well-known results in this direction are:

(6.1) Each nonzero meromorphic function on $\mathbb{P}^1(\mathbb{C})$ has only finitely many zeros and poles. That is, any meromorphic function f on $\mathbb{P}^1(\mathbb{C})$ defines a divisor $\operatorname{div}(f) = \sum m_i P_i$, $|m_i|$ being the multiplicity of the zero ($m_i > 0$) or pole ($m_i < 0$).

(6.2) Each nonzero meromorphic function on $\mathbb{P}^1(\mathbb{C})$ has as many zeros as it does poles. More precisely, $\deg(\operatorname{div}(f)) = 0$.

(6.3) A meromorphic function on $\mathbb{P}^1(\mathbb{C})$ has no zeros and no poles iff it is a nonzero constant. (This implies that the only functions everywhere holomorphic on all $\mathbb{P}^1(\mathbb{C})$ are the constant functions. It also implies that multiplying a function by a nonzero constant does not affect its zeros or poles.)

(6.4) For each divisor D of degree zero on $\mathbb{P}^1(\mathbb{C})$ there is a meromorphic function f on $\mathbb{P}^1(\mathbb{C})$ with $D = \operatorname{div}(f)$; f is unique up to a nonzero constant factor.

Note that the set consisting of the zero function and those meromorphic functions f on $\mathbb{P}^1(\mathbb{C})$ with $\operatorname{div}(f) = 0$, forms the 1-dimensional vector space $\mathbb{C}^1$. More generally, for a fixed point P_0, and a fixed integer $m \leqslant 0$, the set consisting of the zero function and all meromorphic functions f with $mP_0 \leqslant \operatorname{div}(f)$ forms a vector space $\mathbb{C}^{1-m}$. Using (6.1)–(6.4) one can easily extend this to

(6.5) Let D be a divisor on $\mathbb{P}^1(\mathbb{C})$ of nonpositive degree d. The set consisting of the zero function and all meromorphic functions f such that $D \leqslant \operatorname{div}(f)$, forms a vector space $\mathbb{C}^{1-d}$.

To see to what extent (6.1)–(6.5) generalize to nonsingular curves in $\mathbb{P}^2(\mathbb{C})$, we first extend the notion of divisor of a function.

For the remainder of this section, C will denote an irreducible nonsingular curve in $\mathbb{P}^2(\mathbb{C})$. Let K_C be C's function field. Since C is nonsingular, its local ring at any point is a discrete rank one valuation ring (cf. Remark 4.14). Thus any element $f \in K_C\backslash\{0\}$ has a well-defined value in $\mathbb{C} \cup \{\infty\}$ and a well-defined order at each point $P \in C$.

Definition 6.6. Let f be a nonzero element in the function field of a nonsingular curve $C \subset \mathbb{P}^2(\mathbb{C})$. The formal sum $\sum_{P \in C} \operatorname{ord}_P(f) \cdot P$ is called the **divisor of** f, and is denoted $\operatorname{div}(f)$. The sum

$$\sum_{\operatorname{ord}_P(f) > 0} \operatorname{ord}_P(f) \cdot P$$

is the **divisor of zeros of** f, and is denoted by $\operatorname{div}_0(f)$;

$$\sum_{\operatorname{ord}_P(f) < 0} -\operatorname{ord}_P(f) \cdot P$$

is the **divisor of poles of** f, denoted by $\operatorname{div}_\infty(f)$. (Thus $\operatorname{div}(f) = \operatorname{div}_0(f) - \operatorname{div}_\infty(f)$).

We will use this important observation from time to time: $\operatorname{div}(f)$ is the difference of two positive divisors, each one being induced by a curve with multiplicity in $\mathbb{P}^2(\mathbb{C})$. Specifically, we have the following: Let $\mathbb{C}_{XY}$ be an arbitrary affine part of $\mathbb{P}^2(\mathbb{C})$, and let $\mathbb{C}[x, y]$ be the coordinate ring of $C \cap \mathbb{C}_{XY}$. Let $f \in \mathbb{C}(x, y)\backslash\{0\}$; write

$$f = \frac{p\,(x, y, 1)}{q\,(x, y, 1)},$$

where $p(X, Y, Z)$ and $q(X, Y, Z)$ are forms in $\mathbb{C}[X, Y, Z]$ of equal degree. (See the definition, in Section 2, of function field of an irreducible projective variety.) Let P be any point of $C \cap \mathbb{C}_{XY}$. On C, define $\operatorname{ord}_P(p(X, Y, Z))$ to be $\operatorname{ord}_P(p(x, y, 1))$ (Note that $p(x, y, 1) \in \mathfrak{o}(P; C)$). Theorem 5.16, together with the geometric way in which the integer of (5.16.3) is defined, shows that $\operatorname{ord}_P(p(X, Y, Z))$ is independent of the particular P-containing dehomogenization of $\mathbb{P}^2(\mathbb{C})$ chosen. Then let us define $\operatorname{div}(p(X, Y, Z))$ to be $\sum_{P \in C}(\operatorname{ord}_P\, p(X, Y, Z))P$; similarly for $\operatorname{div}(q(X, Y, Z))$. It is clear that

$$\operatorname{ord}_P(f) = \operatorname{ord}_P(p(X, Y, Z)) - \operatorname{ord}_P(q(X, Y, Z));$$

hence

$$\operatorname{div}(f) = \operatorname{div}(p(X, Y, Z)) - \operatorname{div}(q(X, Y, Z)).$$

From Theorem 5.16, we see this can also be written as $\operatorname{div}(f) = \mathsf{C}(p(X, Y, Z)) \cdot C - \mathsf{C}(q(X, Y, Z)) \cdot C$.

One can now try to generalize (6.1)–(6.5) from $\mathbb{P}^1(\mathbb{C})$ to C.

(6.1)* Each $f \in K_C\backslash\{0\}$ has only finitely many zeros and poles. (Hence $\operatorname{div}(f)$ really is a divisor, in the sense of the last section.)

Proof. Write $f = p(x, y, z)/q(x, y, z)$, where $p(X, Y, Z)$ and $q(X, Y, Z)$ are relatively prime and homogeneous of the same degree. Since $p(x, y, z)$ and $q(x, y, z)$ are nonzero, neither $p(X, Y, Z)$ nor $q(X, Y, Z)$ is in $\mathsf{J}(C)$. Now $\operatorname{ord}_P(f) > 0$ implies that $\mathsf{V}(p(X, Y, Z))$ intersects C at P, and $\operatorname{ord}_P(f) < 0$ implies that $\mathsf{V}(q(X, Y, Z))$ intersects C at P. By Bézout's theorem, both these intersections with C are finite, so (6.1)* is proved. □

The above argument allows us to generalize (6.2) to C:

(6.2)* Each nonzero element of K_C has as many zeros as it does poles—that is, $\deg(\operatorname{div}_0(f)) = \deg(\operatorname{div}_\infty(f))$. Equivalently, $\deg(\operatorname{div}(f)) = 0$.

Proof. Let $p(X, Y, Z)$ and $q(X, Y, Z)$ be as in the proof of (6.1)*. By the observation made just after Definition 6.6, we have

$$\deg(\operatorname{div}(f)) = \deg(\mathsf{C}(p) \cdot C) - \deg(\mathsf{C}(q) \cdot C).$$

Define $\deg \mathsf{C}(p)$ to be $\sum m_i \deg \mathsf{V}(p_i)$; similarly for $\mathsf{C}(q)$. Bézout's theorem (Theorem 7.1 of Chapter IV) at once implies that this last difference is

$$\deg \mathsf{C}(p) \cdot \deg C - \deg \mathsf{C}(q) \cdot \deg C.$$

Now surely $\deg \mathbf{C}(p) = \deg p$ and $\deg \mathbf{C}(q) = \deg q$; since $\deg p = \deg q$, this last difference equals 0. □

Generalizing (6.3) to C is easy:

(6.3)* An element of K_C has no zeros and no poles iff it is a nonzero constant.

PROOF

$\Leftarrow$ is obvious.

$\Rightarrow$: Assume $f \in K_C$ has no zeros and no poles, and that f is nonconstant. For any $P_0 \in C$, $f(P_0) \in \mathbb{C} \backslash \{0\}$. Then $f - f(P_0)$ has a zero at P_0, but is still nonconstant. Yet $f - f(P_0)$ surely has no poles, since f has no poles. Hence $\deg(\operatorname{div}(f - f(P_0))) > 0$, a contradiction to (6.2)*. □

Now let us try to extend (6.4) to C; this would say that given any divisor D of degree zero on C, there is a function $f \in K_C$ such that $\operatorname{div}(f) = D$ (f unique up to a nonzero constant factor). Thus if D were of the form $1 \cdot P + (-1)Q$, where $P \neq Q$, then such a function f would assume every value in $\mathbb{C} \cup \{\infty\}$ exactly once. (See Exercise 6.1.) Therefore f would map C to $\mathbb{P}^1(\mathbb{C})$ in a continuous, 1 : 1-onto way; this means C would have to be homeomorphic to the sphere $\mathbb{P}^1(\mathbb{C})$. Of course this is so only if $g = 0$. Hence:

> **(6.7)** For any C of positive genus, there is never a function $f \in K_C$ whose divisor of zeros (or divisor of poles) is a single point of multiplicity one.

One can at once ask, "How about one with two or three zeros or poles?" More generally, it is natural to search for conditions on a divisor D on C so that there is a function in K having D as its divisor. Of course by (6.2)* we know the degree of any such divisor must be zero. Our example then shows the divisors of degree zero fall into two classes—those which are divisors of functions, and those which are not. This leads to a basic definition which we include here for future reference:

Definition 6.8. A divisor on C which is the divisor of an element of $K_C \backslash \{0\}$ is a **principal divisor**.

Definition 6.9. Two divisors D_1 and D_2 are **linearly equivalent** ($D_1 \simeq D_2$) if they differ by a principal divisor ($D_1 = D_2 + \operatorname{div}(g)$, for some $g \in K_C \backslash \{0\}$). The set of principal divisors forms a subgroup of the group of all divisors on C, the quotient group being the set of **linear equivalence classes** of divisors on C.

A search for further conditions will shed light on possible analogues of (6.4) and (6.5). Of course, the above example shows that (6.5) does not generalize verbatim to C. (However, note that if D is a fixed divisor on C, then the zero function together with all functions f of K satisfying $D \leqslant \text{div}(f)$ forms a complex vector space $L(D)$; this follows at once from the definition of $\text{div}(f)$ and the fact that $\text{ord}_P(f + g) \geqslant \min\{\text{ord}_P f, \text{ord}_P g\}$. It follows from Lemma 7.1 that this vector space is finite dimensional.) What are possible natural generalizations? The answers to such questions constitute some of the most central facts about algebraic curves. One generalization of (6.5) will in fact be the Riemann–Roch theorem.

Before beginning a study of this problem, we first look at differentials, which are intimately connected to the above questions. We motivate this discussion by briefly looking at a well-spring of differentials, namely integration.

To integrate on any space S one needs some sort of measure on S, perhaps given directly, perhaps induced by a metric, perhaps by a system of local coordinates, etc. In complex integration on $\mathbb{C}$, one customarily uses the canonical measure induced by the coordinate system $Z = X + iY$. But in contrast to $\mathbb{C}$, it is easy to see that for any nonsingular projective curve, there is never any one coordinate neighborhood covering the whole curve—we always need several neighborhoods and each coordinate neighborhood has its own canonical measure.

For instance, $\mathbb{C}_Z$ covers all of $\mathbb{C} \cup \{\infty\} = \mathbb{P}^1(\mathbb{C})$ except $\{\infty\}$; one can then choose a second copy $\mathbb{C}_W$ of $\mathbb{C}$ covering all of $\mathbb{C} \cup \{\infty\}$ except $\{0\}$, $\mathbb{C}_Z$ and $\mathbb{C}_W$ being related by $W = 1/Z$ in their intersection. In this example, the distance from a point $P \neq 0 \in \mathbb{C}_Z$ to $\{\infty\}$ is infinite in $\mathbb{C}_Z$, but finite in $\mathbb{C}_W$; hence the metrics in the two coordinate neighborhoods are surely different. One can get around this kind of problem by adjusting the metrics in different neighborhoods so they agree on their common part. Thus at each point common to two neighborhoods on C, coordinatized by, say, Z and by W, the metric element dZ may be modified to agree with dW by multiplying by a derivative: $dW = (dW/dZ) \cdot dZ$, where $W = W(Z)$. For instance in the example of $\mathbb{P}^1(\mathbb{C})$ above, where $W = 1/Z$, we have $dW = -(1/Z^2)dZ$.

On $\mathbb{C}$, when one uses a phrase such as *integrating a function* $f(Z)$, the canonical dZ can be left in the background. But it is the differential $f(Z)dZ$ which tells a more complete story since it takes into account the underlying measure; it is the natural object to use when coordinate changes are involved.

Aside from the obvious relation to integration, a study of differentials helps to reveal the connection between the topology of C and the existence of functions with prescribed zeros and poles; this will be our main use of differentials.

We now formally define *differential on an irreducible variety*; our definition is purely algebraic and has certain advantages; it affords a clean algebraic development and can be used in a very general setting. In Remark 6.14 we indicate for nonsingular plane curves how these differentials may be looked at as geometric objects on a variety.

Definition 6.10. Let $k \subset K$ be any two fields of characteristic 0, and let V_1 be the vector space over k generated by the set of indeterminate objects $\{dx \mid x \in K\}$; let V_2 be the subspace of V_1 generated by the set of indeterminate objects.

$$\{d(ax + by) - (adx + bdy), d(xy) - (xdy + ydx) \mid x, y \in K \text{ and } a, b \in k\}.$$

Then V_1/V_2 is the vector space of **differentials of** K **over** k, which we denote by $\Omega(K, k)$. If K_V is the function field of an irreducible variety V over k, then the elements $\omega \in \Omega(K_V, k)$ are the **differentials on** V.

Remark 6.11. One might more accurately call the above differentials *differentials of the first order*, for on varieties of dimension >1 one may consider differentials of higher order. (See, for example, [Lang, Chapter VII].) However we shall use only differentials on curves in this book. Note that the generators of V_2 simply express familiar algebraic properties of a differential.

Remark 6.12. Definition 6.10 immediately implies that $da = 0 \in V_1/V_2$ for each $a \in k$, and that if $f \in K = k(x_1, \ldots, x_n)$, then df is just the usual total differential $d(p/q)(=(qdp - pdq)/q^2$ evaluated at $(x_1, \ldots, x_n)$, where p and q are polynomials in $\mathbb{C}[X_1, \ldots, X_n]$ such that $f = p(x_1, \ldots, x_n)/q(x_1, \ldots, x_n)$.

For our applications in this book, we now assume that $k = \mathbb{C}$ and that K has transcendence degree one over $\mathbb{C}$. Thus K is the function field of an irreducible curve. In this case, for any two functions $f \in K$ and $g \in K \backslash \mathbb{C}$, there is a well-defined derivative $df/dg \in K$ having the properties one would expect of a derivative. We use this fundamental fact (Theorem 6.13) to see the geometric meaning behind Definition 6.10.

Theorem 6.13. *Let a field K have transcendence degree one over $\mathbb{C}$, and let $f \in K$, $g \in K \backslash \mathbb{C}$. Then there is a unique element $\kappa \in K$ such that $df = \kappa dg$. (We denote κ by the symbol df/dg; it is called the* **derivative of** f **with respect to** g.)

PROOF. Let $\dim_K \Omega$ denote the dimension of the K-vector space $\Omega(K, \mathbb{C})$. Theorem 6.13 will follow easily once we show $\dim_K \Omega = 1$. By the theorem of the primitive element, we may write $K = \mathbb{C}(x, y)$; suppose x is transcendental over $\mathbb{C}$, write $x = X$, and let a minimal polynomial of y over $\mathbb{C}[X]$ be $p(X, Y) \in \mathbb{C}[X, Y]$. It follows from Remark 6.12 that $\{dx, dy\}$ K-generates $\Omega(K, \mathbb{C})$, so $\dim_K \Omega \leqslant 2$. Now the element $p(x, y) = 0 \in \mathbb{C}$ has differential zero; but $p(x, y)$ is $p(X, Y)$ evaluated at (x, y), so $0 = d(p(x, y)) = p_X(x, y)dx + p_Y(x, y)dy$ (p_X, p_Y being ordinary partials with respect to the indeterminates X and Y), that is, $p_Y(x, y)dy = -p_X(x, y)dx$. Since p is minimal, $p_Y(x, y) \neq 0$, hence

$$dy = \frac{-p_X(x, y)}{p_Y(x, y)} dx;$$

thus dx generates $\Omega(K, \mathbb{C})$, so $\dim_K \Omega \leqslant 1$. To get equality, it suffices to show that $dx \neq 0$—that is, that dx is not in the subspace V_2 of Definition 6.10. For this, define a map from V_1 to K in this way: Let h, h^* be any two elements of K, and let H be an element of $\mathbb{C}(X, Y)$ such that $h = H(x, y)$. Our map is then

$$\phi: h^* dh \to h^*(H_X(x, y)p_Y(x, y) + H_Y(x, y)p_X(x, y)).$$

It is easily seen that this map is well defined and that any element in V_2 of Definition 6.10 must map to $0 \in K$. Hence ϕ induces a linear function from $\Omega(K, \mathbb{C}) = V_1/V_2$ into K. Now $\phi(dx) = 1 \cdot p_Y(x, y)$ (since $X_X = dX/dX = 1$, and $X_Y = 0$). But we know $p_Y(x, y) \neq 0$, so $\phi(dx) \neq 0$. Hence $dx \neq 0$, and therefore $\dim_K \Omega = 1$.

Since dx K-generates $\Omega(K, \mathbb{C})$, we have $df = \eta_1 dx$ and $dg = \eta_2 dx$ for unique $\eta_1, \eta_2 \in K$. Now $g \in K \backslash \mathbb{C}$, so g is transcendental over $\mathbb{C}$; the same argument that showed $dx \neq 0$ also shows that $dg \neq 0$, so we can write $df = (\eta_1/\eta_2)dg = \kappa\, dg$. κ is unique, so the proof is complete. □

Remark 6.14. For an irreducible nonsingular curve C in $\mathbb{P}^2(\mathbb{C})$ or $\mathbb{C}^2$, Theorem 6.13 allows us to see that differentials on C are actually objects "living on C." Let K_C be C's function field, hdf any element of $\Omega(K_C, \mathbb{C})$, P any point of C, and z any uniformizing parameter of the local ring $\mathfrak{o}(P; C)$ of C at P. Since $z \in K_C \backslash \mathbb{C}$, Theorem 6.13 allows us to write $df = (df/dz)dz$ ($df/dz \in K_C$), so $hdf = h(df/dz)dz = g(z)dz$. There is a Laurent series expansion for $g(z)$ representing g in a neighborhood of P in C, z induces its canonical metric on the neighborhood it coordinatizes, and if z and w are uniformizing parameters of overlapping neighborhoods, z and w being (analytically) related by $z = z(w)$, then $g(z)dz = g(z(w))(dz/dw)dw$. In short, the abstract element $hdf \in \Omega(K_C, \mathbb{C})$ can be made to look and act like the differentials on $\mathbb{R}$ or $\mathbb{C}$ one familiarly meets in calculus or elementary complex analysis.

It is important to note that a differential is quite different from a function. A function assigns a definite value to each point of a set. Though coordinates are often used in defining or representing functions, a function still associates values to points, and is independent of any particular coordinate system. A differential, on the other hand, does not assign a unique value to a point. Rather, it assigns to each point P the product of two different things—a function-value, and a "local measure." Both these in general vary with a change in coordinates; however the product is well defined in a certain sense, a change in metric being offset by a change in functional value via an appropriate derivative, as above, and when one integrates these local products over a path in C, one gets a value independent of any particular choices of coordinates used in evaluating the integral.

Although differentials don't have an intrinsic value at points, they do have a well-defined notion of *order* (and of being *holomorphic*, for instance) and the divisor of a differential can be correspondingly defined. One can then meaningfully seek differential analogues of (6.1)–(6.5)! Since differentials enter in an essential way into the Riemann–Roch theorem, we now explore this idea.

Definition 6.15. Let P be a point of C, z any uniformizing variable of C's local ring at P, and write, for any $\omega \in \Omega(K_C, \mathbb{C})$, $\omega = f dz$ where $f \in K$. The **order of** ω **at** P is the order of f at P, and is written $\mathrm{ord}_P(\omega)$. If $\mathrm{ord}_P(\omega) \geqslant 0$, then (in analogy with functions) we say that ω **is holomorphic at** P.

Note that $\mathrm{ord}_P(\omega)$ is the leading exponent of f's Laurent expansion at P.

Lemma 6.16. *$\mathrm{ord}_P\omega$ is independent of the choice of uniformizing variable.*

Proof. Let z and w be any two uniformizing variables in $\mathfrak{o}(P; C)$; then $f dz = f(dz/dw)dw$. We wish to prove $\mathrm{ord}_P(dz/dw) = 0$. Now $z = uw$ where u is a unit in $\mathfrak{o}(P; C)$, so $dz/dw = u \cdot 1 + w(du/dw)$; it therefore suffices to show that $\mathrm{ord}_P(du/dw) \geqslant 0$—that is, $du/dw \in \mathfrak{o}(P; C)$.

For this, write, in obvious notation, $\mathfrak{o}(P; C) = \mathbb{C}[x, y]_{\mathfrak{m}}$, with $\mathfrak{m} = (w)$. Let $m = \min\{\mathrm{ord}_P(dx/dw), \mathrm{ord}_P(dy/dw)\}$. Then for every element $g \in \mathfrak{o}(P; C)$ we have

$$\mathrm{ord}_P\left(\frac{dg}{dw}\right) \geqslant m. \tag{27}$$

(Write $g = r(x, y)/s(x, y)$, r, $s \in \mathbb{C}[X, Y]$ and $\mathrm{ord}_P(s(x, y)) = 0$. Then the quotient rule shows that dg/dw is

$$\frac{s \cdot \left(r_X \dfrac{dx}{dw} + r_Y \dfrac{dy}{dw}\right) - r \cdot \left(s_X \dfrac{dx}{dw} + s_Y \dfrac{dy}{dw}\right)}{s^2}$$

evaluated at (x, y). Since $\mathrm{ord}_P(r) \geqslant 0$, we have $\mathrm{ord}_P(dg/dw) \geqslant m$.) For any $n \geqslant 0$, there is an element $g_n \in \mathfrak{o}(P; C)$ such that

$$u = c_0 + c_1 w + c_2 w^2 + \ldots + c_{n-1} w^{n-1} + w^n g_n \qquad (c_i \in \mathbb{C}).$$

Then

$$\frac{du}{dw} = c_1 + 2c_2 w + \ldots + n w^{n-1} g_n + w^n \frac{dg_n}{dw}.$$

All terms on the right-hand side except perhaps the last, are surely in $\mathfrak{o}(P; C)$; by (27) also $w^n(dg_n/dw) \in \mathfrak{o}(P; C)$ for n sufficiently large. □

In analogy with the divisor of a function, we have

Definition 6.17. If ω is a differential on C, then we define the **divisor of** ω, denoted by $\mathrm{div}(\omega)$, to be the formal sum (or **chain**) $\sum_P (\mathrm{ord}_P\omega)P$, as P ranges over all points P of C. The **divisor of zeros of** ω, $\mathrm{div}_0(\omega)$, is $\sum_P (\mathrm{ord}_P\omega)P$ as P ranges over points where $\mathrm{ord}_P(\omega) > 0$, and the **divisor of poles of** ω, $\mathrm{div}_\infty(\omega)$, is $\sum_P -(\mathrm{ord}_P\omega)P$, P ranging over points where $\mathrm{ord}_P(\omega) < 0$. (Hence $\mathrm{div}(\omega) = \mathrm{div}_0(\omega) - \mathrm{div}_\infty(\omega)$.) The divisor of any differential is called a **canonical divisor** (in analogy with principal divisors of functions).

We now turn to the differential analogue of the first of (6.1)–(6.5).

(6.1)** Each nonzero differential ω on C has only finitely many zeros and poles—that is, $\mathrm{div}(\omega)$ actually is a divisor.

This will follow at once from Theorem 6.18 (which we prove next); it is basic in its own right because it describes, for a large class of curves, the geometric relation between canonical and principal divisors.

Theorem 6.18. *Let* $C \subset \mathbb{P}^2(\mathbb{C})$ *be an irreducible nonsingular curve. If* $\deg C = d \geqslant 3$ *then:*

(6.18.1) *Every canonical divisor* $\mathrm{div}(\omega)$ *is of the form* $\mathrm{div}(F) + \mathrm{div}(f)$ *for some form (that is, curve with multiplicity)* $F = F(X, Y, Z)$ *of degree* $d - 3$ *and some* $f \in K_C$;

(6.18.2) *Conversely, for every curve with multiplicity* F *in* $\mathbb{P}^2(\mathbb{C})$ *of degree* $d - 3$ *and any* $f \in K_C$, $\mathrm{div}(F) + \mathrm{div}(f)$ *is a canonical divisor.*

PROOF. Since the differentials on C form a 1-dimensional K_C-vector space, any two differentials ω_1 and ω_2 differ by an element of K_C: $\omega_2 = g\omega_1$, for some $g \in K_C$. Hence $\mathrm{div}(\omega_2) = \mathrm{div}(g) + \mathrm{div}(\omega_1)$. Therefore all canonical divisors are linearly equivalent. Clearly any divisor linearly equivalent to a canonical divisor $\mathrm{div}(\omega)$ is canonical, so the canonical divisors are all of the same degree and form a linear equivalence class.

These facts imply, first, that it suffices to prove just one of (6.18.1) and (6.18.2). We prove (6.18.1). Second, they imply that if (u, v, w) is a generic point of the homogeneous variety in $\mathbb{C}^3$ corresponding to C, and if $\mathbb{C}[x, y] = \mathbb{C}[u/w, v/w, 1]$ is the coordinate ring of C's dehomogenization $\mathsf{D}_Z(C) = \mathsf{D}(C)$ at Z, then we may assume without loss of generality that ω is dx. Also, for any two curves with multiplicity F_1 and F_2 of the same degree, $\mathrm{div}(F_1)$ and $\mathrm{div}(F_2)$ differ by a principal divisor, since F_1/F_2 is an element of C's function field; thus we may assume F in our theorem to be of a very simple form: We let $F = Z^{d-3}$. (We may thus regard F as $\mathbb{P}^2(\mathbb{C})$'s line at infinity $\mathsf{V}(Z)$, this line being counted with multiplicity $d - 3$.) With these assumptions, to prove (6.18.1) it suffices to find an element $g \in K_C$ such that

$$\mathrm{div}(dx) = \mathrm{div}(Z^{d-3}) + \mathrm{div}(g). \tag{28}$$

We begin by computing $\mathrm{div}(dx)$; this will tell us how to choose g. To evaluate $\mathrm{div}(dx)$ we choose a uniformizing variable at each point of C. Theorem 2.31 tells us that if a line $c_1X + c_2Y + c_3$ through a point $P \in \mathsf{D}(C)$ is not tangent to $\mathsf{D}(C)$ there, then $c_1x + c_2y + c_3 \in K_C$ is a uniformizing variable for $\mathsf{D}(C)$ at P.

We first compute the part of div(dx) in $\mathsf{D}(C)$—that is, we find $\text{ord}_P dx$ at all finite points P of C. First, for any divisor E on C, let the **support** $|E|$ of E be the union of all points at which E has nonzero coefficient. Now let (a, b) be a typical point of $\mathsf{D}(C)$. Then $x - a$ is a uniformizing variable at (a, b) provided the line $X - a$ is not tangent to $\mathsf{D}(C)$ at (a, b)—that is, provided $G_Y(a, b, 1) \neq 0$, where $G(X, Y, Z)$ is an irreducible d-form defining C. In this case, $1dx = 1(dx/d(x-a))d(x-a) = 1d(x-a)$, so $\text{ord}_{(a,b)}dx = 0$. Hence if $(a, b) \in |\text{div}(dx)| \cap |\mathsf{D}(C)|$, then $G_Y(a, b, 1) = 0$. Thus, suppose that the point $(a, b) \in \mathsf{D}(C)$ satisfies $G_Y(a, b, 1) = 0$. Since $\mathsf{D}(C)$ is nonsingular, $G_Y(a, b, 1) = 0$ implies $G_X(a, b, 1) \neq 0$. Therefore $y - b$ is a uniformizing variable of $\mathsf{D}(C)$ at (a, b), so $dx = (dx/d(y-b))d(y-b)$. What is $dx/d(y-b)$? Note that Definition 6.10 implies that $d(y-b) = dy$; thus $dx/d(y-b) = dx/dy$. We know $G(X, Y, 1)$ is zero at each point of $\mathsf{D}(C)$, so $0 = G_X(x, y, 1)dx + G_Y(x, y, 1)dy$. Hence $dx/dy = -G_Y(x, y, 1)/G_X(x, y, 1) \in K_C$. Since $G_X(a, b, 1) \neq 0$, its order at (a, b) is zero; thus on $\mathsf{D}(C)$ we have $\text{div}(dx) = \text{div}(G_Y)$.

Now we have only to consider the points of C at infinity. We may assume our coordinates have been chosen so that the line at infinity $\mathsf{V}(Z)$ is not tangent to C at any of its points of intersection with C, and we may further assume that the point of $\mathbb{P}^2(\mathbb{C})$ corresponding to $\mathbb{C}_X \subset \mathbb{C}_{XYZ}$ is not in C; then all points of C on $\mathsf{V}(Z)$ are contained in C's dehomogenization at Y. The associated coordinate ring is $\mathbb{C}[u/v, 1, w/v] = \mathbb{C}[x/y, 1/y]$. Since $\mathsf{V}(Z)$ is not tangent to C at any of its points of intersection with C, Z's image $1/y$ in $\mathbb{C}[x/y, 1/y]$ serves as a uniformizing variable *simultaneously* for all the points of $C \cap \mathsf{V}(Z)$. At each such point we have $dx = (dx/d(1/y))d(1/y)$. From $d(1/y) = (-1/y^2)dy$ and $0 = G_X(x, y, 1)dx + G_Y(x, y, 1)(-y^2)d(1/y)$, we get

$$dx = \frac{G_Y(x, y, 1)(y^2)d\left(\dfrac{1}{y}\right)}{G_X(x, y, 1)} = \frac{G_Y(x, y, 1)}{G_X(x, y, 1)}\left(\frac{1}{y}\right)^{-2} d\left(\frac{1}{y}\right).$$

Again, $G_X \neq 0$ at each of these points. (Suppose not: Then $d \cdot G = XG_X + YG_Y + ZG_Z$ implies that at some one of these points, $0 = YG_Y + ZG_Z$. At this point $Z = 0$, hence $YG_Y = 0$. Now $Y \neq 0$ since no point of $C \cap \mathsf{V}(Z)$ corresponds to $\mathbb{C}_X = \mathsf{V}(Y, Z)$, so G_Y would be zero, contradicting C's nonsingularity.) Hence the part of div(dx) on $\mathsf{V}(Z)$ is $\text{div}(G_Y) - 2(C \cdot \mathsf{V}(Z))$; therefore for all C we have

$$\text{div}(dx) = \text{div}(G_Y) - 2(C \cdot \mathsf{V}(Z)).$$

Of course $C \cdot \mathsf{V}(Z) = \text{div}(Z)$, so

$$\text{div}(dx) = \text{div}(G_Y) - \text{div}(Z^{d-1}) + \text{div}(Z^{d-3}).$$

But $\text{div}(G_Y) - \text{div}(Z^{d-1}) = \text{div}(G_Y/Z^{d-1})$. Since G_Y is a form of degree $d - 1$, G_Y/Z^{d-1} is indeed an element of K_C, and the g in (28) we were searching for is simply G_Y/Z^{d-1}; thus (6.18.1) is established, and therefore also Theorem 6.18. □

We now briefly consider differential analogues of (6.2)–(6.5).

As for a possible (6.2)**, $C = \mathbb{P}^1(\mathbb{C})$ above shows that for differentials it is not in general true that "the number of zeros equals the number of poles;" for example $\deg(\operatorname{div}(dx)) = \deg(-2\{\infty\}) = -2$. In fact, more generally, for any nonsingular C of degree 1 or 2 (that is, of genus 0), any nonzero $\omega \in \Omega(K_C, \mathbb{C})$ satisfies $\deg(\operatorname{div}(\omega)) = -2$. (See Exercise 6.2.) As for curves of higher degree, Theorem 6.18 already tells us that for any differential ω on C of degree $d \geqslant 3$, $\deg(\operatorname{div}(\omega)) = d(d-3)$. This may also be expressed in terms of C's genus g; g is $(d-1)(d-2)/2$, and $d(d-3)$ may be written as

$$2\left(\frac{(d-1)(d-2)}{2}\right) - 2 = 2g - 2.$$

Since this last equation also holds for $d = 1$ and $d = 2$, we have:

(6.2)**. If C is of degree d and genus g, then for each nonzero $\omega \in \Omega(K_C, \mathbb{C})$, we have $\deg(\operatorname{div}(\omega)) = d(d-3) = 2g - 2$. (Note that $2g - 2$ is the negative of the familiar *Euler characteristic* $2 - 2g$.)

As for a possible "(6.3)**," note that (6.3)* says that the only functions holomorphic on all of C are the constant functions (which form a 1-dimensional vector space); one differential analogue of (6.3)* is therefore this:

(6.3)**. The $\mathbb{C}$-vector space of all differentials holomorphic at each point of C, has dimension g.

We prove this in Exercise 7.3.

What about an analogue of (6.4)? In analogy with functions on C, given a divisor D of degree $2g - 2$, it turns out that for $C = \mathbb{P}^1(\mathbb{C})$ (in which case $2g - 2 = -2$), there is always a differential ω such that $D = \operatorname{div}(\omega)$, but that for C of higher genus, there is not in general a differential ω with $D = \operatorname{div}(\omega)$. (See Exercise 6.5.) Hence for $g > 0$, just as principal divisors do not exhaust all divisors of degree 0, canonical divisors do not exhaust all divisors of degree $2g - 2$.

As for an analogue of (6.5), if D is a divisor on C, then the zero differential together with the set of all differentials on C satisfying $D \leqslant \operatorname{div}(\omega)$ forms a vector space; the big question of its dimension is answered by the Riemann–Roch theorem, which we turn to in the next section.

It is fair to ask whether the results of this section can be generalized to arbitrary curves—that is, to curves in $\mathbb{P}^n(\mathbb{C})$ which may have singularities. The answer is "yes"; we show this in Exercises 6.11–17. In the following, we supply a little background.

First, we have assumed in this section that C is nonsingular; this allows us to regard elements of K_C as being evaluated at points of C rather than at branches of C (see Exercise 6.1). Also, for a plane curve we have the useful facts that it is definable by a single polynomial, and that it intersects any other curve in $\mathbb{P}^2(\mathbb{C})$.

Although being plane and being nonsingular are very pleasant properties for a curve to have, in general, curves in $\mathbb{P}^n(\mathbb{C})$ of course have neither of these properties. However, it turns out that many notions of this section are "birational" (see Exercise 6.12); as we will see, this has the consequence of allowing us to generalize results of this section to an arbitrary curve by establishing them for a more well-behaved curve birationally equivalent to the original one. For instance, for any irreducible curve $C \subseteq \mathbb{P}^n(\mathbb{C})$, one can find a plane curve (that is, one in $\mathbb{P}^2(\mathbb{C})$) which is birationally equivalent to C (Exercise 6.8). This will be used in generalizing the function-theoretic results of this section. At the other extreme, for any irreducible curve $C \subseteq \mathbb{P}^n(\mathbb{C})$, one can find a nonsingular curve in some $\mathbb{P}^m(\mathbb{C})$ birationally equivalent to C (Exercise 6.9), However, one cannot in general transform C to a curve which is simultaneously plane and nonsingular (Exercise 6.10). To generalize the differential results of this section, we will use a kind of compromise. Namely, one can transform an arbitrary $C \subseteq \mathbb{P}^n(\mathbb{C})$ to a plane curve having singularities of a very simple nature—the ordinary singularities introduced in Exercise IV, 7.5 (c). (The part of a curve near an ordinary singularity is essentially just a finite union of transversally-intersecting smooth analytic arcs.) The "desingularization theorem," which ensures that we can do this, is basic in the theory of algebraic curves. Since this theorem is proved in most books devoted to curves, we only state it here. See, for instance, [Fulton, Chapter 7], or [Walker, Chapter III, Section 7].

Theorem 6.19 (Desingularization theorem for algebraic curves). *Let C be any irreducible curve in $\mathbb{P}^n(\mathbb{C})$. There is a curve $C' \subseteq \mathbb{P}^2(\mathbb{C})$ which is birationally equivalent to C, and which is either nonsingular or which has only ordinary singularities.*

Exercises

In Exercises 6.1–6.6, C denotes a nonsingular curve in $\mathbb{P}^2(\mathbb{C})$ of genus g.

6.1 Let P and Q be two distinct points on C; suppose a function $f \in K_C$ has divisor $1P + (-1)Q$. Show that as P ranges over points of $C, f(P)$ attains each point of $\mathbb{C} \cup \{\infty\}$ exactly once.

6.2 Let C be of degree 1 or 2 (that is, $g = 0$), and let $f \in K_C \backslash \mathbb{C}$ be any nonconstant element of C's function field. Prove that $\deg(\operatorname{div}(df)) = -2$.

6.3 Prove (6.5) in the text.

6.4 Let C be of genus zero, let D be any divisor on C, Z any canonical divisor on C and $L(D)$ the vector space consisting of 0 and all $f \in K_C$ for which $-D \leqslant \operatorname{div}(f)$. Prove that

$$\dim L(D) = \deg D + 1 + \dim L(Z - D)$$

by looking at the two cases when $\deg D \geqslant -1$, and $\deg D \leqslant -2$. (This is the Riemann–Roch theorem for nonsingular curves in $\mathbb{P}^2(\mathbb{C})$ of genus 0. We assume this result in the next section.)

6.5 Show that every divisor of degree $2g - 2$ on C is canonical iff $g = 0$.

6.6 Show that for any divisor D on C, $\dim L(D) > 0$ iff D is linearly equivalent to a positive divisor.

6.7 Let P be a point of a curve $C \subseteq \mathbb{P}^n(\mathbb{C})$. Prove that if C is nonsingular at P, then there is only one branch $B_P\tilde{}$ of C centered at P. (Thus if $C \subseteq \mathbb{P}^n(\mathbb{C})$ is nonsingular, then there is a 1 : 1-onto correspondence between points of the curve and branches of the curve. Therefore, if C is in addition irreducible, one may consider that elements of C's function field may be evaluated at points of C rather than at branches of C.)

Is the converse true? That is, if there is only one branch $B_P\tilde{}$ of C centered at P, must P be nonsingular at P?

6.8 Show that for any irreducible curve C in $\mathbb{P}^n(\mathbb{C})$ or in $\mathbb{C}^n$, there is a curve C' in $\mathbb{P}^2(\mathbb{C})$ (and also one in $\mathbb{C}^2$) birationally equivalent to C.

6.9 Show that any irreducible curve in $\mathbb{P}^n(\mathbb{C})$ or in $\mathbb{C}^n$ is birationally equivalent to a nonsingular curve in some $\mathbb{P}^m(\mathbb{C})$. [Hint: Use Exercise 5.3 and Lemma III, 6.2.]

6.10 Find a nonsingular irreducible curve in $\mathbb{P}^n(\mathbb{C})$ which is not birationally equivalent to any nonsingular curve in $\mathbb{P}^2(\mathbb{C})$. [Hint: Use Exercise IV, 7.8.]

6.11 In this exercise, we generalize some of the concepts of this section.

Let $C \subseteq \mathbb{P}^n(\mathbb{C})$ be any irreducible curve with function field K_C. A **place** of K_C, or of C, is any discrete rank one valuation subring of K_C. (Note that a discrete rank one valuation ring of K_C determines a "place" in the sense indicated before Theorem 2.33.) On C itself, each place may be identified with a branch of C; however, a place is more "birational" in character, in that a discrete rank one valuation subring of K_C determines a branch on every projective curve having K_C as function field. Since a place represents a generalization of a point, we denote places on C by suggestive letters such as Q and Q_i.

A **divisor** D on C is any finite formal sum $\sum n_i Q_i$ ($n_i \in \mathbb{Z}$) of places of C; the **degree** of D, written $\deg D$, is the sum $\sum n_i$ of all the coefficients of $\sum n_i Q_i$. Partial order on the set of all divisors on C is defined in the expected way: $\sum n_i Q_i \leqslant \sum m_i Q_i$ iff $n_i \leqslant m_i$ for each i. Let Q be a place of K_C, and let $\mathfrak{m}$ be the maximal ideal of Q. For any $g \in K_C$, let $\operatorname{ord}_Q(g)$ be the order of g relative to the valuation ring Q. (See Remark 2.17.) With these definitions, definitions of $\operatorname{div}(g)$, $\operatorname{div}_0(g)$, $\operatorname{div}_\infty(g)$, $L(D)$, and $\dim L(D)$ given in the text may now all be extended in the obvious way to C.

(a) Suppose that $C \subseteq \mathbb{P}^2(\mathbb{C})$. Let Q be any place of C, and let F be a homogeneous polynomial in $\mathbb{C}[X, Y, Z]$. Define $\operatorname{ord}_Q(F)$ to be $\operatorname{ord}_Q(f)$, where f is the dehomogenization of F at any hyperplane of $\mathbb{P}^2(\mathbb{C})$ not containing Q's center. Show that this integer is independent of the choice of such a hyperplane.

(b) If an irreducible curve $C' \subseteq \mathbb{P}^2(\mathbb{C})$ properly intersects C, and if C' is defined by an irreducible homogeneous polynomial $F \in \mathbb{C}[X, Y, Z]$, then C' defines on C the divisor $\sum_Q \operatorname{ord}_Q(F)$, where Q ranges over all the places of C. (Since $\operatorname{ord}_Q(F) = 0$ for all but finitely many places Q on C, $\sum_Q \operatorname{ord}_Q(F)Q$ is actually a divisor.) More generally, let $F \in \mathbb{C}[X, Y, Z]$ be any homogeneous polynomial, and suppose that $F = F_1^{n_1} \cdot \ldots \cdot F_r^{n_r}$, where each F_i is irreducible in $\mathbb{C}[X, Y, Z]$. If $\mathsf{V}(F)$ intersects C properly, then F defines on C the divisor $\operatorname{div}(F) = \sum_{i,Q} n_i \operatorname{ord}_Q(F_i)$ (Q ranging over the places of C). Show that when C is nonsingular, the divisor associated with C', and the divisor $\operatorname{div}(F)$, each reduce to the ones defined just after Definition 6.6.

6.12 A property of an irreducible variety which is invariant under an arbitrary birational equivalence is called a **birational property**, or a **birational invariant**. For instance, the genus of an irreducible curve C is a birational invariant, since for any curve C' birationally equivalent to C, the genus of C is equal to that of C'. (See Exercise IV, 7.8.) Similarly, the dimension of a variety is easily checked to be a birational invariant, but nonsingularity is not.

Show that the concepts and integers defined in Exercise 6.11 are all birational.

6.13 Let C be an irreducible curve in $\mathbb{P}^2(\mathbb{C})$, let F be any irreducible homogeneous polynomial in $\mathbb{C}[X, Y, Z]$, and let P be any point of C. Show that if C and $\mathsf{V}(F)$ intersect properly at P, then $i(C, \mathsf{V}(F); P) = \sum_Q \operatorname{ord}_Q(F)$, where Q ranges over those places of C which have center P.

6.14 Making use of the above exercises, generalize (6.1)*, (6.2)*, and (6.3)* to arbitrary irreducible curves in $\mathbb{P}^n(\mathbb{C})$.

In Exercises 6.15–6.17 we generalize some "differential" notions and results of this section to arbitrary irreducible curves in $\mathbb{P}^n(\mathbb{C})$.

6.15 (a) Show how to generalize Definition 6.15 to an arbitrary irreducible curve in $\mathbb{P}^n(\mathbb{C})$ by replacing the notion of point (and local ring at a point) by *place*. Verify that Lemma 6.16 holds in this more general setting.
(b) Extend Definition 6.17 to irreducible curves in $\mathbb{P}^n(\mathbb{C})$.

6.16 Theorem 6.18 can be generalized to a structure theorem for canonical divisors on any irreducible curve C in $\mathbb{P}^2(\mathbb{C})$ having at worst ordinary singularities. On such a curve C, define the divisor Δ to be $\sum_{Q_i} (r_i - 1)Q_i$, where Q_i ranges over all places of C; here r_i denotes the multiplicity of C at the center of Q_i. Then Theorem 6.18 generalizes to a curve C whose only singularities are ordinary, provided that we replace "$\operatorname{div}(F) + \operatorname{div}(f)$" in that theorem by "$\operatorname{div}(F) + \operatorname{div}(f) - \Delta$." (Note that if C is nonsingular, then Δ is the zero-divisor.) State and prove this generalization. [Hint: In computing $\operatorname{div}(dx)$ in $\mathsf{D}(C)$, prove that at a place Q_i having center $P_i \in \mathsf{D}(C)$, $\operatorname{ord}_{Q_i}(G_X(x, y, 1)) = r_i - 1$.]

6.17 Let C be an irreducible curve in $\mathbb{P}^2(\mathbb{C})$ which is either nonsingular or has only ordinary singularities. Let the genus of C be g, and let ω be any nonzero differential C. Show that $\deg(\operatorname{div}(\omega)) = 2g - 2$. (This generalizes (6.2)** in the text.)

7 The Riemann–Roch theorem

Throughout this section, C denotes a nonsingular curve in $\mathbb{P}^2(\mathbb{C})$. Then C is irreducible (Exercise 4.2 of Chapter IV); let it be defined by an irreducible form $F \subset \mathbb{C}[X, Y, Z]$. We observed in the preceding section that for a given divisor D on C, the zero function and the set of all functions $f \in K_C$ such that $-D \leqslant \operatorname{div}(f)$, form a $\mathbb{C}$-vector space, denoted by $L(D)$. By Lemma 7.1 below, we see that $\dim L(D) < \infty$. The Riemann–Roch theorem gives more precise information on how $\dim L(D)$ depends on D. Before stating it, we establish the preliminary Lemmas 7.1 and 7.2.

Lemma 7.1. *For each divisor D on C,* dim $L(D)$ *is finite.*

Proof. Since $\deg \operatorname{div}(f) = 0$, $-D \leqslant \operatorname{div}(f)$ implies that $\deg D \geqslant 0$. Hence if $\deg D \leqslant 0$, then $\dim L(D) = 0$. Now suppose $\deg D \geqslant 0$, and write $D = P_1 + \ldots + P_n - Q_1 - \ldots - Q_{n'}$, with $n \geqslant n'$ (some of the terms P_i and Q_j may be repeated). Clearly $\dim L(D) \leqslant \dim L(P_1 + \ldots + P_n)$, so we may assume $D = P_1 + \ldots + P_n$. Now $L(0) \subset L(P_1) \subset \ldots \subset L(D)$, so one can form successive vector space quotients. Thus

$$\dim L(D) = \dim\left(\frac{L(D)}{L(D - P_1)}\right) + \dim\left(\frac{L(D - P_1)}{L(D - P_1 - P_2)}\right) + \ldots \qquad (29)$$

$$+ \dim\left(\frac{L(P_n)}{L(0)}\right).$$

We complete the proof by showing that each dimension on the right-hand side is finite (in fact, 0 or 1). We prove this for the first quotient; the others are similar. Let u be a uniformizing variable in $\mathfrak{o}(P; C)$ $(P = P_1)$. Suppose that exactly m of $P_1, \ldots, P_n$ are equal to $P = P_1$. Now any $f \in L(D)$ has order $\geqslant -m$ at P, so $f \to (u^m f)(P)$ maps $L(D)$ into $\mathbb{C}$. This map is linear, and its kernel is easily seen to be $L(D - P)$. If $L(D) = L(D - P)$, then the dimension of the quotient is 0. If $L(D) \neq L(D - P)$, then the quotient is isomorphic to $\mathbb{C}$, and its dimension is 1. □

We next show that all divisors in a fixed linear equivalence class have the same degree and determine vector spaces of the same dimension; we can thus work with divisors from a given class which best suit our needs.

Lemma 7.2. $D_1 \simeq D_2$ implies $\dim L(D_1) = \dim L(D_2)$.

Proof. If $D_1 \simeq D_2$, then $D_1 - D_2 = \operatorname{div}(f)$ for some $f \in K_C$, so

$$\deg D_1 - \deg D_2 = \deg \operatorname{div}(f) = 0. \qquad \square$$

Lemma 7.3. $D_1 \simeq D_2$ implies $\dim L(D_1) = \dim L(D_2)$.

Proof. For some $f \in K_C \backslash \{0\}$, $D_1 = D_2 + \operatorname{div}(f)$. Since f is not the zero function, $g \to fg$ is 1:1 and linear from $L(D_1)$ onto $L(D_2)$. Hence $\dim L(D_1) = \dim L(D_2)$. □

Recall that all canonical divisors are linearly equivalent. Also, for divisors D_i and E_i on C $(i = 1, 2)$ we have that $D_1 \simeq D_2$ and $E_1 \simeq E_2$ imply $D_1 + E_1 \simeq D_2 + E_2$. We now state

Theorem 7.4 (Riemann–Roch theorem). *Let D be any divisor and let Z be any canonical divisor on a nonsingular curve $C \subset \mathbb{P}^2(\mathbb{C})$ of genus g. Then*

$$\dim L(D) = \deg D + 1 - g + \dim L(Z - D). \qquad (30)$$

Instead of proving this result right away we first look at a few of its basic features, some of which we establish directly, others following from assuming the theorem itself. These observations will add to our understanding and intuition about the theorem, and some of the facts we prove will form parts of its proof.

First, for any increasing sequence of divisors $\{D\}$ on C, one can consider the graph in $\mathbb{Z} \times \mathbb{Z}$ of deg D versus dim $L(D)$. All these graphs will have certain features in common; for a fixed g, differences in various graphs are due entirely to the term dim $L(Z - D)$.

To begin, note that dim $L(D) = 0$ for all D of negative degree. (If there were a nonzero $f \in L(D)$, then $-D \leqslant \operatorname{div}(f)$. Thus since deg $D < 0$, we would have deg $\operatorname{div}(f) > 0$, in contradiction to (6.2)*.) Thus the graph lies on the "deg D axis" for all negative deg D. This also shows that if deg $D >$ deg Z $(=2g - 2)$, then $\dim(Z - D) = 0$, because $\deg(Z - D) < 0$. Though Z is determined only up to linear equivalence, deg Z is well defined by Lemma 7.2. We thus see that for deg $D \geqslant 2g - 1$ the graph lies on a line of "slope 1." Now assuming the Riemann–Roch theorem, we can see that a rough sketch looks like Figure 2.

When $\dim L(Z - D) = 0$ (that is, when $\dim L(D) = \deg D + 1 - g$), points of the graph are on the lower line of slope 1; the equation of this line is thus $Y = X + 1 - g$. Note that the horizontal and vertical distances between the two lines of slope 1 are both equal to the genus g.

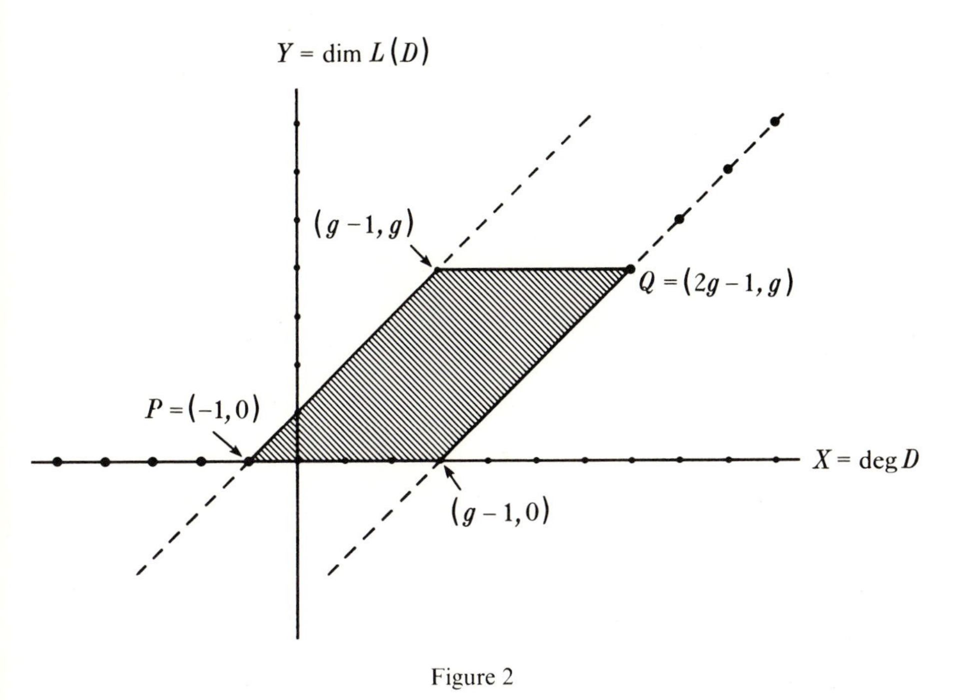

Figure 2

If we start with a divisor of degree -1 and add $2g$ points with multiplicity one at a time, we bridge the gap from the point $(-1, 0)$ on the upper line to the point $(2g - 1, g)$ on the lower line. The way in which this happens depends on the particular points chosen, in great contrast to the points of the graph outside the shaded region, which depend only on C and $\deg D$. For instance if C's genus is $\geqslant 1$, then by (30) $\dim L(0) = 1$ (0 = zero-divisor), but $\dim L(D) = 0$ if $D = 1P_1 - 1P_2$ ($P_1 \neq P_2$, $P_i \in C$). Note that the graph in Figure 2 increases monotonically; in fact the proof of Lemma 7.1 shows that increasing $\deg D$ by 1 increases $\dim L(D)$ by 0 or 1. Hence the graph between $X = -1$ and $X = 2g - 1$ must lie between the lines $Y = 0$ and $Y = g$. It also shows that for $\deg D \geqslant -1$, the graph must lie on or below the upper line of slope 1. And since, for each D, $\dim L(Z - D) \geqslant 0$, the graph must lie on or above the lower line of slope 1. Thus the part between $X = -1$ and $X = 2g - 1$ must lie in the shaded region of Figure 2.

The shape of the parallelogram in Figure 2 implies that for a given strictly increasing sequence $D_{-1}, \ldots, D_{2g-1}$ of $2g + 1$ divisors, starting from one of degree -1 and ending in one of degree $2g - 1$, there are g of these divisors D_i such that $\dim L(D_i) = \dim(L(D_{i-1}))$. In the special case when we simply increase the multiplicity of a *fixed* point P to $(2g - 1)P$, the degrees of these g D_i are then called *Weierstrass gaps*. It turns out that for all but finitely many points of C, the Weierstrass gaps are $1, 2, \ldots, g$. The finitely many points of C where this does not happen are called *Weierstrass points*.

We have just seen some aspects of the Riemann–Roch theorem from the standpoint of graphs. Next, without assuming the Riemann–Roch theorem, we answer some questions along this line: Given $\deg D$, what possible values can $\dim L(D)$ have?

We have seen that $\dim L(D) = 0$ whenever $\deg D < 0$ (from (6.2)*), and that $\dim L(D)$ increases by 0 or 1 as we add one point to D. Hence since $\dim L(D) = 0$ when $\deg D = -1$, for $\deg D \geqslant 0$ all possible values of $\dim L(D)$ must lie on or below the upper line of slope 1 in Figure 2—that is, for $\deg D \geqslant 0$ we have the relation $\dim L(D) \leqslant \deg D + 1$.

Let us now establish the existence of the lower line of slope 1 in Figure 2. We begin by showing there is *some* line of slope 1 serving as lower bound to the graph. That is, we show

Lemma 7.5. *For a given C, there exists a constant c so that*

$$\dim L(D) \geqslant \deg D - c$$

for all divisors D on C.

We show in Theorem 7.12 that the smallest c is actually $g - 1$. The inequality with $c = g - 1$ was first established by B. Riemann, and is known as the *Riemann inequality*. The sharpening of this result to (30) was completed by his student G. Roch.

PROOF OF LEMMA 7.5. The strategy of the proof is this. We first establish

> **(7.6)** Let f be any nonconstant function in K_C, and D any divisor on C. Then there is a divisor $D' \simeq D$ such that for some $n > 0$, $D' \leqslant n\,\mathrm{div}_0(f)$.

Using (7.6), we prove

> **(7.7)** For any fixed $f \in K_C \backslash \mathbb{C}$, there is a constant c' so that for all $n > 0$,
>
> $$\dim L(n(\mathrm{div}_0(f))) \geqslant \deg(n\,\mathrm{div}_0(f)) - c'.$$

We then show that (7.6) and (7.7) imply Lemma 7.5. Actually, this implication is easy to establish—let D be any divisor on C, let D' be as in (7.6), and let f be a fixed element of $K_C \backslash \mathbb{C}$. Choose n large and remove, one at a time, points with multiplicity from $n(\mathrm{div}_0(f))$ to arrive at D'; at each step the degree must decrease by one, while the dimension decreases by only 0 or 1. In this manner the inequality in (7.7) becomes the inequality in Lemma 7.5.

To establish Lemma 7.5 it therefore remains to prove (7.6) and (7.7).

PROOF OF (7.6). First, by adding an appropriate principal divisor to D if necessary, we may assume that no point of D is a pole of f. (This follows from the easily-established fact that for any $Q \in |D| \cap |\mathrm{div}_\infty(f)|$, there is an element in K_C having a zero at Q and no poles on $|D| \backslash \{Q\}$.) Next, we construct a function $g \in K_C \backslash \mathbb{C}$ such that for some $n > 0$ we have $D + \mathrm{div}(g) \leqslant n(\mathrm{div}_0(f))$. Since $\mathrm{div}_0(f) \geqslant 0$, surely one can find an n working for all points of D which either have negative coefficient, or which are points of $\mathrm{div}_0(f)$. It is for the remaining points that one needs the compensating term $\mathrm{div}(g)$. Thus let

$$g = \prod_{P \in S} \left(\frac{f}{f - f(P)} \right)^{m_i},$$

where P runs over the set S of those points having positive coefficient m_i in D but which are not in $\mathrm{div}_0(f)$. Then any point of S having positive coefficient has nonpositive coefficient in $D + \mathrm{div}(g)$. Of course, $\mathrm{div}(g)$ will in general introduce new points with positive coefficients, but these zeros of g are, of course, points of $\mathrm{div}_0(f)$. We can now choose n large enough so $D + \mathrm{div}(g) \leqslant n(\mathrm{div}_0(f))$. □

PROOF OF (7.7). We want a constant c so that for every $n > 0$, we can find at least $\deg(n(div_0(f))) - c$ elements in $L(n(\mathrm{div}_0(f)))$ linearly independent over $\mathbb{C}$. Since for each $n > 0$, increasing n by 1 increases $\deg(n(\mathrm{div}_0(f)))$ by the finite number $\deg(\mathrm{div}_0(f))$, to prove (7.7) it suffices to find an integer constant $N \geqslant 0$ such that for all $n > 0$,

$$\dim L((n + N)\mathrm{div}_0(f)) \geqslant \deg(n(\mathrm{div}_0(f))). \tag{31}$$

We begin by proving a stronger result in the case $n = 1$, namely (7.8). First, note that for any $f \in K_C \backslash \mathbb{C}$, K_C is a vector space over $\mathbb{C}(f)$. (However, no $L((n + N)\text{div}_0(f))$ is a vector space over $\mathbb{C}(f)$.) To prove (31), we first prove:

> **(7.8)** For any fixed $f \in K_C \backslash \mathbb{C}$, there exists an integer $N \geqslant 0$ so that there are $\deg \text{div}_0(f)$ $\mathbb{C}(f)$-linearly independent elements y_i in K_C which are in the subset $L((1 + N)\text{div}_0(f)) \subseteq K_C$.

And to prove (7.8), we in turn first prove this weaker form of (7.8):

> **(7.9)** There are $\deg \text{div}_0(f)$ elements z_i in K_C linearly independent over $\mathbb{C}(f)$.

For the proof of (7.9), first note that if $D = \sum_i m_i P_i$, then

$$\{g \in K_C \,|\, \text{ord}_P(g) \geqslant -m_i \text{ for all } P_i \in |D| \cap |\text{div}_0(f)|\}$$

in a $\mathbb{C}$-vector space W. (Note that conditions on order are imposed on $g \in W$ at only the finitely many points of $|D| \cap |\text{div}_0(f)|$, and W is infinite dimensional; in contrast, with the finite-dimensional vector space $L(D)$, conditions on order are imposed at *every* point of C.) Then, in obvious notation, if $D_1 \leqslant D_2$ are divisors on C, we have $W_1 \subset W_2$. We now show that if $D_1 \leqslant D_2$, then $\dim(W_2/W_1) = \deg_f D_2 - \deg_f D_1$, where for any divisor D, $\deg_f D$ denotes the sum of the coefficients of D over $|D| \cap |\text{div}_0(f)|$.

The argument is much like that of Lemma 7.1's proof. First, we may split up W_2/W_1 as in (29); we may clearly assume that $D_1 = D_2 - P$ for some $P \in |\text{div}_0(f)|$. With this assumption, we now show that $\dim(W_2/W_1) = 1$.

By the proof of Lemma 7.1 (with D_2 in place of D), $\dim(W_2/W_1)$ is 0 or 1; now if P has coefficient m in D_2, then the order at P of any function in W_1 is $\geqslant -m + 1$. So $W_2 \supsetneqq W_1$ (that is, $\dim(W_2/W_1) = 1$) provided there is an $h \in W_2$ of order $-m$ at P, and order $\geqslant -m_i$ at the other points P_i of $D_2 = \sum_i m_i P_i$ which are in $|\text{div}_0(f)|$. Now using Theorem 2.35, one can easily choose a linear form $aX + bY + c \in \mathbb{C}[X, Y]$ whose image $ax + by + c \in R_C = \mathbb{C}[x, y]$ is a uniformizing parameter at any given P_i (that is, $\text{ord}_{P_i}(ax + by + c) = 1$), and has order 0 at the other points of D_2. By multiplying together powers from $\mathbb{Z}$ of such parameters, one then gets an h meeting our requirements.

Now in particular, if D_1 is $-\text{div}_0(f)$ and D_2 is the zero-divisor, then $\dim(W_2/W_1) = \deg \text{div}_0(f)$. For our z_i we choose $\deg \text{div}_0(f)$ elements of W_2 whose images in W_2/W_1 form a $\mathbb{C}$-basis of W_2/W_1. These z_i are linearly independent over $\mathbb{C}(f)$ as follows: Consider $\sum_i f_i z_i = 0$, where $f_i \in \mathbb{C}(f)$; if not all $f_i = 0$, then by multiplying by a common denominator of the f_i (the coefficients of z_i then becoming polynomials in f), we deduce that $\sum_i (c_i + fg_i)z_i = 0$ for some $g_i \in \mathbb{C}[f]$ and $c_i \in \mathbb{C}$, where not all $c_i = 0$. Therefore $\sum_i c_i z_i = -\sum_i fg_i z_i$. Since $\sum_i fg_i z_i \in W_1$, its image in W_2/W_1 is

0; thus the image of $\sum_i c_i z_i$ is 0, which means that the images of the z_i in W_2/W_1 are not linearly independent over $\mathbb{C}$, a contradiction. Thus (7.9) is proved. □

The above $\deg \operatorname{div}_0(f)$ elements z_i may not be elements of any $L((1 + N)\operatorname{div}_0(f))$ (since nonconstant elements z_i have poles off $\mathsf{V}(f)$ while, of course, elements of $L((1 + N)\operatorname{div}_0(f))$ do not); but we can easily modify them so that they are, by scalar multiplying them by an appropriate nonzero element of $\mathbb{C}(f)$. First, since f is nonconstant, it is transcendental over $\mathbb{C}$, so each z_i is algebraic over $\mathbb{C}(f)$. Since $\mathbb{C}(f) = \mathbb{C}(1/f)$, each z_i satisfies

$$a_{i0} z_i^{m_i} + a_{i1} z_i^{m_i - 1} + \dots = 0 \qquad \left(a_{ij} \in \mathbb{C}\left[\frac{1}{f}\right] \text{ and } a_{i0} \neq 0\right).$$

Multiplying this equation by $a_{i0}{}^{m_i - 1}$ shows that $y_i = a_{i0} z_i$ is integral over $\mathbb{C}[1/f]$. That is,

$$y_i^{m_i} = b_{i1} y^{m_i - 1} + \dots \qquad \left(b_{ij} \in \mathbb{C}\left[\frac{1}{f}\right]\right). \tag{32}$$

Next, observe the following:

(7.10) $y_i \in L((1 + N)\operatorname{div}_0(f))$ for some $N \geqslant 0$ iff every pole of y_i is a zero of f.

Now (32) shows that any pole P of y_i must be a zero of f, for if P is not a zero of f, it is not a pole of $1/f$, hence not a pole of any b_{ij}. If P were a pole of y_i, then, from Definition 2.6, we see that the left-hand side of (32) would have order smaller than the right-hand side, a contradiction. Thus, using (7.10), we have (7.8).

The inequality in (31) follows easily: The elements

$$f^{-j} y_i \qquad (i = 1, \dots, \deg \operatorname{div}_0(f); \qquad j = 0, \dots, n - 1)$$

are linearly independent over $\mathbb{C}$ (since the y_i are linearly independent over $\mathbb{C}(f)$ and the f^{-j}, being transcendental over $\mathbb{C}$, are linearly independent over $\mathbb{C}$). There are $n \deg \operatorname{div}_0(f) = \deg(n(\operatorname{div}_0(f)))$ of these elements, and they all belong to $L((n + N)\operatorname{div}_0(f))$. Thus (31) is proved, hence (7.7), and therefore also Lemma 7.5. □

Lemma 7.5 shows there is a line of slope one serving as lower bound for $\dim L(D)$. Since $\dim L(D)$ increases by 0 or 1 as $\deg D$ increases by 1, this result shows the increase must always be 1 for $\deg D$ sufficiently large. Thus not only is there a line of slope 1 serving as a "greatest lower bound," but this line coincides with $\dim L(D)$ for all sufficiently large $\deg D$. What is this line? We are asking for the smallest c working in Lemma 7.5. We answer this in Theorem 7.12 by finding certain divisors D of arbitrarily large degree for which we can compute $\dim L(D)$. To make this computation we shall use a famous result due to Max Noether, namely Theorem 7.11. This theorem is a polynomial translation of this geometric fact: Let curves with multiplicity $\mathsf{C}(G)$ and $\mathsf{C}(H)$ in $\mathbb{P}^2(\mathbb{C})$ define divisors $\operatorname{div} G = \mathsf{C}(G) \cdot C$ and $\operatorname{div} H = \mathsf{C}(H) \cdot C$ on C $(= \mathsf{C}(F))$, and suppose $\operatorname{div}(G) \leqslant \operatorname{div}(H)$. Then it happens that there is a curve with multiplicity $\mathsf{C}(G^*)$ making up the difference—that is, so that $\operatorname{div}(G) + \operatorname{div}(G^*) = \operatorname{div}(H)$ (or what is the same, so that $\operatorname{div}(GG^*) = \operatorname{div}(H)$). Of course adding any (homogeneous) multiple F^*F of F to GG^* still gives us the same divisor on C (that is, $\operatorname{div} GG^* = \operatorname{div}(GG^* + FF^*)$, and in translating this to polynomial form it is necessary to account for this. With the above notation, the result is

Theorem 7.11 (M. Noether). *Let G and H be forms in $\mathbb{C}[X, Y, Z]$ defining divisors $\operatorname{div}(G) \leqslant \operatorname{div}(H)$ on C. Then there are forms G^* and $F^* \in \mathbb{C}[X, Y, Z]$ such that $H = GG^* + FF^*$.*

Before proving Theorem 7.11, we note the following (see the observations made just after Definition 6.6). First, for an arbitrary form $G \in \mathbb{C}[X, Y, Z]$,

$$\operatorname{div}(G) = \mathsf{C}(G) \cdot C = \sum_{P \in C} (\operatorname{ord}_P G)P. \tag{33}$$

Also, on C we have

$$\operatorname{div} f = \operatorname{div}(F_1) - \operatorname{div}(F_2), \tag{34}$$

for appropriate forms F_1, F_2 in $\mathbb{C}[X, Y, Z]$.

PROOF OF THEOREM 7.11. First note that Theorem 7.11 is independent of the choice of linear coordinates X, Y, Z; thus in the proof we may choose coordinates which best suit our needs. We begin by proving a special case which naturally arises in dealing with resultants. It follows from Theorem 5.16 that with respect to appropriate coordinates X, Y, Z, if $R = R(X, Y)$ is the resultant with respect to Z of F and another form $G \in \mathbb{C}[X, Y, Z]$, then on C, $\operatorname{div}(G) = \operatorname{div}(R)$. Then there are forms G_R and $F_R \in \mathbb{C}[X, Y, Z]$ such that

$$R = GG_R + FF_R. \tag{35}$$

We shall use this special case in proving the full theorem.

To prove (35), let

$$F = a_0 Z^m + a_1 Z^{m-1} + \ldots + a_m,$$
$$G = b_0 Z^n + b_1 Z^{n-1} + \ldots + b_n,$$

where a_i and b_i are 0 or forms in X and Y of degree i (but a_0 and b_0 are nonzero). Let $A_1, \ldots, A_n, B_1, \ldots, B_m$ denote the cofactors of the last column of R's representation in (14) of Chapter II. We may then take

$$F_R = A_1 Z^{n-1} + \ldots + A_n,$$
$$G_R = B_1 Z^{m-1} + \ldots + B_m.$$

This is because the Z^0-term of $GG_R + FF_R$ is $A_n a_m + B_m b_n$, which is the expansion down the last column of the determinant in (14) of Chapter II—that is, $A_n a_m + B_m b_n = R$. The coefficient of Z^1 in $GG_R + FF_R$ is $A_n a_{m-1} + A_{n-1} a_m + B_m b_{n-1} + B_{m-1} b_n$, and this is zero since it is the expansion down the last column of an array like the one in (14) of Chapter II, but with the last column changed to agree with the $(m + n - 1)^{\text{st}}$ column. Similarly, the coefficient of Z^i is the expansion down the last column of an array like the one in (14) of Chapter II, but with the last column changed to agree with the $(m + n - i)^{\text{th}}$ column; these coefficients are all zero for $i \geqslant 1$, so $R = GG_R + FF_R$. Thus (35) is proved.

Now let us prove the full theorem. Our basic strategy is this: Starting with $R = GG_R + FF_R$, we multiply by H, getting $RH = GG_R H + FF_R H$. We then show that R can be cancelled from both sides without, however, cancelling any factors of F or G. Note that if we knew R divided, say, $G_R H$, it would follow that R would divide $F_R H$, since R is not a factor of F. (If R were a factor of F, then R would vanish on infinitely many points of $\mathsf{V}(F)$. But R vanishes on $\mathsf{V}(F)$ only at the finitely many (projective) points where G does too.) Now in general R does not divide $G_R H$, but we show that one can take away from $G_R H$ a multiple $\bar{F}F$ of F and give it to $F_R H$ in such a way that the difference $G_R H - \bar{F}F$ is divisible by R. Then so is $F_R H + \bar{F}F$, and one therefore has

$$H = G \cdot \frac{F_R H + \bar{F}F}{R} + F \cdot \frac{G_R H - \bar{F}F}{R}.$$

With respect to appropriate coordinates, a multiple of F which works is given by the division algorithm:

$$G_R H = \bar{F}F + S,$$

for some form $\bar{F}$, and for some S which is either 0 or a form of Z-degree less than m (= degree of F with respect to Z). (See Exercise 7.1(b).) In view of these comments, we shall have proved Theorem 7.11 once we show R divides $S = G_R H - \bar{F}F$.

For this, let $aX - bY$ be an arbitrary nonconstant factor of R, and suppose it occurs with multiplicity s. It suffices to show that each such $(aX - bY)^s$ divides S. For this, we may choose coordinates X, Y, Z so that no point of $C \cap \mathsf{V}(G)$ is on the line $\mathsf{V}(X)$ (which may be thought of as the line at infinity), and so that each line of the form $\mathsf{V}(cX + dY) \subset \mathbb{P}^2(\mathbb{C})$ intersects $C \cap \mathsf{V}(G)$ in at most one point. (For $d \neq 0$, these last lines may be thought of as the completions of those lines in $\mathbb{C}_{YZ}$ which are parallel to $\mathbb{C}_Z$.) We can further stipulate that whenever such a line $\mathsf{V}(cX + dY)$ does intersect $C \cap \mathsf{V}(G)$, it intersects C in $m = \deg F$ distinct points. (See Figure 3.) This is easily seen to imply that the coefficient of Z^m in F is in $\mathbb{C}\backslash\{0\}$; hence we can use the division algorithm of Exercise 7.1(b). Now the line $\mathsf{V}(aX - bY)$ intersects $C \cap \mathsf{V}(G)$ in a single point P, and by Theorem 5.16, $s = \operatorname{ord}_P(R)$ is just $\operatorname{ord}_P(G)$.

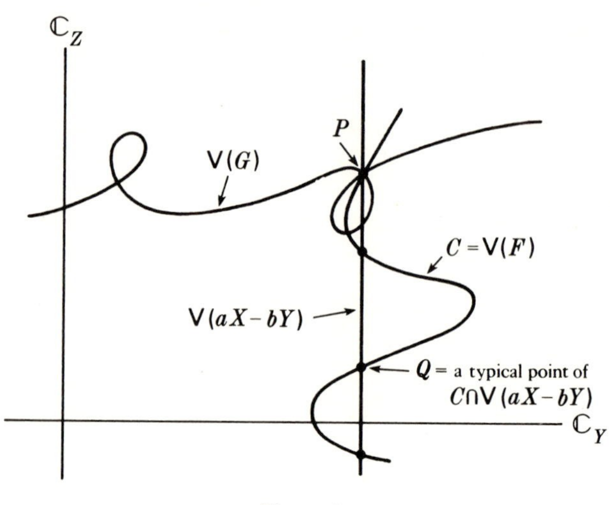

Figure 3

To show $(aX - bY)^s$ divides S, suppose it does not—that is, suppose $S = (aX - bY)^p \cdot T$, where $p < s$ and where $aX - bY$ is not a factor of T. We then show that at least $\deg F = m$ points of $\mathsf{V}(T)$ lie on $\mathsf{V}(aX - bY)$. This will give us a contradiction, for since $\deg F > \deg S \geqslant \deg T$, we have $\deg F > \deg T \cdot \deg(aX - bY) = \deg T$. Corollary 7.8 to Bézout's theorem then tells us that $\mathsf{V}(aX - bY)$ must be a component of $\mathsf{V}(T)$—that is, $aX - bY$ is a factor of T. These $\deg F$ points of $\mathsf{V}(T)$ will be the $\deg F$ points of $C \cap \mathsf{V}(aX - bY)$. The following argument shows that an arbitrary point Q of these $\deg F$ points is in $\mathsf{V}(T) \cap \mathsf{V}(aX - bY)$—that is, that $\operatorname{ord}_Q(T) \geqslant 1$.

First note that $\operatorname{ord}_Q(aX - bY) = 1$, since by our choice of coordinates, $\mathsf{V}(aX - bY)$ intersects C in $\deg F$ distinct points, each with multiplicity one. Then $\operatorname{ord}_Q(T) = \operatorname{ord}_Q(S) - p$, because $S = (aX - bY)^p \cdot T$, and $\operatorname{ord}_Q(aX - bY)^p = p$. Hence to show $\operatorname{ord}_Q(T) \geqslant 1$ we need to show $\operatorname{ord}_Q(S) \geqslant s$ (since we are assuming $p < s$).

For this, we have

$$\begin{aligned} G_R H = \bar{F}F + S \quad \text{implies} \quad & \operatorname{ord}_Q(G_R) + \operatorname{ord}_Q(H) \\ &= \min(\infty, \operatorname{ord}_Q(S)) = \operatorname{ord}_Q(S). \end{aligned}$$

But $\operatorname{ord}_Q(H) \geqslant \operatorname{ord}_Q(G)$, so $\operatorname{ord}_Q(S) \geqslant \operatorname{ord}_Q(G_R) + \operatorname{ord}_Q(G)$. On the other hand, $R = GG_R + FF_R$ implies $\operatorname{ord}_Q(R) = s = \min(\operatorname{ord}_Q(GG_R), \infty) = \operatorname{ord}_Q(G) + \operatorname{ord}_Q(G_R)$, so $s \leqslant \operatorname{ord}_Q(S)$, as required. Thus Theorem 7.11 is proved. □

Notice in the above proof that if $Q \in \mathsf{V}(G)$, then $s = \operatorname{ord}_Q(G)$, and $\operatorname{ord}_Q(G_R) = 0$; when $Q \notin \mathsf{V}(G)$ (representing the other $\deg F - 1$ cases), $s = \operatorname{ord}_Q(G_R)$, and $\operatorname{ord}_Q(G) = 0$.

Now we are ready to prove

Theorem 7.12 (Riemann's theorem). *Let $C \subset \mathbb{P}^2(\mathbb{C})$ be a nonsingular curve of genus g. Then for all divisors D on C,*

$$\dim L(D) \geqslant \deg D - (g - 1);$$

$g - 1$ is the smallest constant c for which $\dim L(D) \geqslant \deg D - c$ *holds for all divisors D on C.*

PROOF. As noted earlier, we want to find divisors D of arbitrary large degree for which we can compute $L(D)$. We choose these as follows: Take any line in $\mathbb{P}^2(\mathbb{C})$ intersecting C in $n = \deg C$ distinct points $P_1, \ldots, P_n$; let D be the divisor $P_1 + \ldots + P_n$. Then we choose for our divisors of arbitrary large degree

$$lD = lP_1 + \ldots + lP_n \qquad (l \in \mathbb{Z}^+).$$

Such divisors do in fact give us Riemann's theorem for arbitrary divisors, since by (7.6), for any divisor E on C, there is a divisor E^* linearly equivalent to E, such that $lD \geqslant E^*$ for some l. Now one can go from lD to E^* by subtracting points with multiplicity one; if Riemann's theorem holds for lD, it surely holds for E^*, since in subtracting one such point, "deg" decreases by exactly one, while "dim" decreases by at most one, so the inequality in Riemann's theorem is preserved.

We can compute $\dim L(lD)$ for arbitrarily large l; we do this by looking at $L(lD)$ as the quotient of two vector spaces, each of whose dimensions is easy to compute.

First, let L be a linear form in $\mathbb{C}[X, Y, Z]$ defining our line which intersects C in $P_1, \ldots, P_n$. Then for any form $A \subset \mathbb{C}[X, Y, Z]$ of degree l, A/L^l (regarded as a function on C) is an element of $L(lD)$. But much more is true. *Every element of $L(lD)$ can be written as A/L^l, for some l-form A.* For if $f \in L(lD)$, then surely $f = F_1/F_2$ for some forms $F_1, F_2 \in \mathbb{C}[X, Y, Z]$ of equal degree. Thus $F_2 = F_1/f$, so $\operatorname{div}(F_2) = \operatorname{div}(F_1) - \operatorname{div}(f)$. But $f \in L(lD)$, i.e., $-\operatorname{div}(f) \leqslant \operatorname{div}(L^l)$; therefore $\operatorname{div}(F_2) \leqslant \operatorname{div}(L^l F_1)$. We can now apply Theorem 7.11 to conclude that there are forms F^* and F_2^* such that

$$L^l F_1 = F_2 F_2^* + F F^* \qquad (C = \mathsf{V}(F)).$$

Since F is zero at each point of C, we have on C

$$L^l F_1 = F_2 F_2^*;$$

hence $f = F_1/F_2 = F_2^*/L^l = A/L^l$, as asserted.

The above characterization of the elements of $L(lD)$ allows us to express the vector space $L(lD)$ as the quotient of two vector spaces of known dimensions. First, the set of all the above vectors A/L^l, as functions on $\mathbb{P}^2(\mathbb{C})$, forms a $\mathbb{C}$-vector space, and it is isomorphic, under $A \leftrightarrow A/L^l$, to the vector space V_1 of all forms in X, Y, Z of degree l. Then $L(lD)$ is isomorphic to the quotient of V_1 by the vector space V_2 of those l-forms identically zero on C. The vector space V_2 is obviously $V_2 = \{F \cdot H \mid H \text{ an } (l-n)\text{-form in } \mathbb{C}[X, Y, Z]\}$ (where $n \geqslant 1$). Thus V_2 is isomorphic, under $H \leftrightarrow F \cdot H$, to the space of $(l-n)$-forms. Since $V = V_1/V_2$, $\dim V = \dim V_1 - \dim V_2$. To evaluate V_1 and V_2, we shall find the dimension of the vector space of m-forms in $\mathbb{C}[X, Y, Z]$; this dimension is evidently the same as the number of monomials of degree m in the indeterminates X, Y, Z. The number of such monomials in which X appears with exponent i ($0 \leqslant i \leqslant m$) is obviously $m - i + 1$. Hence the total number of monomials is

$$(m+1) + (m) + \ldots + 1 = \frac{(m+2)(m+1)}{2}.$$

Thus

$$\dim V = \dim V_1 - \dim V_2 = \frac{(l+2)(l+1)}{2} - \frac{(l-n+2)(l-n+1)}{2}.$$

This simplifies to

$$ln - \frac{(n-1)(n-2)}{2} + 1.$$

The first term ln is $\deg(lD)$; the next term is g (by the genus formula), so we have

$$\dim L(lD) = \deg(lD) - (g-1) \quad \text{for all } l \geqslant 1;$$

hence in view of our earlier discussion, we have, for all divisors D on C,

$$\dim L(D) \geqslant \deg D - (g-1),$$

and $g - 1$ is the smallest constant c for which $\dim L(D) \geqslant \deg D - c$ holds for all divisors D on C. □

At this point we are not far from a proof of the Riemann–Roch theorem. Riemann's theorem is the Riemann–Roch theorem for $\deg D$ sufficiently large and positive, since $\dim L(Z - D) = 0$ for $\deg D$ sufficiently large and positive. And since $\deg Z = 2g - 2$, we have established the Riemann–Roch theorem for $\deg D$ large and negative, too.

Now as $\deg(D)$ changes by ± 1, $\dim L(D)$ changes by 0 or ± 1, and $\dim L(Z - D)$ changes by 0 or ∓ 1. Since we can get from any D to any other D' by successive degree-one changes it suffices to show that the equality $\dim L(D) = \deg D + 1 - g + \dim L(Z - D)$ (which we know holds for D of large degree) continues to hold under an arbitrary degree-one change of D. This amounts to showing that exactly one of $\dim L(D)$, $\dim L(Z - D)$ changes—that is, for any divisor D and any point $P \in C$,

$$\dim L(D) = \dim L(D + P) \quad \text{iff} \quad \dim L(Z - D) \neq \dim L(Z - (D + P)).$$

Actually, proving just the following part of this will allow us to complete the proof of the Riemann–Roch theorem:

Lemma 7.13. *Suppose* $\dim L(D) > 0$. *Then for any* $P \in C$, $\dim L(Z - D) \neq \dim L(Z - (D + P))$ implies $\dim L(D) = \dim L(D + P)$.

PROOF. Since the Riemann–Roch theorem holds for curves of degree one and two (Exercise 6.4), the lemma likewise holds for $n = 1$ or 2. We therefore assume $n \geqslant 3$. Now assume that for some $P \in C$, $\dim L(Z - D) \neq \dim L(Z - (D + P))$—that is, assume that there is an $f \in L(Z - D)$ not in $L(Z - (D + P))$. We may translate this hypothesis into a condition on divisors induced by curves as follows. First, we may assume $D \geqslant 0$ (Exercise 6.6). Thus each pole of f is a point of Z. Also, Z may be assumed to be induced by a curve with multiplicity $\mathsf{C}(F_2)$ of degree $n - 3$ (from Theorem 6.18). Hence we may write $f = F_1/F_2$, for $(n - 3)$-forms F_1, $F_2 \in \mathbb{C}[X, Y, Z]$. Thus the hypothesis implies that $\operatorname{div}(f) \geqslant D - Z$, $\operatorname{div}(f) \ngeqslant D + P - Z$, and $\operatorname{div}(f) = \operatorname{div} F_1 - Z$. From this we see that $\operatorname{div}(F_1) - Z \geqslant D - Z$, and therefore $\operatorname{div}(F_1) \geqslant D$; likewise we have $\operatorname{div}(F_1) \ngeqslant D + P$. Thus $D + E = \operatorname{div}(F_1)$, for some $E \geqslant 0$, where $P \notin E$. We may similarly translate the conclusion: Since always $L(D) \subset L(D + P)$, the conclusion holds iff $L(D + P) \subset L(D)$, which means

$$g \in L(D + P) \quad \text{implies} \quad g \in L(D).$$

That is, $\operatorname{div}(g) + D + P \geqslant 0$ implies $\operatorname{div}(g) + D \geqslant 0$.

Now $\operatorname{div}(g) + D \geqslant 0$ can be translated into a condition on curves with multiplicity, using the fact that $\operatorname{div}(g) + D \geqslant 0$ iff $\operatorname{div}(g) + D + P \geqslant P$; assuming $\operatorname{div}(g) + D + P \geqslant 0$, to prove $\operatorname{div}(g) + D + P \geqslant P$, it suffices to find some $\mathsf{C}(H)$ containing P, and to find a divisor $E' \geqslant 0$ ($E' \leqslant H$) not containing P, such that

$$\operatorname{div}(g) + D + P = \operatorname{div}(H) - E'. \tag{36}$$

Using F_1, we can easily obtain (36) with equality replaced by linear equivalence, for we can add $1P$ to each side of $\operatorname{div}(F_1) = D + E$ by multiplying F_1 by a linear form L defining a line through P intersecting C in n distinct points. Thus $\operatorname{div}(F_1 L) = D + P + E'$, $P \notin E'$. (E' now includes E and the remaining $n - 1$ points.) Note that $P \in \mathsf{V}(F_1 L)$. To get equality in (36), write $g = G_1/G_2$, where G_1 and G_2 are forms. Then $\operatorname{div}(F_1 L G_1) = D + P + E' + \operatorname{div}(G_1) = E' + D + P + \operatorname{div}(g) + \operatorname{div}(G_2) \geqslant \operatorname{div}(G_2)$. (The "$\geqslant$" follows because $E' \geqslant 0$ and $D + P + \operatorname{div}(g) \geqslant 0$.) Then Theorem 7.11 tells us there are forms F^* and H such that $F_1 L G_1 = FF^* + HG_2$; thus $\operatorname{div}(F_1 L G_1) = D + P + E' + \operatorname{div}(G_1) = \operatorname{div}(H) + \operatorname{div}(G_2)$. Since $\operatorname{div}(g) = \operatorname{div}(G_1) - \operatorname{div}(G_2)$, we have $\operatorname{div}(g) + D + P = \operatorname{div}(H) - E'$, which is (36). Note that P is still in the new curve $\mathsf{V}(H)$, for $\deg \mathsf{V}(H) \leqslant \deg H = \deg F_1 L = n - 3 + 1 = n - 2$, and E' still contains the $n - 1$ points of intersection lying on $\mathsf{V}(L)$. Since $n - 1 > n - 2$, Bézout's theorem tells us $\mathsf{V}(L)$ must be a component of $\mathsf{V}(H)$; since $P \in L$, $P \in \mathsf{V}(H)$. □

We now prove the Riemann–Roch theorem. We divide the proof into two cases: (I), $\dim L(Z - D) = 0$, and (II) $\dim L(Z - D) > 0$; we use induction on dim $L(D)$ to prove Case I. We then use Case I to show that the set S of divisors in Case II for which the Riemann–Roch theorem is false, is empty. Lemma 7.13 will be used in a crucial way.

PROOF OF CASE I (dim $L(Z - D) = 0$). In accordance with the literature, we shall call a divisor D **nonspecial** if it satisfies dim $L(Z - D) = 0$. For any nonspecial divisor D it is easily shown that

$$\deg D \geqslant g. \tag{37}$$

Now (37) together with Theorem 7.12 shows that dim $L(D) \geqslant 1$. So the induction in Case I starts with dim $L(D) = 1$.

Thus let D be any nonspecial divisor such that dim $L(D) = 1$. We want the Riemann–Roch equation to read

$$1 = \deg D + 1 - g + 0,$$

that is,

$$\deg D = g.$$

The inequality $\deg D \geqslant g$ is (37); $\deg D \leqslant g$ is Riemann's theorem (Theorem 7.12) when dim $L(D) = 1$, namely

$$1 \geqslant \deg D + 1 - g,$$

that is,

$$\deg D \leqslant g.$$

Hence the first step, dim $L(D) = 1$, is established.

Now assume, as our induction hypothesis, that the Riemann–Roch theorem is true for all nonspecial divisors D for which dim $L(D) \leqslant n$; we want to prove the theorem in the case dim $L(D) = n + 1$. Let D be a typical such divisor. To complete the induction step it suffices to find a nonspecial divisor $D - P$ ($P \in C$) such that dim $L(D - P) = \dim L(D) - 1$. For if we had such a divisor, we could write (using our induction hypothesis):

$$\dim L(D - P) = \deg(D - P) + 1 - g,$$

which is just

$$\dim L(D) - 1 = \deg D - 1 + 1 - g,$$

and this is the Riemann–Roch theorem for D, as desired.

It is not hard to find such a P; choose for P any point which fails to be a zero of at least one function in $L(D)$. Now Lemma 7.13 applied to $D - P$ shows that $D - P$ is nonspecial (that is, dim $L(Z - (D - P)) = 0$), for certainly the hypothesis dim $L(D - P) > 0$ is satisfied (since dim $L(D) \geqslant 2$). But dim $L(D) \neq \dim L(D - P)$—that is, dim $L((D - P) + P) \neq \dim L(D - P)$; hence Lemma 7.13 tells us that

$$\dim L(Z - (D - P) - P) = \dim L(Z - (D - P)).$$

Since $\dim L(Z - (D - P) - P) = \dim L(Z - D) = 0$, we have

$$\dim (Z - (D - P)) = 0,$$

so $D - P$ is indeed nonspecial. Thus Case I is established.

PROOF OF CASE II ($\dim L(Z - D) > 0$). We call any such D **special**. Let S be the set of special divisors for which the Riemann–Roch theorem is false. We show $S = \varnothing$.

If $S \neq \varnothing$, then since $\dim L(Z - D) = 0$ for all divisors of large degree, there must be a *maximal* divisor D_M in S. Now either $\dim L(D_M) = 0$ or $\dim L(D_M) > 0$. Assume first that $\dim L(D_M) > 0$; then the hypothesis of Lemma 7.13 is satisfied for D_M. Let P be a point of C which fails to be a zero of at least one function of $L(Z - D_M)$. Then

$$\dim L(Z - D_M - P) = \dim L(Z - D_M) - 1,$$

so Lemma 7.13 tells us that $\dim L(D_M) = \dim L(D_M + P)$. But the Riemann–Roch theorem is true for $D_M + P$. (If $D_M + P$ is special, it is true since D_M is maximal in S; if $D_M + P$ is nonspecial, it is true by Case I.) Thus $\dim L(D_M + P) = \deg(D_M + P) + 1 - g + \dim L(Z - D_M - P)$. The above facts show this reduces to

$$\dim L(D_M) = \deg D_M + 1 + 1 - g + \dim L(Z - D_M) - 1,$$

which is the Riemann–Roch theorem for D_M. Hence if D_M exists, it must satisfy $\dim L(D_M) = 0$.

Therefore assume $\dim L(D_M) = 0$. In this case, set

$$Z - D_M = E. \tag{38}$$

From Case I we know:

(7.14) If $\dim L(Z - E) = 0$, then

$$\dim L(E) = \deg E + 1 - g + \dim L(Z - E).$$

Using (38), (7.14) becomes:

If $\dim L(D_M) = 0$, then

$$\dim L(Z - D_M) = \deg(Z - D_M) + 1 - g + \dim L(D_M).$$

Since $\deg Z = 2g - 2$ (by (6.2)**), this last reduces to

$$\dim L(Z - D_M) = -\deg D_M - 1 + g + \dim L(D_M)\text{—that is,}$$

$$\dim L(D_M) = \deg D_M + 1 - g + \dim L(Z - D_M).$$

This shows that $S = \phi$, so the Riemann–Roch theorem is true for special divisors, too. □

Exercises

7.1 (a) Prove the division algorithm: Let X be a single indeterminate, and let D be any integral domain. Then for any elements $f, g \in D[X]$ ($g \neq 0$), there are polynomials $q(X), r(X) \in D[X]$ (with $r = 0$ or $\deg r < \deg g$) such that $f(X) = g(X)q(X) + r(X)$.

(b) Prove the following homogeneous analogue of the above division algorithm: If F and G are homogeneous polynomials in $\mathbb{C}[X_1, \ldots, X_n]$ (therefore F, $G \neq 0$), and if the coefficient of the highest power of X_n in G is in $\mathbb{C}\backslash\{0\}$, then there exist polynomials Q and R in $\mathbb{C}[X_1, \ldots, X_n]$ such that (i) Q is homogeneous; (ii) R is either 0 or else it is homogeneous and its degree in X_n is less than that of G; (iii) $F = GQ + R$.

7.2 Prove (37) in the text, assuming only results established before (37).

In Exercises 7.3–7.5, $C \subset \mathbb{P}^2(\mathbb{C})$ denotes a nonsingular curve of genus g.

7.3 Recall that in analogy with functions, a differential ω on C is said to be **holomorphic** if the order of ω at each point $P \in C$ is nonnegative. Use the Riemann–Roch theorem to show that the set of all homomorphic differentials on C forms a $\mathbb{C}$-vector space of dimension g.

7.4 Prove that a divisor D on C is canonical iff $\deg D = 2g - 2$ and $\dim L(D) = g$.

7.5 (a) Let Z be any canonical divisor on C. For any divisor D on C, D and $D^* = Z - D$ are dual in the sense that $D^{**} = D$. Show that the dual of any principal divisor is a canonical divisor, and vice versa. Incidentally, note that $(\deg Z)/2 = g - 1$ is the "center of symmetry" of $\deg D$ and $\deg D^*$—that is, for any D, $\deg Z/2$ is midway between $\deg D$ and $\deg Z - D$. (Cf. Figure 2.)

(b) Show that replacing D in the Riemann–Roch formula by $Z - D$ gives us back exactly the same formula.

(c) Let D and $D^* = Z - D$ be dual divisors. Prove the Brill–Noether reciprocity theorem—namely,

$$\dim L(D) - 2 \deg D = \dim L(D^*) - 2 \deg D^*.$$

(d) With D and D^* as above, prove that if $\dim L(D) > 0$ and $\dim L(D^*) > 0$, then

$$\dim L(D) - 1 \leqslant \frac{\deg D}{2}.$$

7.6 By making the necessary changes in the proof of Theorem 7.11, prove the following more general form of Noether's theorem:

Let $C \subseteq \mathbb{P}^2(\mathbb{C})$ be an irreducible curve which is either nonsingular, or has only ordinary singularities.

Let forms $G, H \in \mathbb{C}[X, Y, Z]$ define divisors $\operatorname{div}(G)$, $\operatorname{div}(H)$ on C, and suppose that these divisors satisfy $\operatorname{div} G + \Delta \leqslant \operatorname{div} H$, where Δ is as in Exercise 6.16. Then there are forms G^* and $F^* \in \mathbb{C}[X, Y, Z]$ such that $H = GG^* + FF^*$.

7.7 Prove this more general form of the Riemann–Roch theorem:

Let $C \subseteq \mathbb{P}^n(\mathbb{C})$ be any irreducible curve of genus g. Let D be any divisor on C, and Z, any canonical divisor on C. Then

$$\dim L(D) = \deg D + 1 - g + \dim L(Z - D).$$

[Suggestions: Reduce the theorem to the case of a curve in $\mathbb{P}^2(\mathbb{C})$ having at worst ordinary singularities (Exercise 6.12 and Theorem 6.19). Use the definition of genus in Exercise IV, 7.6 and the birational invariance of g proved in Exercise IV, 7.8. Note that the statements of the lemmas in this section generalize verbatim to arbitrary irreducible curves in $\mathbb{P}^n(\mathbb{C})$.]

Bibliography

Ahlfors, L. V. *Complex Analysis*. New York: McGraw-Hill, 1953.

Bochner, S., and Martin, W. T. *Several Complex Variables*. Princeton, N.J.: Princeton University Press, 1948.

Borevich, Z. I., and Shafarevich, I. R. *Number Theory*. New York: Academic Press, 1966.

Cairns, S. S. *Introductory Topology*. New York: The Ronald Press Company, 1961.

Eisenbud, D., and Evans, E. G., Jr. Every algebraic set in n-space is the intersection of n hypersurfaces. *Inventiones Math.*, **19** (1973), 107–112.

Fulton, W. *Algebraic Curves*. New York: W. A. Benjamin, Inc., 1969.

Lang, S. *Introduction to Algebraic Geometry*. New York: Interscience Publishers, Inc., 1958.

Markov, A. A. The problem of homeomorphy. *Proceedings of the International Congress of Mathematicians* (1958) [Russian]. Cambridge: The Cambridge University Press, 1960.

Massey, W. S. *Algebraic Topology: An Introduction*. New York: Harcourt, Brace & World, Inc., 1967.

Narasimhan, R. *Analysis on Real and Complex Manifolds*. Amsterdam: North-Holland Publishing Company, 1968.

Reid, C. *Hilbert*. New York: Springer-Verlag, 1970.

Spivak, M. *Calculus on Manifolds*. New York: W. A. Benjamin, Inc., 1965.

van der Waerden, B. L. *Modern Algebra*. 2 vols. New York: Frederick Ungar Publishing Co., 1949.

Vick, J. W. *Homology Theory*. New York: Academic Press, 1973.

Walker, R. J. *Algebraic Curves*. New York: Dover Publications, Inc., 1962.

Wells, R. O., Jr. *Differential Analysis on Complex Manifolds*. Englewood Cliffs, N.J.: Prentice-Hall, Inc., 1973.

Zariski, O., and Samuel, P. *Commutative Algebra*. 2 vols. New York: Springer-Verlag, 1976.

Notation index

Subject index

A

B

C

D

E

F

G

H

I

J

K

L

M

N

Q

R

S

T

U

V

W

Z

Graduate Texts in Mathematics

Soft and hard cover editions are available for each volume up to Vol. 14, hard cover only from Vol. 15

1. TAKEUTI/ZARING. Introduction to Axiomatic Set Theory. vii, 250 pages. 1971.
2. OXTOBY. Measure and Category. viii, 95 pages. 1971.
3. SCHAEFFER. Topological Vector Spaces. xi, 294 pages. 1971.
4. HILTON/STAMMBACH. A Course in Homological Algebra. ix, 338 pages. 1971. (Hard cover edition only)
5. MACLANE. Categories for the Working Mathematician. ix, 262 pages. 1972.
6. HUGHES/PIPER. Projective Planes. xii, 291 pages. 1973.
7. SERRE. A Course in Arithmetic. x, 115 pages. 1973.
8. TAKEUTI/ZARING. Axiomatic Set Theory. viii, 238 pages. 1973.
9. HUMPHREYS. Introduction to Lie Algebras and Representation Theory. xiv, 169 pages. 1972.
10. COHEN. A course in Simple Homotopy Theory. xii, 114 pages. 1973.
11. CONWAY. Functions of One Complex Variable. 2nd corrected reprint. xiii, 313 pages. 1975. (Hard cover edition only.)
12. BEALS. Advanced Mathematical Analysis. xi, 230 pages. 1973.
13. ANDERSON/FULLER. Rings and Categories of Modules. ix, 339 pages. 1974.
14. GOLUBITSKY/GUILLEMIN. Stable Mappings and Their Singularities. x, 211 pages. 1974.
15. BERBERIAN. Lectures in Functional Analysis and Operator Theory. x, 356 pages. 1974.
16. WINTER. The Structure of Fields. xiii, 205 pages. 1974.
17. ROSENBLATT. Random Processes. 2nd ed. x, 228 pages. 1974.
18. HALMOS. Measure Theory. xi, 304 pages. 1974.
19. HALMOS. A Hilbert Space Problem Book. xvii, 365 pages. 1974.
20. HUSEMOLLER. Fibre Bundles. 2nd ed. xvi, 344 pages. 1975.
21. HUMPHREYS. Linear Algebraic Groups. xiv. 272 pages. 1975.
22. BARNES/MACK. An Algebraic Introduction to Mathematical Logic. x, 137 pages. 1975.

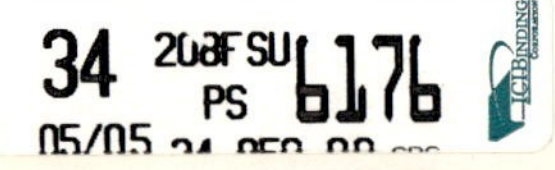